2D AND 3D IMAGE ANALYSIS BY MOMENTS

2D AND 3D IMAGE ANALYSIS BY MOMENTS

Jan Flusser, Tomáš Suk and Barbara Zitová

Institute of Information Theory and Automation,
Czech Academy of Sciences,
Prague,
Czech Republic

Library of Congress Cataloging-in-Publication data

Names: Flusser, Jan, author. | Suk, Tomáš, author. | Zitová, Barbara,
  author.
Title: 2D and 3D image analysis by moments / Jan Flusser, Tomáš Suk, Barbara
  Zitová.
Description: Chichester, UK ; Hoboken, NJ : John Wiley & Sons, 2016. |
  Includes bibliographical references and index.
Identifiers: LCCN 2016026517| ISBN 9781119039358 (cloth) | ISBN 9781119039372
  (Adobe PDF) | ISBN 9781119039365 (epub)
Subjects: LCSH: Image analysis. | Moment problems (Mathematics) | Invariants.
Classification: LCC TA1637 F58 2016 | DDC 621.36/7015159–dc23 LC record available at
  https://lccn.loc.gov/2016026517

Set in 10/12pt, TimesLTStd by SPi Global, Chennai, India.
Printed and bound in Malaysia by Vivar Printing Sdn Bhd

10 9 8 7 6 5 4 3 2 1

To my wife, Vlasta, my sons, Michal and Martin,
and my daughters, Jana and Veronika.

Jan Flusser

To my wife, Lenka, my daughter, Hana, and my son, Ondřej.

Tomáš Suk

To my husband, Pavel, my sons, Janek and Jakub,
and my daughter, Babetka.

Barbara Zitová

Contents

Authors' biographies

Prof. Jan Flusser, PhD, DrSc, received the MSc degree in mathematical engineering from the Czech Technical University, Prague, Czech Republic, in 1985, the PhD degree in computer science from the Czechoslovak Academy of Sciences in 1990, and the DrSc degree in technical cybernetics in 2001. Since 1985 he has been with the Institute of Information Theory and Automation, Czech Academy of Sciences, Prague. In 1995–2007, he held the position of head of Department of Image Processing. Since 2007 he has been a Director of the Institute. He is a full professor of computer science at the Czech Technical University, Faculty of Nuclear Science and Physical Engineering, and at the Charles University, Faculty of Mathematics and Physics, Prague, Czech Republic, where he gives undergraduate and graduate courses on Digital Image Processing, Pattern Recognition, and Moment Invariants and Wavelets. Jan Flusser's research interests cover moments and moment invariants, image registration, image fusion, multichannel blind deconvolution, and super-resolution imaging. He has authored and coauthored more than 200 research publications in these areas, including the monograph *Moments and Moment Invariants in Pattern Recognition* (Wiley, 2009), approximately 60 journal papers, and 20 tutorials and invited/keynote talks at major conferences (ICIP'05, ICCS'06, COMPSTAT'06, ICIP'07, DICTA'07, EUSIPCO'07, CVPR'08, CGIM'08, FUSION'08, SPPRA'09, SCIA'09, ICIP'09, CGIM'10, AIA'14, and CI'15). His publications have received about approximately 1000 citations.

In 2007 Jan Flusser received the Award of the Chairman of the Czech Science Foundation for the best research project and won the Prize of the Czech Academy of Sciences for his contribution to image fusion theory. In 2010, he was awarded the SCOPUS 1000 Award presented by Elsevier. He received the Felber Medal of the Czech Technical University for excellent contribution to research and education in 2015. Personal webpage: `http://www.utia.cas.cz/people/flusser`

Tomáš Suk, PhD, received the MSc degree in technical cybernetics from the Czech Technical University, Prague, Czech Republic, in 1987 and the PhD degree in computer science from the Czechoslovak Academy of Sciences in 1992. Since 1992 he has been a research fellow with the Institute of Information Theory and Automation, Czech Academy of Sciences, Prague. His research interests include invariant features, moment and point-based invariants, color spaces, geometric transformations, and applications in botany, remote sensing, astronomy, medicine, and computer vision.

Tomáš Suk has authored and coauthored more than thirty journal papers and fifty conference papers in these areas, including tutorials on moment invariants held at the conferences ICIP'07 and SPPRA'09. He also coauthored the monograph *Moments and Moment Invariants in Pattern Recognition* (Wiley, 2009). His publications have received about approximately citations. In 2002 he received the Otto Wichterle Premium of the Czech Academy of Sciences for young scientists. Personal webpage: http://zoi.utia.cas.cz/suk

Barbara Zitová, PhD, received the MSc degree in computer science from the Charles University, Prague, Czech Republic, in 1995 and the Ph.D degree in software systems from the Charles University, Prague, Czech Republic, in 2000. Since 1995, she has been with the Institute of Information Theory and Automation, Czech Academy of Sciences, Prague. Since 2008 she has been the head of Department of Image Processing. She gives undergraduate and graduate courses on Digital Image Processing and Wavelets in Image Processing at the Czech Technical University and at the Charles University, Prague, Czech Republic. Barbara Zitová's research interests include geometric invariants, image enhancement, image registration, image fusion, and image processing in medical and in cultural heritage applications. She has authored or coauthored more than seventy research publications in these areas, including the monograph *Moments and Moment Invariants in Pattern Recognition* (Wiley, 2009) and tutorials at several major conferences. In 2003 Barbara Zitová received the Josef Hlavka Student Prize, in 2006 the Otto Wichterle Premium of the Czech Academy of Sciences for young scientists, and in 2010 she was awarded the prestigious SCOPUS 1000 Award for receiving more than 1000 citations of a single paper. Personal webpage: http://zoi.utia.cas.cz/zitova

Preface

Seven years ago we published our first monograph on moments and their applications in image recognition: J. Flusser, T. Suk, and B. Zitová, *Moments and Moment Invariants in Pattern Recognition*, Wiley, 2009.

That book (referred to as MMIPR) was motivated by the need for a monograph covering theoretical aspects of moments and moment invariants and their relationship to practical image recognition problems. Long before 2009, object recognition had become an established discipline inside image analysis. Moments and moment invariants, introduced to the image analysis community in the early 1960s, have played a very important role as features in invariant recognition. Nevertheless, such a book had not been available before 2009[1].

The development of moment invariants after 2009 was even more rapid than before. In SCOPUS, which is probably the most widely-used publication database, we have received 16,000 search results as the response to the "image moment" keyword and 6,000 results of the "moment invariants" search[2]. There has been an overlap of about 2,000 papers, which results in 20,000 relevant papers in total. This huge number of entries illustrates how a large and important area of computer science has been formed by the methods based on image moments. In Figure 1 we can observe the development in time. A relatively slow growth in the last century was followed by a rapid increase of the number of publications in 2009–2010 (we believe that the appearance of MMIPR at least slightly contributed to the growing interest in moment invariants). Since then, the annual number of publications has slightly fluctuated, reaching another local maximum in 2014. In 2014, a new multi-authored book edited by G. A. Papakostas[3] appeared on the market. Although the editor did a very good job, this book suffers from a common weakness of multi-authored books – the topics of individual chapters had been selected by their authors according to their interest, which made some areas overstressed, while some others, remained unmentioned despite their importance. The Papakostas book reflects recent developments in some areas but can hardly be used as a course textbook.

---

[1] The very first moment-focused book by R. Mukundan and K. R. Ramakrishnan, *Moment Functions in Image Analysis*, World Scientific, 1998, is just a short introduction to this field. The second book by M. Pawlak, *Image Analysis by Moments: Reconstruction and Computational Aspects*, Wroclaw, Poland, 2006, is focused narrowly on numerical aspects of image moments, without providing a broader context of invariant image recognition and of practical applications.

[2] The search was performed within the title and abstract of the papers.

[3] G. A. Papakostas ed., *Moments and Moment Invariants – Theory and Applications*, Science Gate Publishing, 2014.

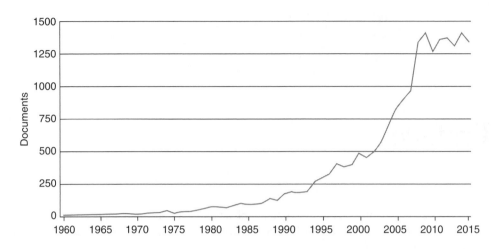

Figure 1   The number of moment-related publications as found in SCOPUS.

The great number of publications that have appeared since 2009 led us to the idea of writing a new, more comprehensive book on this topic. The abundant positive feedback we have received from the MMIPR readers and from our students was another factor which has strengthened our intentions. In 2014, the MMIPR was even translated into Chinese[4].

The main goal of the book you are now holding in your hands is to be a comprehensive monograph covering the current state of the art of moment-based image analysis and presenting the latest developments in this field in a consistent form. In this book, the reader will find a survey of all important theoretical results as well as a description of how to use them in practical image analysis tasks. In particular, our aims were

- To review the development of moments and moment invariants after 2009;
- To cover the area of 3D moments and invariants, which were mostly omitted in the MMIPR but have become important in the last few years;
- To present some of the latest unpublished results, especially in the field of blur invariants; and
- To provide readers with an introduction to a broader context, showing the role of the moments in image analysis and recognition and reviewing other relevant areas and approaches.

At the same time, we aimed to write a self-contained book. This led us to the decision (with the kind permission of Wiley) to include also the core parts of the MMIPR, of course in enhanced and extended/up-to-date form. Attentive readers may realize that about one half of this book was already treated in the MMIPR in some form, while the other half is original. In particular, Chapters 1, 2, 4, and 6 are completely or mostly original. Chapters 7, 8, 9, and 10 are substantially extended versions of their ancestors in the MMIPR, and Chapters 3 and 5 were adopted from the MMIPR after minor updates.

---

[4] Published by John Wiley & Univ. of Science and Technology of China Press, see http://library.utia.cas.cz/separaty/2015/ZOI/flusser-0444327-cover.jpg.

The book is based on our deep experience with moments and moment invariants gained from twenty-five years of research in this area, from teaching graduate courses on moment invariants and related fields at the Czech Technical University and at the Charles University, Prague, Czech Republic, and from presenting several tutorials on moments at major international conferences.

The target readership includes academic researchers and R&D engineers from all application areas who need to recognize 2D and 3D objects extracted from binary/graylevel/color images and who look for invariant and robust object descriptors, as well as specialists in moment-based image analysis interested in a new development on this field. Last but not least, the book is also intended for university lecturers and graduate students of image analysis and pattern recognition. It can be used as textbook for advanced graduate courses on Invariant Pattern Recognition. The first two chapters can be even utilized as supplementary reading to undergraduate courses on Pattern Recognition and Image Analysis.

We created an accompanying website at http://zoi.utia.cas.cz/moment_ invariants2 containing selected Matlab codes, the complete lists of the invariants, the slides for those who wish to use this book for educational purposes, and Errata (if any). This website is free for the book readers (the password can be found in the book) and is going to be regularly updated.

Acknowledgements

First of all, we would like to express gratitude to our employer, the Institute of Information Theory and Automation (ÚTIA) and to its mother organization, the Czech Academy of Sciences, for creating an inspiring and friendly environment and for continuous support of our research. We also thank both universities where we teach, the Czech Technical University, Faculty of Nuclear Science and Physical Engineering, and the Charles University, Faculty of Mathematics and Physics, for their support and for administration of all our courses.

Many our colleagues, students, and friends contributed to the book in various ways. We are most grateful to all of them, especially to

Jiří Boldyš for his significant contribution to Chapters 4 and 6,
Bo Yang for his valuable contribution to Chapters 4 and 7,
Sajad Farokhi for participation in the experiments with Gaussian blur in Chapter 6,
Roxana Bujack for sharing her knowledge on vector field invariants,
Matteo Pedone for his contribution to blur-invariant image registration,
Jaroslav Kautsky for sharing his long-time experience with orthogonal polynomials,
Filip Šroubek for his valuable contribution to Chapter 6 and for providing test images and
    deconvolution codes,
Cyril Höschl IV for implementing the graph-based decomposition in Chapter 8,
Jitka Kostková for creating illustrations in Chapter 2,
Michal Šorel for his help with the optical flow experiment in Chapter 9,
Babak Mahdian for his contribution to the image forensics experiment in Chapter 9,
Michal Breznický for his comments on the blur invariants,
Stanislava Šimberová for providing the astronomical images,
Digitization Center of Cultural Heritage, Xanthi, Greece, for providing the 3D models of the
    ancient artifacts,
Alexander Kutka for his help with the focus measure experiment in Chapter 9,
Jan Kybic for providing us with his code for elastic matching, and
Jarmila Pánková for the graphical design of the book front cover.

Special thanks go to the staff of the publisher John Wiley & Sons Limited, who encouraged us to start writing the book and who provided us with help and assistance during the publication process. We would also like to thank the many students in our courses and our tutorial attendees who have asked numerous challenging and stimulating questions.

The work on this book was partially supported by the Czech Science Foundation under the projects No. GA13-29225S in 2014 and No. GA15-16928S in 2015–16.

Last but not least, we would like to thank our families. Their encouragement and patience was instrumental in completing this book.

Jan Flusser
Tomáš Suk
Barbara Zitová
Institute of Information Theory and Automation of the CAS,
Prague, Czech Republic, November 2015.

About the companion website

Don't forget to visit the companion website for this book:

http://www.wiley.com/go/flusser/2d3dimageanalysis

There you will find valuable material designed to enhance your learning, including:

- 300 slides
- Matlab codes

Scan this QR code to visit the companion website

1

Motivation

In our everyday life, each of us constantly receives, processes, and analyzes a huge amount of information of various kinds, significance and quality, and has to make decisions based on this analysis. More than 95% of information we perceive is of a visual character. An image is a very powerful information medium and communication tool capable of representing complex scenes and processes in a compact and efficient way. Thanks to this, images are not only primary sources of information, but are also used for communication among people and for interaction between humans and machines. Common digital images contain enormous amounts of information. An image you can take and send to your friends using a smart phone in a few seconds contains as much information as hundreds of text pages. This is why in many application areas, such as robot vision, surveillance, medicine, remote sensing, and astronomy, there is an urgent need for automatic and powerful image analysis methods.

1.1 Image analysis by computers

Image analysis in a broad sense is a many-step process the input of which is an image while the output is a final (usually symbolic) piece of information or a decision. Typical examples are localization of human faces in the scene and recognition (against a list or a database of persons) who is who, finding and recognizing road signs in the visual field of the driver, and identification of suspect tissues in a CT or MRI image. A general image analysis flowchart is shown in Figure 1.1 and an example what the respective steps look like in the car licence plate recognition is shown in Figure 1.2.

The first three steps of image analysis—image acquisition, preprocessing, and object segmentation/detection—are comprehensively covered in classical image processing textbooks [1–5], in recent specialized monographs [6–8] and in thousands of research papers. We very briefly recall them below, but we do not deal with them further in this book. The topic of this book falls into the fourth step, the feature design. The book is devoted to one particular family of features which are based on image moments. The last step, object classification, is shortly reviewed in Chapter 2, but its deeper description is beyond the scope of this book.

In *image acquisition*, the main theoretical questions are how to choose the sampling scheme, the sampling frequency and the number of the quantization levels such that the artifacts caused by aliasing, moire, and quantization noise do not degrade the image much while keeping the

2D and 3D Image Analysis by Moments, First Edition. Jan Flusser, Tomáš Suk and Barbara Zitová.
© 2017 John Wiley & Sons, Ltd. Published 2017 by John Wiley & Sons, Ltd.
Companion website: www.wiley.com/go/flusser/2d3dimageanalysis

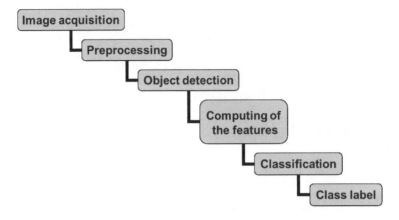

Figure 1.1   General image analysis flowchart

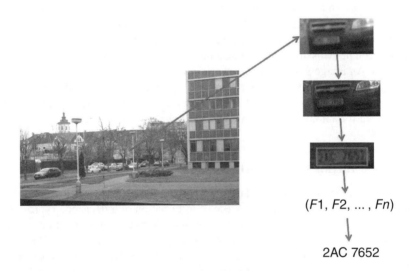

Figure 1.2   An example of the car licence plate recognition

image size reasonably low (there are of course also many technical questions about the appropriate choice of the camera and the spectral band, the objective, the memory card, the transmission line, the storage format, and so forth, which we do not discuss here).

Since real imaging systems as well as imaging conditions are usually imperfect, the acquired image represents only a degraded version of the original scene. Various kinds of degradations (geometric as well as graylevel/color) are introduced into the image during the acquisition process by such factors as imaging geometry, lens aberration, wrong focus, motion of the scene, systematic and random sensor errors, noise, etc. (see Figure 1.3 for the general scheme and an illustrative example). Removal or at least suppression of these degradations is a subject of *image preprocessing*. Historically, image preprocessing was one of the first topics systematically studied in digital image processing (already in the very first monograph [9] there was a chapter

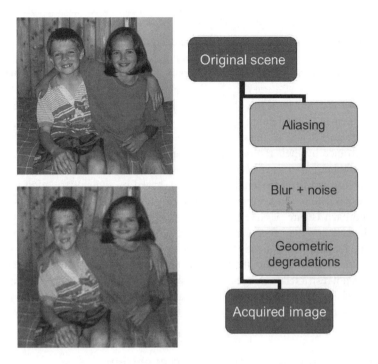

Figure 1.3 Image acquisition process with degradations

devoted to this topic) because even simple preprocessing methods were able to enhance the visual quality of the images and were feasible on old computers. The first two steps, image acquisition and preprocessing, are in the literature often categorized into *low-level processing*. The characteristic feature of the low-level methods is that both their input and output are digital images. On the contrary, in *high-level processing*, the input is a digital image (often an output of some preprocessing) while the output is a symbolic (i.e. high-level) information, such as the coordinates of the detected objects, the list of boundary pixels, etc.

Object detection is a typical example of high-level processing. The goal is to localize the objects of interest in the image and separate (segment) them from the other objects and from the background[1]. Hundreds of segmentation methods have been described in the literature. Some of them are universal, but most of them were designed for specific families of objects such as characters, logos, cars, faces, human silhouettes, roads, etc. A good basic survey of object detection and segmentation methods can be found in [5] and in the references therein.

Feature definition and computing is probably the most challenging part of image analysis. The features should provide an accurate and unambiguous quantitative description of the

---

[1] Formally, the terms such as "object", "boundary", "interior", "background", and others are subject of the discrete topology we are working in. Various discrete topologies differ from one another by the notion of neighboring pixels and hence by the notion of connectivity. However, here we use the above terms intuitively without specifying the particular topology.

objects[2]. The feature values are elements of the *feature space* which should be for the sake of efficient computation of low dimensionality. The design of the features is highly dependent on the type of objects, on the conditions under which the images have been acquired, on the type and the quality of preprocessing, and on the application area. There is no unique "optimal" solution.

Classification/recognition of the object is the last stage of the image analysis pipeline. It is entirely performed in the feature space. Each unknown object, now being represented by a point in the feature space, is classified as an element of a certain class. The classes can be specified in advance by their representative samples, which create a *training set*; in such a case we speak about *supervised classification*. Alternatively, if no training set is available, the classes are formed from the unknown objects based on their distribution in the feature space. This case is called *unsupervised classification* or *clustering*, and in visual object recognition this approach is rare. Unlike the feature design, the classification algorithms are application independent—they consider neither the nature of the original data nor the physical meaning of the features. Classification methods are comprehensively reviewed in the famous Duda-Hart-Stork book [10] and in the most recent monograph [11]. The use of the classification methods is not restricted to image analysis. We can find numerous applications in artificial intelligence, decision making, social sciences, statistical data analysis, and in many other areas beyond the scope of this book.

1.2   Humans, computers, and object recognition

Why is image analysis so easy and natural for human beings and so difficult for machines? There are several reasons for this performance difference. Incorporating lifetime experience, applying information fusion, the ability of contextual perception, and perceptual robustness are the key elements. Current research directions are often motivated by these factors, but up to now the results have been achieved only in laboratory conditions and have not reached the quality of human beings yet. Probably the most serious difference is that humans incorporate their lifetime knowledge into the recognition process and do so in a very efficient way. If you see an object you have already met (even a long time ago), you are able to "retrieve" this information from your brain almost immediately (it appears that the "retrieval time" does not depend on the number of the objects stored in the brain) and use it as a hint for the classification. Your brain is also able to supply the missing information if only a part of the object is visible. This ability helps us a lot when recognizing partially occluded objects. Computers could in principle do the same, but it would require massive learning on large-scale databases, which would lead not only to a time-consuming learning stage but also to relatively slow search/retrieval of the learned objects. To overcome that, several projects have been started recently which are trying to substitute human life experience by shared knowledge available at search engines (see [12] for example) but this research is still at the beginning stages.

Our ability to combine supplementary information of other kinds in parallel with vision makes human perception efficient, too. We also have other senses (hearing, smell, taste, and

---

[2] There exists an alternative way of object description by its structure. The respective recognition methods are then called structural or syntactic ones. They use the apparatus of formal language theory. Each object class is given by a language, and each object is a word. An object is assigned to the class if its word belongs to the class language. This approach is useful if the structural information is more appropriate than a quantitative description.

touch), and we are very good in combining them with the visual input and making a decision by fusing all these input channels. This helps us a lot, especially if the visual information is insufficient such as in the darkness. Modern-day robots are often equipped with such sensors, even working in modalities we are not able to perceive. However the usage of such multi-modal robots is limited by their cost and, what is more important, by the non-triviality of efficient fusion of the acquired information.

Contextual perception used in object recognition gives humans yet another advantage comparing to computers. Computer algorithms either do not use the context at all because they work with segmented isolated objects or classify the objects in the neighborhood, and the contextual relations are used as additional features or prior probability estimators (for instance, if we recognize a car in front of the unknown object and another car behind it, the prior probability that the object in question is also a car is high). Such investigation of the context is, however, time expensive. Humans perceive, recognize, and evaluate the context very quickly which gives them the ability of "recognizing" even such objects that are not visible at all at the moment. Imagine you see a man with a leash in his hand on the picture. You can, based on context (a man in the a park) and your prior knowledge, quickly deduce that the animal on the other end of the leash, that cannot be seen in the photo, is a dog (which actually may not be true in all cases). This kind of thinking is very complex and thus difficult to implement on a computer.

Yet another advantage of human vision over computer methods is its high robustness to the change of the orientation, scale, and pose of the object. Most people can easily recognize a place or a person on a photograph that is upside down or mirrored. The algorithms can do that as well, but it requires the use of special features, which are insensitive to such modifications. Standard features may change significantly under spatial image transformations, which may lead to misclassifications.

From this short introduction, we can see that automatic object recognition is a challenging and complex task, important for numerous application areas, which requires sophisticated algorithms in all its stages. This book contributes to the solution of this problem by systematic research of one particular type of features, which are known as *image moments*.

1.3 Outline of the book

This book deals systematically with moments and moment invariants of 2D and 3D images and with their use in object description, recognition, and in other applications.

Chapter 2 introduces basic terms and concepts of object recognition such as feature spaces and related norms, equivalences, and space partitions. Invariant based approaches are introduced together with normalization methods. Eight basic categories of invariants are reviewed together with illustrative examples of their usage. We also recall main supervised and unsupervised classifiers as well as related topics of classifier fusion and reduction of the feature space dimensionality.

Chapters 3 – 6 are devoted to four classes of moment invariants. In Chapter 3, we introduce 2D moment invariants with respect to the simplest spatial transformations—translation, rotation, and scaling. We recall the classical Hu invariants first, and then we present a general method for constructing invariants of arbitrary orders by means of complex moments.

We prove the existence of a relatively small basis of invariants that is complete and independent. We also show an alternative approach—constructing invariants via normalization. We discuss the difficulties which the recognition of symmetric objects poses and present moment invariants suitable for such cases.

In Chapter 4 we introduce 3D moment invariants with respect to translation, rotation, and scaling. We present the derivation of the invariants by means of three approaches—the tensor method, the expansion into spherical harmonics, and the object normalization. Similarly as in 2D, the symmetry issues are also discussed there.

Chapter 5 deals with moment invariants to the affine transformation of spatial coordinates. We present four main approaches showing how to derive them in 2D—the graph method, the method of normalized moments, the transvectants, and the solution of the Cayley-Aronhold equation. Relationships between the invariants produced by different methods are mentioned, and the dependency among the invariants is studied. We describe a technique used for elimination of reducible and dependent invariants. Numerical experiments illustrating the performance of the affine moment invariants are carried out, and a generalization to color images, vector fields, and 3D images is proposed.

Chapter 6 deals with a completely different kind of moment invariants, with invariants to image blurring. We introduce the theory of projection operators, which allows us to derive invariants with respect to image blur regardless of the particular convolution kernel, provided it has a certain type of symmetry. We also derive so-called combined invariants, which are invariant to composite geometric and blur degradations. Knowing these features, we can recognize objects in the degraded scene without any restoration/deblurring.

Chapter 7 presents a survey of various types of orthogonal moments. The moments orthogonal on a rectangle/cube as well as the moments orthogonal on a unit disk/sphere are described. We review Legendre, Chebyshev, Gegenbauer, Jacobi, Laguerre, Gaussian-Hermite, Krawtchouk, dual Hahn, Racah, Zernike, pseudo-Zernike, and Fourier-Mellin polynomials and moments. The use of the moments orthogonal on a disk in the capacity of rotation invariants is discussed. The second part of the chapter is devoted to image reconstruction from its moments. We explain why orthogonal moments are more suitable for reconstruction than geometric ones, and a comparison of reconstructing power of different orthogonal moments is presented.

In Chapter 8, we focus on computational issues. Since the computing complexity of all moment invariants is determined by the computing complexity of the moments, efficient algorithms for moment calculations are of prime importance. There are basically two major groups of methods. The first one consists of methods that attempt to decompose the object into non-overlapping regions of a simple shape, the moments of which can be computed very fast. The moment of the object is then calculated as a sum of the moments of all regions. The other group is based on Green's theorem, which evaluates the integral over the object by means of a less-dimensional integration over the object boundary. We present efficient algorithms for binary and graylevel objects and for geometric as well as for selected orthogonal moments.

Chapter 9 is devoted to various applications of moments and moment invariants in image analysis. We demonstrate their use in image registration, object recognition, medical imaging, content-based image retrieval, focus/defocus measurement, forensic applications, robot navigation, digital watermarking, and others.

Chapter 10 contains a summary and a concluding discussion.

References

[1] A. Rosenfeld and A. C. Kak, *Digital Picture Processing*. Academic Press, 2nd ed., 1982.
[2] H. C. Andrews and B. R. Hunt, *Digital Image Restoration*. Prentice Hall, 1977.
[3] W. K. Pratt, *Digital Image Processing*. Wiley Interscience, 4th ed., 2007.
[4] R. C. Gonzalez and R. E. Woods, *Digital Image Processing*. Prentice Hall, 3rd ed., 2007.
[5] M. Šonka, V. Hlaváč, and R. Boyle, *Image Processing, Analysis and Machine Vision*. Thomson, 3rd ed., 2007.
[6] P. Campisi and K. Egiazarian, *Blind Image Deconvolution: Theory and Applications*. CRC Press, 2007.
[7] P. Milanfar, *Super-Resolution Imaging*. CRC Press, 2011.
[8] A. N. Rajagopalan and R. Chellappa, *Motion Deblurring: Algorithms and Systems*. Cambridge University Press, 2014.
[9] A. Rosenfeld, *Picture Processing by Computer*. Academic Press, 1969.
[10] R. O. Duda, P. E. Hart, and D. G. Stork, *Pattern Classification*. Wiley Interscience, 2nd ed., 2001.
[11] S. Theodoridis and K. Koutroumbas, *Pattern Recognition*. Academic Press, 4th ed., 2009.
[12] X. Chen, A. Shrivastava, and A. Gupta, "NEIL: Extracting visual knowledge from web data," in *The 14th International Conference on Computer Vision ICCV'13*, pp. 1409–1416, 2013.

2

Introduction to Object Recognition

This chapter is a brief introduction to the principles of automatic object recognition. We introduce the basic terms, concepts, and approaches to feature-based classification. For understanding the rest of the book, the most important part of this chapter is the introduction of the term *invariant* and an overview of the invariants which have been proposed for visual object description and recognition. In addition to that, we concisely review the existing classifiers. The chapter starts with a short introduction to feature metric spaces[1].

2.1 Feature space

As we mentioned in Chapter 1, the features are measurable quantitative characteristics of the objects and images. From a mathematical point of view, the features are elements of a *feature space*.

Let us now look closer at the feature space, at its required and desirable properties, and at some connections with metric spaces, equivalence relations, set partition, and group theory. At this moment, we are not going to define any particular feature set. We stay on a general level such that the following considerations are valid regardless of the specific choice of the features.

All object recognition algorithms that act in a feature space use a measure of object similarity or dissimilarity as the central tool. Hence, most of the feature spaces are constructed such that they are *metric spaces*. For certain special features where the construction of a metric is not appropriate or possible, we weaken this requirement and create *pseudometric, quasimetric, semimetric*, or *premetric* spaces. In some other cases, the similarity/dissimilarity measure is not explicitly defined, but it is still present implicitly.

---

[1] Readers who are familiar with these topics at least on an intermediate level may skip this chapter and proceed directly to Chapter 3, where the theory of moment invariants begins.

2.1.1 *Metric spaces and norms*

A metric space is an ordered pair (P, d), where P is a non-empty set and d is a *metric* on P.

Definition 2.1 A real-valued function d defined on $P \times P$ is a *metric* if it satisfies the four following axioms for any $\mathbf{a}, \mathbf{b}, \mathbf{c} \in P$.

1. $d(\mathbf{a}, \mathbf{b}) \geq 0$ (positivity axiom),
2. $d(\mathbf{a}, \mathbf{b}) = 0 \Leftrightarrow \mathbf{a} = \mathbf{b}$ (identity axiom),
3. $d(\mathbf{a}, \mathbf{b}) = d(\mathbf{b}, \mathbf{a})$ (symmetry axiom),
4. $d(\mathbf{a}, \mathbf{b}) \leq d(\mathbf{a}, \mathbf{c}) + d(\mathbf{b}, \mathbf{c})$ (triangle inequality).

The first axiom follows from the other three ones and could be omitted. The function d is also called *distance function* or simply *distance* between two points.

The conditions 1–4 express our intuitive notion about the distance between two points. If axiom 2 is weakened such that only $d(\mathbf{a}, \mathbf{a}) = 0$ is required but possibly also $d(\mathbf{a}, \mathbf{b}) = 0$ for some distinct values $\mathbf{a} \neq \mathbf{b}$, then d is called *pseudometric*. If d satisfies 1, 2, and 4 but it is not necessarily symmetric, we call it *quasimetric*; *semimetric* is such d that satisfies the first three axioms, but not necessarily the triangle inequality. The weakest version is a *premetric*, which only requires $d(\mathbf{a}, \mathbf{b}) \geq 0$ and $d(\mathbf{a}, \mathbf{a}) = 0$. Some of these modifications were inspired by real-life situations. Imagine the "distance" which is measured by the time you need to get from place $\mathbf{a}$ to place $\mathbf{a}$ by bike. Obviously, if the road is uphill, then $d(\mathbf{a}, \mathbf{b}) > d(\mathbf{b}, \mathbf{a})$ and d is a quasimetric. If the distance between two countries is defined as the shortest distance between the points on their borders, then the distance between any two neighboring countries is zero and we get a (symmetric) premetric. The reader can find many other examples easily.

For the given set P, there may exist many (even infinitely many) various metrics which all fulfill the above definition but induce different distance relations between the points of P. Hence, when speaking about a metric space, it is always necessary not only to mention the set P itself but also to specify the metric d.

The definition of a metric space does not require any algebraic structure on P. However, most feature spaces are *normed vector spaces* with the operations addition and multiplication by a constant and with a norm $\| \cdot \|$. Let us recall that any norm must satisfy certain conditions.

Definition 2.2 A real-valued function $\| \cdot \|$ defined on a vector space P is a *norm* if it satisfies the four following axioms for any $\mathbf{a}, \mathbf{b} \in P$ and a real or complex number α.

1. $\|\mathbf{a}\| \geq 0$,
2. $\|\mathbf{a}\| = 0 \Rightarrow \mathbf{a} = \mathbf{0}$,
3. $\|\alpha \cdot \mathbf{a}\| = |\alpha| \cdot \|\mathbf{a}\|$ (homogeneity axiom),
4. $\|\mathbf{a} + \mathbf{b}\| \leq \|\mathbf{a}\| + \|\mathbf{b}\|$ (triangle inequality).

Similarly to the case of metric, the first axiom follows from the homogeneity and triangle inequality and could be omitted.

If we now define d as

$$d(\mathbf{a}, \mathbf{b}) = \|\mathbf{a} - \mathbf{b}\|,$$

then, thanks to the properties of the norm, such d actually fulfills all axioms of a metric. We speak about the metric that is *induced* by the norm[2]. Any norm induces a metric, but only some metrics are induced by a norm. The norm is a generalization of our intuitive notion of the vector "length" and can be also understood as a distance of $\mathbf{a}$ from the origin. If P is a space with a scalar product, then the norm is induced by this scalar product as $\|\mathbf{a}\| = \sqrt{(\mathbf{a}, \mathbf{a})}$.

Feature spaces of finite dimensions[3] are mostly (but not necessarily) equivalent to n-dimensional vector space of real numbers $\mathbb{R}^n$, and we use the notation $\mathbf{a} = (a_1, a_2, \cdots, a_n)^T$. The most common norms in such spaces are ℓ_p norms

$$\|\mathbf{a}\|_p = \left(\sum_{i=1}^{n} |a_i|^p \right)^{1/p}$$

and weighted ℓ_p norms

$$\|\mathbf{a}\|_p^w = \left(\sum_{i=1}^{n} w_i |a_i|^p \right)^{1/p},$$

where $p \geq 1$ and w_i are positive constants. The ℓ_p norms are well known from linear algebra and for certain p's they are of particular interest. For $p = 1$, it turns to the *city-block norm* popular in digital geometry. Euclidean norm, probably the most intuitive and the most frequently used norm ever, is a special case if $p = 2$ (we mostly drop the index p for simplicity when using Euclidean norm and write just $\|\mathbf{a}\|$). If $p \to \infty$, the ℓ_p norm converges to the *maximum norm* ℓ_∞

$$\|\mathbf{a}\|_\infty = \max_i (|a_i|)$$

which is also known as the *chessboard norm*. If $0 < p < 1$, then the triangle inequality is violated, and the corresponding ℓ_p-norm is actually only a seminorm, but it still finds applications: for instance, in robust fitting the feature points with a curve/surface.

Introducing the weights w_i into the ℓ_p norm is of great importance when constructing the feature space and often influences significantly the classification results. The goal of the weights is to bring all the individual features (i.e. components a_i of the feature vector) to approximately the same range of values. It is equivalent to a non-uniform scaling of the feature space. Why is it so important? The features may be quantities of very different character and range. Suppose that, for example, $a_1 \in (-1, 1)$ while $a_2 \in (-10^{10}, 10^{10})$. When using the pure Euclidean norm, we actually lose the information that is contained in the feature a_1 because the a_2 values are so big that $\|\mathbf{a}\| \doteq |a_2|$. However, if we use the weights $w_1 = 1$ and $w_2 = 10^{-20}$, then small relative changes of a_1 have approximately the same impact on the norm as small relative changes of a_2.

There are of course many vector norms that are not ℓ_p-norms. An example is a Hamming norm (sometimes referred to as ℓ_0-norm), which is simply the number of the non-zero components of $\mathbf{a}$.

---

[2] Although this is the most common way to induce a metric by the norm, it is not the only one. However, the other options are rather obscure and of low importance when working in feature spaces. An example is the rail metric defined as $d(\mathbf{a}, \mathbf{b}) = \|\mathbf{a}\| + \|\mathbf{b}\|$ for distinct $\mathbf{a}, \mathbf{b}$ and $d(\mathbf{a}, \mathbf{a}) = 0$. Its name somehow ironically reflects the fact that a railway trip usually passes through the main hub regardless of where the final destination is.

[3] Feature spaces of infinite dimensions are of very little practical interest, and we do not discuss them in this book.

Although the (weighted) metrics induced by ℓ_p norms are almost a standard choice, there exist feature spaces where these metrics are not appropriate. Consider a feature space that consists of individual features, which are very different from one another in their nature. The first feature is real-valued, the second is a periodic quantity (for instance, the orientation angle or the phase of a complex number), the third feature is a logical variable, and the last one expresses a subjective rating on the scale "excellent-good-fair-average-poor". Here the ℓ_p norm would not be a good choice, and we have to construct the metric carefully such that it reflects the information contained in all features and weights it according to its actual importance for the given task.

2.1.2   Equivalence and partition

The most transparent way to explain the principles of object classification is through two elementary terms of the set theory—*partition* and *equivalence*. Given a non-empty set M, a set of its subsets $\{M_1, \cdots, M_C\}$ such that[4]

1. $M_i \neq \emptyset$ $\forall i$,
2. $M_i \cap M_j = \emptyset$ $\forall i \neq j$,
3. $\bigcup\limits_{i=1}^{C} M_i = M$

is called partition of M.

Partition is closely connected with the equivalence relation on $M \times M$. Let us recall that a binary relation $\approx$ is called equivalence if and only if it is reflexive, symmetric, and transitive, that is,

1. $a \approx a$ for any $a \in M$,
2. If $a \approx b$, then $b \approx a$ for any $a, b \in M$,
3. If $a \approx b$ and $b \approx c$, then $a \approx c$ for any $a, b, c \in M$.

Any equivalence on $M \times M$ unambiguously determines a partition of M and vice versa. The components M_i are then called *equivalence classes*. Clearly, if equivalence $\approx$ is given on $M \times M$, we can choose an arbitrary $a \in M$ and find a set $M_a = \{x \in M | a \approx x\}$. This is our first partition component. Then we find $b \in M$ such that $b \notin M_a$ and construct the partition component M_b. We repeat this process until the whole M is covered. The sets $M_a, M_b, \cdots$ fulfill the definition of the partition. The backward implication is even more apparent. If partition $\{M_1, \cdots, M_C\}$ of M is given, then the relation

$$a \approx b \quad \Leftrightarrow \quad a, b \text{ lie in the same component } M_i$$

is an equivalence on $M \times M$.

If we want to abstract away from the properties that may change inside the equivalence classes, we may work with a set of the classes that is called the *quotient set* and is denoted as $(M/\approx)$. Working with quotient sets is very common in mathematics because it helps to

---

[4] We use a notation for finite number of components C because object classification is always to a finite number of classes. A set partition in general may have an infinite (countable or uncountable) number of components.

concentrate on the substantial properties of the objects and to ignore the insignificant ones. If there has been some algebraic and/or metric structure defined on M, it usually smoothly propagates into $(M/\approx)$.

In object classification, the classes (determined by the user) are in fact equivalence classes in the object space. In face recognition, for instance, the person should be identified regardless of the head pose, facial expression, and hair style; in character recognition, the type and the color of the pen, the font, size and orientation of the character are absolutely insignificant parameters, which should in no way influence the recognition (provided that we are interested in "reading" the characters, not in font/size recognition). Hence, the equivalence classes are defined as "all pictures of the person A" and "all possible appearance of the letter B", etc. The variability within each class, which is insignificant for the given task, is called *intraclass variability*. The intraclass variability can be described by an operator D, which maps any object to another object *of the same class*. The relation between two objects

$$f \approx g \quad \Leftrightarrow \quad \text{There exists } D \text{ such that } g = D(f) \text{ or } f = D(g)$$

should be an equivalence that induces the same partition into classes as the one specified by the user. Note that the operator D is *not* defined "automatically" by selecting the object space. In one object space, we may have several meaningful task formulations, each of them leading to distinct space partitions. As an example, consider again the space of portrait photographs. One task could be person recognition, the other one gender recognition, and yet another one classification into age categories.

Very often, D describes all possible object degradations that may arise in the given task: if f is an "ideal" image of an object from the class M_f, then $g = D(f)$ can be understood as a degraded version of f but still belonging to M_f. The ideal classification algorithm should assign g into the class M_f regardless of particular form and parameters of D because they are usually unknown.

From the above considerations, we can clearly see one of the fundamental requirements of object recognition. The operator of the intraclass variability D must be *closed under a composition* in each equivalence class. If D is applied twice (possibly with different parameters), then the composition $D_c = D_1 D_2$ must be of the same type. Without this *closure property* the task would be ill-posed because the "classes" would not be equivalence classes and would not define a partition, so an object could be an element of multiple classes. Since the closure property is often coupled with *associativity* and *invertibility* of D (which is, however, much less important here), the intraclass transformations form a *group* (or a *monoid* if D is not invertible) and the action of D is a *group action*. Typical examples of intraclass variations that form a group are object rotation, translation, and scaling (similarity group), affine transformation (affine group), perspective projection (projective group). Blurring the image by a Gaussian kernel does not form a group because it is not invertible but forms a monoid. On the other hand, quadratic transformation of the image coordinates is not closed and does not form equivalence classes.

2.1.3   Invariants

Now we approach the explanation of the central term of the feature-based object recognition—the notion of invariants. The *invariant feature* or simply the *invariant* is a functional that maps the image space into the feature space such that $I(f)$ depends on the class f belongs to but does not depend on particular appearance of f. In other words, $I(f) = I(D(f))$ for any f and any instance of D. In yet other words, I is actually defined on a quotient object space factorized by D. Simple examples of the invariants under a similarity transformation are angles and ratios, the object size is invariant to rotation, and the property of two lines "to be parallel" is invariant under affine transformation.

The functional I that satisfies the above definition is sometimes called the *absolute* invariant, to emphasize the difference from *relative* invariants. The relative invariant satisfies a weaker and more general condition $I(f) = \Lambda(D)I(D(f))$, where Λ is a scalar function of the parameters of D. If D is for instance an affine transformation, $\Lambda(D)$ is often a function of its Jacobian. Relative invariants cannot be used directly for classification since they vary within the class. It is usually not difficult to normalize a relative invariant by another relative invariant to cancel the factor $\Lambda(D)$ and to get an absolute invariant.

The invariance property is the necessary but not sufficient condition for I to be really able to classify objects (for example, the trivial feature $I(f) = 0$ is perfectly invariant but completely useless). Any functional I defines another equivalence (and hence another partition and also another quotient space) in the object space

$$f \sim g \iff I(f) = I(g).$$

The partition induced by equivalence $\sim$ cannot be arbitrary. Since I is supposed to be an invariant, the quotient space $(M/\sim)$ can only be the same or coarser than $(M/\approx)$. In an ideal case both partitions are equal, i.e. $(M/\sim) = (M/\approx)$, and I is said to be *discriminative* because it has distinct values on objects that belong to distinct user classes. If it is not the case, then there exist at least two objects f and g belonging to different classes but fulfilling $I(f) = I(g)$, which means they are not discriminable by means of I.

Invariance and discriminability are somehow opposed properties. One should keep in mind that the invariance property is not something "always advantageous".

Making the invariance broader than necessary decreases the recognition power (for instance, using affine invariants instead of similarity invariants leads in the feature space to mixing all parallelograms together). The user must tune these two properties of the features with respect to the given data and to the purpose. Choosing a proper trade-off between the invariance and the discrimination power is a very important task in feature-based object recognition (see Figure 2.1 for an example of a desirable situation).

Usually, a single real-valued invariant does not provide enough discrimination power, and several invariants $I_1, \ldots, I_n$ must be used simultaneously. Then we speak about an *invariant vector* and the feature space has n dimensions. The invariant vector should be *independent*, which means that no its component is a function of the others[5]. This is a natural and important requirement. Dependent invariants do not contribute to the discrimination power of the vector, they only increase the dimensionality of the feature space. This leads not only to increasing the time complexity of the classification algorithms and of the feature measurement (and in practice often to a growth of the cost), but also may cause a drop of performance.

Apart from the object recognition, invariants often play an important role in the reconstruction of the original object from its feature vector. We face this situation when the invariants

---

[5] The feature dependence/independence in this sense is not necessarily connected with the correlation, which is evaluated pair-wise on the training set and reflects a linear component of the dependency only.

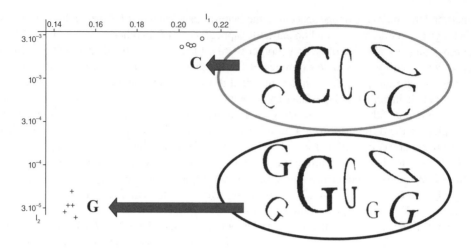

Figure 2.1　Two-dimensional feature space with two classes, almost an ideal example. Each class forms a compact cluster (the features are invariant to translation, rotation, scaling, and skewing of the characters) and the clusters are well separated from one another (the features are discriminative although the characters are visually similar to each other)

are considered an alternative representation of the objects and are stored in a database instead of the original images. The loss-less reconstruction is possible only if the vector is *complete*, which means that other independent features do not exist. Such a reconstruction is of course possible only modulo D since there is no way to recover the particular appearance from the invariants. In case of digital images, complete invariant vectors usually have almost the same number of components as the number of the pixels in the input image. Hence, the requirement of completeness is for classification purposes often highly excessive and useless.

2.1.4　Covariants

Unlike the invariants, there exist features $C(f)$ that change within the classes, i.e. $C(f) \neq C(D(f))$. They are called *covariants*[6]. A covariant usually does not have the ability to discriminate the classes (if it had such ability, it could be turned into an invariant by thresholding) but carry the information about the particular intra-class parameters of f, which may be very useful when transforming the object into certain canonical (standard) positions. Such transformation is called *object normalization*. If, for instance, the intra-class variation is a rotation, then the directions of the principal axes of the objet are covariants. When defining the standard position such that the principal axes are horizontal and vertical, we need to know their actual orientation in order to properly rotate the object. If we want to reconstruct the object including its particular position in the class, we also need, in addition to a complete invariant vector, a complete covariant vector.

---

[6] The terminology has not been unified here. Some authors use the term "variant feature". The term "covariant" has also other meanings in tensor algebra, in programming languages, and in quantum mechanics.

2.1.5   *Invariant-less approaches*

For the sake of completeness, let us mention that along with the invariant approach, there exist two other approaches to visual object recognition—*brute force* and *image normalization*. In the brute force approach, we search the parametric space of all possible image degradations. That means the training set of each class should contain not only all class representatives but also all their rotated, scaled, blurred, and deformed versions. Clearly, this approach would lead to extreme time complexity and is practically inapplicable. In the normalization approach, the objects are transformed into a certain standard position before they enter the classifier. This is very efficient in the classification stage, but the object normalization itself usually requires solving difficult inverse problems that are often ill-conditioned or even ill-posed. For instance, in case of image blurring, "normalization" means in fact blind deconvolution [1], and in case of spatial image deformation, "normalization" requires performing registration of the image to some reference frame [2].

2.2   Categories of the invariants

The existing invariant features for object description and recognition can be categorized from various points of view (see the survey papers [3] and [4]). The main categorization criteria, which have appeared in literature, are the following.

1. The dimension of the objects the invariants describe,
2. The type of the invariance (i.e., the type of the intra-class variations),
3. The range of intensity values of the object,
4. The part of the object which is necessary for the invariant computation,
5. The mathematical tools used for the construction of the invariants.

The first three categorizations are very natural and straightforward; they, however, sort the invariants into only a few categories. When sorting according to the dimensionality[7], we speak about 2D invariants, 3D invariants, and d-D invariants for $d > 3$. According to the type of the invariance, we have geometric invariants or intensity-based ones. In this first category, we recognize translation, rotation, scaling, affine, and projective geometric invariants. Invariants to intensity variations exist with respect to linear contrast stretching, non-linear intensity transformations, and convolution. Sorting according to the range of the intensity values distinguishes invariants of binary, gray-level, and color objects and of vector-valued images.

The fourth viewpoint reflects what part of the object is needed to calculate the invariant. *Global* invariantsare calculated from the whole image (including background if no segmentation has been performed). Most of them include projections of the image onto certain basis functions and are calculated by integration. Comparing to local invariants, global invariants are much more robust with respect to noise, inaccurate boundary detection, and similar factors. On the other hand, their serious drawback is the fact that any local change of the image influences the values of all invariants and is not "localized" in a few components only. This is why global invariants cannot be used when the object in question is partially occluded by another object and/or when a part of it is out of the visual field. *Local* invariants are, on the

---

[7] The term "dimension" here refers to the dimension of the object/images, not to the dimension of the invariant vector.

contrary, calculated from a certain neighborhood of dominant points only. Their primary purpose is to recognize objects which are visible only partially. Differential invariants are typical representatives of this category. *Semi-local* invariants attempt to keep the strengths of the two above groups and to overcome their weaknesses. They divide the object into stable parts and describe each part by some kind of global invariants.

Categorization according to the mathematical tools used is probably the most detailed type. It provides a grouping, where the common factor is the mathematical background, which at the same time often implicitly determines the feature categorization according to the previous four criteria. One can find numerous categories of this kind in the literature, depending on the "depth" of sorting. Here we present eight basic ones—simple shape features, complete visual features, transformation coefficient features, wavelet-based features, textural features, differential invariants, point set invariants, and moment invariants.

2.2.1 Simple shape features

Simple shape features, sometimes referred to as *visual features*, are low-dimensional features of very limited discriminability. On the other hand, they are easy to compute (or even visually estimate) and interpret, which implies their popularity. They have been designed for binary objects (shapes) and are defined for 2D objects only, although some of them can be readily extended to 3D as well. They are mostly based on the measurement of object similarity to a certain reference shape—a circle, an ellipse, or a rectangle.

One of the oldest and widely used is *compactness* defined as

$$c = \frac{4\pi S(A)}{O(A)^2},$$

where $S(A)$ is the area and $O(A)$ is the perimeter of object A. It holds $0 < c \leq 1$, where $c = 1$ comes only if A is a circle. The name of this feature originates from our intuitive notion of the circle as the "most compact" object. That is why this feature is sometimes called *circularity*, although other definitions of circularity which come from the fit of the object boundary by a circle can be found in the literature (see [5], for instance).

The feature, which reflects the similarity between the object and its convex hull is called *convexity*, defined as

$$v = \frac{S(A)}{S(C_A)},$$

where C_A is the convex hull of A (see Figure 2.2). Clearly, $0 < v \leq 1$, and $v = 1$ for convex shapes only.

Another feature called *rectangularity* measures the similarity to the minimum bounding rectangle R_A (see Figure 2.3)

$$r = \frac{S(A)}{S(R_A)}.$$

Again, $0 < r \leq 1$ and the equality arises for rectangles only. There is another popular feature of this kind called *elongation*. It is again derived from the minimum bounding rectangle as the ratio of its side lengths.

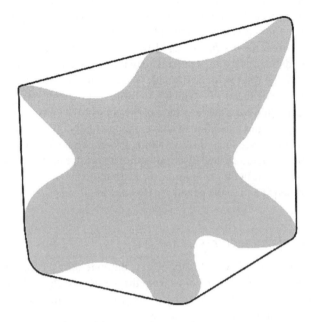

Figure 2.2   The object and its convex hull

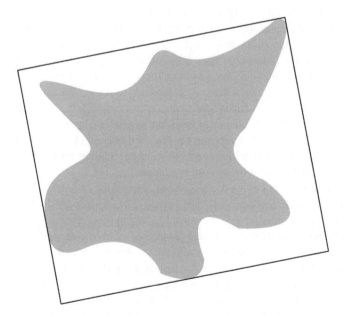

Figure 2.3   The object and its minimum bounding rectangle

Two other simple features refer to the properties of ellipses. The *ellipticity* is given by an error of the fitting the object by an ellipse. The *eccentricity* is simply the eccentricity of the fitted ellipse (other definitions can be found in literature).

Another scalar feature is the *bending energy* [6] of the object boundary. The bending energy is in the discrete case defined as a sum of the square of the boundary curvature over all boundary points, normalized by the total boundary length. We can imagine this feature as a potential energy of an ideally flexible metal wire which is deformed into the shape of the boundary. The bending energy is minimum for a circle, which corresponds to the physical interpretation.

All the simple shape features mentioned above are inherently invariant to translation, rotation, and uniform scaling. Individual discrimination power of each of them is poor, but when used together they may provide enough discriminability to resolve at least easy tasks or to serve as pre-selectors in a more difficult recognition. To learn more about simple shape features we refer to the monograph [5] and to the papers by Rosin et al. [7–12].

2.2.2 Complete visual features

A natural step forward is to stay with binary objects but to require a completeness of the features. This cannot be fulfilled by the features from the previous section. Probably the simplest way to reach the completeness while keeping the translation, rotation, and scaling invariance is to use polar encoding of the shape.

We explain this possibility for *star-shaped* objects first (the object is called star-shaped if any half line originating in the object centroid intersects the object boundary just once). Let us put the coordinate origin into the object centroid and construct boundary radial function $r(\theta), \theta \in \langle 0, 2\pi \rangle$. This function fully describes the object (see Figure 2.4). Instead of computing features of the object, it is sufficient to compute features of the radial function. The radial function is invariant to object translation and undergoes a cyclic shift when the object rotates. The change of the starting point at the boundary has the same impact. If we now sample the radial function in a constant angular step, we obtain the *radial shape vector* [13] (see Figure 2.5). The radial shape vector can be used directly as the feature vector. We can control the discrimination power by increasing/decreasing the shape vector length (i.e., by changing the angular sampling step). If the similarity (distance) between two radial shape vectors $\mathbf{a}, \mathbf{b}$ is defined as the minimum of $d(\mathbf{a}, \mathbf{b})$, where the minimum is searched over all cyclic shifts of $\mathbf{b}$, then we get a reasonable metric which ensures invariants to rotation. Scaling invariance can be achieved by normalizing the shape vector by its largest element.

The concept of the polar encoding can be generalized into the *shape matrix*, which is a binary matrix of user-defined dimensions [14] (see Figure 2.6). A polar grid of concentric circles and radial lines is placed into the object centroid such that the largest circle circumscribes the object. Each segment of the grid corresponds to one element of the shape matrix. If the segment is covered by the object, the corresponding value in the matrix is 1 and is zero otherwise. Since the segments have different sizes, appropriate weighting of the matrix elements was proposed in [15]. The shape matrix can be used for any object of a finite extent regardless of its particular shape. It works also for the shapes with holes and with multiple isolated components.

The *chain code*, which was introduced by Freeman in [16] and further developed in [17, 18] has become very popular in loss-less binary image compression and can also be considered a complete visual feature. The discrete object boundary is represented by a

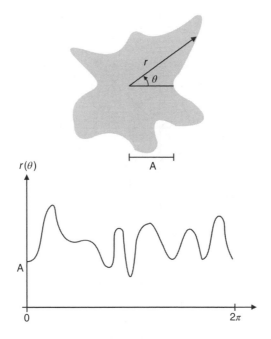

Figure 2.4   Radial function of the object

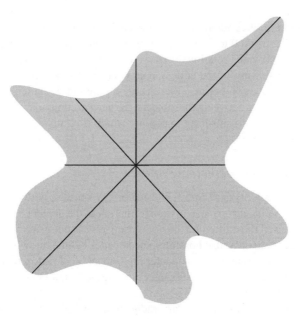

Figure 2.5   Star-shaped object and its radial shape vector

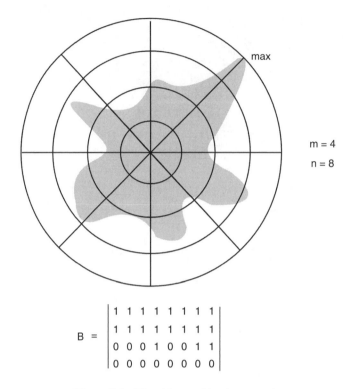

$$B = \begin{vmatrix} 1 & 1 & 1 & 1 & 1 & 1 & 1 & 1 \\ 1 & 1 & 1 & 1 & 1 & 1 & 1 & 1 \\ 0 & 0 & 0 & 1 & 0 & 0 & 1 & 1 \\ 0 & 0 & 0 & 0 & 0 & 0 & 0 & 0 \end{vmatrix}$$

Figure 2.6   The object and its shape matrix

sequence of elementary directional unit vectors, each direction being encoded by an integer (usually from 0 to 3 or from 0 to 7, depending on the topology). If implemented as a relative (differential) code, it provides rotational invariance while the invariance to the starting point is achieved by cyclic shifting of the code. Chain codes are, however, not frequently used as features for object recognition because there is no simple metric which would be consistent with the actual difference between the objects. Hence, a comparison of two chain codes should be done via maximum substring matching, which is relatively slow.

2.2.3   Transformation coefficient features

These features have been primarily designed for binary objects, both 2D and 3D. They are calculated from the coefficients of certain transformation of the object boundary. Fourier transformation has been used most often for this purpose [19–21] but using Walsh-Hadamard and other similar transformations is also possible. The key idea behind these features is to use polar or log-polar transformations in order to transfer rotation and scale into a (cyclic) shift, which is much easier to handle because it preserves the Fourier transformation magnitude as is implied by the Fourier Shift Theorem.

   If an object has a well-defined boundary radial function $r(\theta)$, we compute its 1D Fourier transformation. Since the object rotation results in a cyclic shift of $r(\theta)$ and the scaling with a factor s results in the radial function $s \cdot r(\theta)$, the Fourier transformation magnitude is under these two transformations only multiplied by s. Hence, the magnitude normalized by the first coefficient is invariant w.r.t. TRS (translation, rotation, and uniform change of scale) transformation. In practice, we take only the first n coefficients of discrete Fourier transformation; by increasing n we increase the discriminability. These features are known as the *Fourier descriptors*.

In the general case, the function $r(\theta)$ does not exist, but we still can construct the Fourier descriptors in another way. Consider the boundary pixels (x_k, y_k) as points in the complex plane $z_k = x_k + iy_k$. Now we can calculate discrete Fourier transformation of $(z_1, z_2, \cdots)$. The position of the object in the plane is encoded solely in the Fourier coefficient Z_0. Discarding them, we get a translation invariance. Rotation invariance along with the invariance to the origin of the sampling is achieved by taking the magnitudes $|Z_k|$'s, similarly to the previous case. Finally, the normalization by $|Z_1|$ yields the scaling invariance. Alternatively, we can use the phase of Z_1 for normalization of the other phases to rotation and sampling origin. The phases are not changed under scaling.

Fourier descriptors are very popular and powerful features in the presence of TRS transformation. In case of binary images, they compete with moment invariants. In a different form, they can be applied to graylevel images as well. If only an image translation was considered, we could use the Fourier transformation magnitude directly as an invariant. If only a rotation and scaling (but no translation) were present, then we could make a log-polar transformation of the image and again to use its Fourier transformation magnitude. However, if a full TRS transformation is present, we have to insert an intermediate step. First, we make a Fourier transformation of the original image and take its magnitude (which is invariant to shift). The log-polar transformation is now applied in the frequency domain to the magnitude. Then another Fourier transformation of this magnitude yields the TRS invariants.

Radon transformation offers another, yet similar possibility, of how to handle object rotations. The rotation (and the choice of the origin of angular projections) generates a one-dimensional cyclic shift in the Radon spectrum, which can be again eliminated by 1D Fourier transformation or by calculating the distance over all possible shifts.

2.2.4   Textural features

A particular group of features has been designed for description of images with prominent *textures*. The texture is a repeating pattern in graylevels or colors which may not be strictly periodic—it may contain irregularities and random perturbations. For objects made from materials such as wood, sand, skin, and many others, their texture is a dominant property; more class-specific than the shape and the color (see Figure 2.7 for some examples).

All textural features try to somehow capture the spatial appearance and its possible repetition frequency in different directions. In principle, any local directional features can be used for texture description. Autocorrelation function and its spectrum are straightforward general solutions. The review of textural features can be found in [22] and in the monograph [23], where the reader will also find many other relevant topics such as texture image preprocessing and texture modeling.

One of the first specialized textural features were proposed by Haralic et al. [24] in the form of spatial-dependence matrix. This idea has been re-used many times and has led to co-occurrence matrices. Another approach models the texture as a random process

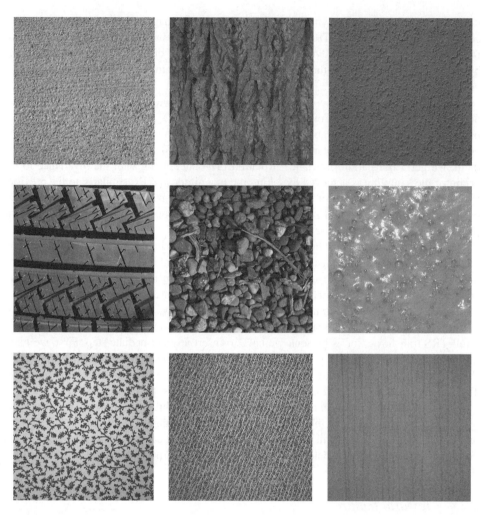

Figure 2.7   Examples of textures. The texture is often a more discriminative property than the shape and the color

(AR, ARMA, and Markov models have been used for this purpose), the parameters of which serve as the features. This approach requires estimating the process parameters from the image, which may not be trivial.

Very powerful textural features are the *Local binary patterns* (LBPs) introduced by Ojala et al. [25]. The LBPs are high-dimensional features the core idea of which is to encode, for each pixel, whether or not its value is higher than that of the neighboring pixels. The LBPs exist in many versions and extensions, the most important are multiresolution LBPs [26] and 3D LBPs [27].

Most recently, the attention of researchers has been drawn to the *Bidirectional texture function* (BTF), which describes a kind of texture the appearance of which depends not only on the position in the scene but also on the view and illumination spherical angles. BTF is highly redundant description, so for recognition it is usually used after a lossy compression. The notion of BTF was introduced by Dana et al. [28]. Later on, the methods for modelling, recognition, and synthesis of the BTF were developed by Haindl et al. [29, 30].

2.2.5 Wavelet-based features

In fact, this category could be included in the previous two sections, but due to its importance it is handled separately. Wavelet-based features make use of the *wavelet transformation* (WT) [31], which provides a spatial-frequency representation of the analyzed image (see Figure 2.8 for an example), and thus it is able to capture both the shape and textural information of the object. The two most frequent ways to incorporate wavelet transformation in the object description are the following. For binary objects, the features are constructed from 1D WT of the boundary [32, 33] which is analogous to Fourier descriptors. Using a proper boundary parametrization, these features can be made even affine invariant [34, 35]. A comparison of the efficiency of the Fourier and wavelet boundary coefficients can be found in [36] (the authors claimed wavelets slightly outperform Fourier descriptors).

Graylevel images are described by features which are calculated from individual bands of 2D WT and capture the object texture. The texture-oriented descriptors are based on the representation of an energy distribution over wavelet subbands, such as overall energy, entropy, or covariance between decompositions of individual color channels (in the case of color images) [37]. The decomposition can be done by means of wavelet transformation often using Gabor or Daubechies wavelets, but other approaches are used too, such as wavelet packet transformation [38] and wavelet frames. One of the first applications was described in [39]. Great attention has been paid to proper selection of the bands used for the descriptor construction [40].

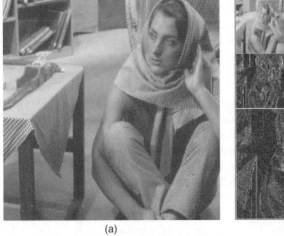
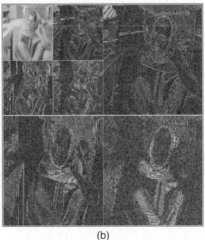

(a) (b)

Figure 2.8 The original Barabara image (a) and its wavelet decomposition into depth two (b)

Recently, with increasing computational power, the third approach to wavelet-based object description has begun to appear. The overcomplete representation of objects is created by means of all coefficients at chosen coarser levels of the object wavelet decomposition [41]. Such feature vectors are then used as object descriptors.

2.2.6   Differential invariants

Differential invariants are local features based on the derivatives of the image intensity function in case of graylevel objects or, when dealing with binary objects, on the derivatives of the object boundary.

In the case of binary objects with a smooth boundary, the invariants are calculated for each boundary point as functions of the boundary derivatives from order two to eight. In that way they map the boundary onto a so-called *signature curve* which is an invariant to affine or even projective transformation. The signature curve depends at any given point only on the shape of the boundary in its immediate vicinity. Any local change or deformation of the object has only a local impact on the signature curve. This property of locality, along with the invariance, makes them a seemingly perfect tool for recognition of partially occluded objects. Recognition under occlusion can be performed by substring matching in the space of the signatures. Differential invariants of this kind were discovered by Wilczynski [42], who proposed to use derivatives up to the eighth order. Weiss introduced differential invariants to the computer vision community. He published a series of papers [43–45] on various invariants of orders from four to six. Although differential invariants may seem to be promising from a theoretical point of view, they have been experimentally proven to be extremely sensitive to inaccurate segmentation, sampling errors, and noise.

Such high vulnerability of "pure" differential invariants led to development of *semi-differential* invariants. This method divides the object into affine-invariant parts. Each part is described by some kind of global invariants, and the whole object is then characterized by a string of vectors of invariants. Recognition under occlusion is again performed by maximum substring matching. Since inflection points of the boundary are invariant to affine (and even projective) deformation of the shape, they have become a popular tool for the definition of the affine-invariant parts. This approach was used by Ibrahim and Cohen [46], who described the object by the area ratios of two neighboring parts. Horacek et al. [47] used the same division of the object by means of inflection points but described the cuts by affine-invariant modification of the shape vector. Other similar variants can be found in [48] and [49].

As a modification which does not use inflection points, concave residua of a convex hull can be used. For polygon-like shapes, however, inflection points cannot be used since they do not exist. Instead, one can construct the cuts defined by three or four neighboring vertices. Yang and Cohen [50] and Flusser [51] used the area ratios of the cuts to construct affine invariants. A similar method was successfully tested for perspective projection by Rothwell et al. [52].

A slightly different approach, but still belonging to semi-differential invariants, is the *Curvature scale space* (CSS) proposed in [53]. It detects inflection points of the shape and investigates the movement trajectory and the "lifetime" of each inflection point under progressive smoothing of the boundary by a Gaussian filter; see Figure 2.9. Under smoothing, neighboring inflection points converge to each other and then vanish. The number of inflection points on each smoothing level, the speed of their convergence and

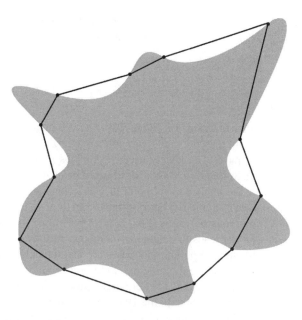

Figure 2.9   Semi-differential invariants. The object is divided by inflection points. Both convex and concave cuts can be used for a description by global invariants

the number of smoothings that each inflection point survives are encoded into a curve which characterizes the object (although not completely). This curve can be made rotationally (and even affine) invariant and for objects with a number of inflection points provides enough discriminability.

Local differential invariants exist also for graylevel and color images. Probably the most famous representative is the *Scale-invariant feature transformation* (SIFT) [54]. This method detects dominant points in the image at different scales, looking for the extrema of the scale space created by means of the *Difference of Gaussians* (DoG) and describes their neighborhood by means of orientation histograms. The descriptors of individual dominant points are then concatenated into a long feature vector, which is invariant to image translation, scaling, and rotation, and robust to illumination changes. SIFT method has inspired numerous authors to improvements, modifications, and developing of other methods of the same nature, among which the *Histogram of oriented gradients* (HOG) [55] and *Speeded up robust features* (SURF) [56] belong to the most popular ones.

2.2.7   Point set invariants

Invariants of point sets form a relatively narrow, specialized category. They were developed for binary objects with polygonal boundary (i.e., for objects where the differential invariants are not well defined). The vertices of the polygon fully describe the complete shape, so it is meaningful to use them for feature generation. Point set invariants were usually constructed to be invariant w.r.t. projective transformation and also to the vertex labeling order. They are defined either as products or ratios of the distances between two points or as products or ratios of various triangle areas which are formed by point triplets [57–60]. The simplest projective invariant of this kind—the *cross ratio* —was known already to ancient mathematicians Euclid and Pappus.

2.2.8   Moment invariants

Moment invariants are special functions of *image moments*. Moments are scalar quantities, which have been used for more than hundred years to characterize a function and to capture its significant features. They have been widely used in statistics for description of the shape of a probability density function and in classic rigid-body mechanics to measure the mass distribution of a body. From the mathematical point of view, moments are "projections" of a function onto a polynomial basis (similarly, Fourier transformation is a projection onto a basis of harmonic functions). Numerous polynomial bases have been used to construct moments. The standard power basis $\{x^p y^q\}$ leads to geometric moments; various orthogonal bases lead to orthogonal moments. Although all polynomial bases are theoretically equivalent, it makes sense to use different bases (and different moments) in different situations, either for the sake of simplicity or because some bases provide better numerical properties than others.

The history of moment invariants began in the 19th century, many years before the appearance of the first computers, under the framework of group theory and of the theory of algebraic invariants. The theory of algebraic invariants was thoroughly studied by famous German mathematicians P. A. Gordan and D. Hilbert [61] and was further developed in the 20th century in [62, 63] and [64], among others.

Moment invariants were first introduced to the pattern recognition and image processing community in 1962, when Hu employed the results of the theory of algebraic invariants and derived his seven famous invariants to rotation of 2-D objects [65]. Since that time, thousands of papers[8] have been devoted to various improvements, extensions and generalizations of moment invariants and also to their use in many areas of applications. Moment invariants were extensively reviewed in four monographs. The first one by Mukundan [67], published as early as in 1998, covered only selected subtopics. The second one by Pawlak [68] is focused mainly on computational aspects of moments and detailed error analysis in the discrete domain rather than on their invariant properties. Various moment invariants in 2D have been the main topics of the monograph by Flusser et al. [69]. The most recent book edited by Papakostas [66] reflects the latest development on the field.

Moment invariants have become important and frequently used shape descriptors. Nowadays, they exist for 2D as well as for 3D objects with a possible extension into arbitrary dimensions in some cases. We have moment invariants for binary, gray-level, color, and even vector-valued images. Concerning the type of invariance, there exist moment invariants to similarity and affine object transformations as well as invariants w.r.t. image blurring with certain types of the blurring filters.

Even though moment invariants suffer from certain intrinsic limitations (the worst of which is their globalness, which prevents a direct utilization for occluded object recognition), they frequently serve as the "first-choice descriptors" and as the reference method for evaluating the performance of other shape features.

---

[8] According to [66], about 18,000 research papers relevant to moments and/or moment invariants in image analysis have appeared in SCOPUS.

2.3   Classifiers

A classifier is an algorithm which "recognizes" objects by means of their representations in the feature space and assigns each object to one of the pre-defined classes $\omega_1, \cdots, \omega_C$. The input of a classifier is the feature vector $(I_1(f), \cdots, I_n(f))^T$ of an unknown object f, and the output is the label of the class the object has been assigned to. The classifiers are totally independent of what the original objects actually look like and also of the particular kind of features I_k.

In a *supervised classification*, the classes are specified beforehand by means of a *training set*. The training set is a set of objects the class membership of which is supposed to be known. It should contain enough typical representatives of each class including intra-class variations. Selection of the training set is the role of a human expert in the application domain. This is not a mathematical/computer science problem and cannot be resolved by any automatic algorithm because it requires a deep knowledge of the application area. Since the training set is selected manually, it often contains misclassified objects. This is caused not only by human mistakes but also because some objects may be so deformed or of such non-typical appearance that the expert may be in doubt which class they should be assigned to.

In an *unsupervised classification*, the training set is not available; all objects are of unknown classification at the beginning. Sometimes even the number of classes is unknown. The classes are formed iteratively or hierarchically during the classification process such that they create compact clusters in the feature space. Since the unsupervised classification is rarely used in visual object recognition, we mention this approach for the sake of completeness but do not go into details.

Constructing a classifier is equivalent to a partition of the feature space (see Figure 2.10). Each partition component is assigned to certain class (multiple partition components can be assigned to the same class but not vice versa), and the unknown object f is classified according to the position of $(I_1(f), \cdots, I_n(f))^T$. Hence, constructing a classifier means defining the boundaries of the partition. The partition boundaries are set up by means of the training set.

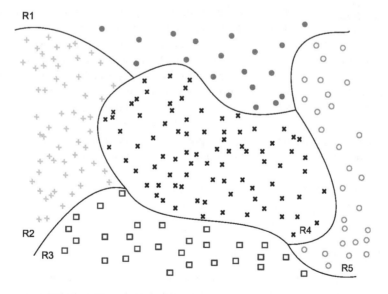

Figure 2.10   Partition of the feature space defines a classifier

This stage is called *training* or *learning* of the classifier, and it is the most difficult and challenging part of the classifier design and usage. The main challenge follows from the fact that the training set is always of a limited size (sometimes of a very small one), but we try to set up the classification rules, which should work well on a high number of unknown objects. So, it is a kind of generalization or extrapolation from a small sample set, where the result is always uncertain. As soon as the classifier has been trained, the rest (i.e., the recognition of unknown objects) is a routine task.

Theoretically, if all the assumptions were perfectly fulfilled, the training should be very simple. Assuming that $I_1, \cdots, I_n$ are invariant and discriminatory, then *all* members of each class are mapped into a single point in the feature space irrespective of the intra-class variations and the points representing different classes are distinct. Then a one-element training set per class would be sufficient, and the classifier would just check which training sample the incoming object coincides in the feature space. This model is, however, very far from reality. The features I_k, although theoretically invariant, may not provide a perfect invariance within the classes simply because the intra-class variations are broader than they have been expected to be. Moreover, in all stages of object recognition we encounter errors of various kinds, which all contribute to the violation of the theoretical assumptions. In such a case, the training samples of the same class may fall into distinct points in the feature space.

In the literature, the classifiers are sometimes formalized by means of *decision functions*. Each class ω_i is assigned to its real-valued decision function g_i which is defined in the feature space. Then an object f represented by a feature point $\mathbf{a} = (I_1(f), \cdots, I_n(f))^T$ is classified into class ω_k if and only if $g_k(\mathbf{a}) > g_i(\mathbf{a})$ for all $i \neq k$. This concept is equivalent to the previous one. If the decision functions have been given, then the partition boundaries can be found as solutions of the equations $g_j(\mathbf{a}) = g_i(\mathbf{a})$. On the other hand, given the partition, we can easily construct the decision functions as characteristic functions of the partition components.

There have been many approaches to the classifier training and to how the partition is set up. In some classifiers, the partition boundaries are defined explicitly, and their parameters are found via training; some others do not use any parametric curves/surfaces, and the partition, is defined implicitly or by means of another quantity. For the same training set, we can construct infinitely many classifiers (by means of different training algorithms) which will most probably differ one another by its performance on independent data. If we use an over-simplified model of the partition, then we only roughly approximate the "ground-truth" partition, and the number of misclassified objects is high even inside the training set. If, on the other hand, we allow relatively complex boundaries and minimize the classification error on the training set, we end up with an *over-trained* classifier, which is optimal on the training set, but there is no guarantee of its optimality on independent data (more precisely, it is highly probable that such classifier will perform poorly). One feels that the "optimal" classifier should be somewhere in between (see Figure 2.11).

In the rest of the section we briefly review the basic classifier types. For details and for other classifiers, we refer to the specialized monographs [70, 71].

2.3.1 Nearest-neighbor classifiers

The *nearest-neighbor* (NN) classifier, sometimes also called the *minimum distance* classifier, is the most intuitive classifier. Vector $\mathbf{a}$ is assigned to the class which minimizes the distance

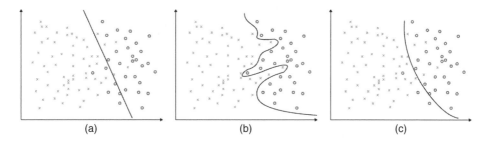

Figure 2.11 Three different classifiers as the results of three training algorithms on the same training set. (a) Over-simplified classifier, (b) over-trained classifier, and (c) close-to-optimal classifier

$\varrho(\mathbf{a}, \omega_i)$ between $\mathbf{a}$ and respective classes. The feature metric space with the metric d offers a variety of possibilities of how the distance ϱ between a point and a set (class) can be defined. The one which is common in metric space theory

$$\varrho_1(\mathbf{a}, \omega) = \min_{\mathbf{b} \in \omega} d(\mathbf{a}, \mathbf{b})$$

can of course be used, but it is very sensitive to outliers in the training set. To avoid this sensitivity, the "mean distance"

$$\varrho_2(\mathbf{a}, \omega) = d(\mathbf{a}, \mathbf{m}_\omega),$$

where $\mathbf{m}_\omega$ is the centroid of the class ω, is sometimes used instead. Using the mean distance is also justified by a known paradigm from statistics, which says that no algorithm working with real data should use extreme points or values, because they tend to be unstable and most often are influenced by measurement errors and noise. On the other hand, ϱ_1 reflects the size and the shape of the classes in a better way than ϱ_2 (see Figure 2.12).

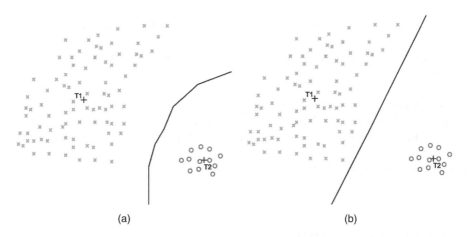

Figure 2.12 Decision boundary of the NN classifier depends on the used distance: (a) the nearest distance ϱ_1 and (b) the mean distance ϱ_2

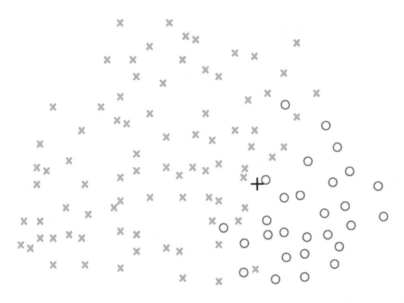

Figure 2.13 Robustness of the k-NN classifier. The unknown sample "+" is classified as a circle by the NN classifier and as a cross by the 2-NN classifier. The later choice better corresponds to our intuition

If each class is represented by a single training sample, then of course $\varrho_1 = \varrho_2$. In the case of two classes, this is the simplest situation ever, which leads to a linear decision boundary. In the case of more classes, the feature space partition is given by Voronoi tessellation, where each training sample is a seed point of a Voronoi cell that corresponds to the respective class.

In case of multiple training samples in each class, there is of course a difference between the NN classifiers which use ϱ_1 and ϱ_2 distances. While ϱ_2 again leads to a Voronoi tessellation the seeds of which are the class centers, the use of ϱ_1 generates complex curved decision boundaries.

The sensitivity of the NN classifier with ϱ_1 to outliers can be suppressed by introducing a more general k-NN classifier, with k being a (usually small) integer user-defined parameter (see Figure 2.13). We can find two versions of this algorithm in the literature.

Version 1

1. Unknown feature vector **a** is given.
2. Find k training samples which are the closest (in the sense of metric d) to sample **a**.
3. Check for each class how many times it is represented among the k samples found in Step 1.
4. Assign **a** to the class with the maximum frequency.

Version 2

1. Unknown feature vector **a** is given.
2. Denote the number of the closest training samples from ω_i as k_i. Set $k_i = 0$ for $i = 1, 2, \cdots, C$.

3. Until $k_j = k$ for some j do

   Find training sample **c** which is the closest (in the sense of metric d) to sample **a**.

   If $\mathbf{c} \in \omega_i$, then set $k_i = k_i + 1$.

   Exclude **c** from the training set.
4. Assign **a** to the class ω_j.

Version 1 is simpler because we need to find just k closest neighbors, but it may not provide an unambiguous decision. Version 2 requires finding between k and $C(k - 1) + 1$ closest neighbors (their exact number cannot be predicted) but mostly yields a decision[9]. Although the meaning of k is slightly different in both versions, its choice is always a heuristic often done on a trial and error basis. Small k provides less robustness to outliers, but the classification is faster. The corresponding decision boundary may be curved and complex, and the classifier tends to get overtrained. For $k = 1$, we obtain a plain NN classifier. When k is higher, the decision boundary is smooth and less distinct. More training samples may be classified incorrectly (which says nothing about the overall quality of the classifier). Anyway, the upper limit on k is that it should be by an order smaller than the number of the training samples in each class. A typical choice of k in practice is from 2 to 10, depending on the size and reliability of the training set.

The k-NN classifier is relatively time expensive, especially for high k. Several efficient implementations can be found in the literature, and some are even available on the internet.

2.3.2   Support vector machines

Classifiers called the *Support vector machines* (SVMs) are generalizations of a classical notion of linear classifiers. The SVMs were invented by Vapnik [72]. In the training stage, the SVM classifier looks for two parallel hyperplanes which separate the training samples such that the distance (which is called *margin*) between these hyperplanes is maximized. The decision boundary is then another hyperplane parallel with these two and lying in the middle between them (see Figure 2.14). The margin hyperplanes always pass through some training samples; these samples are called *support vectors*. The support vectors can lie on the convex hull of the training set only. The SVM classifier satisfying this constraint is called the *hard margin* SVM.

Training of the SVM is a constrained optimization problem, where we minimize the norm of the hyperplane normal vector subject to the condition that the margin hyperplanes separate the data. This can be efficiently solved by quadratic programming. The hard margin constraint is applicable only to linearly separable classes with almost error-free samples. If there are some outliers in the training set, we can apply the *soft margin* constraint proposed by Cortes [73]. Soft margin constraint allows some training samples to lie inside the margin or even on its other side. It again leads to a constrained optimization problem, but the objective function contains an additional penalty term. For the difference between the hard- and soft-margin classifiers, see Figure 2.14.

Boser et al. [74] generalized the idea of the maximum margin hyperplanes to linearly non-separable classes. They proposed a mapping of the current feature space into a new one,

---

[9] The decision still may not be unique because there may exist multiple minima, but this ambiguity is rare comparing to the ambiguity of Version 1.

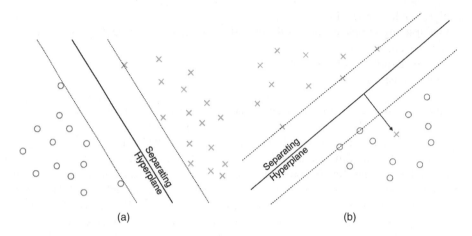

(a)   (b)

Figure 2.14   SVM classifiers. The hard margin (a) and the soft margin (b) constraints

in which the classes are linearly separable. This mapping is defined by means of various radial basis functions and is known as the *kernel trick*.

The SVM's have become very popular namely because their relatively fast training. They provide a good trade-off between speed and accuracy, namely for mid-sized training sets (although they can be applied to large training sets as well). However, they also exhibit several potential drawbacks. The hard margin version ignores the distribution of the samples inside the training set; it considers only the samples on the convex hull. Unlike probabilistic classifiers, the SVM decision is deterministic and does not take into account prior probability of individual classes. SVMs were originally proposed for a two-class (dichotomic) problem. Generalization to more classes is commonly done by decomposition into several dichotomies [75], which may be time-expensive if C is large.

2.3.3   Neural network classifiers

Artificial neural networks (ANNs) are "biologically inspired" classifiers. The original idea was to employ a massive parallelism offered by the most powerful computers and to construct a "network" of many "neurons", which can perform simple operations in parallel. These neurons are organized into layers. The first (input) layer contains as many neurons as is the dimensionality of the feature vector. It reads the object features, makes a simple operation(s) with them, and sends these results to the second layer. This process is repeated until the last (output) layer has been reached; the output layer provides a decision about the class membership. The layers between the input and output layers are called *hidden* layers. The neurons of the same layer do not communicate among themselves, so they can actually work in parallel.

The ANNs were firstly used for classification purposes probably by Rosenblatt [76], who proposed the first single-layer *perceptron*. It was, in fact, a linear dichotomic classifier. Each neuron was assigned a weight, and the perceptron calculated the linear combination of the features (possibly with an offset) weighted by these weights. The set where this linear combination was zero, defined a decision hyperplane. So, the perceptron only checked the sign of the output and classified the object according to this sign. The training of the perceptron consists of setting up the neuron weights such that the training set (provided it is linearly separable) is classified correctly. The training algorithm is iterative and upgrades the weights after the processing of each training sample.

Later, the multilayer ANNs and especially the popular *radial basis function* (RBF), networks broke the limitation to linearly separable tasks. The RBF networks have at least one hidden layer, where the features are used as arguments of certain RBFs. So, the evaluation of the RBF is used instead of doing linear combinations. The RBFs are popular, smooth, and "infinitely flexible" functions known from approximation theory (multidimensional thin-plate splines and Gaussian functions are examples of the RBFs). When used in the ANN classifier, they can create an arbitrary smooth decision boundary [77].

The popularity of the ANN's in the 1970s and 1980s originated from two common beliefs. The first one was that artificial intelligence algorithms should copy the "algorithms" in a human brain (even though the computer and brain architectures are different) and the second one that the future of computers is in very powerful mainframes equipped by thousands of CPUs. As we know today, the recent development of computer technologies followed a different way to the distributed computing and personal computers.

The ANN classifiers have been compared in many studies to the SVM with ambiguous results. There is a belief these two types of classifiers are more or less comparable in performance (although they use different terminology and mathematical background). We can find many parallels between them. The single-layer perceptron corresponds to the hard-margin linear SVM; the RBF networks use the same "linearization" principle as the kernel trick in SVM (both even use the same functions to create the decision boundary). The differences in performance reported in individual experiments were caused probably by the properties of the particular data rather than by general properties of these classifiers. In the 1980s, the ANN classifiers were believed to overcome a construction of small sophisticated feature sets, which is sometimes difficult. In many tasks, such as character recognition, the ANN was trained directly on images, taking each pixel value as a feature. This was, however, a dead-end road because these "features" did not provide any invariance to elementary image transformations. All the above facts implied that neural network classifiers were gradually overtaken in popularity and in the number of reported applications by support vector machines in 1990–2005. After 2005, the new concept of *deep learning* and *convolution networks* resurrected interest in the ANNs.

Deep convolution neural networks (CNN) are modern classifiers which go beyond the conventional framework of separated feature design and classifier training. Although they were firstly proposed as early as 1980 [78], they have attracted noticeable attention very recently when they repeatedly achieved an excellent performance in the *Large Scale Visual Recognition Challenge* [79].

Instead of working with features "manufactured" beforehand, the CNN generates the features by a cascade of convolutions and downsampling. The parameters of each convolution kernel are learned by a backpropagation algorithm. There are many convolution kernels in each layer, and each kernel is replicated over the entire image with the same parameters. The function of the convolution operators is to extract different features of the input. The capacity of a neural net varies, depending on the number of layers. The first convolution layers obtain the low-level features, such as edges, lines, and corners. The more layers the network has, the higher-level features it produces. The CNNs virtually skip the feature extraction step and require only basic preprocessing, which makes them, if enough computing power is available, very powerful [80].

The absence of sophisticated features defined in advance is not only an advantage but also a drawback of the CNNs. For CNNs it is very difficult (if not impossible) to generate features (by their inner layers), invariant to rotation or affine transformations (at least no such methods have been reported in the literature). In general, it is possible to reach translation and scale invariance, but probably the only way of dealing with other intraclass variabilities is a brute force approach or prior image normalization.

2.3.4   Bayesian classifier

The Bayesian classifier is a statistical classifier which considers the features to be random variables. It is suitable when the training set is large. It makes a decision by maximizing the posterior probability $p(\omega_i|\mathbf{a})$ of assigning an unknown feature vector $\mathbf{a}$ to class ω_i. This is a reasonable idea because $p(\omega_i|\mathbf{a})$ is the probability that an object whose feature vector is $\mathbf{a}$ belongs to the class ω_i. Maximizing the posterior probability assures that the probability of misclassification is minimized [70]. This is why the Bayesian classifier is sometimes called the "optimal" classifier.

Since the posterior probabilities cannot be estimated directly from the training set, we take advantage of the Bayes formula, which express the posterior probability in terms of prior probability and class-conditional probability

$$p(\omega_i|\mathbf{a}) = \frac{p(\mathbf{a}|\omega_i) \cdot p(\omega_i)}{\sum_{j=1}^{C} p(\mathbf{a}|\omega_j) \cdot p(\omega_j)}.$$

Since the denominator is constant over i, maximizing posterior probability is equivalent to maximizing the numerator.

The prior probability $p(\omega_i)$ express the relative frequency with which the class ω_i appears in reality. In other words, $p(\omega_i)$ is a probability that the next object, which we have not seen/measured yet, will be from the class ω_i. The priors are unknown and must be estimated during the classifier training. There are basically three approaches how to estimate the priors. The best possibility is to use prior knowledge, which might be available from previous studies. In medical diagnostics, for instance, we know the incidence of certain diseases in the population. In character recognition, the probabilities of the frequency of occurrence of each character is well known from extensive linguistic studies (in English, for instance, the most probable letter is E with $p(E) = 0.13$ while the least probable is Z with $p(Z) = 0.0007$ [81]).

If such information is not available, we may estimate $p(\omega_i)$ by the relative frequency of occurrence of ω_i in the training set. This is, however, possible only if the training set is large enough and has been selected in accordance with reality. If it is not the case and no prior knowledge is available, we set $p(\omega_i) = 1/C$ for any i.

The class-conditional probability $p(\mathbf{a}|\omega_i)$, sometimes referred to as the *likelihood* of the class ω_i, is a probability of occurrence of point $\mathbf{a}$ in ω_i. It is given by the respective probability density function (pdf), which is to be estimated from the training data independently for each class. This estimation is usually a parametric one, where a particular form of the pdf is assumed and we estimate only their parameters. Although any parametric pdf can be employed, the Gaussian pdf is used most frequently. Hence, we assume the n-dimensional pdf in the form

$$p(\mathbf{a}|\omega_i) = \frac{1}{\sqrt{(2\pi)^n|\Sigma_i|}} \exp\left(-\frac{1}{2}(\mathbf{a} - \mathbf{m}_i)^T \Sigma_i^{-1}(\mathbf{a} - \mathbf{m}_i)\right), \qquad (2.1)$$

where mean vector $\mathbf{m}_i$ and covariance matrix Σ_i are estimated by a sample mean and a sample covariance matrix from the training data[10]. In this way, the classifier no longer works with the individual training samples, it works only with the pdf parameters. In the case of equal priors, the classifier maximizes the likelihood. The decision boundary in the feature space between two classes is given by the equation

$$p(\mathbf{a}|\omega_i) = p(\mathbf{a}|\omega_j),$$

which always leads to a hyperquadric (a conic if $n = 2$). If the covariance matrices are equal, $\Sigma_i = \Sigma_j$, then the decision boundary between ω_i and ω_j is a hyperplane (a line if $n = 2$), and vice versa (in this case, the joint covariance matrix Σ can be robustly estimated from the training samples from both ω_i and ω_j simultaneously). In such a case, the classifier can be understood as a minimum distance classifier which minimizes the *Mahalanobis distance*

$$\varrho_M(\mathbf{a},\omega_i) = (\mathbf{a} - \mathbf{m}_i)^T \Sigma^{-1}(\mathbf{a} - \mathbf{m}_i).$$

If Σ is diagonal with equal variances, we end up with a standard Euclidean minimum distance classifier.

If the classes are not normally distributed, their pdf's can be estimated using other parametric models or, alternatively, by non-parametric techniques (the estimation by means of Parzen window uses to be applied, see [70] for details).

Since the Bayesian classifier requires a large number of training samples in each class, it is not convenient for face and fingerprint recognition, where typically each class is represented by a single (or very few) sample(s). On the other hand, it is widely used and performs very well in the pixel-wise classification of multispectral and hyperspectral data (see Figure 2.15).

2.3.5 Decision trees

Decision trees are simple classifiers designed particularly for logical "yes/no" features and for categorial features[11], where no "natural" metric exists. The classifier is arranged into a (usually but not necessarily binary) tree, where the unknown object enters the root and passes the tree until it reaches a leaf. In each node, a simple question is asked, and the next move is chosen according to the answer. Each leaf is associated with just one class, and the object is classified accordingly (several leaves may be associated with the same class). Decision trees can be applied to real-valued features as well. The queries have the form of simple inequalities (is the feature a_i less or greater than threshold t?) or compound inequalities, where a function of several features is compared to the threshold.

---

[10] In some cases when the features are independent, we can assume all Σ_i being diagonal. Such classifier is called the *naive* Bayesian classifier.

[11] Categorial features can take only few distinct values, for instance "good", "average", and "poor". They are not very common in image analysis but are of great importance in medical diagnostics and social science statistics.

Figure 2.15 Multispectral satellite image. The objects are single pixels, the features are their intensities in the individual spectral bands. This kind of data is ideal for the Bayesian classifier

In the training phase, we must design the tree shape, choose the queries in all nodes and select the thresholds (and the compound functions) the real-valued features (if any) are to be compared with. Obviously, our goal is to "grow" a small tree with few nodes and simple decision queries. This is not a trivial task, and there have been several approaches to tree training. The most popular one relies on the concept of *purity* or *impurity* of the node. The node is called *pure* if it contains only the objects belonging to one class (clearly, the only pure nodes are the leaves); the node with a maximum impurity contains different classes with the same frequency.

The principle underlaying the tree creation is that we want to maximize the decrease of impurity between the node and the immediate descendent nodes. Hence, a proper query at the node should split the data into the parts which are as pure as possible. There have been several definitions of the impurity. Most of them are inspired by information theory, such as *entropy impurity, information impurity, and Gini impurity*, some others follow from our intuitive understanding of the problem (*variance impurity, misclassification impurity*) [70]. The chosen impurity measure is optimized in a "greedy" manner, which means a maximum drop is required in each node. It is well known that this locally optimal approach may not lead to the simplest tree but global tree optimization is not feasible due to its complexity.

2.3.6   Unsupervised classification

As we already mentioned, unsupervised classification (also called *clustering*) has in object recognition less importance than the supervised one, because in most tasks we do have a training set. Still, clustering may be applied in the preliminary stage before the training set has been selected, for instance in multispectral pixel-wise classification, to see how many potential classes there are.

Clustering is a traditional discipline of statistics and data analysis. Unlike the supervised classification, the desired number of clusters may be unknown. Intuitively, the cluster is a subset of the data, which is "compact" and "far from other clusters" in the feature space. The meaning of these terms, of course, depends on the chosen metric and may be defined in various ways.

Clustering methods can be divided into two major groups—*iterative* and *hierarchical* methods. Iterative methods typically require the number of the clusters as an input parameter. In each iteration, they try to re-assign the data points such that a given quality criterion is maximized. A typical representative of iterative methods is the famous *C*-means clustering.

Hierarchical methods built the cluster structure "from scratch". *Agglomerative clustering* starts from the initial stage, where each data point is considered to be a cluster. On each level of the hierarchy, the two "most similar" clusters are merged together. Various criteria can be used to select the clusters to be merged. The stopping level and the final configuration are usually chosen by means of the *dendrogram*, a symbolic tree graph showing the clustering process. A complementary approach is provided by *divisive* algorithms, which on the initial level consider all data to be contained in a single cluster. In each level, the algorithm selects the cluster to be split and then divides this cluster into two parts, trying to optimize the same or similar criteria as are used in agglomerative methods. Although agglomerative and divisive methods may seem to converge to the same clustering, they are not completely "inverse" and may lead to different results. Agglomerative methods are usually faster and more frequently used, particularly if the desired/expected number of clusters is high. To learn more about clustering techniques, we refer to [71].

2.4   Performance of the classifiers

Before we proceed to the explanation of how we can increase the classifier performance, let us mention how the classifier performance should be evaluated. After that, we briefly review two popular techniques used for improving the classification—*classifier fusion* and *dimensionality reduction*.

2.4.1   Measuring the classifier performance

Before we proceed to the explanation of how to increase the classifier performance, let us mention how the classifier performance should be evaluated. Intuitively, one may expect that the classifier performance could be measured by a single scalar indicator called the *success rate*, which is a ratio of the number of successfully recognized objects to all trials (it is often presented as a percentage). The success rate, however, shows the quality of the classifier in a limited way, and sometimes it is even completely misleading. Consider a medical prophylactical screening for a disease whose statistical frequency in the population is 0.001. If any tested person was automatically "classified" as a healthy person, then the classifier would reach an excellent success rate 0.999 but such classifier is totally useless since it does not detect any disease occurrence. Hence, the success rate can be used only if the relative frequency of all classes is the same, but that is not the only limitation.

Let us again imagine a medical test whether or not the test for person has a cancer. Since only suspected persons pass this test, the relative frequency of the cancer occurrence can be considered 0.5. Still, the success rate itself does not say anything about the quality of the test because the two possible misclassifications have completely different weights. While the "false positive" results will be later disproved by other tests (which may be expensive and unpleasant for the patient but does not do any harm), the "false negative" result leads to a delay in the treatment and may be fatal for the patient. The success rate does not distinguish these two cases.

The two above examples illustrate why the classifier performance should be reported by a full *confusion table*, which is a $C \times C$ matrix with the ij-th element being the number of the ω_i members which have been assigned to ω_j.

The confusion table provides a very good insight into the quality of the classifier if it has been evaluated on a representative set of objects. It is seriously misleading to use the training set for this purpose. The error estimates would be highly optimistic, and if the confusion table was diagonal, we could not conclude that the classifier is perfect, but it rather shows it is heavily overtrained. The correct way is to evaluate the classifier on the *test set*, which should be selected in the same way as the training set but has not been used for the training. If the user has not provided us with the separate test set, we can just leave out 10% of the training set and use it for the evaluation. By repeating this evaluation several times with randomly selected 10% of the training samples and calculating the means and standard deviations, we obtain a very reliable and informative confusion table.

2.4.2   Fusing classifiers

Fusing or combining the results of different classifiers together in order to reach better performance is a modern trend in classifier design. This idea was proposed by Kittler et al. [82], who presented several fusion schemes—algorithms, how to combine the decisions of several classifiers together. By the term "different classifiers" we understand either two classifiers of the same nature but being trained on different training sets and/or in different feature spaces, or two classifiers of different nature trained on the same or different data. For deterministic classifiers such as the k-NN and SVM, there is not much freedom in constructing the fusion rule. The absolute majority and the simple majority vote are probably the only reasonable choices. In case of probabilistic classifiers, such as the Bayesian classifier, the individual probabilities of the assignments are fused instead of fusing final decisions. Then we maximize the product, the sum, the median or the maximum of individual posterior probabilities.

The classifier fusion has been successfully used in handwritten character recognition [83, 84], in medical diagnostics [85, 86], and in many other problems. It is, however, true that classifier fusion does not always guarantee better accuracy than the best-performing individual classifier. The best chance of improvement is if the input classifiers have a comparable (and not very high) accuracy. The fusion rules may be not only fixed but may be also adaptive w.r.t. the dataset, which improves the performance particularly if more classifiers and hierarchical fusion schemes are used [87]. For an exhaustive overview of classifier fusion techniques we refer to the recent monograph [88].

2.4.3   Reduction of the feature space dimensionality

We should keep the dimensionality of the feature space as low as possible. The reason for that is twofold—working in fewer dimensions is faster, and the classifier might be even more accurate than in higher dimensions. While the former proposition is evident, the latter one might be a bit surprising because one could intuitively expect that the more features, the better classification results. To explain this paradox, let us imagine we are using a Bayesian classifier with normally distributed classes. In n dimensions, we have to estimate about $n^2/2$ parameters in each class. If we double the dimensionality, then the number of the unknown parameters increases four times but the number of the training samples is still the same. Hence, increasing the dimensionality yields less accurate parameter estimation. If we consider, in addition to that, that the features are often correlated, the need for dimensionality reduction is straightforward.

There are basically two goals (criteria) of the dimensionality reduction, which determine the used approaches. The first goal is to remove the correlation (or other dependencies) between the features globally, regardless of whether or not any classes have been specified. The most famous algorithm of this kind is the *principal component transformation* (PCT)[12]. The joint covariance matrix, estimated from all available data, is diagonalized. This is always possible because the covariance matrix is symmetric; their eigenvectors are orthogonal and create a new coordinate system in which the features are uncorrelated. The transformation from the old to the new coordinates is a rotation. Then, the features are sorted according to their variance, and p features called *principal components* with the highest variance are kept while the others are removed from the system. Hence, the PCT creates new synthetic features, which are linear combinations of the original ones and are not correlated (see Figure 2.16). It should be, however, noted that the PCT suppresses only the linear component of the dependency between the features and cannot identify and handle higher-order dependencies.

The PCT is an efficient tool suitable for many purposes, such as for multichannel image compression, but its application to classification tasks is limited. As we have seen, the PCT evaluates the "quality" of the features solely according to their variance, which may be different from their actual discrimination power (see Figure 2.17).

Another criterion is to select such features, which maximize the between-class separability on the training set. There are two conceptual differences from the PCT—the new features are selected among the old ones (no features are artificially created) and the labeled training set is required. The separability can be measured in various ways. For normally distributed classes,

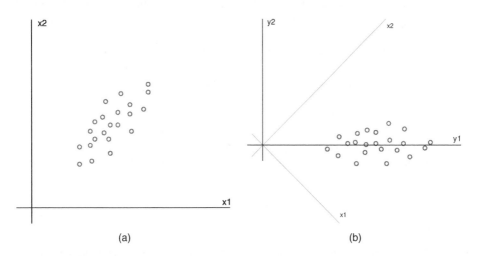

(a)   (b)

Figure 2.16   Principal component transformation: (a) unstructured original data in (X_1, X_2) feature space, correlated (b) transformation into new feature space (Y_1, Y_2), decorrelated. The first principal component is Y_1

[12] The term *principal component analysis* (PCA) has been equivalently used in the literature for this technique.

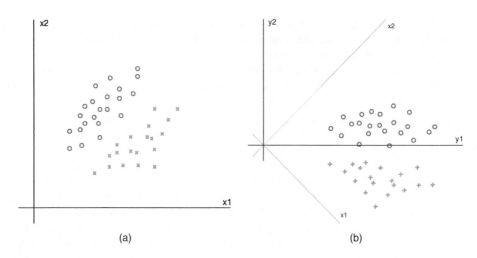

Figure 2.17 PCT of data consisting of two classes: (a) the original (X_1, X_2) feature space, (b) new feature space (Y_1, Y_2) after the PCT. The first principal component is Y_1 thanks to higher variance but the between-class separability is provided solely by Y_2

the simplest separability measure is the *Mahalanobis distance* between two classes

$$s_M(\omega_i, \omega_j) = (\mathbf{m}_i - \mathbf{m}_j)^T (\Sigma_i + \Sigma_j)^{-1} (\mathbf{m}_i - \mathbf{m}_j).$$

More sophisticated (but not necessarily better) separability measures can be derived from between-class ant within-class scatter matrices [89], trying to select the feature subspace such that the classes are "compact" and "distant" from one another. In this way, the feature selection problem has been reformulated as a maximization of the chosen separability measure. To resolve it, several optimization algorithms have been proposed, both optimal [90] as well as suboptimal [91–93] ones.

The advantage of the latter approach is that it considers the between-class separability, preserves the original meaning of the features, and makes the choice on the training set before the unknown objects are actually measured, so the features which have not been selected can be excluded from the measurements.

Most recently, the *wrappers* feature selection methods have appeared [94]. They do not maximize any separability measure but directly optimize the performance of the classifier on a test set. Since this approach requires training and evaluating the classifier repeatedly many times, it is significantly slower than the previous one, but it may be better for the given classifier.

2.5 Conclusion

This chapter presented a glance at the whole object recognition process. Our aim was not to go into particular details but rather show briefly the context in which moment invariants appear, their role in object description, and what we expect from them. In the next chapters we study moments and moment invariants in detail.

References

[1] D. Kundur and D. Hatzinakos, "Blind image deconvolution," *IEEE Signal Processing Magazine*, vol. 13, no. 3, pp. 43–64, 1996.

[2] B. Zitová and J. Flusser, "Image registration methods: A survey," *Image and Vision Computing*, vol. 21, no. 11, pp. 977–1000, 2003.

[3] S. Lončarić, "A survey of shape analysis techniques," *Pattern Recognition*, vol. 31, no. 8, pp. 983–1001, 1998.

[4] D. Zhang and G. Lu, "Review of shape representation and description techniques," *Pattern Recognition*, vol. 37, no. 1, pp. 1–19, 2004.

[5] F. B. Neal and J. C. Russ, *Measuring Shape*. CRC Pres, 2012.

[6] I. T. Young, J. E. Walker, and J. E. Bowie, "An analysis technique for biological shape. I," *Information and Control*, vol. 25, no. 4, pp. 357–370, 1974.

[7] J. Žunić, K. Hirota, and P. L. Rosin, "A Hu moment invariant as a shape circularity measure," *Pattern Recognition*, vol. 43, no. 1, pp. 47–57, 2010.

[8] P. L. Rosin and J. Žunić, "2D shape measures for computer vision," in *Handbook of Applied Algorithms : Solving Scientific, Engineering and Practical Problems* (A. Nayak and I. Stojmenovic, eds.), pp. 347–372, Wiley, 2008.

[9] P. L. Rosin and C. L. Mumford, "A symmetric convexity measure," *Artificial Intelligence*, vol. 103, no. 2, pp.101–111, 2006.

[10] P. L. Rosin and J. Žunić, "Measuring rectilinearity," *Computer Vision and Image Understanding*, vol. 99, no. 2, pp. 175–188, 2005.

[11] P. L. Rosin, "Measuring sigmoidality," *Pattern Recognition*, vol. 37, no. 8, pp. 1735–1744, 2004.

[12] J. Žunić and P. L. Rosin, "A new convexity measure for polygons," *IEEE Transactions on Pattern Analysis and Machine Intelligence*, vol. 26, no. 7, pp. 923–934, 2004.

[13] T. Peli, "An algorithm for recognition and localization of rotated and scaled objects," *Proceedings of the IEEE*, vol. 69, no. 4, pp. 483–485, 1981.

[14] A. Goshtasby, "Description and discrimination of planar shapes using shape matrices," *IEEE Transactions on Pattern Analysis and Machine Intelligence*, vol. 7, no. 6, pp. 738–743, 1985.

[15] A. Taza and C. Y. Suen, "Description of planar shapes using shape matrices," *IEEE Transactions on Systems, Man, and Cybernetics*, vol. 19, no. 5, pp. 1281–1289, 1989.

[16] H. Freeman, "On the encoding of arbitrary geometric configurations," *IRE Transactions on Electronic Computers*, vol. 10, no. 2, pp. 260–268, 1961.

[17] H. Freeman, "Computer processing of line-drawing images," *ACM Computing Surveys*, vol. 6, no. 1, pp. 57–97, 1974.

[18] H. Freeman, "Lines, curves, and the characterization of shape," in *International Federation for Information Processing (IFIP) Congress* (S. H. Lavington, ed.), pp. 629–639, Elsevier, 1980.

[19] C. T. Zahn and R. Z. Roskies, "Fourier descriptors for plane closed curves," *IEEE Transactions on Computers*, vol. C-21, no. 3, pp. 269–281, 1972.

[20] C. C. Lin and R. Chellapa, "Classification of partial 2-D shapes using Fourier descriptors," *IEEE Transactions on Pattern Analysis and Machine Intelligence*, vol. 9, no. 5, pp. 686–690, 1987.

[21] K. Arbter, W. E. Snyder, H. Burkhardt, and G. Hirzinger, "Application of affine-invariant Fourier descriptors to recognition of 3-D objects," *IEEE Transactions on Pattern Analysis and Machine Intelligence*, vol. 12, no. 7, pp. 640–647, 1990.

[22] J. Zhang and T. Tan, "Brief review of invariant texture analysis methods," *Pattern Recognition*, vol. 35, no. 3, pp. 735–747, 2002.

[23] M. Petrou and P. G. Sevilla, *Image Processing: Dealing with Texture*. Wiley, 2006.

[24] R. M. Haralick, K. Shanmugam, and I. Dinstein, "Textural features for image classification," *IEEE Transactions on Systems, Man and Cybernetics*, vol. 3, no. 6, pp. 610–621, 1973.

[25] T. Ojala, M. Pietikäinen, and D. Harwood, "A comparative study of texture measures with classification based on feature distributions," *Pattern Recognition*, vol. 19, no. 3, pp. 51–59, 1996.

[26] T. Ojala, M. Pietikäinen, and T. Mäenpää, "Multiresolution gray-scale and rotation invariant texture classification with local binary patterns," *IEEE Transactions on Pattern Analysis and Machine Intelligence*, vol. 24, no. 7, pp. 971–987, 2002.

[27] J. Fehr and H. Burkhardt, "3D rotation invariant local binary patterns," in *19th International Conference on Pattern Recognition ICPR'08*, pp. 1–4, IEEE, 2008.

[28] K. J. Dana, B. van Ginneken, S. K. Nayar, and J. J. Koenderink, "Reflectance and texture of real-world surfaces," *ACM Transactions on Graphics*, vol. 18, no. 1, pp. 1–34, 1999.

[29] M. Haindl and J. Filip, *Visual Texture. Advances in Computer Vision and Pattern Recognition*, London, UK: Springer-Verlag, 2013.

[30] J. Filip and M. Haindl, "Bidirectional texture function modeling: A state of the art survey," *IEEE Transactions on Pattern Analysis and Machine Intelligence*, vol. 31, no. 11, pp. 1921–1940, 2009.

[31] S. Mallat, *A Wavelet Tour of Signal Processing*. Academic Press, 3rd ed., 2008.

[32] G. C.-H. Chuang and C.-C. J. Kuo, "Wavelet descriptor of planar curves: Theory and applications," *IEEE Transactions on Image Processing*, vol. 5, no. 1, pp. 56–70, 1996.

[33] P. Wunsch and A. F. Laine, "Wavelet descriptors for multiresolution recognition of handprinted characters," *Pattern Recognition*, vol. 28, no. 8, pp. 1237–1249, 1995.

[34] Q. M. Tieng and W. W. Boles, "An application of wavelet-based affine-invariant representation," *Pattern Recognition Letters*, vol. 16, no. 12, pp. 1287–1296, 1995.

[35] M. Khalil and M. Bayeoumi, "A dyadic wavelet affine invariant function for 2D shape recognition," *IEEE Transactions on Pattern Analysis and Machine Intelligence*, vol. 23, no. 10, pp. 1152–1163, 2001.

[36] S. Osowski and D. D. Nghia, "Fourier and wavelet descriptors for shape recognition using neural networks –a comparative study," *Pattern Recognition*, vol. 35, no. 9, pp. 1949–1957, 2002.

[37] G. V. de Wouwer, P. Scheunders, S. Livens, and D. V. Dyck, "Wavelet correlation signatures for color texture characterization," *Pattern Recognition*, vol. 32, no. 3, pp. 443–451, 1999.

[38] A. Laine and J. Fan, "Texture classification by wavelet packet signatures," *IEEE Transactions on Pattern Analysis and Machine Intelligence*, vol. 15, no. 11, pp. 1186–1191, 1993.

[39] S. G. Mallat, "A theory for multiresolution signal decomposition: The wavelet representation," *IEEE Transactions on Pattern Analysis and Machine Intelligence*, vol. 11, no. 1, pp. 674–693, 1989.

[40] K. Huang and S. Aviyente, "Wavelet feature selection for image classification," *IEEE Transactions on Image Processing*, vol. 17, no. 9, pp. 1709–1720, 2008.

[41] C. Papageorgiou and T. Poggio, "A trainable system for object detection," *International Journal of Computer Vision*, vol. 38, no. 1, pp. 15–33, 2000.

[42] E. Wilczynski, *Projective Differential Geometry of Curves and Ruled Surfaces*. Leipzig, Germany: B. G. Teubner, 1906.

[43] I. Weiss, "Noise resistant invariants of curves," in *Geometric Invariance in Computer Vision* (J. L. Mundy and A. Zisserman, eds.), (Cambridge, Massachusetts, USA), pp. 135–156, MIT Press, 1992.

[44] I. Weiss, "Projective invariants of shapes," in *Proceedings of the Computer Vision and Pattern Recognition CVPR '88*, pp. 1125–1134, IEEE, 1988.

[45] A. M. Bruckstein, E. Rivlin, and I. Weiss, "Scale space semi-local invariants," *Image and Vision Computing*, vol. 15, no. 5, pp. 335–344, 1997.

[46] W. S. Ibrahim Ali and F. S. Cohen, "Registering coronal histological 2-D sections of a rat brain with coronal sections of a 3-D brain atlas using geometric curve invariants and B-spline representanion," *IEEE Transactions on Medical Imaging*, vol. 17, no. 6, pp. 957–966, 1998.

[47] O. Horáček, J. Kamenický, and J. Flusser, "Recognition of partially occluded and deformed binary objects," *Pattern Recognition Letters*, vol. 29, no. 3, pp. 360–369, 2008.

[48] Y. Lamdan, J. Schwartz, and H. Wolfson, "Object recognition by affine invariant matching," in *Proceedings of the Computer Vision and Pattern Recognition CVPR'88*, pp. 335–344, IEEE, 1988.

[49] F. Krolupper and J. Flusser, "Polygonal shape description for recognition of partially occluded objects," *Pattern Recognition Letters*, vol. 28, no. 9, pp. 1002–1011, 2007.

[50] Z. Yang and F. Cohen, "Image registration and object recognition using affine invariants and convex hulls," *IEEE Transactions on Image Processing*, vol. 8, no. 7, pp. 934–946, 1999.

[51] J. Flusser, "Affine invariants of convex polygons," *IEEE Transactions on Image Processing*, vol. 11, no. 9, pp. 1117–1118, 2002.

[52] C. A. Rothwell, A. Zisserman, D. A. Forsyth, and J. L. Mundy, "Fast recognition using algebraic invariants," in *Geometric Invariance in Computer Vision* (J. L. Mundy and A. Zisserman, eds.), pp. 398–407, MIT Press, 1992.

[53] F. Mokhtarian and S. Abbasi, "Shape similarity retrieval under affine transforms," *Pattern Recognition*, vol. 35, no. 1, pp. 31–41, 2002.

[54] D. Lowe, "Object recognition from local scale-invariant features," in *Proceedings of the International Conference on Computer Vision ICCV'99*, vol. 2, pp. 1150–1157, IEEE, 1999.

[55] N. Dalal and B. Triggs, "Histograms of oriented gradients for human detection," in *Proceedings of the Conference on Computer Vision and Pattern Recognition CVPR'05*, vol. 1, pp. 886–893, IEEE, 2005.

[56] H. Bay, A. Ess, T. Tuytelaars, and L. V. Gool, "Speeded-up robust features (SURF)," *Computer Vision and Image Understanding*, vol. 110, no. 3, pp. 346–359, 2008.

[57] J. L. Mundy and A. Zisserman, *Geometric Invariance in Computer Vision*. Cambridge, Massachusetts, USA: MIT Press, 1992.

[58] T. Suk and J. Flusser, "Vertex-based features for recognition of projectively deformed polygons," *Pattern Recognition*, vol. 29, no. 3, pp. 361–367, 1996.

[59] R. Lenz and P. Meer, "Point configuration invariants under simultaneous projective and permutation transformations," *Pattern Recognition*, vol. 27, no. 11, pp. 1523–1532, 1994.

[60] N. S. V. Rao, W. Wu, and C. W. Glover, "Algorithms for recognizing planar polygonal configurations using perspective images," *IEEE Transactions on Robotics and Automation*, vol. 8, no. 4, pp. 480–486, 1992.

[61] D. Hilbert, *Theory of Algebraic Invariants*. Cambridge, U.K.: Cambridge University Press,1993.

[62] J. H. Grace and A. Young, *The Algebra of Invariants*. Cambridge University Press, 1903.

[63] G. B. Gurevich, *Foundations of the Theory of Algebraic Invariants*. Groningen, The Netherlands: Nordhoff, 1964.

[64] I. Schur, *Vorlesungen über Invariantentheorie*. Berlin, Germany: Springer, 1968. (in German).

[65] M.-K. Hu, "Visual pattern recognition by moment invariants," *IRE Transactions on Information Theory*, vol. 8, no. 2, pp. 179–187, 1962.

[66] G. A. Papakostas, ed., *Moments and Moment Invariants - Theory and Applications*. Science Gate Publishing, 2014.

[67] R. Mukundan and K. R. Ramakrishnan, *Moment Functions in Image Analysis*. Singapore: World Scientific, 1998.

[68] M. Pawlak, *Image Analysis by Moments: Reconstruction and Computational Aspects*. Wrocław, Poland: Oficyna Wydawnicza Politechniki Wrocławskiej, 2006.

[69] J. Flusser, T. Suk, and B. Zitová, *Moments and Moment Invariants in Pattern Recognition*. Chichester, U.K.: Wiley, 2009.

[70] R. O. Duda, P. E. Hart, and D. G. Stork, *Pattern Classification*. Wiley Interscience, 2nd ed., 2001.

[71] S. Theodoridis and K. Koutroumbas, *Pattern Recognition*. Academic Press, 4th ed., 2009.

[72] V. N. Vapnik, "Pattern recognition using generalized portrait method," *Automation and Remote Control*, vol. 24, pp. 774–780, 1963.

[73] V. N. Vapnik and C. Cortes, "Support-vector networks," *Machine Learning*, vol. 20, no. 3, pp. 273–297, 1995.

[74] B. E. Boser, I. M. Guyon, and V. N. Vapnik, "A training algorithm for optimal margin classifiers," in *Proceedings of the fifth annual workshop on Computational learning theory COLT'92*, pp. 144–152, ACM Press, 1992.

[75] K.-B. Duan and S. S. Keerthi, "Which is the best multiclass SVM method? an empirical study," in *Multiple Classifier Systems MCS'05* (N. C. Oza, R. Polikar, J. Kittler, and F. Roli, eds.), vol. 3541 of *Lecture Notes in Computer Science*, (Berlin, Heidelberg, Germany), pp. 278–285, Springer, 2005.

[76] F. Rosenblatt, "The perceptron: A probabilistic model for information storage and organization in the brain," *Psychological Review*, vol. 65, no. 6, pp. 386–408, 1958.

[77] B. D. Ripley, *Pattern Recognition and Neural Networks*. Cambridge University Press, 3rd ed., 1996.

[78] K. Fukushima, "Neocognitron: A self-organizing neural network model for a mechanism of pattern recognition unaffected by shift in position," *Biological Cybernetics*, vol. 36, no. 4, pp. 193–202, 1980.

[79] Stanford Vision Lab, "Imagenet large scale visual recognition challenge (ILSVRC)," 2015. http://www.image-net.org/challenges/LSVRC/.

[80] D. C. Cireşan, U. Meier, J. Masci, L. M. Gambardella, and J. Schmidhuber, "Flexible, high performance convolutional neural networks for image classification," in *Proceedings of the 22nd international joint conference on Artificial Intelligence IJCAI'11*, vol. 2, pp. 1237–1242, AAAI Press, 2011.

[81] Algoritmy.net, "Letter frequency (English)," 2008–2015. http://en.algoritmy.net/article/40379/Letter-frequency-English.

[82] J. Kittler, M. Hatef, R. P. Duin, and J. Matas, "On combining classifiers," *IEEE Transactions on Pattern Analysis and Machine Intelligence*, vol. 20, no. 3, pp. 226–239, 1998.

[83] L. Prevost and M. Milgram, "Static and dynamic classifier fusion for character recognition," in *Proceedings of the Fourth International Conference on Document Analysis and Recognition ICDAR'97*, vol. 2, pp. 499–506, IEEE, 1997.

[84] Q. Fu, X. Q. Ding, T. Z. Li, and C. S. Liu, "An effective and practical classifier fusion strategy for improving handwritten character recognition," in *Ninth International Conference on Document Analysis and Recognition ICDAR'07*, vol. 2, pp. 1038–1042, IEEE, 2007.

[85] I. N. Dimou, G. C. Manikis, and M. E. Zervakis, "Classifier fusion approaches for diagnostic cancer models," in *Engineering in Medicine and Biology Society EMBS'06.*, pp. 5334–5337, IEEE, 2006.

[86] D. Wang, J. M. Keller, C. A. Carson, K. K. McAdoo-Edwards, and C. W. Bailey, "Use of fuzzy-logic-inspired features to improve bacterial recognition through classifier fusion," *IEEE Transactions on Systems, Man, and Cybernetics, Part B: Cybernetics*, vol. 28, no. 4, pp. 583–591, 1998.

[87] C.-X. Zhang and R. P. W. Duin, "An experimental study of one- and two-level classifier fusion for different sample sizes," *Pattern Recognition Letters*, vol. 32, no. 14, pp. 1756–1767, 2011.

[88] L. I. Kuncheva, *Combining Pattern Classifiers: Methods and Algorithms*. Wiley, 2004.

[89] K. Fukunaga, *Introduction to Statistical Pattern Recognition*. Computer Science and Scientific Computing, Morgan Kaufmann, Academic Press, 2nd ed., 1990.

[90] P. Somol, P. Pudil, and J. Kittler, "Fast branch & bound algorithms for optimal feature selection," *IEEE Transactions on Pattern Analysis and Machine Intelligence*, vol. 26, no. 7, pp. 900–912, 2004.

[91] P. Pudil, J. Novovičová, and J. Kittler, "Floating search methods in feature selection," *Pattern Recognition Letters*, vol. 15, no. 11, pp. 1119–1125, 1994.

[92] P. Somol, P. Pudil, J. Novovičová, and P. Paclík, "Adaptive floating search methods in feature selection," *Pattern Recognition Letters*, vol. 20, no. 11–13, pp. 1157–1163, 1999.

[93] J. Novovičová, P. Pudil, and J. Kittler, "Divergence based feature selection for multimodal class densities," *IEEE Transactions on Pattern Analysis and Machine Intelligence*, vol. 18, no. 1, pp. 218–223, 1996.

[94] R. Kohavi and G. H. John, "Wrappers for feature subset selection," *Artificial Intelligence*, vol. 97, no. 1–2, pp. 273–324, 1997.

3

2D Moment Invariants to Translation, Rotation, and Scaling

3.1 Introduction

In this chapter, we introduce 2D moment invariants with respect to the simplest spatial in-plane transformations–translation, rotation, and scaling (TRS). Invariance with respect to TRS is widely required in almost all practical applications, because the object should be correctly recognized regardless of its particular position and orientation in the scene and of the object to camera distance (see Figure 3.1). On the other hand, the TRS model is a sufficient approximation of the actual image deformation if the scene is flat and (almost) perpendicular to the optical axis. Due to these reasons, much attention has been paid to TRS invariants. While translation and scaling invariants can be mostly derived in an intuitive way, derivation of invariants to rotation is far more complicated.

Before we proceed to the design of the invariants, we start this chapter with a few basic definitions and with an introduction to moments. The notation introduced in the next section will be used throughout the book if not specified otherwise.

3.1.1 Mathematical preliminaries

Spatial coordinates in the image domain are denoted as $\mathbf{x} = (x_1, x_2, \cdots, x_d)^T$, where d is the dimension of the space. In 2D and 3D domains, if there is no danger of misunderstanding, we sometimes use a simpler (and more common in the literature) notation $\mathbf{x} = (x, y)^T$ and $\mathbf{x} = (x, y, z)^T$, respectively. The superscript $(\cdots)^T$ means a transposition, so our coordinates are arranged into a column vector.

Definition 3.1 By an *image function* (or *image*) we understand any piece-wise continuous real function $f(\mathbf{x})$ defined on a compact support $\Omega \subset \mathbb{R}^d$, which has a finite nonzero integral.

According to this definition, our "images" need not be non-negative. The piece-wise continuity and the compact support are assumed to make the operations, we are going to apply on the images, well defined, which enables a comfortable mathematical treatment without a tedious verifying of the existence of the operations in each individual case. From a purely mathematical point of view, these requirements may seem to be too restrictive because certain

2D and 3D Image Analysis by Moments, First Edition. Jan Flusser, Tomáš Suk and Barbara Zitová.
© 2017 John Wiley & Sons, Ltd. Published 2017 by John Wiley & Sons, Ltd.
Companion website: www.wiley.com/go/flusser/2d3dimageanalysis

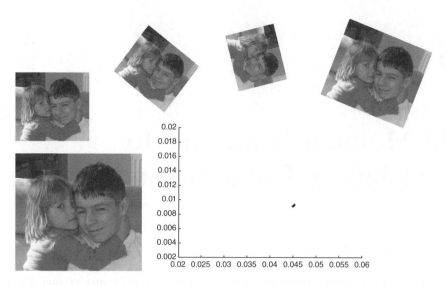

Figure 3.1   The desired behavior of TRS moment invariants–all instances of a rotated and scaled image have almost the same values of the invariants (depicted for two invariants)

operations are well defined on broader classes of functions such as integrable or square integrable functions or infinitely supported functions of fast decay. However, these nuances make absolutely no difference from a practical point of view when working with digital images. The non-zero integral is required because its value will be frequently used as a normalization factor[1].

If the image is mathematically described by the image function, we sometimes speak about the *continuous representation*. Although the continuous representation may not reflect certain properties of digital images, such as sampling and quantization errors, we will adopt the continuous formalism because of its mathematical transparency and simplicity. If the discrete character of the image is substantial (such as, for instance, in Chapter 8 where we explain numerical algorithms for moment computations), we will work with a *discrete representation* of the image in a form of a finite-extent 2D matrix $\mathbf{f} = (f_{ij})$ or 3D matrix $\mathbf{f} = (f_{ijk})$, which is supposed to be obtained from the continuous representation $f(\mathbf{x})$ by sampling and quantization.

The image function from Definition 3.1 represents *monochromatic* images (sometimes also called graylevel or scalar images). If f has only two possible values (which are usually encoded as 0 and 1), we speak about a *binary* image. Color images and vector-valued images are represented as a vector image function, each component of which satisfies Definition 3.1.

Convolution is an operation between two image functions[2], the result of which is another image function defined as

$$(f * g)(\mathbf{x}) = \int_{\mathbb{R}^d} f(\mathbf{t})g(\mathbf{x} - \mathbf{t})\mathrm{d}\mathbf{t}. \tag{3.1}$$

---

[1] If we relaxed this assumption, the invariants still could be constructed provided that at least one moment is non-zero.
[2] Note that the convolution is not a point-wise operation between the function values, so the notation $f(x) * g(x)$, commonly used in many engineering textbooks, is misleading and mathematically incorrect.

Fourier transformation (FT) of image function f is defined as

$$\mathcal{F}(f)(\mathbf{u}) \equiv F(\mathbf{u}) = \int_{\mathbb{R}^d} e^{-2\pi i \mathbf{u}\mathbf{x}} f(\mathbf{x})\mathrm{d}\mathbf{x}, \tag{3.2}$$

where i is the imaginary unit, the components of $\mathbf{u}$ are spatial frequencies, and $\mathbf{u}\mathbf{x} = u_1 x_1 + \cdots + u_d x_d$ means scalar product[3]. Note that $\mathcal{F}(f)$ always exists thanks to the integrability of f, and its support cannot be bounded[4]. Fourier transformation is invertible through the *inverse Fourier transformation* (IFT)

$$\mathcal{F}^{-1}(F)(\mathbf{x}) = \int_{\mathbb{R}^d} e^{2\pi i \mathbf{u}\mathbf{x}} F(\mathbf{u})\mathrm{d}\mathbf{u} = f(\mathbf{x}). \tag{3.3}$$

We recall two important properties of FT, which we employ in this book. The first one, also known as the *Fourier shift theorem*, tells that the FT of a translated image function equals the FT of the original up to a phase shift

$$\mathcal{F}(f(\mathbf{x}-\mathbf{t}))(\mathbf{u}) = e^{-2\pi i \mathbf{u}\mathbf{t}} \mathcal{F}(f(\mathbf{x}))(\mathbf{u}). \tag{3.4}$$

The second property, known as the *convolution theorem*, shows why FT is so useful in signal processing–it transfers the convolution into a point-wise multiplication in the frequency domain

$$\mathcal{F}(f * g)(\mathbf{u}) = F(\mathbf{u})G(\mathbf{u}). \tag{3.5}$$

3.1.2 Moments

Moments are scalar real or complex-valued features which have been used to characterize a given function. From the mathematical point of view, moments are "projections"[5] of function f onto a polynomial basis (similarly, Fourier transformation is a projection onto a basis of the harmonic functions).

Definition 3.2 Let $\{\pi_{\mathbf{p}}(\mathbf{x})\}$ be a d-variable polynomial basis of the space of image functions defined on Ω and let $\mathbf{p} = (p_1, \ldots, p_d)$ be a multi-index of non-negative integers which show the highest power of the respective variables in $\pi_{\mathbf{p}}(\mathbf{x})$. Then the *general moment* $M_{\mathbf{p}}^{(f)}$ of image f is defined as

$$M_{\mathbf{p}}^{(f)} = \int_{\Omega} \pi_{\mathbf{p}}(\mathbf{x})f(\mathbf{x})\mathrm{d}\mathbf{x}. \tag{3.6}$$

The number $|\mathbf{p}| = \sum_{k=1}^{d} p_k$ is called the *order* of the moment. We omit the superscript $^{(f)}$ whenever possible without confusion.

---

[3] A more correct notation of the scalar product is $\mathbf{u}^T\mathbf{x}$; we drop the superscript ^{T} for simplicity.
[4] More precisely, $\mathcal{F}(f)$ cannot vanish on any open subset of $\mathbb{R}^d$.
[5] The moments in general are not the coordinates of f in the given basis in the algebraic sense. That is true for orthogonal bases only.

Depending on the polynomial basis $\{\pi_\mathbf{p}(\mathbf{x})\}$, we recognize various systems of moments. The most common choice is a standard power basis $\pi_\mathbf{p}(\mathbf{x}) = \mathbf{x}^\mathbf{p}$ which leads to *geometric moments*

$$m_\mathbf{p}^{(f)} = \int_\Omega \mathbf{x}^\mathbf{p} f(\mathbf{x}) d\mathbf{x}. \tag{3.7}$$

In the literature, one can find various extensions of Definition 3.2. Some authors allow non-polynomial bases (more precisely, they allow basis functions which are products of a polynomial and some other–usually harmonic–functions) and/or include various scalar factors and weighting functions in the integrand. Some other authors even totally replaced the polynomial basis by some other basis but still call such features moments–we can find *wavelet moments* [1] and *step-like moments* [2], where wavelets and step-wise functions are used in a combination with harmonic functions instead of the polynomials. These modifications broadened the notion of moments but have not brought any principle differences in moment usage.

3.1.3   Geometric moments in 2D

In case of 2D images, the choice of the standard power basis $\pi_{pq}(x, y) = x^p y^q$ yields *2D geometric moments*

$$m_{pq} = \int_{-\infty}^{\infty} \int_{-\infty}^{\infty} x^p y^q f(x, y) dx dy. \tag{3.8}$$

Geometric moments have been widely used in statistics for description of the shape of a probability density function and in classic rigid-body mechanics to measure the mass distribution of a body. Geometric moments of low orders have an intuitive meaning–m_{00} is a "mass" of the image (on binary images, m_{00} is an area of the object), m_{10}/m_{00} and m_{01}/m_{00} define the *center of gravity* or *centroid* of the image. Second-order moments m_{20} and m_{02} describe the "distribution of mass" of the image with respect to the coordinate axes. In mechanics, they are called the *moments of inertia*. Another popular mechanical quantity, the *radius of gyration* with respect to an axis, can be also expressed in terms of moments as $\sqrt{m_{20}/m_{00}}$ and $\sqrt{m_{02}/m_{00}}$, respectively.

If the image is considered to be a joint probability density function (PDF) of two random variables (i.e., its values are non-negative and normalized such that $m_{00} = 1$), then the horizontal and vertical projections of the image are the 1D marginal densities, and m_{10} and m_{01} are the mean values of the variables. In case of zero means, m_{20} and m_{02} are their *variances*, and m_{11} is a *covariance* between them. In this way, the second-order moments define the principal axes of the image. As will be seen later, the second-order geometric moments can be used to find the normalized position of the image. In statistics, two higher-order moment characteristics of a probability density function have been commonly used – the *skewness* and the *kurtosis*. These terms are mostly used for 1D marginal densities only. Skewness of the horizontal marginal density is defined as $m_{30}/\sqrt{m_{20}^3}$ and that of the vertical marginal distribution as $m_{03}/\sqrt{m_{02}^3}$. The skewness measures the deviation of the PDF from symmetry. If the PDF is symmetric with respect to the mean (i.e., to the origin in this case), then the corresponding skewness equals zero. The kurtosis measures the "peakedness" of the probability density function and is again defined separately for each marginal distribution–the horizontal kurtosis as m_{40}/m_{20}^2 and the vertical kurtosis as m_{04}/m_{02}^2.

Characterization of the image by means of the geometric moments is complete and unambiguous in the following sense. For any image function, its geometric moments of all orders do exist and are finite. The image function can be exactly reconstructed from the set of all its moments (this assertion is known as the *uniqueness theorem* and holds thanks to the Weirstrass theorem on infinitely accurate polynomial approximation of continuous functions)[6].

Geometric moments of function f are closely related to its Fourier transformation. If we expand the kernel of the Fourier transformation $e^{-2\pi i(ux+vy)}$ into a power series, we realize that the geometric moments form Taylor coefficients of F

$$F(u, v) = \sum_{p=0}^{\infty} \sum_{q=0}^{\infty} \frac{(-2\pi i)^{p+q}}{p!q!} m_{pq} u^p v^q.$$

This link between the geometric moments and Fourier transformation is employed in many tasks, for instance in image reconstruction from moments as we will see in detail in Chapter 7.

3.1.4   *Other moments*

Geometric moments are very attractive thanks to the formal simplicity of the basis functions. This is why many theoretical considerations about moment invariants have been based on them. On the other hand, they have also certain disadvantages. One of them is their complicated transformation under rotation. This has led to introducing a class of *circular moments*, which change under rotation in a simple way and allow to design rotation invariants systematically, as we will show later in this chapter for the 2D case and in Chapter 4 for the 3D case.

Another drawback of the geometric moments are their poor numerical properties when working in a discrete domain. Standard powers are nearly dependent both for small and large values of the exponent and increase rapidly in range as the order increases. This leads to correlated geometric moments and to the need for high computational precision. Using lower precision results in unreliable computation of geometric moments. This has led several authors to employing *orthogonal* (OG) moments, that is, moments the basis polynomials of which are orthogonal on Ω.

In theory, all polynomial bases of the same degree are equivalent because they generate the same space of functions. Any moment with respect to a certain basis can be expressed in terms of moments with respect to any other basis. From this point of view, OG moments of any type are equivalent to geometric moments. However, a significant difference appears when considering stability and other computational issues. OG moments have the advantage of requiring lower computing precision because we can evaluate them using recurrent relations, without expressing them in terms of the standard powers. OG moments are reviewed in Chapter 7.

---

[6] A more general moment problem is well known in statistics: can a given sequence be a set of moments of some compactly supported function? The answer is yes if the sequence is completely monotonic.

3.2   TRS invariants from geometric moments

Translation, rotation, and scaling in 2D is a four-parameter transformation, which can be
described in a matrix form as

$$\mathbf{x}' = s\mathbf{R}_\alpha \mathbf{x} + \mathbf{t},$$

where $\mathbf{t}$ is a translation vector, s is a positive scaling factor (note that here we consider *uniform*
scaling only, that is, s is the same both in horizontal and vertical directions), and $\mathbf{R}_\alpha$ is a rotation
matrix by the angle α

$$\mathbf{R}_\alpha = \begin{pmatrix} \cos\alpha & -\sin\alpha \\ \sin\alpha & \cos\alpha \end{pmatrix}.$$

Note that the TRS transformations actually form a group because

$$\mathbf{R}_0 = \mathbf{I},$$

$$\mathbf{R}_\alpha \mathbf{R}_\beta = \mathbf{R}_{\alpha+\beta}$$

and

$$\mathbf{R}_\alpha^{-1} = \mathbf{R}_{-\alpha} = \mathbf{R}_\alpha^T.$$

3.2.1   Invariants to translation

Geometric moments are not invariant to translation. If f is shifted by vector $\mathbf{t} = (a, b)^T$, then
its geometric moments change as

$$
\begin{aligned}
m'_{pq} &= \int_{-\infty}^{\infty}\int_{-\infty}^{\infty} x^p y^q f(x-a, y-b)\mathrm{d}x\mathrm{d}y \\
&= \int_{-\infty}^{\infty}\int_{-\infty}^{\infty} (x+a)^p (y+b)^q f(x,y)\mathrm{d}x\mathrm{d}y \qquad (3.9) \\
&= \sum_{k=0}^{p}\sum_{j=0}^{q} \binom{p}{k}\binom{q}{j} a^k b^j m_{p-k,q-j}.
\end{aligned}
$$

Invariance to translation can be achieved simply by seemingly shifting the object into a certain
well-defined position before the moments are calculated. The best way is to use its centroid–if
we shift the object such that its centroid coincides with the origin of the coordinate system, then
all the moments are translation invariants. This is equivalent to keeping the object fixed and
shifting the polynomial basis into the object centroid[7]. We obtain so-called *central* geometric
moments

$$\mu_{pq} = \int_{-\infty}^{\infty}\int_{-\infty}^{\infty} (x-x_c)^p (y-y_c)^q f(x,y)\mathrm{d}x\mathrm{d}y, \qquad (3.10)$$

---

[7] We recall that the use of the centralized coordinates is a common way of reaching translation invariance also in case
of many other than moment-based features.

where

$$x_c = m_{10}/m_{00}, \quad y_c = m_{01}/m_{00}$$

are the coordinates of the object centroid. Note that it always holds $\mu_{10} = \mu_{01} = 0$ and $\mu_{00} = m_{00}$. Translation invariance of the central moments is straightforward.

The central moments can be expressed in terms of geometric moments as

$$\mu_{pq} = \sum_{k=0}^{p} \sum_{j=0}^{q} \binom{p}{k} \binom{q}{j} (-1)^{k+j} x_c^k y_c^j m_{p-k,q-j}.$$

Although this relation has little importance for theoretic consideration, it is sometimes used when we want to calculate the central moments by means of some fast algorithm for geometric moments computation.

3.2.2 Invariants to uniform scaling

Scaling invariance is obtained by a proper normalization of each moment. To find a normalization factor, let us first look at how the moment is changed under image scaling[8] by a factor s

$$\mu'_{pq} = \int_{-\infty}^{\infty} \int_{-\infty}^{\infty} (x - x_c)^p (y - y_c)^q f(x/s, y/s) dx dy \tag{3.11}$$

$$= \int_{-\infty}^{\infty} \int_{-\infty}^{\infty} s^p (x - x_c)^p s^q (y - y_c)^q f(x, y) s^2 dx dy = s^{p+q+2} \mu_{pq}. \tag{3.12}$$

In particular,

$$\mu'_{00} = s^2 \mu_{00}.$$

In principle, any moment can be used as a normalizing factor provided that it is non-zero for all images in the experiment. Since low-order moments are more stable to noise and easier to calculate, we normalize most often by a proper power of μ_{00}

$$\nu_{pq} = \frac{\mu_{pq}}{\mu_{00}^w}. \tag{3.13}$$

To eliminate the scaling parameter s, the power must be set as

$$w = \frac{p+q}{2} + 1. \tag{3.14}$$

---

[8] There is a difference between scaling the image and scaling the coordinates–upscaling the image is the same as downscaling the coordinates and vice versa, the scaling factors are inverse. However, for the purpose of deriving invariants, it does not matter which particular scaling we consider.

The moment v_{pq} is called the *normalized* central geometric moment[9] and is invariant to uniform scaling:

$$v'_{pq} = \frac{\mu'_{pq}}{(\mu'_{00})^w} = \frac{s^{p+q+2}\mu_{pq}}{(s^2\mu_{00})^w} = v_{pq}.$$

The moment that has been used for scaling normalization can no longer be used for recognition because the value of the corresponding normalized moment is always one (in the above normalization, $v_{00} = 1$). If we want to keep the zero-order moment valid, we have to normalize by another moment. Such kind of scaling normalization is used very rarely; probably the only meaningful normalization is by $(\mu_{20} + \mu_{02})^{w/2}$. Numerical stability of v_{20} is illustrated in Figure 3.2.

3.2.3 Invariants to non-uniform scaling

Non-uniform scaling is a transformation beyond the TRS framework that maps a unit square onto a rectangle. It is defined as

$$x' = ax,$$

$$y' = by,$$

where $a \neq b$ are positive scaling factors. Invariants to non-uniform scaling are sometimes called *aspect-ratio invariants*.

Geometric central moments change under non-uniform scaling simply as

$$\mu'_{pq} = \int\limits_{-\infty}^{\infty}\int\limits_{-\infty}^{\infty} a^p(x-x_c)^p b^q(y-y_c)^q f(x,y)ab \ dx\,dy = a^{p+1}b^{q+1}\mu_{pq}.$$

To eliminate the scaling factors a and b, we need at least two normalizing moments. When using for instance μ_{00} and μ_{20}, we get the invariants

$$A_{pq} = \frac{\mu_{pq}}{\mu_{00}^\alpha \mu_{20}^\beta},$$

where

$$\alpha = \frac{3q-p}{2} + 1$$

and

$$\beta = \frac{p-q}{2}.$$

One can derive many different invariants to non-uniform scaling. For instance, Pan and Keane [4] proposed more "symmetric" normalization by three moments μ_{00}, μ_{20} and μ_{02},

---

[9] This normalization was proposed already in Hu's paper [3], but the exponent was stated incorrectly as $w = \frac{p+q+1}{2}$. Many authors adopted this error, and some other authors introduced a new one claiming $w = [\frac{p+q}{2}] + 1$, where $[.]$ denotes an integer part.

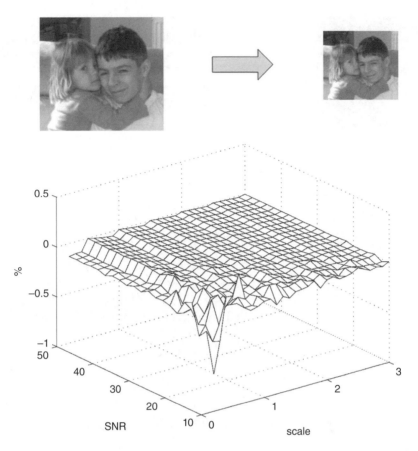

Figure 3.2 Numerical test of the normalized moment v_{20}. Computer-generated scaling of the test image ranged form $s = 0.2$ to $s = 3$. To show robustness, each image was corrupted by additive Gaussian white noise. Signal-to-noise ratio (SNR) ranged from 50 (low noise) to 10 (heavy noise). Horizontal axes: scaling factor s and SNR, respectively. Vertical axis–relative deviation (in %) between v_{20} of the original and that of the scaled and noisy image. The test proves the invariance of v_{20} and illustrates its high robustness to noise

which leads to

$$S_{pq} = \frac{\mu_{00}^{(p+q+2)/2}}{\mu_{20}^{(p+1)/2} \cdot \mu_{02}^{(q+1)/2}} \cdot \mu_{pq}.$$

Invariance to non-uniform scaling cannot be combined in a simple way with rotation invariance. The reason is that rotation and non-uniform scaling are not closed operations. When applied repeatedly, they generate an affine group of transformations, which implies that if we want to have invariance to non-uniform scaling and rotation, we must use affine invariants (they will be introduced in Chapter 5). There is no transformation group "between" the TRS, and affine groups and, consequently, no special set of invariants "between" the TRS and affine moment invariants may exist. This is why the invariants to non-uniform scaling described above

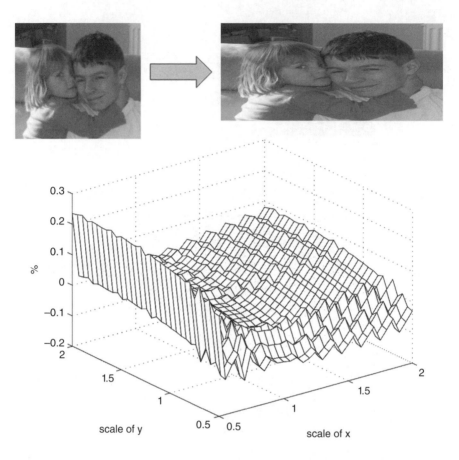

Figure 3.3 Numerical test of the aspect-ratio invariant A_{22}. Computer-generated scaling of the test image ranged form 0.5 to 2 in both directions independently. Horizontal axes: scaling factors a and b, respectively. Vertical axis–relative deviation (in %) between A_{22} of the original and that of the scaled image. The test illustrates the invariance of A_{22}. Higher relative errors for low scaling factors and typical jagged surface of the graph are the consequences of the image resampling

are of only little importance in practical applications, and we presented them here merely for illustration. Numerical stability of A_{22} is illustrated in Figure 3.3.

3.2.4 Traditional invariants to rotation

Rotation moment invariants were firstly introduced in 1962 by Hu [3], who employed the results of the theory of algebraic invariants and derived his seven famous invariants to an in-plane rotation around the origin

$$\phi_1 = m_{20} + m_{02},$$
$$\phi_2 = (m_{20} - m_{02})^2 + 4m_{11}^2,$$

$$\phi_3 = (m_{30} - 3m_{12})^2 + (3m_{21} - m_{03})^2,$$

$$\phi_4 = (m_{30} + m_{12})^2 + (m_{21} + m_{03})^2, \tag{3.15}$$

$$\phi_5 = (m_{30} - 3m_{12})(m_{30} + m_{12})((m_{30} + m_{12})^2 - 3(m_{21} + m_{03})^2)$$
$$+ (3m_{21} - m_{03})(m_{21} + m_{03})(3(m_{30} + m_{12})^2 - (m_{21} + m_{03})^2),$$

$$\phi_6 = (m_{20} - m_{02})((m_{30} + m_{12})^2 - (m_{21} + m_{03})^2)$$
$$+ 4m_{11}(m_{30} + m_{12})(m_{21} + m_{03}),$$

$$\phi_7 = (3m_{21} - m_{03})(m_{30} + m_{12})((m_{30} + m_{12})^2 - 3(m_{21} + m_{03})^2)$$
$$- (m_{30} - 3m_{12})(m_{21} + m_{03})(3(m_{30} + m_{12})^2 - (m_{21} + m_{03})^2).$$

If we replace geometric moments by central or normalized moments in these relations, we obtain invariants not only to rotation but also to translation and/or scaling, which at the same time ensures invariance to rotation around an arbitrary point. Hu's derivation was rather complicated, and that is why only these seven invariants were derived explicitly and no hint how to derive invariants from higher-order moments was given in [3]. However, once we have the formulae, the proof of rotation invariance is easy. Let us demonstrate it for ϕ_1 and ϕ_2.

The second-order moments after rotation by angle α can be expressed as

$$\mu'_{20} = \cos^2\alpha \cdot \mu_{20} + \sin^2\alpha \cdot \mu_{02} - \sin 2\alpha \cdot \mu_{11}$$

$$\mu'_{02} = \sin^2\alpha \cdot \mu_{20} + \cos^2\alpha \cdot \mu_{02} + \sin 2\alpha \cdot \mu_{11}$$

$$\mu'_{11} = \frac{1}{2}\sin 2\alpha \cdot (\mu_{20} - \mu_{02}) + \cos 2\alpha \cdot \mu_{11}.$$

Thus,

$$\phi'_1 = \mu'_{20} + \mu'_{02} = (\sin^2\alpha + \cos^2\alpha)(\mu_{20} + \mu_{02}) = \phi_1$$

and similarly for ϕ'_2, applying the formula $\cos 2\alpha = \cos^2\alpha - \sin^2\alpha$.   □

Although Hu's invariants suffer from the limited recognition power, mutual dependence, and the restriction to the second- and third- order moments only, they have become classics, and, despite of their drawbacks, they have found numerous successful applications in various areas. The major weakness of Hu's theory is that it does not provide for a possibility of any generalization. By means of it, we could not derive invariants from higher-order moments. To illustrate that, let us consider how a general moment is changed under rotation:

$$m'_{pq} = \int\limits_{-\infty}^{\infty}\int\limits_{-\infty}^{\infty} (x\cos\alpha - y\sin\alpha)^p(x\sin\alpha + y\cos\alpha)^q f(x,y)dxdy$$

$$= \int\limits_{-\infty}^{\infty}\int\limits_{-\infty}^{\infty} \sum_{k=0}^{p}\sum_{j=0}^{q}\binom{p}{k}\binom{q}{j} x^{k+j}y^{p+q-k-j}(-1)^{p-k}(\cos\alpha)^{q+k-j}(\sin\alpha)^{p-k+j} f(x,y)dxdy$$

$$\tag{3.16}$$

$$= \sum_{k=0}^{p}\sum_{j=0}^{q}(-1)^{p-k}\binom{p}{k}\binom{q}{j}(\cos\alpha)^{q+k-j}(\sin\alpha)^{p-k+j}m_{k+j,p+q-k-j}.$$

This is a rather complicated expression from which the rotation parameter α cannot be easily eliminated. An attempt was proposed by Jin and Tianxu [5], but compared to the approach described in the sequel it is difficult and not very transparent.

In the next section, we present a general approach to deriving rotation invariants, which uses circular moments.

3.3 Rotation invariants using circular moments

After Hu, various approaches to the theoretical derivation of moment-based rotation invariants, which would not suffer from a limitation to low orders, have been published. Although nowadays the key idea seems to be bright and straightforward, its first versions appeared in the literature only two decades after Hu's paper.

The key idea is to replace the cartesian coordinates (x, y) by (centralized) *polar coordinates* (r, θ)

$$
\begin{aligned}
x &= r\cos\theta & r &= \sqrt{x^2 + y^2} \\
y &= r\sin\theta & \theta &= \arctan(y/x)
\end{aligned}
\tag{3.17}
$$

In the polar coordinates, the rotation is a translation in the angular argument. To eliminate the impact of this translation, we were inspired by the Fourier shift theorem–if the basis functions in angular direction were harmonics, the translation would result just in a phase shift of the moment values, which would be easy to eliminate. This brilliant general idea led to the birth of a family of so-called *circular moments*[10]. The circular moments have the form

$$
C_{pq} = \int\limits_0^\infty \int\limits_0^{2\pi} R_{pq}(r)e^{i\xi(p,q)\theta}f(r,\theta)r\mathrm{d}\theta\mathrm{d}r
\tag{3.18}
$$

where $R_{pq}(r)$ is a radial univariate polynomial and $\xi(p,q)$ is a function (usually a very simple one) of the indices. Complex, Zernike, pseudo-Zernike, Fourier-Mellin, radial Chebyshev, and many other moments are particular cases of the circular moments.

Under coordinate rotation by angle α (which is equivalent to image rotation by $-\alpha$), the circular moment is changed as

$$
\begin{aligned}
C'_{pq} &= \int\limits_0^\infty \int\limits_0^{2\pi} R_{pq}(r)e^{i\xi(p,q)\theta}f(r,\theta+\alpha)r\mathrm{d}\theta\mathrm{d}r \\
&= \int\limits_0^\infty \int\limits_0^{2\pi} R_{pq}(r)e^{i\xi(p,q)(\theta-\alpha)}f(r,\theta)r\mathrm{d}\theta\mathrm{d}r \\
&= e^{-i\xi(p,q)\alpha}C_{pq}.
\end{aligned}
\tag{3.19}
$$

---

[10] They are sometimes referred to as radial or radial-circular moments.

To eliminate the parameter α, we can take the moment magnitude $|C_{pq}|$ (note that the circular moments are generally complex-valued even if $f(x, y)$ is real) or apply some more sophisticated method to reach the phase cancelation (such a method of course depends on the particular function $\xi(p, q)$).

The above idea can be traced in several papers, which basically differ from each other by the choice of the radial polynomials $R_{pq}(r)$, the function $\xi(p, q)$, and by the method by which α is eliminated.

Teague [6] and Wallin [7] proposed to use Zernike moments, which are a special case of circular moments, and to take their magnitudes. Li [8] used Fourier-Mellin moments to derive invariants up to the order 9, Wong [9] used complex monomials up to the fifth order that originate from the theory of algebraic invariants. Mostafa and Psaltis [10] introduced the idea to use complex moments for deriving invariants, but they focused on the evaluation of the invariants rather than on constructing higher-order systems. This approach was later followed by Flusser [11, 12], who proposed a general theory of constructing rotation moment invariants. His theory, presented in the next section, is based on the complex moments. It is formally simple and transparent, allows deriving invariants of any orders, and enables studying mutual dependence/independence of the invariants in a readable way. The usage of the complex moments is not essential; this theory can be easily modified for other circular moments as was for instance demonstrated by Derrode and Ghorbel [13] for Fourier-Mellin moments and Yang et al. [14] for Gaussian-Hermite moments.

3.4   Rotation invariants from complex moments

In this section, we present a general method of deriving complete and independent sets of rotation invariants of any orders. This method employs *complex moments* of the image.

3.4.1   Complex moments

The complex moment c_{pq} is obtained when we choose the polynomial basis of complex monomials $\pi_{pq}(x, y) = (x + iy)^p (x - iy)^q$

$$c_{pq} = \int\limits_{-\infty}^{\infty} \int\limits_{-\infty}^{\infty} (x + iy)^p (x - iy)^q f(x, y) \mathrm{d}x \mathrm{d}y. \tag{3.20}$$

It follows from the definition that only the indices $p \geq q$ are independent and worth considering because $c_{pq} = c_{qp}^*$ (the asterisk denotes complex conjugate).

Complex moments carry the same amount of information as the geometric moments of the same order. Each complex moment can be expressed in terms of geometric moments as

$$c_{pq} = \sum_{k=0}^{p} \sum_{j=0}^{q} \binom{p}{k} \binom{q}{j} (-1)^{q-j} \cdot i^{p+q-k-j} \cdot m_{k+j, p+q-k-j} \tag{3.21}$$

and vice versa[11]

$$m_{pq} = \frac{1}{2^{p+q}i^q} \sum_{k=0}^{p} \sum_{j=0}^{q} \binom{p}{k} \binom{q}{j} (-1)^{q-j} \cdot c_{k+j,p+q-k-j}. \tag{3.22}$$

We can also see the link between the complex moments and Fourier transformation. The complex moments are almost (up to a multiplicative constant) the Taylor coefficients of the Fourier transformation of $\mathcal{F}(f)(u+v, i(u-v))$. If we denote $U = u + v$ and $V = i(u - v)$, then we have

$$\mathcal{F}(f)(U, V) \equiv \int_{-\infty}^{\infty} \int_{-\infty}^{\infty} e^{-2\pi i(Ux+Vy)} f(x, y) dx dy$$

$$= \sum_{j=0}^{\infty} \sum_{k=0}^{\infty} \frac{(-2\pi i)^{j+k}}{j!k!} c_{jk} u^j v^k. \tag{3.23}$$

When expressed in polar coordinates, the complex moments take the form

$$c_{pq} = \int_{0}^{\infty} \int_{0}^{2\pi} r^{p+q} e^{i(p-q)\theta} f(r, \theta) r d\theta dr. \tag{3.24}$$

Hence, the complex moments are special cases of the circular moments with a choice of $R_{pq} = r^{p+q}$ and $\xi(p, q) = p - q$. The following lemma is a consequence of the rotation property of circular moments.

Lemma 3.1 Let f' be a rotated version (around the origin) of f, that is, $f'(r, \theta) = f(r, \theta + \alpha)$. Then

$$c'_{pq} = e^{-i(p-q)\alpha} \cdot c_{pq}. \tag{3.25}$$

3.4.2 Construction of rotation invariants

The simplest method proposed by many authors (see [15] for instance) is to use as invariants the moment magnitudes themselves. However, they do not generate a complete set of invariants–by taking the magnitudes only we miss many useful invariants. In the following theorem, phase cancelation is achieved by multiplication of appropriate moment powers.

Theorem 3.1 Let $n \geq 1$, and let $k_i, p_i \geq 0$, and $q_i \geq 0$ $(i = 1, \ldots, n)$ be arbitrary integers such that

$$\sum_{i=1}^{n} k_i(p_i - q_i) = 0.$$

---

[11] While the proof of (3.21) is straightforward, the proof of (3.22) requires first to express x and y as $x = ((x + iy) + (x - iy))/2$ and $y = ((x + iy) - (x - iy))/2i$.

Then

$$I = \prod_{i=1}^{n} c_{p_i q_i}^{k_i} \tag{3.26}$$

is invariant to rotation.

Proof. Let us consider rotation by angle α. Then

$$I' = \prod_{i=1}^{n} c_{p_i q_i}'^{k_i} = \prod_{i=1}^{n} e^{-ik_i(p_i - q_i)\alpha} \cdot c_{p_i q_i}^{k_i} = e^{-i\alpha \sum_{i=1}^{n} k_i(p_i - q_i)} \cdot I = I.$$

$\square$

According to Theorem 3.1, some simple examples of rotation invariants are c_{11}, $c_{20}c_{02}$, $c_{20}c_{12}^2$, and so on. Most invariants (3.26) are complex. If real-valued features are required, we consider real and imaginary parts of each of them separately. To achieve also translation invariance, we use central coordinates in the definition of the complex moments (3.20), and we discard all invariants containing c_{10} and c_{01}. Scaling invariance can be achieved by the same normalization as we used for the geometric moments.

3.4.3 Construction of the basis

In this section, we will pay attention to the construction of a basis of invariants up to a given order. Theorem 3.1 allows us to construct an infinite number of invariants for any order of moments, but only few of them being mutually independent. By the term *basis* we intuitively understand the smallest set required to express all other invariants. More precisely, the basis must be *independent*, which means that none of its elements can be expressed as a function of the other elements, and also *complete*, meaning that any rotation invariant can be expressed by means of the basis elements only.

The knowledge of the basis is a crucial point in all pattern recognition tasks because the basis provides the same discriminative power as the set of all invariants and thus minimizes the computational cost. For instance, the set

$$\{c_{20}c_{02}, c_{21}^2 c_{02}, c_{12}^2 c_{20}, c_{21}c_{12}, c_{21}^3 c_{02}c_{12}\}$$

is a dependent set whose basis is $\{c_{12}^2 c_{20}, c_{21}c_{12}\}$.

To formalize these terms, we first introduce the following definitions.

Definition 3.3 Let $k \geq 1$, let $\mathcal{I} = \{I_1, \dots, I_k\}$ be a set of rotation invariants. Let J be also a rotation invariant. J is said to be *dependent on* $\mathcal{I}$ if and only if there exists a function F of k variables such that

$$J = F(I_1, \dots, I_k).$$

J is said to be *independent* on $\mathcal{I}$ otherwise.

Definition 3.4 Let $k > 1$ and let $\mathcal{I} = \{I_1, \dots, I_k\}$ be a set of rotation invariants. The set $\mathcal{I}$ is said to be *dependent* if and only if there exists $k_0 \leq k$ such that I_{k_0} depends on $\mathcal{I} \div \{I_{k_0}\}$. The set $\mathcal{I}$ is said to be *independent* otherwise.

According to this definition, $\{c_{20}c_{02}, c_{20}^2c_{02}^2\}$, $\{c_{21}^2c_{02}, c_{21}c_{12}, c_{21}^3c_{02}c_{12}\}$, and $\{c_{20}c_{12}^2, c_{02}c_{21}^2\}$ are examples of dependent invariant sets.

Definition 3.5 Let $\mathcal{I}$ be a set of rotation invariants, and let $\mathcal{B}$ be its subset. $\mathcal{B}$ is called a *complete* subset if and only if any element of $\mathcal{I} \doteq \mathcal{B}$ depends on $\mathcal{B}$. The set $\mathcal{B}$ is a *basis* of $\mathcal{I}$ if and only if it is independent and complete.

Now we can formulate a fundamental theorem that tells us how to construct an invariant basis of a given order.

Theorem 3.2 Let us consider complex moments up to the order $r \geq 2$. Let a set of rotation invariants $\mathcal{B}$ be constructed as follows:

$$\mathcal{B} = \{\Phi(p, q) \equiv c_{pq}c_{q_0p_0}^{p-q} | p \geq q \wedge p + q \leq r\},$$

where p_0 and q_0 are arbitrary indices such that $p_0 + q_0 \leq r$, $p_0 - q_0 = 1$ and $c_{p_0q_0} \neq 0$ for all admissible images. Then $\mathcal{B}$ is a basis of all rotation invariants[12] created from the moments of any kind up to the order r.

Theorem 3.2 is very strong because it claims $\mathcal{B}$ is a basis of *all possible* rotation moment invariants, not only of those constructed according to (3.26) and not only of those which are based on complex moments. In other words, $\mathcal{B}$ provides at least the same discrimination power as any other set of moment invariants up to the given order $r \geq 2$ because any possible invariant can be expressed in terms of $\mathcal{B}$. (Note that this is a theoretical property; it may be violated in the discrete domain where different polynomials have different numerical properties.)

Proof. Let us prove the independence of $\mathcal{B}$ first. Let us assume $\mathcal{B}$ is dependent, that is, there exists $\Phi(p, q) \in \mathcal{B}$, such that it depends on $\mathcal{B} \doteq \{\Phi(p, q)\}$. As follows from the linear independence of the polynomials $(x + iy)^p(x - iy)^q$ and, consequently, from independence of the complex moments themselves, it must hold that $p = p_0$ and $q = q_0$. That means, according to the above assumption, there exist invariants $\Phi(p_1, q_1), \ldots, \Phi(p_n, q_n)$ and $\Phi(s_1, t_1), \ldots, \Phi(s_m, t_m)$ from $\mathcal{B} \doteq \{\Phi(p_0, q_0)\}$ and positive integers $k_1, \ldots, k_n$ and $\ell_1, \ldots, \ell_m$ such that

$$\Phi(p_0, q_0) = \frac{\prod_{i=1}^{n_1} \Phi(p_i, q_i)^{k_i} \cdot \prod_{i=n_1+1}^{n} \Phi^*(p_i, q_i)^{k_i}}{\prod_{i=1}^{m_1} \Phi(s_i, t_i)^{\ell_i} \cdot \prod_{i=m_1+1}^{m} \Phi^*(s_i, t_i)^{\ell_i}}. \tag{3.27}$$

Substituting into (3.27) and grouping the factors $c_{p_0q_0}$ and $c_{q_0p_0}$ together, we get

$$\Phi(p_0, q_0) = \frac{c_{q_0p_0}^{\sum_{i=1}^{n_1} k_i(p_i-q_i)} \cdot c_{p_0q_0}^{\sum_{i=n_1+1}^{n} k_i(q_i-p_i)} \cdot \prod_{i=1}^{n_1} c_{p_iq_i}^{k_i} \cdot \prod_{i=n_1+1}^{n} c_{q_ip_i}^{k_i}}{c_{q_0p_0}^{\sum_{i=1}^{m_1} \ell_i(s_i-t_i)} \cdot c_{p_0q_0}^{\sum_{i=m_1+1}^{m} \ell_i(t_i-s_i)} \cdot \prod_{i=1}^{m_1} c_{s_it_i}^{\ell_i} \cdot \prod_{i=m_1+1}^{m} c_{t_is_i}^{\ell_i}}. \tag{3.28}$$

---

[12] The correct notation should be $\mathcal{B}(p_0, q_0)$ because the basis depends on the choice of p_0 and q_0; however, we drop these indexes for simplicity.

Comparing the exponents of $c_{p_0 q_0}$ and $c_{q_0 p_0}$ on both sides, we get the constraints

$$K_1 = \sum_{i=1}^{n_1} k_i(p_i - q_i) - \sum_{i=1}^{m_1} \ell_i(s_i - t_i) = 1 \qquad (3.29)$$

and

$$K_2 = \sum_{i=n_1+1}^{n} k_i(q_i - p_i) - \sum_{i=m_1+1}^{m} \ell_i(t_i - s_i) = 1. \qquad (3.30)$$

Since the rest of the right-hand side of eq. (3.28) must be equal to 1, and since the moments themselves are mutually independent, the following constraints must be fulfilled for any index i:

$$n_1 = m_1, \qquad n = m, \qquad p_i = s_i, \qquad q_i = t_i, \qquad k_i = \ell_i.$$

Introducing these constraints into (3.29) and (3.30), we get $K_1 = K_2 = 0$ which is a contradiction.

To prove the completeness of B, it is sufficient to resolve the so-called *inverse problem*, which means to recover all complex moments (and, consequently, all geometric moments) up to the order r when knowing the elements of B. Thus, the following nonlinear system of equations must be resolved for the c_{pq}'s:

$$
\begin{aligned}
\Phi(p_0, q_0) &= c_{p_0 q_0} c_{q_0 p_0}, \\
\Phi(0, 0) &= c_{00}, \\
\Phi(1, 0) &= c_{10} c_{q_0 p_0}, \\
\Phi(2, 0) &= c_{20} c_{q_0 p_0}^2, \\
\Phi(1, 1) &= c_{11}, \\
\Phi(3, 0) &= c_{30} c_{q_0 p_0}^3, \\
&\cdots \\
\Phi(r, 0) &= c_{r0} c_{q_0 p_0}^r, \\
\Phi(r - 1, 1) &= c_{r-1,1} c_{q_0 p_0}^{r-2}, \\
&\cdots
\end{aligned}
\qquad (3.31)
$$

Since B is a set of rotation invariants, it does not reflect the actual orientation of the object. Thus, there is one degree of freedom when recovering the object moments that correspond to the choice of the object orientation. Without loss of generality, we can choose such orientation in which $c_{p_0 q_0}$ is real and positive. As can be seen from eq. (3.25), if $c_{p_0 q_0}$ is nonzero then such orientation always exists. The first equation of (3.31) can be then immediately resolved for $c_{p_0 q_0}$:

$$c_{p_0 q_0} = \sqrt{\Phi(p_0, q_0)}.$$

Consequently, using the relationship $c_{q_0 p_0} = c_{p_0 q_0}$, we obtain the solutions

$$c_{pq} = \frac{\Phi(p, q)}{c_{q_0 p_0}^{p-q}}$$

and

$$c_{pp} = \Phi(p,p)$$

for any p and q. Recovering the geometric moments is straightforward from eq. (3.8). Since any polynomial is a linear combination of standard powers $x^p y^q$, any moment (and any moment invariant) can be expressed as a function of geometric moments, which completes the proof.   □

Theorem 3.2 allows us not only to create the basis but also to calculate the number of its elements in advance. Let us denote it as $|\mathcal{B}|$. If r is odd then

$$|\mathcal{B}| = \frac{1}{4}(r+1)(r+3),$$

if r is even then

$$|\mathcal{B}| = \frac{1}{4}(r+2)^2.$$

(These numbers refer to complex-valued invariants.) We can see that the authors who used the moment magnitudes only actually lost about half of the information containing in the basis $\mathcal{B}$.

The basis defined in Theorem 3.2 is generally not unique. It depends on the particular choice of p_0 and q_0, which is very important. How shall we select these indices in practice? On the one hand, we want to keep p_0 and q_0 as small as possible because lower-order moments are less sensitive to noise than the higher-order ones. On the other hand, the close-to-zero value of $c_{p_0 q_0}$ may cause numerical instability of the invariants. Thus, we propose the following algorithm. We start with $p_0 = 2$ and $q_0 = 1$ and check if $|c_{p_0 q_0}|$ exceeds a pre-defined threshold for all objects (in practice this means for all given training samples or database elements). If this condition is met, we accept the choice; in the opposite case we increase both p_0 and q_0 by one and repeat the above procedure.

3.4.4   Basis of the invariants of the second and third orders

In this section, we present a basis of the rotation invariants composed of the moments of the second and third orders, that is constructed according to Theorem 3.2 by choosing $p_0 = 2$ and $q_0 = 1$. The basis is

$$\Phi(1,1) = c_{11},$$

$$\Phi(2,1) = c_{21}c_{12},$$

$$\Phi(2,0) = c_{20}c_{12}^2,$$ (3.32)

$$\Phi(3,0) = c_{30}c_{12}^3.$$

In this case, the basis is determined unambiguously and contains six real-valued invariants. It is worth noting that formally, according to Theorem 3.2, the basis should contain also invariants $\Phi(0,0) = c_{00}$ and $\Phi(1,0) = c_{10}c_{12}$. We did not include these two invariants in the basis because $c_{00} = \mu_{00}$ is in most recognition applications used for normalization to scaling, and $c_{10} = m_{10} + im_{01}$ is used to achieve translation invariance. Then $\Phi(0,0) = 1$ and $\Phi(1,0) = 0$ for any object and it is useless to consider them. Numerical stability of $\Phi(2,0)$ is illustrated in Figure 3.4.

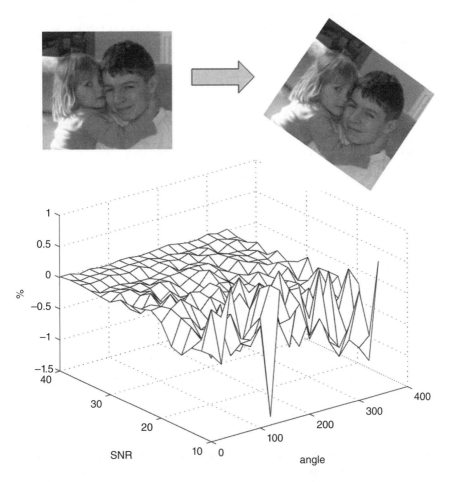

Figure 3.4 Numerical test of the basic invariant $\Phi(2,0)$. Computer-generated rotation of the test image ranged form 0 to 360 degrees. To show robustness, each image was corrupted by additive Gaussian white noise. Signal-to-noise ratio (SNR) ranged from 40 (low noise) to 10 (heavy noise). Horizontal axes: rotation angle and SNR, respectively. Vertical axis–relative deviation (in %) between $\mathcal{R}e(\Phi(2,0))$ of the original and that of the rotated and noisy image. The test proves the invariance of $\mathcal{R}e(\Phi(2,0))$ and illustrates its high robustness to noise

3.4.5 Relationship to the Hu invariants

In this section, we highlight the relationship between the Hu invariants (3.15) and the proposed invariants (3.32). We show the Hu invariants are incomplete and mutually dependent, which can explain some practical problems connected with their usage.

It can be seen that the Hu invariants are nothing but particular representatives of the general form (3.26)

$$\phi_1 = c_{11},$$

$$\phi_2 = c_{20}c_{02},$$

$$\phi_3 = c_{30}c_{03},$$

$$\phi_4 = c_{21}c_{12},$$

$$\phi_5 = \mathcal{R}e(c_{30}c_{12}^3),$$

$$\phi_6 = \mathcal{R}e(c_{20}c_{12}^2),$$

$$\phi_7 = \mathcal{I}m(c_{30}c_{12}^3)$$

$$(3.33)$$

and that they can be expressed also in terms of the basis (3.32)

$$\phi_1 = \Phi(1,1)$$

$$\phi_2 = \frac{|\Phi(2,0)|^2}{\Phi(2,1)^2}$$

$$\phi_3 = \frac{|\Phi(3,0)|^2}{\Phi(2,1)^3}$$

$$\phi_4 = \Phi(2,1)$$

$$\phi_5 = \mathcal{R}e(\Phi(3,0))$$

$$\phi_6 = \mathcal{R}e(\Phi(2,0))$$

$$\phi_7 = \mathcal{I}m(\Phi(3,0)).$$

Using (3.33), we can demonstrate the dependency of the Hu invariants. It holds

$$\phi_3 = \frac{\phi_5^2 + \phi_7^2}{\phi_4^3},$$

which means that either ϕ_3 or ϕ_4 is useless and can be excluded from Hu's system without any loss of discrimination power.

Moreover, we show the Hu's system is incomplete. Let us try to recover complex and geometric moments when knowing $\phi_1, \ldots, \phi_7$ under the same normalization constraint as in the previous case, i.e. c_{21} is required to be real and positive. Complex moments $c_{11}, c_{21}, c_{12}, c_{30}$, and c_{03} can be recovered in a straightforward way:

$$c_{11} = \phi_1,$$

$$c_{21} = c_{12} = \sqrt{\phi_4},$$

$$\mathcal{R}e(c_{30}) = \mathcal{R}e(c_{03}) = \frac{\phi_5}{c_{12}^3},$$

$$\mathcal{I}m(c_{30}) = -\mathcal{I}m(c_{03}) = \frac{\phi_7}{c_{12}^3}.$$

Unfortunately, c_{20} cannot be fully recovered. We have, for its real part,

$$\mathcal{R}e(c_{20}) = \frac{\phi_6}{c_{12}^2}$$

but for its imaginary part we get

$$(\mathcal{I}m(c_{20}))^2 = |c_{20}|^2 - (\mathcal{R}e(c_{20}))^2 = \phi_2 - \left(\frac{\phi_6}{c_{12}^2}\right)^2.$$

There is no way of determining the sign of $\mathcal{I}m(c_{20})$. In terms of geometric moments, it means the sign of m_{11} cannot be recovered.

The incompleteness of the Hu invariants implicates their lower discrimination power compared to the proposed invariants of the same order. Let us consider two objects $f(x,y)$ and $g(x,y)$, both in the normalized central positions, having the same geometric moments up to the third order except m_{11}, for which $m_{11}^{(f)} = -m_{11}^{(g)}$. In the case of artificial data, such object $g(x,y)$ exists for any given $f(x,y)$ and can be designed as

$$g(x,y) = (f(x,y) + f(-x,y) + f(x,-y) - f(-x,-y))/2, \tag{3.34}$$

see Figure 3.5 for an example. It is easy to prove that under (3.34) the moment constraints are always fulfilled. While the basic invariants (3.32) distinguish these two objects by the imaginary part of $\Phi(2,0)$, the Hu invariants are not able to do so (see Table 3.1), even if the objects are easy to discriminate visually.

This property can be demonstrated also on real data. In Figure 3.6 (left), one can see the photograph of a pan. A picture of a virtual "two-handle" pan (Figure 3.6 right) was created from the original image according to (3.34). Although these two objects are apparently different, all their Hu invariants are exactly the same. On the other hand, the new invariant $\Phi(2,0)$ distinguishes these two objects clearly thanks to the opposite signs of its imaginary part.

This ambiguity cannot be avoided by a different choice of the normalization constraint when recovering the moments. Let us, for instance, consider another normalization constraint, which

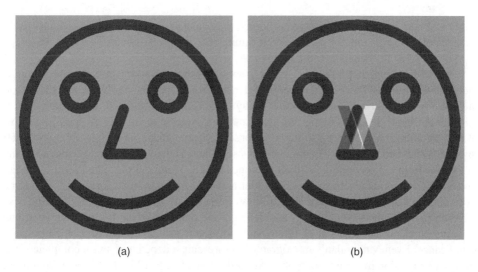

(a)   (b)

Figure 3.5 The smiles: (a) original and (b) another figure created from the original according to Eq. (3.34). For the values of the respective invariants see Table 3.1

Table 3.1 The values of the Hu invariants and the basic invariants (3.32) of "The smiles" in Figure 3.5. The only invariant discriminating them is $Im(\Phi(2,0))$. (The values shown here were calculated after bringing Figure 3.5 into normalized central position $c_{21} > 0$ and nonlinearly scaled to the range from -10 to 10 for display.)

Hu	Figure 3.5(a)	Figure 3.5(b)	Basic	Figure 3.5(a)	Figure 3.5(b)
ϕ_1	4.2032	4.2032	$\Phi(1,1)$	4.2032	4.2032
ϕ_2	1.7563	1.7563	$\Phi(2,1)$	6.5331	6.5331
ϕ_3	3.6395	3.6395	$\mathcal{R}e(\Phi(2,0))$	8.6576	8.6576
ϕ_4	6.5331	6.5331	$\mathbf{Im(\Phi(2,0))}$	**7.6437**	**−7.6437**
ϕ_5	1.0074	1.0074	$\mathcal{R}e(\Phi(3,0))$	1.0074	1.0074
ϕ_6	8.6576	8.6576	$\mathcal{I}m(\Phi(3,0))$	−6.9850	−6.9850
ϕ_7	−6.9850	−6.9850			

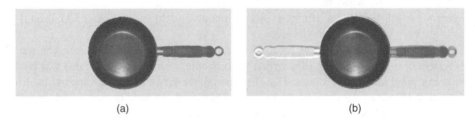

(a)   (b)

Figure 3.6 (a) Original image of a pan and (b) a virtual "two-handle" pan. These objects are distinguishable by the basic invariants but not by the Hu invariants

requires c_{20} to be real and positive. In terms of geometric moments, it corresponds to the requirement $m_{11} = 0$ and $m_{20} > m_{02}$. Then, similarly to the previous case, the signs of $(m_{30} + m_{12})$ and $(m_{03} + m_{21})$ cannot be unambiguously determined.

It is worth noting that also other earlier sets of rotation invariants are dependent and/or incomplete. This was, however, not mentioned in the original papers, partly because the authors did not pay attention to these issues and partly because their methods did not allow them to find out the dependencies among the invariants in a systematic way. Li [8] published a set of invariants from moments up to the ninth order. Unfortunately, his system includes the Hu invariants and therefore it also cannot be a basis. Jin and Tianxu [5] derived twelve invariants in explicit form but only eight of them are independent. Wong [9] presented a set of sixteen invariants from moments up to the third order and a set of "more than forty-nine" invariants from moments up to the fourth order. It follows immediately from Theorem 3.2 that the basis of the third-order invariants has only six elements and the basis of the fourth-order invariants has nine elements. Thus, most of the Wong's invariants are dependent and of no practical significance. Even Mukundan's monograph [16] presents a dependent and incomplete set of twenty-two rotation invariants up to the sixth order (see [16], p. 125). The construction of the invariants from the complex moments has resolved the independence and completeness issues in an elegant way.

3.5 Pseudoinvariants

In this section, we investigate the behavior of rotation invariants under mirror reflection of the image (see Figure 3.7). Since the mirror reflection is not a special case of the TRS transformation, the rotation invariants may change when the image is mirrored. The mirror reflection can occur in some applications only, but it is useful to know how the rotation invariants behave in that case. We show they cannot change arbitrarily. There are only two possibilities–a (real-valued) invariant can either stay constant or change its sign. The invariants, which preserve their value under reflection are traditionally called *true invariants* while those that change the sign are called *pseudoinvariants* [6] or, misleadingly, *skew invariants* [3]. Pseudoinvariants discriminate between mirrored images of the same object, which is useful in some applications but may be undesirable in other cases. We show which invariants from the basis introduced in Theorem 3.2 are pseudoinvariants and which are the true ones.

Let us consider a basic invariant, and let us investigate its behavior under a reflection across an arbitrary line. Due to the rotation and shift invariance, we can limit ourselves to the reflection across the *x*-axis.

Figure 3.7   The test image and its mirrored version. Basic invariants of the mirrored image are complex conjugates of those of the original

Let $\bar{f}(x,y)$ be a mirrored version of $f(x,y)$, i.e. $\bar{f}(x,y) = f(x,-y)$. As follows from the definition,

$$\overline{c_{pq}} = c_{qp} = c_{pq}^*.$$

Thus, it holds for any basic invariant $\Phi(p,q)$

$$\overline{\Phi(p,q)} = \overline{c_{pq} c_{q_0 p_0}^{p-q}} = c_{pq}^* \cdot (c_{q_0 p_0}^*)^{p-q} = \Phi^*(p,q).$$

This proves that the real parts of the basic invariants are true invariants. On the other hand, the imaginary parts of them change their signs under reflection and hence are pseudoinvariants. This brings a consequence for the recognition of objects with axial symmetry. The pseudoinvariants cannot differentiate among any axially symmetric objects because their values must always be zero.

3.6   Combined invariants to TRS and contrast stretching

So far we have considered invariants to spatial transformations only. However, if the features used in a recognition system are calculated from a graylevel/color image and not just from a binary image, they should be invariant also to various changes of intensity. These changes may be caused by the change of illumination conditions, by the camera setup, by the influence of the environment, and by many other factors. In Chapter 6 we describe invariants to graylevel degradations caused by linear filtering. In this section we consider *contrast stretching* only, which is a very simple graylevel transformation given as

$$f'(x,y) = a \cdot f(x,y),$$

where a is a positive stretching factor. This model approximately describes the change of the illumination of the scene between two acquisitions (in case of digital images, this model may not be valid everywhere because $f'(x,y)$ might overflow the 8-bit intensity range at some places, but we ignore this issue here for simplicity). The invariants to contrast stretching in connection with Hu's rotation invariants were firstly proposed by Maitra [17] and later improved by Hupkens [18]. Here we will demonstrate how to include the invariance to contrast stretching into the TRS invariants described earlier in this chapter.

Since $\mu'_{pq} = a \cdot \mu_{pq}$ and $c'_{pq} = a \cdot c_{pq}$, pure contrast (and translation) invariance and can be achieved simply by normalizing each central or complex moment by μ_{00}. Now consider a contrast stretching *and* a rotation together. It holds for any basic invariant $\Phi(p,q)$

$$\Phi'(p,q) = a^{p-q+1} \Phi(p,q),$$

and it is sufficient to use the normalization

$$\frac{\Phi(p,q)}{\mu_{00}^{p-q+1}}.$$

This approach unfortunately cannot be further extended to scaling. If μ_{00} has been used for normalization to contrast, it cannot be at the same time used for scaling normalization. We have to normalize by some other invariant, preferably by c_{11} for the sake of simplicity.

Let $\widetilde{c_{pq}}$ be defined as follows

$$\widetilde{c_{pq}} = \frac{c_{pq}\mu_{00}^{\frac{p+q}{2}-1}}{c_{11}^{\frac{p+q}{2}}}.$$

Then $\widetilde{c_{pq}}$ is *both* contrast and scaling normalized moment, which can be used for constructing rotation invariants in a usual way as

$$\Psi(p,q) = \widetilde{c_{pq}}\,\widetilde{c_{q_0p_0}}^{p-q}.$$

This is possible because c_{11} itself is a rotation invariant. Hence, $\Psi(p,q)$ is a combined TRS and contrast invariant. Note that $\Psi(1,1) = 1$ for any object due to the normalization. Numerical stability of $\Psi(2,0)$ is illustrated in Figure 3.8.

3.7   Rotation invariants for recognition of symmetric objects

In many applied tasks, we want to classify man-made objects or natural shapes from their silhouettes (i.e., from binary images) only. In most cases such shapes have some kind of symmetry. Different classes may have the same symmetry, which makes the task even more difficult. While humans can use the symmetry as a cue that helps them to recognize objects, moment-based classifiers suffer from the fact that some moments of symmetric shapes are zero and corresponding invariants do not provide any discrimination power. This is why we must pay attention to these situations and why it is necessary to design special invariants for each type of symmetry [19].

For example, all odd-order moments (geometric as well as complex) of a centrosymmetric object are identically equal to zero. If an object is circularly symmetric, all its complex moments, whose indices are different, vanish. Thus, Theorem 3.2 either cannot be applied at all, or many rotation invariants might be useless. Let us imagine an illustrative example. We want to recognize three shapes–a square, a cross, and a circle–independently of their orientation. Because of the symmetry, all complex moments of the second and third orders except c_{11} are zero. If the shapes are appropriately scaled, c_{11} can be the same for all of them. Consequently, neither the Hu invariants nor the basic invariants from Theorem 3.2 provide any discrimination power, even if the shapes can be readily recognized visually. Appropriate invariants in this case would be $c_{22}, c_{40}c_{04}, c_{51}c_{04}, c_{33}, c_{80}c_{04}^2, c_{62}c_{04}, c_{44}$, etc. The above simple example shows the necessity of having different systems of invariants for objects with different types of symmetry.

We have only two types of object symmetry which are relevant to consider in this context[13]–*N-fold rotation symmetry* (*N*-FRS) and *N-fold dihedral symmetry* (*N*-FDS).

Object f is said to have *N*-FRS if it is "rotationally periodic", that is, if it repeats itself when it rotates around the origin (which is supposed to be shifted into the object centroid) by $\alpha_j = 2\pi j/N$ for all $j = 1, \ldots, N$. In polar coordinates this means that

$$f(r,\theta) = f(r, \theta + \alpha_j) \qquad j = 1, \ldots, N.$$

---

[13] In the theory of point groups in 2D, it is well known that there exist only cyclic groups C_N, dihedral groups D_N, and translational symmetry groups which are irrelevant here.

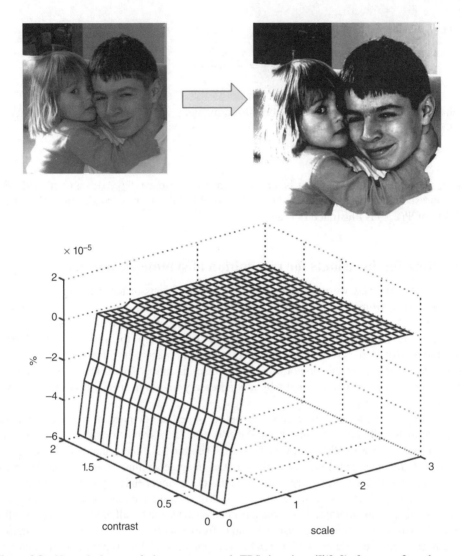

Figure 3.8 Numerical test of the contrast and TRS invariant $\Psi(2,0)$ for $p_0 = 2$ and $q_0 = 1$. Computer-generated scaling of the test image ranged from $s = 0.2$ to $s = 3$, and the contrast stretching factor ranged from $a = 0.1$ to $a = 2$. Horizontal axes: scaling factor s and contrast stretching factor a, respectively. Vertical axis–relative deviation (in %) between $\mathcal{R}e(\Psi(2,0))$ of the original and that of the scaled and stretched image. The test proves the invariance of $\mathcal{R}e(\Psi(2,0))$ with respect to both factors. However, for down-scaling with $s = 0.2$ and $s = 0.3$, the resampling effect leads to higher relative errors

In particular, $N = 1$ means no symmetry in a common sense, and $N = 2$ denotes the central symmetry $f(-x, -y) = f(x, y)$. We use this definition not only for finite N but also for $N = \infty$. Thus, in our terminology, the objects with a *circular symmetry* $f(r, \theta) = f(r)$ are said to have an ∞-FRS.

Object f is said to have N-FDS if it has N-FRS and at least one symmetry axis. The number of symmetry axes cannot be arbitrary–the object with N-FRS may only have N symmetry axes

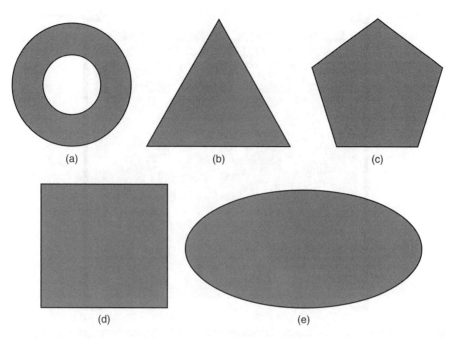

Figure 3.9 Sample objects with an N-fold rotation symmetry. From (a) to (e): $N = \infty$, 3, 5, 4, and 2, respectively. All depicted cases have also an axial symmetry; however this is not a rule

or no axis at all. On the other hand, if an object has N axes of symmetry ($N > 0$), then it is also rotationally symmetric, and N is exactly its number of folds [20]. In case of $N = \infty$, the rotation and dihedral symmetries coincide. Examples of N-fold symmetric shapes are shown in Figure 3.9.

We show that both rotation and dihedral symmetries contribute to vanishing of some moments and invariants.

Lemma 3.2 If $f(x, y)$ has an N-fold rotation symmetry (N finite) and if $(p - q)/N$ is not an integer, then $c_{pq} = 0$.

Proof. Let us rotate f around the origin by $2\pi/N$. Due to its symmetry, the rotated object f' must be the same as the original. In particular, it must hold $c'_{pq} = c_{pq}$ for any p and q. On the other hand, it follows from eq. (3.25) that

$$c'_{pq} = e^{-2\pi i(p-q)/N} \cdot c_{pq}.$$

Since $(p - q)/N$ is assumed not to be an integer, this equation can be fulfilled only if $c_{pq} = 0$.[14]   □

Lemma 3.3a If $f(x, y)$ has ∞-fold rotation symmetry and if $p \neq q$, then $c_{pq} = 0$.

---

[14] This lemma can be generalized for any circular moment; the moment vanishes if $\xi(p, q)/N$ is not an integer.

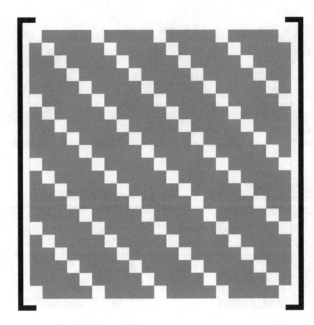

Figure 3.10   The matrix of the complex moments of an N-fold symmetric object. The gray elements are always zero. The distance between neighboring non-zero diagonals is N

Proof. Let us rotate f around the origin by an arbitrary angle α. The rotated version f' must be the same as the original for any α, and, consequently, its moments cannot change under rotation. Eq. (3.25) implies

$$c'_{pq} = e^{-i(p-q)\alpha} \cdot c_{pq}.$$

Since $p \neq q$ and α may be arbitrary, this equation can be fulfilled only if $c_{pq} = 0$.   □

The above lemma shows that the matrix of the complex moments $C_{ij} = c_{i-1,j-1}$ of an N-fold symmetric object always has a specific structure – it is a sparse *multidiagonal* (and of course Hermitian) matrix with non-zero entries on the major diagonal and on the minor diagonals with the equidistant spacing of N (see Figure 3.10).

The other elements are always zero, so we have to avoid their usage as the factors in the invariants. In order to derive invariants for recognition of objects with N-fold rotation symmetry, Theorem 3.2 can be generalized in the following way.

Theorem 3.2 Let $N \geq 1$ be a finite integer. Let us consider complex moments up to the order $r \geq N$. Let a set of rotation invariants $\mathcal{B}_N$ be constructed as follows:

$$\mathcal{B}_N = \{\Phi_N(p,q) \equiv c_{pq}c_{q_0 p_0}^k \mid p \geq q \,\wedge\, p+q \leq r \,\wedge\, k \equiv (p-q)/N \text{ is an integer}\},$$

where p_0 and q_0 are arbitrary indices such that $p_0 + q_0 \leq r$, $p_0 - q_0 = N$, and $c_{p_0 q_0} \neq 0$ for all admissible objects. Then $\mathcal{B}_N$ is a basis of all non-trivial rotation invariants for objects with N-FRS, created from the moments up to the order r.

Proof. Rotation invariance of all elements of B_N follows immediately from Theorem 3.1. The independence and completeness of B_N can be proven exactly in the same way as in Theorem 3.2, when only non-zero moments are recovered. □

If $2 \leq r < N$, no such $c_{p_0 q_0}$ exists, and the basis is

$$B_N = \{c_{pp} \mid p \leq r/2\}.$$

For $N = 1$, which means no rotation symmetry, Theorem 3.2 is reduced exactly to Theorem 3.2; $B_1 = B$ and $\Phi_1(p,q) = \Phi(p,q)$. The following modification of Theorem 3.2 deals with the case of $N = \infty$.

Theorem 3.2a Let us consider complex moments up to the order $r \geq 2$. Then the basis B_∞ of all non-trivial rotation invariants for objects with ∞-FRS is

$$B_\infty = \{c_{pp} \mid p \leq r/2\}.$$

The proof of Theorem 3.2a follows immediately from Theorem 3.1 and Lemma 3.3a. Note that matrix $\mathbf{C}$ is diagonal for $N = \infty$.

Theorems 3.2 and 3.2a have several interesting consequences. Some of them are summarized in the following lemma.

Lemma 3.4 Let us denote all rotation invariants that can be expressed by means of elements of basis B as $\langle B \rangle$. Then it holds for any order r

1. If M and N are finite and L is their least common multiple, then

$$\langle B_M \rangle \cap \langle B_N \rangle = \langle B_L \rangle.$$

In particular, if M/N is an integer, then $\langle B_M \rangle \subset \langle B_N \rangle$.

2.

$$\bigcap_{N=1}^{\infty} \langle B_N \rangle = \langle B_\infty \rangle.$$

3. If N is finite, the number of elements of B_N is

$$|B_N| = \sum_{j=0}^{[r/N]} \left[\frac{r - jN + 2}{2} \right],$$

where $[a]$ means an integer part of a. For $N = \infty$ it holds

$$|B_\infty| = \left[\frac{r+2}{2} \right].$$

From the last point of Lemma 3.4 we can see that the higher the number of folds, the fewer non-trivial invariants exist. The number of them ranges from about $r^2/4$ for non-symmetric objects to about $r/2$ for circularly symmetric ones.

If an object has an N-FDS, then its moment matrix has the same multidiagonal structure, but the axial symmetry introduces additional dependence between the real and imaginary parts of each moment. The actual relationship between them depends on the orientation of the symmetry axis. If we, however, consider the basic rotation invariants from Theorem 3.2, the dependence on the axis orientation disappears; the imaginary part of each invariant is zero (we already showed this property in Section 3.5).

In practical pattern recognition experiments, the number of folds N may not be known *a priori*. In that case we can apply a fold detector (see [21–24] for sample algorithms[15] detecting the number of folds) to all elements of the training set before we choose an appropriate system of moment invariants. In case of equal fold numbers of all classes, the proper invariants can be chosen directly according to Theorem 3.2 or 3.2a. Usually we start selecting from the lowest orders, and the appropriate number of the invariants is determined such that they should provide a sufficient discriminability of the training set.

If there are shape classes with different numbers of folds, the previous theory does not provide a universal solution to this problem. If we want to find such invariants which are non-trivial on *all* classes, we cannot simply choose one of the fold numbers detected as the appropriate N for constructing invariant basis according to Theorem 3.2 (although one could intuitively expect the highest number of folds to be a good choice, this is not that case). A more sophisticated choice is to take the least common multiple of all finite fold numbers and then to apply Theorem 3.2. Unfortunately, taking the least common multiple often leads to high-order instable invariants. This is why in practice one may prefer to decompose the problem into two steps–first, pre-classification into "groups of classes" according to the number of folds is performed, and then final classification is done by means of moment invariants, which are defined separately within each group. This decomposition can be performed explicitly in a separate pre-classification stage or implicitly during the classification. The word "implicitly" here means that the number of folds of an unknown object is not explicitly tested, however, at the beginning we must test the numbers of folds in the training set. We explain the latter version.

Let us have C classes altogether such that C_k classes have N_k folds of symmetry; $k = 1, \ldots, K$; $N_1 > N_2 > \ldots N_K$. The set of proper invariants can be chosen as follows. Starting from the highest symmetry, we iteratively select those invariants providing (jointly with the invariants that have been already selected) a sufficient discriminability between the classes with the fold numbers N_F and higher, but which may equal zero for some (or all) other classes. Note that for some F the algorithm need not select any invariant because the discriminability can be assured by the invariants selected previously or because $C_F = 1$.

In order to illustrate the importance of a careful choice of the invariants in pattern recognition tasks, we carried out the following experimental studies.

---

[15] These fold detectors look for the angular periodicity of the image, either directly by calculating the circular auto-correlation function and searching for its peaks or by investigating circular Fourier transformation and looking for dominant frequency. Alternatively, some methods estimate the fold number directly from the zero values of certain complex moments or other circular moments. Such approach may however be misleading in some cases because the moment values could be small even if the object is not symmetric.

3.7.1   Logo recognition

In the first experiment, we tested the capability of recognizing objects having the same number of folds, particularly $N = 3$. As a test set we used three logos of major companies (Mercedes-Benz, Mitsubishi, and Fischer) and two commonly used symbols ("recycling" and "woolen product"). All logos were downloaded from the respective web-sites, resampled to 128×128 pixels and binarized. We decided to use logos as the test objects because most logos have a certain degree of symmetry. This experiment is also relevant to the state of the art because several commercial logo/trademark recognition systems reported in the literature [25–28] used rotation moment invariants as features, and all of them face the problem of symmetry.

As can be seen in Figure 3.11, all our test logos have three-fold rotation symmetry. Each logo was rotated ten times by randomly generated angles. Since the spatial resolution of the images was relatively high, the sampling effect was insignificant. Moment invariants from Theorem 3.2 ($N = 3$, $p_0 = 3$, and $q_0 = 0$) provide an excellent discrimination power even if only the two simplest ones have been used (see Figure 3.12), while the invariants from Theorem 3.2 are not able to distinguish the logos at all (see Figure 3.13).

3.7.2   Recognition of shapes with different fold numbers

In the second experiment, we used nine simple binary patterns with various numbers of folds: capitals F and L ($N = 1$), rectangle and diamond ($N = 2$), equilateral triangle and tripod ($N = 3$), cross ($N = 4$), and circle and ring ($N = \infty$) (see Figure 3.14). As in the previous case, each pattern was rotated ten times by random angles.

First, we applied rotation invariants according to Theorem 3.2 choosing $p_0 = 2$ and $q_0 = 1$. The positions of our test patterns in the feature space are plotted in Figure 3.15. Although only a 2-D subspace showing the invariants $c_{21}c_{12}$ and $\mathcal{R}e(c_{20}c_{12}^2)$ is visualized there, we can easily observe that the patterns form a single dense cluster around the origin (the only exception is the tripod, which is slightly biased because of its non-symmetry caused by the sampling effect). Two non-symmetric objects–letters F and L–are far from the origin, beyond the displayed area. The only source of non-zero variance of the cluster are spatial sampling errors. All other invariants of the form $c_{pq}c_{12}^{p-q}$ behave in the same way. Thus, according to our theoretical expectation, we cannot discriminate among symmetric objects (even if they are very different) by means of the invariants defined in Theorem 3.2.

Second, we employed the invariants introduced in Theorem 3.2 choosing $N = 4$ (the highest finite number of folds among the test objects), $p_0 = 4$, and $q_0 = 0$ to resolve the above

Figure 3.11   The test logos (from left to right): Mercedes-Benz, Mitsubishi, Recycling, Fischer, and Woolen product

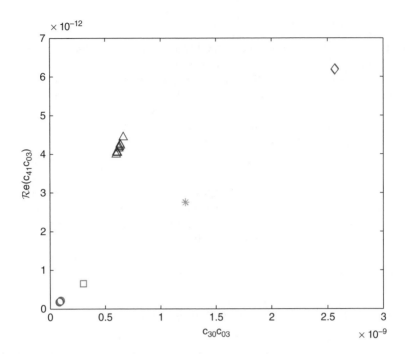

Figure 3.12   The logo positions in the space of two invariants $c_{30}c_{03}$ and $\mathcal{R}e(c_{41}c_{03})$ showing good discrimination power. The symbols: □–Mercedes-Benz, ◇–Mitsubishi, △–Recycling, *–Fischer, and ○–Woolen product. Each logo was randomly rotated ten times

recognition experiment. The situation in the feature space appears to be different from the previous case (see the plot of the two simplest invariants $c_{40}c_{04}$ and $\mathcal{R}e(c_{51}c_{04})$ in Figure 3.16). Five test patterns formed their own very compact clusters that are well separated from each other. However, the patterns circle, ring, triangle, and tripod still made a mixed cluster around the origin and remained non-separable. This result is fully in accordance with the theory, because the number of folds used here is not optimal for our test set.

Third, we repeated this experiment again with invariants according to Theorem 3.2 but selecting N as the least common multiple of all finite fold numbers involved, that is, $N = 12$. One can learn from Figure 3.17 that now all clusters are well separated (because of the high dynamic range, logarithmic scale was used for visualization purposes). The only exception are two patterns having circular symmetry–the circle and the ring–that still created a mixed cluster. If also these two patterns were to be separated from one another, we could use the invariants c_{pp}. On the other hand, using *only* these invariants for the whole experiment is not a good choice from the practical point of view –since there is only one such invariant for each order, we would be pushed into using high-order noise-sensitive moments.

Finally, we used the algorithm described at the end of Section 3.7. In this case, two invariants $c_{30}c_{03}$ and $c_{40}c_{04}$ are sufficient to separate all classes (of course, again with the exception of the circle and the ring), see Figure 3.18. Compared to the previous case, note less correlation of the invariants, their higher robustness, and lower dynamic range. On the other hand, neither $c_{30}c_{03}$ nor $c_{40}c_{04}$ provide enough discrimination power when used individually while the twelfth-order invariants are able to distinguish all classes.

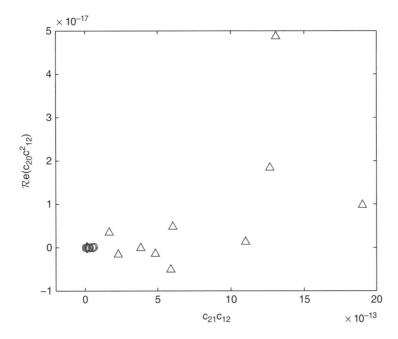

Figure 3.13 The logo positions in the space of two invariants $c_{21}c_{12}$ and $\mathcal{R}e(c_{20}c_{12}^2)$ introduced in Theorem 3.2. These invariants have no discrimination power with respect to this logo set. The symbols: □–Mercedes-Benz, ◇–Mitsubishi, △–Recycling, ∗–Fischer, and ○–Woolen product

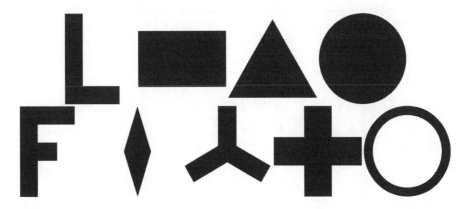

Figure 3.14 The test patterns: capital L, rectangle, equilateral triangle, circle, capital F, diamond, tripod, cross, and ring

3.7.3 Experiment with a baby toy

We demonstrate the performance of invariants in an object matching task. We used a popular baby toy (see Figure 3.19) that is also commonly used in testing computer vision algorithms

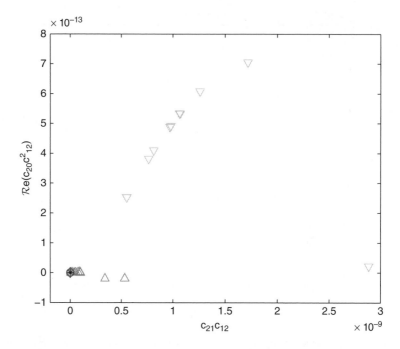

Figure 3.15 The space of two invariants $c_{21}c_{12}$ and $\mathcal{R}e(c_{20}c_{12}^2)$ introduced in Theorem 3.2. The symbols: ×–rectangle, ◇–diamond, △–equilateral triangle, ▽–tripod +–cross, •–circle, and ○–ring. The discriminability is very poor

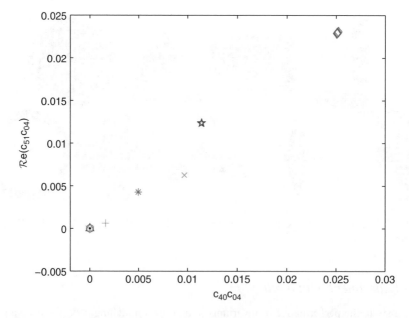

Figure 3.16 The space of two invariants $c_{40}c_{04}$ and $\mathcal{R}e(c_{51}c_{04})$ introduced in Theorem 3.2, $N = 4$. The symbols: ×–rectangle, ◇–diamond, △–equilateral triangle, ▽–tripod +–cross, •–circle, and ○–ring, ∗–capital F, and ☆–capital L. Some clusters are well separated

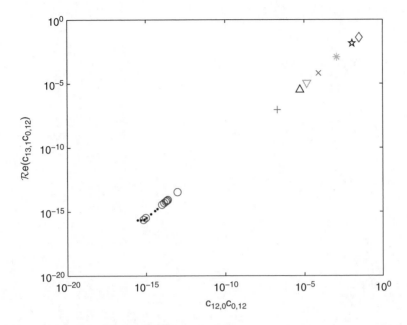

Figure 3.17 The space of two invariants $c_{12,0}c_{0,12}$ and $\mathcal{R}e(c_{13,1}c_{0,12})$ introduced in Theorem 3.2, $N = 12$ (logarithmic scale). The symbols: ×–rectangle, ◇–diamond, △–equilateral triangle, ▽–tripod +–cross, ●–circle, and ○–ring, *–capital F, and ☆–capital L. All clusters except the circle and the ring are separated

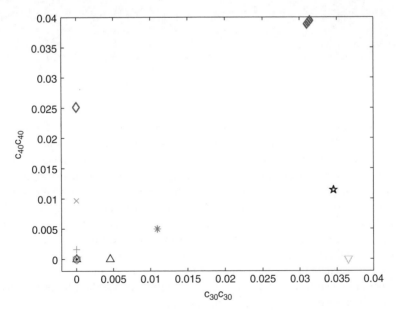

Figure 3.18 The space of two invariants $c_{30}c_{03}$ and $c_{40}c_{04}$. The symbols: ×–rectangle, ◇–diamond, △–equilateral triangle, ▽–tripod +–cross, ●–circle, and ○–ring, *–capital F, and ☆–capital L. All clusters except the circle and the ring are separated. Comparing to Figure 3.17, note less correlation of the invariants and a lower dynamic range

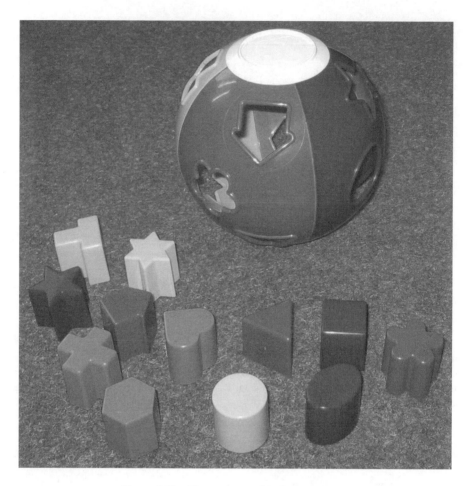

Figure 3.19   The toy set used in the experiment

and robotic systems. The toy consists of a hollow sphere with twelve holes and of twelve objects of various shapes. Each object matches up with just one particular hole. The baby (or the algorithm) is supposed to assign the objects to the corresponding holes and insert them into the sphere. The baby can employ both the color and shape information; however, in our experiment we completely disregarded the colors to make the task more challenging.

First, we binarized the pictures of the holes (one picture per each hole) by simple thresholding. Binarization was the only pre-processing; no sphere-to-plane corrections were applied.

To select proper invariants, we applied the algorithm from Section 3.7 on the images of the holes. As a discriminability measure we took weighted Euclidean distance, where the weights were set up to normalize the dynamic range of the invariants. As one can observe, the highest finite number of folds is 6. The algorithm terminated after passing three loops and selected the three following invariants: $c_{60}c_{06}$, $c_{50}c_{05}$, and $c_{40}c_{04}$.

Then we took ten pictures of each object with random rotations, binarized them, and run the classification. This task is not so easy as it might appear because the holes are a bit larger than the objects, but this relation is morphological rather than linear and does not preserve the shapes exactly. Fortunately, all 120 unknown objects were recognized correctly and assigned to proper holes. It should be emphasized that only three invariants without any other information yielded 100% recognition rate for twelve classes, which is a very good result even though the shapes are relatively simple.

We repeated the classification once again with invariants constructed according to Theorem 3.2 setting $p_0 = 2, q_0 = 1$. Only three objects having 1-fold symmetry were recognized correctly, the others were classified randomly.

3.8   Rotation invariants via image normalization

The TRS invariants can also be derived in another way than by means of Theorem 3.2. This alternative method called *image normalization* first brings the image into some "standard" or "canonical" position, which is defined by setting certain image moments to pre-defined values (usually 0 or 1). Then the moments of the normalized image (except those that have been used to define the normalization constraints) can be used as TRS invariants of the original image. It should be noted that no actual geometric transformation and resampling of the image is required; the moments of the normalized image can be calculated directly from the original image by means of normalization constraints.

While normalization to translation and scaling is trivial by setting $m'_{10} = m'_{01} = 0$ and $m'_{00} = 1$, there are many possibilities of defining the normalization constraints to rotation. The traditional one known as the *principal axis* method [3] requires for the normalized image the constraint $\mu'_{11} = 0$.[16] It leads to diagonalization of the second-order moment matrix

$$\mathbf{M} = \begin{pmatrix} \mu_{20} & \mu_{11} \\ \mu_{11} & \mu_{02} \end{pmatrix}. \tag{3.35}$$

It is well-known that any symmetric matrix can be diagonalized on the basis of its (orthogonal) eigenvectors. After finding the eigenvalues of $\mathbf{M}$ (they are guaranteed to be real) by solving

$$|\mathbf{M} - \lambda \mathbf{I}| = 0,$$

we obtain the diagonal form

$$\mathbf{M}' = \mathbf{G}^T \mathbf{M} \mathbf{G} = \begin{pmatrix} \lambda_1 & 0 \\ 0 & \lambda_2 \end{pmatrix} = \begin{pmatrix} \mu'_{20} & 0 \\ 0 & \mu'_{02} \end{pmatrix},$$

where $\mathbf{G}$ is an orthogonal matrix composed of the eigenvectors of $\mathbf{M}$. It is easy to show that we obtain for the eigenvalues of $\mathbf{M}$

$$\mu'_{20} \equiv \lambda_1 = ((\mu_{20} + \mu_{02}) + \sqrt{(\mu_{20} - \mu_{02})^2 + 4\mu_{11}^2})/2,$$

$$\mu'_{02} \equiv \lambda_2 = ((\mu_{20} + \mu_{02}) - \sqrt{(\mu_{20} - \mu_{02})^2 + 4\mu_{11}^2})/2,$$

---

[16] The reader may notice a clear analogy with the principal component analysis.

and that the angle between the first eigenvector and the x-axis, which is actually the normalization angle, is[17]

$$\alpha = \frac{1}{2} \cdot \arctan\left(\frac{2\mu_{11}}{\mu_{20} - \mu_{02}}\right).$$

Provided that $\lambda_1 > \lambda_2$, there are four different normalization angles (see Figure 3.21) which all make $\mu'_{11} = 0$: $\alpha, \alpha + \pi/2, \alpha + \pi$, and $\alpha + 3\pi/2$. To remove the ambiguity between "horizontal" and "vertical" position, we impose the constraint $\mu'_{20} > \mu'_{02}$. Removing the "left-right" ambiguity can be achieved by introducing an additional constraint $\mu'_{30} > 0$ (provided that μ'_{30} is nonzero; a higher-order constraint must be used otherwise).

If $\lambda_1 = \lambda_2$, then *any* nonzero vector is an eigenvector of $\mathbf{M}$, and the principal axis method would yield infinitely many solutions. This is possible only if $\mu_{11} = 0$ and $\mu_{20} = \mu_{02}$; the principal axis normalization cannot be used in that case.

Now each moment of the normalized image μ'_{pq} is a rotation invariant of the original image. For the second-order moments, we can see clear correspondence with the Hu invariants

$$\mu'_{20} = (\phi_1 + \sqrt{\phi_2})/2,$$

$$\mu'_{02} = (\phi_1 - \sqrt{\phi_2})/2.$$

The principal axis normalization is linked with a notion of the *reference ellipse* (see Figure 3.20). There exist two concepts in the literature[18]. The first one, which is suitable namely for binary images, requires the ellipse to have a unit intensity. The other one, which is more general and more convenient for graylevel images, requires a constant-intensity ellipse, but its intensity may be arbitrary. The former approach allows constraining the first- and the second-order moments of the ellipse such that they equal the moments of the image, but the zero-order moment of the ellipse is generally different from that of the image.

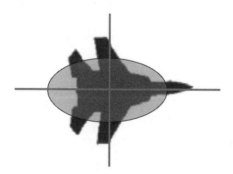

Figure 3.20 Principal axis normalization to rotation–an object in the normalized position along with its reference ellipse superimposed

---

[17] Some authors define the normalization angle with an opposite sign. Both ways are, however, equivalent–we may "rotate" either the image or the coordinates.

[18] Whenever one uses the reference ellipse, it must be specified which definition has been applied.

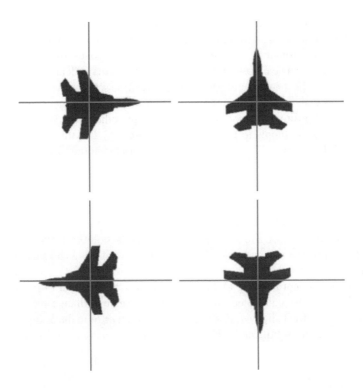

Figure 3.21   Ambiguity of the principal axis normalization. These four positions of the object satisfy $\mu'_{11} = 0$. Additional constraints $\mu'_{20} > \mu'_{02}$ and $\mu'_{30} > 0$ make the normalization unique

After the principal axis normalization, the reference ellipse is in axial position, and we have the following constraints for its moments,

$$\mu_{11}^{(e)} = 0,$$

$$\mu_{20}^{(e)} \equiv \frac{\pi a^3 b}{4} = \mu'_{20},$$

$$\mu_{02}^{(e)} \equiv \frac{\pi a b^3}{4} = \mu'_{02},$$

where a and b are the length of the major and minor semiaxis, respectively. By the last two constraints, a and b are unambiguously determined.

If we apply the latter definition of the reference ellipse, the constraints also contain the ellipse intensity ℓ, which allows us to constraint also the zero-order moment

$$\mu_{00}^{(e)} \equiv \pi \ell ab = \mu'_{00},$$

$$\mu_{11}^{(e)} = 0,$$

$$\mu_{20}^{(e)} \equiv \frac{\pi \ell a^3 b}{4} = \mu'_{20},$$

$$\mu_{02}^{(e)} \equiv \frac{\pi \ell a b^3}{4} = \mu'_{02}.$$

These constraints allow an unambiguous calculation of a, b, and ℓ.

All images having the same second-order moments have identical reference ellipses. This can also be understood from the opposite side–knowing the second-order image moments, we can reconstruct the reference ellipse only, without any finer image details.

Mostafa and Psaltis [29] proposed a general normalization scheme by means of the complex moments. Their normalization constraint requires one complex moment c_{st}, where $s \neq t$, of the image to be real and positive (we could of course use any other well-defined orientation, but this one was chosen for simplicity). As follows from eq. (3.25), this is always possible to achieve provided that $c_{st} \neq 0$. Then the constraint leads to the normalization angle

$$\alpha = \frac{1}{s-t} \cdot \arctan\left(\frac{\mathcal{I}m(c_{st})}{\mathcal{R}e(c_{st})}\right). \tag{3.36}$$

There exist exactly $s - t$ distinct angles satisfying (3.36), see Figure 3.22. This normalization is unambiguous only if $s - t = 1$. For $s - t > 1$, additional constraints are required to remove the ambiguity. If we choose $s = 2$ and $t = 0$, the constraint (3.36) brings us back to the principal axis normalization.

The selection of the normalizing moment c_{st} is a critical step, and in practice it must be done very carefully. If we calculated the normalization angle with an error, this error would propagate and would affect all invariants. Even a very small error in the normalizing angle can influence some invariants significantly. The non-zero constraint is easy to check in a continuous

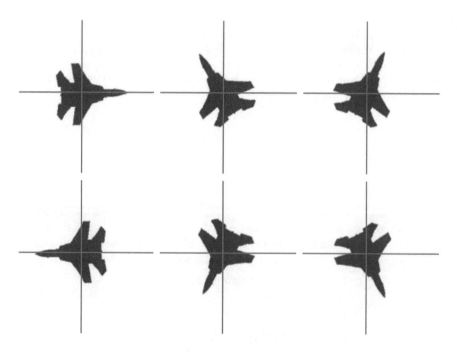

Figure 3.22 An example of the ambiguity of the normalization by complex moments. In all these six positions, c'_{60} is real and positive as required

domain, but in a discrete case some moments that are in fact zero due to the object symmetry may appear in certain positions of the object as nonzero because of the sampling effect. We have to identify such moments and avoid their usage in the normalization. At the same time, we want to keep the moment order $s + t$ as low as possible in order to make the normalization robust.

The normalizing moment can be found as follows. We sort all complex moments (except those that were already used for translation and scaling normalization) up to the given order r according to the index difference $p - q$, and those moments with the same index differences we sort according to their order. We obtain an ordered sequence $c_{21}, c_{32}, c_{43}, \ldots, c_{20}, c_{31}, c_{42}$, $c_{53}, \ldots$, etc. The first nonzero moment in this sequence is then used for normalization. To test if the moment is nonzero, we compare its value with a user-defined threshold. The choice of the threshold must reflect the magnitudes of the moments in question, and, especially when considering high r, it may be useful to define different thresholds for different moment orders. If all moments in the sequence are zero, we consider the object circularly symmetric, and no normalization is necessary.

One can, of course, use another ordering scheme when looking for the normalizing moment. An alternative is to sort the moments in the other way round–first according to their orders and the moments of the same order according to the index differences. Both methods have certain advantages and drawbacks. For non-symmetric objects there is no difference because we always choose c_{21}. For symmetric and close-to-symmetric objects, when we sort the moments according to their orders first, then the order of the chosen moment is kept as low as possible. This is a favorable property because low-order moments are more robust to noise than higher-order ones. On the other hand, one can find objects where this approach fails. Sorting moments according to the index difference first generally leads to a higher-order normalization constraint that is numerically less stable, but from the theoretical point of view this works for any object.

There have been many discussions on the advantages and drawbacks of both approaches–moment invariants and image normalization. Here we show that from the theoretical point of view they are fully equivalent and that the normalizing moment c_{st} and the basic moment $c_{p_0q_0}$ play very similar roles.

Let us consider the basis of the rotation invariants designed according to Theorem 3.2. We can apply the Mostafa and Psaltis normalization method with $c_{st} = c_{p_0q_0}$. Since each $\Phi(p, q)$ is a rotation invariant, it holds

$$\Phi(p, q) = c'_{pq} c'^{p-q}_{q_0p_0},$$

where c'_{pq} denotes the moments of the normalized image. Consequently, since $c'_{q_0p_0}$ is real and positive, we get

$$\Phi(p, q) = c'_{pq} |c_{p_0q_0}|^{p-q}. \tag{3.37}$$

Eq. (3.37) shows that there is a one-to-one mapping between moment invariants $\Phi(p, q)$ of the original image and complex moments c'_{pq} of the normalized image. An analogous result can be obtained for symmetric objects and the invariants from Theorem 3.2. Thus, normalization approach and invariant-based approach are equivalent on the theoretical level. Computational issues in particular applications might be, however, different.

3.9   Moment invariants of vector fields

A vector-valued image $\mathbf{f}(\mathbf{x})$ can be viewed as a set of m scalar images $\mathbf{f}(\mathbf{x}) = (f_1(\mathbf{x}), \cdots, f_m(\mathbf{x}))^T$.
Depending on its character, there may be various relationships among individual components
f_k, which influence the behavior of $\mathbf{f}$ under spatial transformations and hence influence the design of the invariants. Examples of vector-valued images are conventional color/multispectral
images and *vector fields*. The former are not very difficult to handle because their components
(even if they are usually correlated) can be treated as independent graylevel images, which
allows us to apply the rotation invariants derived earlier in this chapter separately to each
component. However, the vector fields behave differently and require a different treatment.

A 2D vector field is a vector-valued image $\mathbf{f}(\mathbf{x})$ with $d = m = 2$. Typical examples of vector
fields are gradient fields, particle velocity fields, optical flow fields, fluid flow fields, and others,
see Figures 3.23–3.25 and 3.27. A common feature is that at each point (x, y), the value of $\mathbf{f}(x, y)$
shows the orientation and the magnitude of certain vector.

If a 2D vector-valued image $\mathbf{f}(\mathbf{x})$ is rotated, we have to distinguish four kinds of rotation.
Inner rotation is a traditional rotation known from scalar images

$$\mathbf{f}'(\mathbf{x}) = \mathbf{f}(\mathbf{R}_{-\alpha}\mathbf{x}).$$

Outer rotation does not affect the spatial coordinates but acts on the vector values only

$$\mathbf{f}'(\mathbf{x}) = \mathbf{R}_\alpha\mathbf{f}(\mathbf{x}).$$

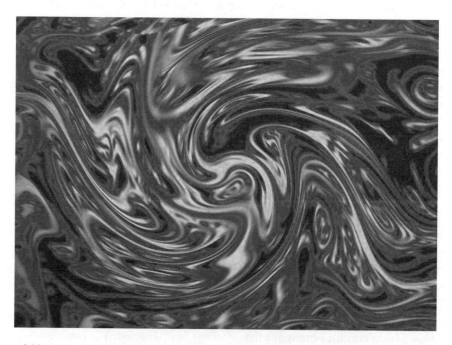

Figure 3.23   Turbulence in a fluid (the graylevels show the local velocity of the flow, the direction is
not displayed)

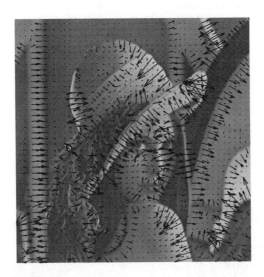

Figure 3.24   Image gradient as a vector field. For visualization purposes, the field is depicted by arrows on a sparse grid and laid over the original Lena image

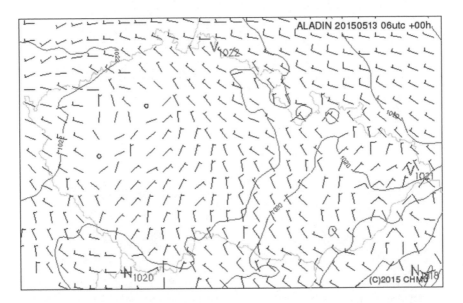

Figure 3.25   The wind velocity forecast for the Czech Republic (courtesy of the Czech Hydrometeoro-logical Institute, the numerical model Aladin). The longer dash is constant part of the wind, the shorter dash expresses perpendicular squalls. The figure actually represents two vector fields

It appears for instance in color images when the color space rotates but the picture itself does not move. *Total rotation* is a combination of both inner and outer rotations of the same rotation matrices

$$\mathbf{f}'(\mathbf{x}) = \mathbf{R}_\alpha \mathbf{f}(\mathbf{R}_{-\alpha}\mathbf{x}).$$

Finally, the most general *independent total rotation* is a total rotation for which the outer and inner rotation matrices are different

$$\mathbf{f}'(\mathbf{x}) = \mathbf{R}_\alpha \mathbf{f}(\mathbf{R}_\beta \mathbf{x}).$$

Let us illustrate these terms in Figure 3.26 for $\alpha = 22.5°$. In the case of the inner rotation, each arrow sustains its direction, but it is rotated around the image center to the new position.

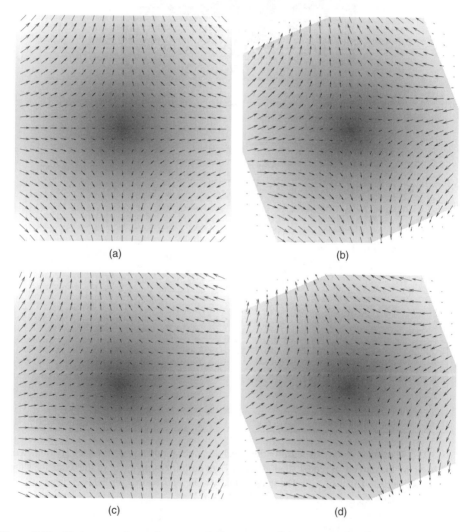

(a)   (b)

(c)   (d)

Figure 3.26   Various rotations of a vector field: (a) the original vector field, (b) the inner rotation, (c) the outer rotation, (d) the total rotation. The graylevels corresponds to the vector sizes

In the outer rotation, each arrow sustains its position, but its direction is rotated by 22.5°. In the total rotation, each arrow is rotated around the image center to the new position, and its direction is also rotated by the same angle.

Traditional spatial-domain rotation of a color image is inner rotation since the colors do not change. Rotation of a vector field is typically a total rotation because if a vector field is rotated in the coordinate space, then its values must be rotated accordingly such that the vector directions are preserved relative to the image content. An independent total rotation may appear when spatially rotating a color image at the same time that we independently rotate the color space.

Analogously, to various kinds of rotations, we recognize *inner scaling* $\mathbf{f}'(\mathbf{x}) = \mathbf{f}(\mathbf{x}/s)$, *outer scaling* $\mathbf{f}'(\mathbf{x}) = s\mathbf{f}(\mathbf{x})$, *total scaling* $\mathbf{f}'(\mathbf{x}) = s\mathbf{f}(\mathbf{x}/s)$, and *independent total scaling* $\mathbf{f}'(\mathbf{x}) = s_1\mathbf{f}(\mathbf{x}/s_2)$. For vector fields, typical scaling in practice is the total scaling. We could also analogously define an inner, outer, total, and independent total translations, but it does not make much sense in this context because the translation of a vector field in practice is always the inner one.

Although some vector field images may arrive as the results of image or video processing, such as gradients and optical flow images, we meet vector field images more often in visualization of fluid flow and of particle motion. These images may come from the solution of Navier-Stokes equations as well as from real measurements and may show, for instance, flowing water in a pipe or an air flow around an aircraft wing or around a coachwork. For engineers and designers, it is very important to identify singularities in the flow such as sinks and vortexes because they increase the friction, decrease the speed of the medium, and consequently increase the power and cost which is necessary to transport the medium through the pipe or the object through the air or water. Searching for the singularities is a typical template matching problem—we have a database of template singularities, and we look for these patterns in the vector field. Since the orientation of the template is unknown and irrelevant, we cannot just correlate the templates but need invariants to total rotation.

This problem was addressed for the first time by Schlemmer et al. [30] who realized that the invariants from complex moments can be used here in a similar way as for the graylevel images. If $d = m = 2$, we can treat the vector field as a field of complex numbers

$$\mathbf{f}(x, y) = f_1(x, y) + if_2(x, y)$$

which allows us to use the standard definition of complex moments. It holds

$$c_{pq}^{(\mathbf{f})} = c_{pq}^{(f_1)} + ic_{pq}^{(f_2)}.$$

Unlike the moments of a scalar image, it holds for the vector fields moments in general

$$c_{pq}^{(\mathbf{f})} \neq c_{qp}^{(\mathbf{f})*}.$$

Now let us investigate how $c_{pq}^{(\mathbf{f})}$ is changed under a total rotation by angle α

$$c_{pq}^{(\mathbf{f}')} = e^{-i\alpha}e^{-i(p-q)\alpha} \cdot c_{pq}^{(\mathbf{f})} = e^{-i(p-q+1)\alpha} \cdot c_{pq}^{(\mathbf{f})}. \tag{3.38}$$

By a multiplication of proper powers we can cancel the rotation parameter. Let $\ell \geq 1$ and let $k_i, p_i \geq 0$, and $q_i \geq 0$ $(i = 1, \ldots, \ell)$ be arbitrary integers such that

$$\sum_{i=1}^{\ell} k_i(p_i - q_i + 1) = 0.$$

Then

$$I = \prod_{i=1}^{\ell} c_{p_i q_i}^{k_i} \tag{3.39}$$

is invariant to total rotation. To obtain a basis, we apply a construction analogous to that one we used for graylevel images. We set $p_0 - q_0 = 2$ to get the desired effect[19]. The basis invariants for vector fields then have the form

$$\Phi(p, q) \equiv c_{pq} c_{q_0 p_0}^{p-q+1}. \tag{3.40}$$

There are, however, slight differences from the basis of the scalar invariants. In the case of vector fields, it is meaningful to consider also such $\Phi(p, q)$ where $p < q$ because they are independent. $\Phi(q_0, p_0) = 1$ and should be discarded from the basis. The invariants (3.40) are not complete because $|c_{q_0, p_0}|$ is another independent invariant. If we incorporate it, we obtain a complete system. Note that always $\Phi(p, p + 1) = c_{p,p+1}$.

Under a total scaling, the complex moment of a vector field changes as

$$c_{pq}^{(\mathbf{f}')} = s^{(p+q+3)} c_{pq}^{(\mathbf{f})} \tag{3.41}$$

which means we can achieve the scaling invariance when normalizing by means of $c_{00}^{(p+q+3)/3}$.

Note that for vector fields, c_{00} is not invariant w.r.t. a total rotation. To combine total scaling and rotation invariance, it cannot be used as a sole normalization factor because such normalization would violate the rotation invariance. Instead, we have to use some rotation invariant for normalization. We can, for instance, normalize each moment by $(c_{00} c_{02})^{(p+q+3)/8}$, which preserves the rotation invariance after normalization and yields the invariance to total scaling and rotation.

Schlemmer [30] used a few invariants of the lowest orders of the type (3.40) (without an explicit mention of this general formula) to detect specific patterns in a turbulent swirling jet flow. He also noticed a symmetry problem[20] similar to that one we discussed in the Section 2.6. Specific templates we search for are often N-fold symmetric with respect to total rotation. Hence, any vector field moment c_{pq} such that $(p - q + 1)/N$ is not an integer vanishes. We have to take this into account when choosing the basis of the invariants, and we should avoid vanishing invariants. Similarly to the scalar case, the *total dihedral symmetry* of the template makes all imaginary parts of the rotation invariants zero (note that the total axial symmetry is defined as $\mathbf{f}(x, -y) = \mathbf{f}^*(x, y)$).

We present here a very simple experiment, which demonstrates the basic property of rotation invariance of the above-mentioned features. We detected optical flow in a video sequence

---

[19] Other choices of p_0, q_0 are also possible, we may for instance choose $p_0 = q_0$. The most general solution has been proposed by Bujack [31], who uses any indices such that $p_0 - q_0 \neq 1$. This choice is the most flexible one. It allows avoiding the use of close-to-zero moments, but leads to rational exponents of $c_{p_0 q_0}$.

[20] Schlemmer mentioned this issue for the second-order moments only; we describe it here in a general form.

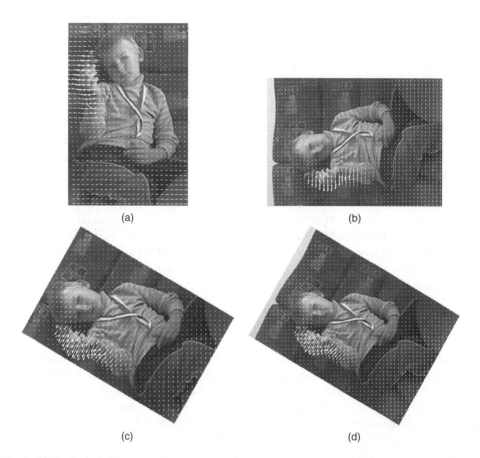

(a)   (b)

(c)   (d)

Figure 3.27   Optical flow as a vector field: (a) the original field, (b) the optical flow computed from the video sequence rotated by 90°, (c) the original optical flow field after total rotation by 60°, (d) the optical flow computed from the video sequence rotated by 60°. All rotations are counterclockwise. The arrows show the direction and velocity of the movement between two consecutive frames of the video sequence

(see Figure 3.27a) by the method described in [32], using the authors' code [33]. Then we calculated five invariants c_{01}, $c_{00}c_{02}$, $c_{11}c_{02}$, $c_{10}c_{02}^2$, and $c_{20}c_{02}^3$ of the optical flow vector field. Thereafter we rotated the flow field by 90° (this angle was chosen to avoid resampling errors) and calculated the invariants again. The average relative error was $2 \cdot 10^{-14}\%$, which shows that the only source of errors is numerical inaccuracy. Another comparison was done such that we rotated the video sequence, then we detected the optical flow again (see Figure 3.27b–there are no visible differences between it and the rotated vector field from previous case) and calculated the invariants. This scenario corresponds to the real situations, and one can expect bigger errors since the two vector fields may be generally different if the optical flow detection is not perfectly isotropic. The average relative error was 2.79% which is still very good.

Finally, we tested the influence of resampling of the field when it has been rotated by 60° (see Figure 3.27c). The average relative error was only 0.0025%. On the other hand, if these two factors–image resampling and independent optical flow detection–are combined together, the relative error of the invariants may range from 10 to 100% depending on the image texture. In our case it was 46.6%, the corresponding vector field is in Figure 3.27d.

The idea of vector field invariants has found several applications. Liu and Ribeiro [34] used it, along with a local approximation of the vector field by a polynomial, to detect singularities on meteorological satellite images where the respective field was a wind velocity map. Basically the same kind of rotation invariants was used by Liu and Yap [35] for indexing and recognition of fingerprint images. The respective vector field was a field of local orientation of the fingerprint ridges and valleys; the directional information was discarded (Liu and Yap call this the *orientation field*).

Similarly to scalar invariants, the vector field invariants can also be alternatively derived via normalization of the vector field w.r.t. total rotation and scaling. This approach has been developed by Bujack et al. [36, 37], who proposed the normalization constraints based on the zero and second-order moments. The authors demonstrated the usage of the normalized moments in template matching, where the template vortexes were searched in the image showing the *Karman vortex street* simulation[21]. They also applied the normalized moments to a segmentation of vector fields via clustering [38].

3.10 Conclusion

In this chapter, we introduced 2D moment invariants with respect to the translation, rotation and scaling. We described a general theory showing how to generate invariants of any orders. We defined the basis of the invariants as the smallest complete subset, we showed its importance and described an explicit method how to find it. As a consequence of this theory, we proved that some traditional invariant sets are dependent and/or incomplete, which explains certain failures reported in the literature. It was shown that the proposed invariants outperform the widely used Hu moment invariants both in discrimination power and dimensionality requirements.

Furthermore, we discussed the difficulties with recognition of symmetric objects, where some moment invariants vanish, and we presented moment invariants suitable for such cases. We briefly reviewed an alternative approach to constructing invariants via normalization and showed that both methods are theoretically equivalent.

At the end, we addressed a recent development on the field of moment invariants–rotation and scale invariants of vector fields.

References

[1] Z. Feng, L. Shang-Qian, W. Da-Bao, and G. Wei, "Aircraft recognition in infrared image using wavelet moment invariants," *Image and Vision Computing*, vol. 27, no. 4, pp. 313–318, 2009.

[2] S. Dominguez, "Image analysis by moment invariants using a set of step-like basis functions," *Pattern Recognition Letters*, vol. 34, no. 16, pp. 2065–2070, 2013.

[3] M.-K. Hu, "Visual pattern recognition by moment invariants," *IRE Transactions on Information Theory*, vol. 8, no. 2, pp. 179–187, 1962.

---

[21] In fluid dynamics, the Karman vortex street is a repeating pattern of swirling vortexes caused by the flow of a fluid around blunt bodies.

[4] F. Pan and M. Keane, "A new set of moment invariants for handwritten numeral recognition," in *Proceedings of the International Conference on Image Processing ICIP'94*, pp. 154–158, IEEE, 1994.

[5] L. Jin and Z. Tianxu, "Fast algorithm for generation of moment invariants," *Pattern Recognition*, vol. 37, no. 8, pp. 1745–1756, 2004.

[6] M. R. Teague, "Image analysis via the general theory of moments," *Journal of the Optical Society of America*, vol. 70, no. 8, pp. 920–930, 1980.

[7] Å. Wallin and O. Kübler, "Complete sets of complex Zernike moment invariants and the role of the pseudoinvariants," *IEEE Transactions on Pattern Analysis and Machine Intelligence*, vol. 17, no. 11, pp. 1106–1110, 1995.

[8] Y. Li, "Reforming the theory of invariant moments for pattern recognition," *Pattern Recognition*, vol. 25, no. 7, pp. 723–730, 1992.

[9] W.-H. Wong, W.-C. Siu, and K.-M. Lam, "Generation of moment invariants and their uses for character recognition," *Pattern Recognition Letters*, vol. 16, no. 2, pp. 115–123, 1995.

[10] Y. S. Abu-Mostafa and D. Psaltis, "Recognitive aspects of moment invariants," *IEEE Transactions on Pattern Analysis and Machine Intelligence*, vol. 6, no. 6, pp. 698–706, 1984.

[11] J. Flusser, "On the independence of rotation moment invariants," *Pattern Recognition*, vol. 33, no. 9, pp. 1405–1410, 2000.

[12] J. Flusser, "On the inverse problem of rotation moment invariants," *Pattern Recognition*, vol. 35, no. 12, pp. 3015–3017, 2002.

[13] S. Derrode and F. Ghorbel, "Robust and efficient Fourier–Mellin transform approximations for gray-level image reconstruction and complete invariant description," *Computer Vision and Image Understanding*, vol. 83, no. 1, pp. 57–78, 2001.

[14] B. Yang, J. Flusser, and T. Suk, "Design of high-order rotation invariants from Gaussian-Hermite moments," *Signal Processing*, vol. 113, no. 1, pp. 61–67, 2015.

[15] D. Bhattacharya and S. Sinha, "Invariance of stereo images via theory of complex moments," *Pattern Recognition*, vol. 30, no. 9, pp. 1373–1386, 1997.

[16] R. Mukundan and K. R. Ramakrishnan, *Moment Functions in Image Analysis*. Singapore: World Scientific, 1998.

[17] S. Maitra, "Moment invariants," *Proceedings of the IEEE*, vol. 67, no. 4, pp. 697–699, 1979.

[18] T. M. Hupkens and J. de Clippeleir, "Noise and intensity invariant moments," *Pattern Recognition Letters*, vol. 16, no. 4, pp. 371–376, 1995.

[19] J. Flusser and T. Suk, "Rotation moment invariants for recognition of symmetric objects," *IEEE Transactions on Image Processing*, vol. 15, no. 12, pp. 3784–3790, 2006.

[20] D. Shen, H. H.-S. Ip, and E. K. Teoh, "A novel theorem on symmetries of 2D images," in *Proceedings of the 15th International Conference on Pattern Recognition ICPR'00*, vol. 3, pp. 1014–1017, IEEE, 2000.

[21] J. Lin, W. Tsai, and J. Chen, "Detecting number of folds by a simple mathematical property," *Pattern Recognition Letters*, vol. 15, no. 11, pp. 1081–1088, 1994.

[22] J. Lin, "A simplified fold number detector for shapes with monotonic radii," *Pattern Recognition*, vol. 29, no. 6, pp. 997–1005, 1996.

[23] D. Shen, H. H.-S. Ip, K. K. T. Cheung, and E. K. Teoh, "Symmetry detection by generalized complex (GC) moments: A close-form solution," *IEEE Transactions on Pattern Analysis and Machine Intelligence*, vol. 21, no. 5, pp. 466–476, 1999.

[24] S. Derrode and F. Ghorbel, "Shape analysis and symmetry detection in gray-level objects using the analytical Fourier–Mellin representation," *Signal Processing*, vol. 84, no. 1, pp. 25–39, 2004.

[25] J. Chen, L. Wang, and D. Chen, *Logo Recognition: Theory and Practice*. Boca Raton, FL: CRC Press, 2011.

[26] Y.-S. Kim and W.-Y. Kim, "Content-based trademark retrieval system using a visually salient feature," *Image and Vision Computing*, vol. 16, no. 12–13, pp. 931–939, 1998.

[27] A. Jain and A. Vailaya, "Shape-based retrieval: A case study with trademark image databases," *Pattern Recognition*, vol. 31, no. 9, pp. 1369–1390, 1998.

[28] M. S. Hitam, W. Nural, J. Hj Wan Yussof, and M. M. Deris, "Hybrid Zernike moments and color-spatial technique for content-based trademark retrieval," in *Proceedings of the International Symposium on Management Engineering, ISME'06*, 2006.

[29] Y. S. Abu-Mostafa and D. Psaltis, "Image normalization by complex moments," *IEEE Transactions on Pattern Analysis and Machine Intelligence*, vol. 7, no. 1, pp. 46–55, 1985.

[30] M. Schlemmer, M. Heringer, F. Morr, I. Hotz, M.-H. Bertram, C. Garth, W. Kollmann, B. Hamann, and H. Hagen, "Moment invariants for the analysis of 2D flow fields," *IEEE Transactions on Visualization and Computer Graphics*, vol. 13, no. 6, pp. 1743–1750, 2007.

[31] R. Bujack, "General basis of rotation invariants of vector fields," private communication (unpublished), 2015.

[32] C. Liu, *Beyond Pixels: Exploring New Representations and Applications for Motion Analysis*. PhD thesis, Northeastern University, Boston, Massachusetts, USA, 2009.

[33] C. Liu, 2011. `http://people.csail.mit.edu/celiu/OpticalFlow/`.

[34] W. Liu and E. Ribeiro, "Detecting singular patterns in 2-D vector fields using weighted Laurent polynomial," *Pattern Recognition*, vol. 45, no. 11, pp. 3912–3925, 2012.

[35] M. Liu and P.-T. Yap, "Invariant representation of orientation fields for fingerprint indexing," *Pattern Recognition*, vol. 45, no. 7, pp. 2532–2542, 2012.

[36] R. Bujack, I. Hotz, G. Scheuermann, and E. Hitzer, "Moment invariants for 2D flow fields using normalization," in *Pacific Visualization Symposium, PacificVis'14*, pp. 41–48, IEEE, 2014.

[37] R. Bujack, M. Hlawitschka, G. Scheuermann, and E. Hitzer, "Customized TRS invariants for 2D vector fields via moment normalization," *Pattern Recognition Letters*, vol. 46, no. 1, pp. 46–59, 2014.

[38] R. Bujack, J. Kasten, V. Natarajan, G. Scheuermann, and K. Joy, "Clustering moment invariants to identify similarity within 2D flow fields," in *EG / VGTC Conference on Visualization, EuroVis'15* (J. Kennedy and E. Puppo, eds.), pp. 31–35, The Eurographics Association, 2015.

4

3D Moment Invariants to Translation, Rotation, and Scaling

4.1 Introduction

In the last two decades, 3D image analysis has attracted significant attention. The main reason is a dynamic development of 3D imaging devices and technologies, which have become widely accessible. The most prominent area where 3D image analysis is in great demand is medicine. Current medical imaging devices, such as MRI, f-MRI, CT, SPECT, and PET, provide us with full 3D images of the patient's body. Many 3D imaging sensors can be found outside medical area. Even in the world of consumer electronics and computer games, 3D sensors are quite common[1]. Another reason for increasing interest in 3D image analysis algorithms is the existence of powerful multi-core computers equipped with GPUs that can realize 3D algorithms, which are typically of high complexity, in an acceptable (often close to real) time.

In 2D imaging, the scanner or the camera provides mostly a color image. Graylevel or binary image is a result of intensity transformation or image segmentation. On the contrary, only some 3D imaging devices actually produce the full graylevel images. Such image is in the discrete domain represented by a "data cube", that is, by a 3D matrix of *voxels*[2] the values of which correspond to the X-ray absorbance, magnetic susceptibility, or another quantity. Many other 3D sensors do not actually capture an image; they only measure the distance of each point on the object surface from the sensor. Such measurement can be visualized as a distance map or depth map, which is also sometimes called 2.5D image. These devices cannot see "inside" the object. If the object is scanned from various sides such that each surface part is visible on at least one scan, the individual scans (views) can be combined together, and the complete surface of the object can be reconstructed. Hence, we obtain 3D binary images. All range finders, including the above-mentioned Kinect, work in this way. To save memory, binary 3D

---

[1] The Kinect sensor, which is used for playing on Xbox, is a typical example of such device.

[2] The word "voxel" is an abbreviation of "volume picture element". The voxel is a 3D analogue of the pixel.

2D and 3D Image Analysis by Moments, First Edition. Jan Flusser, Tomáš Suk and Barbara Zitová.
© 2017 John Wiley & Sons, Ltd. Published 2017 by John Wiley & Sons, Ltd.
Companion website: www.wiley.com/go/flusser/2d3dimageanalysis

objects are of course not stored as data cubes because they are fully determined by their surface. Storing a list of the coordinates of all surface voxels would be possible but not very efficient. Most of the commercial software for 3D object reconstruction from multiple views generates a *triangulated* surface, which is an approximation of the actual surface. The accuracy of this approximation is determined by the density of the mesh nodes and can be controlled by the user. Denser mesh usually yields a better approximation.

Many public databases of 3D objects, including the famous Princeton Shape Benchmark[3] (PSB) [1] contain objects that were created artificially in a proper software for 3D modelling such as Autodesk 3D Studio MAX and similar software. This software also produces triangulated surfaces and stores the objects efficiently with minimum memory requirement.

Some examples of various 3D data are shown in Figure 4.1: full volumetric CT data depicted as individual slices, real binary object scanned by Kinect, binary object with triangulated surface from the Princeton Shape Benchmark, and a synthetic object.

Numerous 3D image processing algorithms are just extensions of the 2D versions by an additional dimension. Even if they are computationally more demanding, they do not require mathematical tools different from those used in 2D (typical examples of this kind of methods are convolution and Fourier transform). On the other hand, there are operations and algorithms which are distinct in 2D and 3D. Spatial rotation belongs to this category, and that is why 3D rotation moment invariants cannot be obtained by a simple generalization of 2D invariants. As we will see in this chapter, the design of 3D rotation invariants is much more difficult than in 2D. Although we can observe certain analogies, in 3D we face new problems which are easy to resolve in 2D – an example is a construction of the basis of invariants.

Surprisingly, there exist some tasks and situations which are easier to resolve in 3D than in 2D. The underlaying reason is that 2D images are projections of a 3D world. In the z-dimension, the object may be hidden with another object which leads to the appearance of partially occluded objects on the image and requires local invariants for their description. This cannot happen in 3D images since there is no fourth dimension in the real world. Hence, the demand for local invariants is less than in 2D, and most of the 3D recognition tasks can be resolved by global features. This is why moment invariants, which are typical global features, are even more important in 3D than in 2D.

Although many fewer papers have been published on 3D moment invariants than on 2D moment invariants, the first attempts to extend 2D moment invariants to 3D are quite old. The very first 3D moment invariants were proposed by Sadjadi and Hall [2]. Their results were later rediscovered (with some modifications) by Guo [3]. Both derived only three TRS invariants from geometric moments of the second order without any possibility of further extension. Cyganski and Orr [4] proposed a tensor method for derivation of rotation invariants from geometric moments. Xu and Li [5, 6] used a method of geometric primitives that yields the same results, obtained however by different mathematical tools. Suk and Flusser [7] proposed an algorithm for an automated generation of the 3D rotation invariants from geometric moments.

There also exist 3D analogues of complex moments, where the *spherical harmonics* are involved in the basis functions. As in 2D, they can be advantageously used for construction of rotation invariants. Lo and Don [8] employed group representation theory and derived 3D TRS invariants up to the third order from the complex moments. Their approach was later used by Flusser et al. [9], where a coupling of the rotation with image blurring was considered. Galvez

---

[3] The PSB database has become a world-leading benchmark for testing 3D object recognition algorithms.

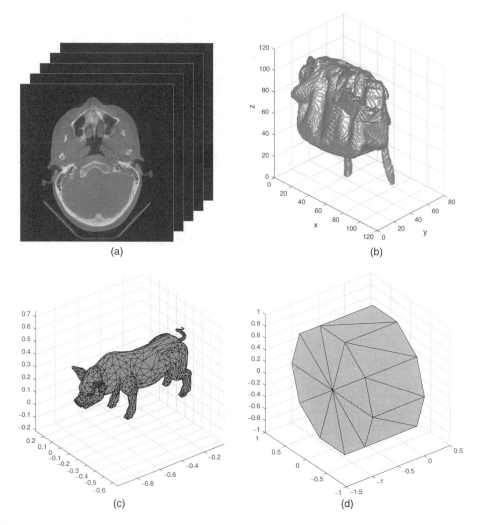

Figure 4.1 3D images of various nature: (a) real volumetric data – CT of a human head, (b) real binary volumetric data measured by Kinect (the bag), (c) binary object with a triangulated surface (the pig) from the PSB, and (d) artificial binary object with a triangulated surface (decagonal prism)

and Canton [10] constructed the invariants via the normalization approach. The use of 3D complex moments for designing rotation invariants was recently systematically studied by Suk et al. [11]. Special invariants for recognition of symmetric bodies were proposed by the same authors in [12]. In order to avoid numerical problems with geometric and complex moments, some authors proposed 3D invariants from orthogonal moments, such as Zernike moments [13] and Gaussian–Hermite moments [14]. Orthogonal moments and related invariants will be thoroughly discussed in Chapter 7.

Other authors have used slightly different approaches to the construction of the 3D rotation invariants. The main alternative to the moments of the complete body is a decomposition of the body to concentric spherical layers. Then the features of various kinds are calculated separately for the individual layers. Kazhdan et al. [15] proposed so-called *spherical harmonic representation*, which is essentially equivalent to moments of the layers. Kakarala [16] computed *bispectrum* of each layer, and Fehr and Burkhardt [17] proposed to construct 3D invariants from *local binary patterns* (LBP), which had been originally introduced in 2D version by Ojala et al. [18].

Several application papers that successfully used 3D rotation moment invariants have been published. In [19], the authors used 3D TRS invariants to test handedness and gender from brain MRI snaps. Two other papers [20] and [21] employed invariants for 3D image registration. Millán et al. [22] used Lo and Don's invariants for prediction of rupture of intracranial saccular aneurysms. They also compared their performance to Zernike moment magnitudes [23].

4.2   Mathematical description of the 3D rotation

In 2D, the rotation is characterized by a single rotation angle or, equivalently, by a 2×2 rotation matrix (see the previous chapter). The 3D rotation has three degrees of freedom. The *3D rotation matrix* is a matrix of directional cosines

$$\mathbf{R} = \begin{pmatrix} \cos\alpha_{xx} & \cos\alpha_{xy} & \cos\alpha_{xz} \\ \cos\alpha_{yx} & \cos\alpha_{yy} & \cos\alpha_{yz} \\ \cos\alpha_{zx} & \cos\alpha_{zy} & \cos\alpha_{zz} \end{pmatrix}, \tag{4.1}$$

where α_{zy} is the angle between the original position of the z-axis and the new position of the y-axis, etc. We can freely choose three of these angles which are neither in the same row nor in the same column. The rotation matrix is *orthonormal*, that is, the dot product of two different rows or columns equals zero and that of the same row/column equals one

$$\forall i \neq j : \sum_{k=1}^{3} r_{ik} r_{jk} = 0, \quad \forall i \sum_{k=1}^{3} r_{ik}^2 = 1$$

$$\forall i \neq j : \sum_{k=1}^{3} r_{ki} r_{kj} = 0, \quad \forall i \sum_{k=1}^{3} r_{ki}^2 = 1. \tag{4.2}$$

It implies $\mathbf{R}^{-1} = \mathbf{R}^T$, and $\det(\mathbf{R}) = 1$, which is analogous to 2D. The rotations in 3D

$$\mathbf{x}' = \mathbf{R}\mathbf{x} \tag{4.3}$$

create a group often denoted as SO(3).

The description by a rotation angle, which is easy in 2D, has also an analogy in 3D but the geometry is more complicated. We can imagine an axis of rotation and attitude given by an angle of rotation around it, but we must describe the position of the axis of rotation. We can measure, for instance, the angle between the axis of rotation and the $x - y$ plane. This angle defines a plane. The other angle is between this plane and $x - z$ plane, that is, we have ordinary spherical coordinates of intersection of the axis of rotation and a unit sphere. The radius of the sphere is not important; therefore we have three angles together for description of the 3D rotation.

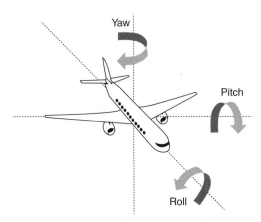

Figure 4.2   Yaw, pitch, and roll angles in the Tait-Bryan convention

Another concept is a decomposition of the rotation to three single-parametric rotations around x, y, and z axes. There are twelve possibilities of the axes ordering – for the first rotation, we can choose an arbitrary axis; for the second and third rotations we must choose at least one of the two remaining axes. It is possible to use some axes twice. Such a convention is called the *Euler convention* or description by *Euler angles*[4]. If three different axes are used, the description is called the *Tait–Bryan convention*[5], and we refer to the angles as *yaw*, *pitch*, and *roll* or as *nautical angles* (see Figure 4.2 for illustration).

After the object has been rotated around the first axis, the question of how to measure the subsequent rotations arises. In the first rotation, the position of the two other axes is changed, and in the second rotation we can rotate the body either around the new axis or around the original one. The same applies in the third rotation, too. The rotations around the moving axes are called *intrinsic rotations*, and those around the static (i.e. original) axes are called *extrinsic rotations*. Both intrinsic and extrinsic rotations can be used for Euler and Tait-Bryan conventions.

The rotation matrix can be calculated both from Euler angle and Tait-Bryan angle conventions by matrix multiplication. We can easily derive the particular rotation matrices around individual axes. The rotation matrix around x axis by angle α is

$$\mathbf{A} = \begin{pmatrix} 1 & 0 & 0 \\ 0 & \cos\alpha & -\sin\alpha \\ 0 & \sin\alpha & \cos\alpha \end{pmatrix}, \tag{4.4}$$

the rotation matrix around y axis by angle β is

$$\mathbf{B} = \begin{pmatrix} \cos\beta & 0 & -\sin\beta \\ 0 & 1 & 0 \\ \sin\beta & 0 & \cos\beta \end{pmatrix} \tag{4.5}$$

---

[4] Leonhard Euler (1707–1783) was a Swiss mathematician and physicist.
[5] Peter Guthrie Tait (1831–1901) was a Scottish mathematical physicist, George Hartley Bryan (1864–1928) was a mathematician working in Wales.

and the rotation matrix around z axis by angle γ is

$$\mathbf{C} = \begin{pmatrix} \cos\gamma & -\sin\gamma & 0 \\ \sin\gamma & \cos\gamma & 0 \\ 0 & 0 & 1 \end{pmatrix}. \tag{4.6}$$

So, the rotation matrix in extrinsic Tait-Bryan convention $(x - y - z)$ is

$$\mathbf{R} = \mathbf{CBA}. \tag{4.7}$$

If we use the intrinsic convention $(x - y - z)$, then the matrices are multiplied in the reverse order

$$\mathbf{R} = \mathbf{ABC}. \tag{4.8}$$

For certain historical reasons, the extrinsic Euler angles $(z - y - z)$ and intrinsic Tait-Bryan convention $(x - y - z)$ are used most frequently.

For the sake of completeness, let us mention that there exists yet another formalism for how to express 3D rotations, which uses *quaternions*. The quaternions are "generalized complex numbers". They consist of a real part and three imaginary parts. The real part is often zero, and the imaginary parts can express the coordinates of a point in 3D space as well as the parameters of 3D rotation. Then both rotation of the point and combination of two rotations can be described as a quaternion multiplication; see for instance [24]. Although the quaternion formalism can be utilized with a connection to moments and to moment invariants, it does not bring any advantages over the other approaches (in most cases, it is even more complicated). That is why we do not follow this approach in this book. The quaternions are more often used for encoding of color images [25–29].

4.3   Translation and scaling invariance of 3D geometric moments

The generalization of translation and uniform scaling invariance to 3D is straightforward. *3D geometric moment* of image f is defined as

$$m_{pqr} = \int\limits_{-\infty}^{\infty} \int\limits_{-\infty}^{\infty} \int\limits_{-\infty}^{\infty} x^p\, y^q\, z^r\, f(x, y, z)\ \mathrm{d}x\ \mathrm{d}y\ \mathrm{d}z. \tag{4.9}$$

The translation invariance can easily be achieved by shifting the object (or the coordinates) such that the coordinate origin coincides with the object centroid. This leads us to the *3D central geometric moments*

$$\mu_{pqr} = \int\limits_{-\infty}^{\infty} \int\limits_{-\infty}^{\infty} \int\limits_{-\infty}^{\infty} (x - x_c)^p\, (y - y_c)^q\, (z - z_c)^r\, f(x, y, z)\ \mathrm{d}x\ \mathrm{d}y\ \mathrm{d}z, \tag{4.10}$$

where $x_c = m_{100}/m_{000}$, $y_c = m_{010}/m_{000}$ and $z_c = m_{001}/m_{000}$ are the centroid coordinates of image f. The central moments are obviously invariant to translation.

Achieving scaling invariance is also basically the same as in 2D. Let us assume a uniform scaling with parameter s. The geometric as well as the central moments are changed under this scaling as

$$m'_{pqr} = m_{pqr} s^{p+q+r+3}.$$

Hence, the normalized moments

$$v_{pqr} = \mu_{pqr} / \mu_{000}^{(p+q+r)/3+1} \tag{4.11}$$

are both translation and scaling invariants.

However, the design of rotation moment invariants in 3D is much more difficult than in 2D. We present three alternative approaches in the sequel. The first one uses geometric moments, and the invariance is achieved by means of tensor algebra. The second one is based on spherical harmonics and can be viewed as an analogue to 2D complex moment invariants. The last one is a normalization approach. Even if all these three approaches are theoretically equivalent (which means there exist a one-to-one mapping among the three sets of invariants), their suitability in particular tasks and their numerical behavior may differ from one another.

4.4 3D rotation invariants by means of tensors

The tensor method could be in principle used in 2D as well, but in 2D complex moments provide a far simpler solution. In 3D, the difficulty of deriving (and understanding) the invariants by these two approaches is comparable. We introduce a few basic terms of tensor algebra first. Then we show how to create moment tensors and how to use them to construct rotation invariants.

4.4.1 Tensors

A *tensor* is a multidimensional array with an operation of multiplication. The ordinary tensor $T^{\beta_1, \beta_2, \ldots, \beta_{k_2}}_{\alpha_1, \alpha_2, \ldots, \alpha_{k_1}}$ has two types of indices – *covariant* written as subscripts and *contravariant* written as superscripts. The tensor behaves under the affine transformation according to the rule

$$\hat{T}^{\beta_1, \beta_2, \ldots, \beta_{k_2}}_{\alpha_1, \alpha_2, \ldots, \alpha_{k_1}} = q^{\beta_1}_{j_1} q^{\beta_2}_{j_2} \cdots q^{\beta_{k_2}}_{j_{k_2}} p^{i_1}_{\alpha_1} p^{i_2}_{\alpha_2} \cdots p^{i_{k_1}}_{\alpha_{k_1}} T^{j_1, j_2, \ldots, j_{k_2}}_{i_1, i_2, \ldots, i_{k_1}}, \tag{4.12}$$

where p^i_j are parameters of the direct affine transformation and q^i_j are parameters of the inverse affine transformation (without translation). The multiplication of two tensors includes two types of behavior – *enumeration*, where we multiply each component of one tensor with each component of the other tensor and *contraction*, where the sum of the corresponding components is computed.

If we use two vectors $\mathbf{u}$ and $\mathbf{v}$ as tensors u_i and v^i, then their matrix product $\mathbf{M} = \mathbf{u}\mathbf{v}^T$ is an example of the enumeration $m^j_i = u_i v^j$, and their dot product $d = \mathbf{u}^T \mathbf{v}$ is an example of the contraction $d = u_i v^i$. If an index is used twice in the product, the contraction is carried out. If it is used just once, the underlaying operation is enumeration. The traditional matrix product $\mathbf{C} = \mathbf{AB}$ can be written as $C_i^j = A_i^k B_k^j$ in the Einstein tensor notation[6]. The matrix product

$$C_i^j = A_i^k B_k^j = \sum_{k=1}^{d} A_i^k B_k^j,$$

where d is the dimension of the space which must equal the size of the matrices.

We can also do the contraction of one tensor only, for instance, A_i^i is the trace of the matrix, that is, the sum of its diagonal elements. The *total contraction* is performed over all indices of some tensor; the result is just a single number. More details about this type of tensors can be found in the chapter about affine invariants.

If we face a rotation only, *Cartesian tensors* are more suitable for derivation of the rotational invariants. The Cartesian tensor behaves under the rotation according to the rule

$$\hat{T}_{\alpha_1,\alpha_2,\ldots,\alpha_k} = r_{\alpha_1 i_1} r_{\alpha_2 i_2} \cdots r_{\alpha_k i_k} T_{i_1,i_2,\ldots,i_k}, \tag{4.13}$$

where r_{ij} is an arbitrary orthonormal matrix. In this case the distinction between covariant and contravariant indices is not necessary, because the inversion of the orthonormal matrix equals its transposition. We can substitute the rotation matrix here.

4.4.2 Rotation invariants

First, we establish the relation between moments and tensors. Then we use the tensor theory for derivation of the invariants. The *moment tensor* [4] is defined as

$$M^{i_1 i_2 \cdots i_k} = \int_{-\infty}^{\infty} \int_{-\infty}^{\infty} \int_{-\infty}^{\infty} x^{i_1} x^{i_2} \cdots x^{i_k} f(x^1, x^2, x^3) \; dx^1 \; dx^2 \; dx^3, \tag{4.14}$$

where $x^1 = x$, $x^2 = y$ and $x^3 = z$. If p indices equal 1, q indices equal 2, and r indices equal 3, then $M^{i_1 i_2 \cdots i_k} = m_{pqr}$. The definition in 2D is analogous. The behavior of the moment tensor under an affine transformation is, in the Einstein notation,

$$\begin{aligned} M^{i_1 i_2 \cdots i_r} &= |J| p_{\alpha_1}^{i_1} p_{\alpha_2}^{i_2} \cdots p_{\alpha_r}^{i_r} \hat{M}^{\alpha_1 \alpha_2 \cdots \alpha_r} \\ \hat{M}^{i_1 i_2 \cdots i_r} &= |J|^{-1} q_{\alpha_1}^{i_1} q_{\alpha_2}^{i_2} \cdots q_{\alpha_r}^{i_r} M^{\alpha_1 \alpha_2 \cdots \alpha_r}. \end{aligned} \tag{4.15}$$

This means that the moment tensor is a relative contravariant tensor with the weight -1, what will be later useful in a derivation of affine invariants. The moment tensor can also be used as a Cartesian tensor for derivation of the rotation invariants, then we write the indices as subscripts.

Each total contraction of a Cartesian tensor or of a product of Cartesian tensors is a rotation invariant. We can compute the total contraction of the moment tensor product, when each index is included just twice in it.

---

[6] A. Einstein introduced this notation to simplify expressions with multi-dimensional coordinates, see, e.g., [30] for an explanation.

The simplest example is the total contraction of the second-order moment tensor M_{ij}. In 2D it is

$$M_{11} + M_{22} : \Phi_1^{2D} = (\mu_{20} + \mu_{02})/\mu_{00}^2.$$

We can easily express it in 3D and obtain a rotation invariant in 3D. It is sufficient to compute the total contraction in three-dimensional space

$$M_{11} + M_{22} + M_{33} : \Phi_1^{3D} = (\mu_{200} + \mu_{020} + \mu_{002})/\mu_{000}^{5/3}.$$

Another example is $M_{ij}M_{ij}$, which is in 2D

$$M_{11}M_{11} + M_{12}M_{12} + M_{21}M_{21} + M_{22}M_{22} : \Phi_2^{2D} = (\mu_{20}^2 + \mu_{02}^2 + 2\mu_{11}^2)/\mu_{00}^4,$$

while in 3D it yields

$$
\begin{aligned}
M_{11}M_{11} &+ M_{12}M_{12} &+ M_{13}M_{13}+ \\
+M_{21}M_{21} &+ M_{22}M_{22} &+ M_{23}M_{23}+ \\
+M_{31}M_{31} &+ M_{32}M_{32} &+ M_{33}M_{33} :
\end{aligned}
$$

$$\Phi_2^{3D} = (\mu_{200}^2 + \mu_{020}^2 + \mu_{002}^2 + 2\mu_{110}^2 + 2\mu_{101}^2 + 2\mu_{011}^2)/\mu_{000}^{10/3}.$$

The last example of the second order is $M_{ij}M_{jk}M_{ki}$. In 2D it is

$$\Phi_3^{2D} = (\mu_{20}^3 + 3\mu_{20}\mu_{11}^2 + 3\mu_{11}^2\mu_{02} + \mu_{02}^3)/\mu_{00}^6,$$

and in 3D it becomes

$$
\begin{aligned}
\Phi_3^{3D} = (&\mu_{200}^3 + 3\mu_{200}\mu_{110}^2 + 3\mu_{200}\mu_{101}^2 + 3\mu_{110}^2\mu_{020} + 3\mu_{101}^2\mu_{002} + \mu_{020}^3 \\
&+ 3\mu_{020}\mu_{011}^2 + 3\mu_{011}^2\mu_{002} + \mu_{002}^3 + 6\mu_{110}\mu_{101}\mu_{011})/\mu_{000}^5.
\end{aligned}
$$

The scaling normalization can be derived from (4.11). The required exponent of the zero-order moment is

$$\omega = \frac{w}{3} + s, \qquad (4.16)$$

where w is the sum of the indices of all moments in one term and s is the number of the moments in a single term.

4.4.3   Graph representation of the invariants

Each invariant generated by the tensor method can be represented by a connected graph, where each moment tensor corresponds to a node and each index corresponds to an edge of the graph (in graph theory, such graphs with multiple edges are called *multigraphs*). From the graph one can learn about the orders of the moments the invariant is composed of and about the structure of the invariant. Since the number of indices of the moment tensor equals its order, it also equals the number of the edges originating from the corresponding node. Each rotation moment invariant is in fact a sum, where each term is a product of a certain number of moments. This number, called the *degree* of the invariant, is constant for all terms of one particular invariant and is equal to the total number of the graph nodes.

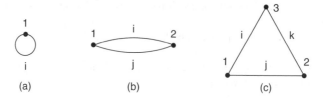

Figure 4.3   The generating graphs of (a) both Φ_1^{2D} and Φ_1^{3D}, (b) both Φ_2^{2D} and Φ_2^{3D} and (c) both Φ_3^{2D} and Φ_3^{3D}

Another parameter is the total number w of the edges in the graph that equals the sum of the indices of all moments in one term. The graphs can include self-loops, see Figure 4.3a.

Finding all invariants up to a certain weight is equivalent to constructing all connected graphs with a maximum of w edges. There is an additional restriction. Since we use zero-order moments for scaling normalization and first-order moments for translation normalization, we cannot use them in the invariants for other purposes. That is why we must avoid such graphs, in which some nodes have just one adjacent edge. Let us denote the set of all graphs with a maximum of w edges as G_w. Generating all graphs of G_w is a combinatorial task with exponential complexity but formally easy to implement.

4.4.4   The number of the independent invariants

Before we proceed to the generation of the invariants and selection of the independent ones, it is worth analyzing how many invariants may exist. An intuitive rule suggests that the number n_i of independent invariants created from n_m independent measurements (i.e., moments in our case) is

$$n_i = n_m - n_p, \tag{4.17}$$

where n_p is the number of independent constraints that must be satisfied (see, e.g., reference [31]). This is true, but the number of independent constraints is often hard to determine.

In most cases we estimate n_p as the number of transformation parameters. In the case of 2D TRS, we have one parameter of scaling, two parameters of translation and one parameter of rotation, i.e. four parameters together. In 3D, there is one parameter of scaling, three parameters of translation, and three parameters of rotation, that is, seven parameters together. Generally, the number of moments of orders from 0 to r equals

$$n_m = \binom{r+d}{d}, \tag{4.18}$$

where d is the number of dimensions of the space, i.e. $d = 2$ in 2D and $d = 3$ in 3D.

In the case of rotation, there are no hidden dependencies, neither among the constraints nor among the moments. We have $\binom{2+2}{2} = 6$ moments minus 4 parameters, which equals 2 independent TRS invariants up to the second order in 2D and $\binom{2+3}{3} - 7 = 10 - 7 = 3$ independent TRS invariants in 3D.

While Φ_3^{2D} is a dependent invariant due to the relation

$$\Phi_3^{2D} = (3\Phi_1^{2D}\Phi_2^{2D} - (\Phi_1^{2D})^3)/2$$

its counterpart Φ_3^{3D} is independent. The main difference between the tensor method for derivation of rotation invariants in 2D and in 3D is in their complexity. We use the same graphs in both dimensions, but we must analyze the dependency among the invariants separately. In 2D, the tensor method yields linear combinations of the invariants obtained by the method of complex moments (see Chapter 3), such as

$$\Phi_1^{2D} = c_{11},$$

$$\Phi_2^{2D} = (c_{20}c_{02} - c_{11}^2)/2,$$

and

$$\Phi_3^{2D} = c_{11}(c_{20}c_{02} + 3c_{11}^2)/4.$$

This is basically true in 3D, too, but the 3D tensor method is easier than the 3D method of complex moments and may serve as a primary method for finding invariants.

4.4.5   Possible dependencies among the invariants

Let us look at the possible dependencies among the invariants in detail. We categorize them into five groups and explain how they can be eliminated. This analysis is valid not only for the tensor method, but for all invariants in the form of polynomials of moments (as we will see in Chapter 5, invariants to affine transform also fall into this category).

1. *Zero invariants.* Some invariants may be identically zero regardless of the image which they are calculated from. When generating the invariants, we have to check if there is at least one non-zero coefficient.
2. *Identical invariants.* Sometimes, an invariant can be generated repeatedly. In the tensor method, different graphs may lead to the same invariant. Elimination can be done by comparing the invariants termwise.
3. *Products.* Some invariants may be products of other invariants generated earlier.
4. *Linear combinations.* Some invariants may be linear combinations of other invariants generated earlier.
5. *Polynomial dependencies.* If there exists a linear combination of products of the invariants (including their integer powers) that equals zero, the invariants involved are said to be polynomially dependent.

The invariants suffering from the dependencies 1–4 are called *reducible* invariants. After eliminating all of them, we obtain a set of so-called *irreducible* invariants. However, irreducibility does not mean independence, as we will see later. In this section we show how to detect reducible and polynomially dependent invariants.

The dependencies 1 and 2 are trivial and easy to find. To remove the products (type 3), we perform an incremental exhaustive search. All possible pairs of the admissible invariants (the sum of their individual weights must not exceed w) are multiplied, and the independence of the results is checked.

To remove the dependencies of type 4 (linear combinations), all possible linear combinations of the admissible invariants (the invariants must have the same structure) are considered, and their independence is checked. The algorithm should not miss any possible combination and can be implemented as follows.

Since only invariants of the same structure can be linearly dependent, we sort the invariants (including all their products) according to the structure first. For each group of the same structure, we construct a matrix of coefficients of all invariants. The "coefficient" here means the multiplicative constant of each term; if the invariant does not contain all possible terms, the corresponding coefficients are zero. Thanks to this, all coefficient vectors are of the same length, and we can arrange them into a matrix. If this matrix has a full rank, all invariants are linearly independent. To calculate the rank of the matrix, we use singular value decomposition (SVD). The rank of the matrix is given by the number of nonzero singular values and equals the number of linearly independent invariants of the given structure. Each zero singular value corresponds to a linearly dependent invariant that should be removed.

At the end of the above procedure, we obtain a set of all irreducible invariants up to the given weight or order. However, this procedure is of exponential complexity and thus very expensive even for low weights/orders. On the other hand, this procedure is run only once. As soon as the invariants are known in explicit forms, they can be used in all future experiments without repeating the derivation. Note that the dependence/independence is checked on the theoretical level, and it is not linked to any particular dataset[7].

Elimination of the polynomially dependent invariants (type 5) is an even more difficult task, which in general has not been completely resolved yet. We leave the detailed explanation to Chapter 5, since historically the elimination of dependent invariants has been linked to affine moment invariants rather than to the 3D rotation invariants even if the technique is the same.

4.4.6 Automatic generation of the invariants by the tensor method

The generation of all invariants up to the given weight/order is equivalent to creating all graphs from G_w. The number of edges from one node equals the order of the corresponding moment. Let us suppose we want to construct all graphs with w edges. The graphs can have w nodes at maximum, we label them by integers from 1 to w (there are graphs of w edges and $w + 1$ nodes but we need not consider them because they contain a node of degree one which is not admissible). We describe the graph by the list of edges arranged into a $2 \times w$ matrix, where each column represents an edge connecting the nodes the labels of which are in the first and the second row, respectively. For the purpose of automatic graph generation, we consider the natural ordering of the graphs such that the "first" graph is, in the above-mentioned representation,

$$
\begin{matrix}
1 & 1 & \cdots & 1 & 1 \\
1 & 1 & \cdots & 1 & 1.
\end{matrix}
\tag{4.19}
$$

This graph consists of a single node only and w selfedges[8]. The "last" graph is

$$
\begin{matrix}
1 & 2 & \cdots & w-1 & w \\
w & w & \cdots & w & w,
\end{matrix}
\tag{4.20}
$$

which is a star-shaped graph of w nodes with $w - 1$ regular edges and one selfedge.

---

[7] This is a significant difference from data-driven feature selection algorithms known from pattern recognition, that check the feature values on a particular dataset and should be run in each experiment again.

[8] A selfedge is an edge connecting a node with itself.

Starting from the first graph, we iterate the following algorithm, which creates the next graph until the last graph has been reached.

1. Find the first element from the end of the second row, which can be increased. The element can be increased, if it is less than the corresponding element of the last graph.
2. If it exists, then increase it by one to v_1. Fill the rest of the row by $\max(v_1, a + 1)$ where a is the value in the first row above the filled one.
3. If it does not exist, then find the first element from the end of the first row, which can be increased.
4. If it exists, then increase it by one to v_2. Fill the rest of the row by v_2 and copy the first row to the second row.
5. If it does not exist, then stop.

At the end, this algorithm yields the set G_w which unfortunately contains many useless graphs that correspond to reducible invariants. We have to apply the above-described procedure of dependency elimination to obtain irreducible invariants. If we work with central moments (as we usually do), all invariants containing first-order moments must be discarded as well, since they are identically zero (they can be easily recognized because the corresponding graph contains at least one node with a single adjacent regular edge).

Taking into account the computing complexity on one hand and the capability of current computers on the other hand, this construction seems to be feasible only for $w < 10$. We performed it for all w from 1 to 8. Table 4.1 summarizes the numbers of the generated invariants.

The first row of the table shows the orders of the invariants, the second row contains the cumulative number of the irreducible invariants up to the given order, which were constructed by the proposed algorithm. Note that the algorithm always works with the limitation of $w \leq 8$, so the actual numbers of irreducible invariants for orders $r > 8$ are higher than those listed in the table. Particularly, for the orders $r > 12$ the weight limitation is so restrictive that the increase of the number of irreducible invariants found by the algorithm is almost negligible. Unfortunately, releasing the weight constraint is not computationally feasible. The third row of the table contains the theoretical maximum number of the independent invariants of the given order, which was estimated by (4.17).

We generated explicit forms of all 1185 invariants; they are available (along with the codes for their evaluation) online at [32][9]. Ten independent invariants of the third order are presented

Table 4.1 The numbers of the 3D irreducible and independent rotation invariants of weight $w \leq 8$.

Order	2	3	4	5	6	7	8	9
irreducible	3	42	242	583	840	1011	1098	1142
independent	3	13	28	49	77	113	158	213

Order	10	11	12	13	14	15	16
irreducible	1164	1174	1180	1182	1184	1184	1185
independent	279	357	448	553	673	809	962

[9] At the time when of this book was prepared, this was by far the largest collection of 3D rotation moment invariants in the world. The corresponding pdf file contains more than 10,000 pages.

in Appendix A (the three independent invariants of the second order, Φ_1, Φ_2, and Φ_3, were already presented earlier in this chapter).

4.5   Rotation invariants from 3D complex moments

3D complex moments provide an alternative to the tensor method when deriving rotation invariants. This approach was inspired by 2D complex moments (see Chapter 3) and was for the first time consistently presented in [11].

The definition of 2D complex moments in polar coordinates (3.17) contains the harmonic angular function $e^{i(p-q)\theta}$. In 3D, analogous functions are called *spherical harmonics*. Their detailed description can be found in the references [33–35]. The spherical harmonics of degree ℓ and order m is defined as

$$Y_\ell^m(\theta, \varphi) = \sqrt{\frac{(2\ell + 1)}{4\pi} \frac{(\ell - m)!}{(\ell + m)!}} P_\ell^m(\cos\theta) e^{im\varphi}, \tag{4.21}$$

where $\ell = 0, 1, 2, \ldots, m = -\ell, -\ell + 1, \ldots, \ell$, and P_ℓ^m is an *associated Legendre function*

$$P_\ell^m(x) = (-1)^m (1 - x^2)^{(m/2)} \frac{d^m}{dx^m} P_\ell(x). \tag{4.22}$$

$P_\ell(x)$ is a *Legendre polynomial*

$$P_\ell(x) = \frac{1}{2^\ell \ell!} \frac{d^\ell}{dx^\ell} (x^2 - 1)^\ell. \tag{4.23}$$

The relation

$$P_\ell^{-m}(x) = (-1)^m \frac{(\ell - m)!}{(\ell + m)!} P_\ell^m(x) \tag{4.24}$$

is used for definition of the associated Legendre function of the negative order.

The first ten spherical harmonics have the forms

$$Y_0^0(\theta, \varphi) = \frac{1}{2}\sqrt{\frac{1}{\pi}},$$

$$Y_1^0(\theta, \varphi) = \frac{1}{2}\sqrt{\frac{3}{\pi}} \cos\theta, \qquad\qquad Y_1^1(\theta, \varphi) = -\frac{1}{2}\sqrt{\frac{3}{2\pi}} \sin\theta\, e^{i\varphi},$$

$$Y_2^0(\theta, \varphi) = \frac{1}{4}\sqrt{\frac{5}{\pi}}(3\cos^2\theta - 1), \qquad Y_2^1(\theta, \varphi) = -\frac{1}{2}\sqrt{\frac{15}{2\pi}} \sin\theta\cos\theta\, e^{i\varphi},$$

$$Y_2^2(\theta, \varphi) = \frac{1}{4}\sqrt{\frac{15}{2\pi}}\sin^2\theta\, e^{2i\varphi},$$

$$Y_3^0(\theta, \varphi) = \frac{1}{4}\sqrt{\frac{7}{\pi}}(5\cos^3\theta - 3\cos\theta), \quad Y_3^1(\theta, \varphi) = -\frac{1}{8}\sqrt{\frac{21}{\pi}} \sin\theta(5\cos^2\theta - 1)\, e^{i\varphi},$$

$$Y_3^2(\theta, \varphi) = \frac{1}{4}\sqrt{\frac{105}{2\pi}}\sin^2\theta\cos\theta\, e^{2i\varphi}, \quad Y_3^3(\theta, \varphi) = -\frac{1}{8}\sqrt{\frac{35}{\pi}}\sin^3\theta\, e^{3i\varphi}.$$

$$\tag{4.25}$$

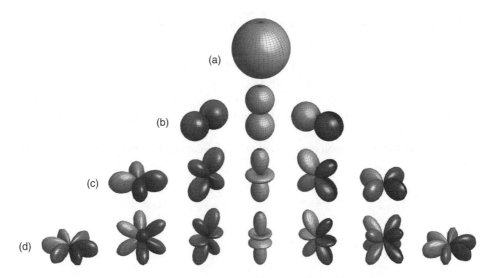

Figure 4.4 The spherical harmonics. (a) $\ell = 0$, $m = 0$, (b) $\ell = 1$, $m = -1, 0, 1$, (c) $\ell = 2$, $m = -2, -1, 0, 1, 2$, (d) $\ell = 3$, $m = -3, -2, -1, 0, 1, 2, 3$. Imaginary parts are displayed for $m < 0$ and real parts for $m \geq 0$

The spherical harmonics of a negative order can be computed as $Y_\ell^{-m} = (-1)^m (Y_\ell^m)^*$. The first few spherical harmonics can be seen in Figure 4.4, where the absolute real value is expressed by the distance of the surface from the origin of the coordinates. Since the spherical harmonics of negative orders are (negative) complex conjugates, the imaginary parts are shown instead. The gray level describes the complex phase. Y_ℓ^0 are real with a zero phase.

If we substitute Cartesian coordinates x, y, z for the spherical coordinates ϱ, θ, φ

$$
\begin{aligned}
x &= \varrho \sin\theta \cos\varphi & \varrho &= \sqrt{x^2 + y^2 + z^2} \\
y &= \varrho \sin\theta \sin\varphi & \sin\theta \; e^{i\varphi} &= (x + iy)/\varrho \\
z &= \varrho \cos\theta & \cos\theta &= z/\varrho
\end{aligned}
\tag{4.26}
$$

in (4.25), we obtain the spherical harmonics in Cartesian coordinates, e.g.

$$
Y_1^1(x, y, z) = -\frac{1}{2}\sqrt{\frac{3}{2\pi}} \; \frac{x + iy}{\varrho} \; .
\tag{4.27}
$$

3D complex moments are defined as

$$
c_{s\ell}^m = \int_0^{2\pi} \int_0^{\pi} \int_0^{\infty} \varrho^{s+2} Y_\ell^m(\theta, \varphi) f(\varrho, \theta, \varphi) \; \sin\theta \; d\varrho \; d\theta \; d\varphi,
\tag{4.28}
$$

where s is the order of the moment, ℓ is called *latitudinal repetition*, and m is called *longitudinal repetition*. The indices ℓ and m correspond to the index difference $p - q$ in 2D. The factor $\varrho^2 \sin\theta$ equals the Jacobian of the transformation from the Cartesian to the spherical coordinates.

Complex moments can be expressed also in Cartesian coordinates as

$$c_{s\ell}^m = \int\limits_{-\infty}^{\infty} \int\limits_{-\infty}^{\infty} \int\limits_{-\infty}^{\infty} \varrho^s Y_\ell^m(x, y, z) f(x, y, z) \, dx \, dy \, dz. \tag{4.29}$$

If $s = 0, 1, 2, \ldots$, then $\ell = 0, 2, 4, \ldots, s - 2, s$ for even s and $\ell = 1, 3, 5, \ldots, s - 2, s$ for odd s, $m = -\ell, -\ell + 1, \ldots, \ell$. The number of complex moments corresponds to the number of geometric moments.

The centralized complex moments of low orders expressed in terms of the geometric moments are

$$c_{00}^0 = \frac{1}{2} \sqrt{\frac{1}{\pi}} \mu_{000}$$

$$c_{20}^0 = \frac{1}{2} \sqrt{\frac{1}{\pi}} (\mu_{200} + \mu_{020} + \mu_{002})$$

$$c_{22}^{-2} = \frac{1}{4} \sqrt{\frac{15}{2\pi}} (\mu_{200} - \mu_{020} - 2i\mu_{110})$$

$$c_{22}^{-1} = \frac{1}{2} \sqrt{\frac{15}{2\pi}} (\mu_{101} - i\mu_{011})$$

$$c_{22}^0 = \frac{1}{4} \sqrt{\frac{5}{\pi}} (-\mu_{200} - \mu_{020} + 2\mu_{002})$$

$$c_{22}^1 = -\frac{1}{2} \sqrt{\frac{15}{2\pi}} (\mu_{101} + i\mu_{011})$$

$$c_{22}^2 = \frac{1}{4} \sqrt{\frac{15}{2\pi}} (\mu_{200} - \mu_{020} + 2i\mu_{110})$$

$$c_{31}^{-1} = \frac{1}{2} \sqrt{\frac{3}{2\pi}} (\mu_{300} + \mu_{120} + \mu_{102} - i(\mu_{030} + \mu_{210} + \mu_{012}))$$

$$c_{31}^0 = \frac{1}{2} \sqrt{\frac{3}{\pi}} (\mu_{201} + \mu_{021} + \mu_{003})$$

$$c_{31}^1 = -\frac{1}{2} \sqrt{\frac{3}{2\pi}} (\mu_{300} + \mu_{120} + \mu_{102} + i(\mu_{030} + \mu_{210} + \mu_{012}))$$

$$c_{33}^{-3} = \frac{1}{8} \sqrt{\frac{35}{\pi}} (\mu_{300} - 3\mu_{120} - i(3\mu_{210} - \mu_{003}))$$

$$c_{33}^{-2} = \frac{1}{4} \sqrt{\frac{105}{2\pi}} (\mu_{201} - \mu_{021} - 2i\mu_{111})$$

$$c_{33}^{-1} = \frac{1}{8} \sqrt{\frac{21}{\pi}} (-\mu_{300} - \mu_{120} + 4\mu_{102} - i(-\mu_{030} - \mu_{210} + 4\mu_{012}))$$

$$c_{33}^0 = \frac{1}{4} \sqrt{\frac{7}{\pi}} (-3\mu_{201} - 3\mu_{021} + 2\mu_{003})$$

$$c_{33}^1 = -\frac{1}{8}\sqrt{\frac{21}{\pi}}(-\mu_{300} - \mu_{120} + 4\mu_{102} + i(-\mu_{030} - \mu_{210} + 4\mu_{012}))$$

$$c_{33}^2 = \frac{1}{4}\sqrt{\frac{105}{2\pi}}(\mu_{201} - \mu_{021} + 2i\mu_{111})$$

$$c_{33}^3 = -\frac{1}{8}\sqrt{\frac{35}{\pi}}(\mu_{300} - 3\mu_{120} + i(3\mu_{210} - \mu_{030}))$$

The first-order central complex moments are zero.

The behavior of the spherical harmonics and 3D complex moments under rotation is not as simple as the behavior of their 2D counterparts, but still it allows designing 3D invariants. The spherical harmonics are transformed under rotation by the formula [36]

$$Y_\ell^m(\mathbf{R}^{-1}\mathbf{x}) = \sum_{m'=-\ell}^{\ell} D_{m'm}^\ell(\mathbf{R})Y_\ell^{m'}(\mathbf{x}), \tag{4.30}$$

that is, the spherical harmonics after applying the rotation matrix $\mathbf{R}$ are linear combinations of the spherical harmonics with the same degree. The coefficients of these linear combinations are so-called *Wigner D-functions*

$$D_{m'm}^\ell(\alpha, \beta, \gamma) = e^{im'\alpha}d_{m'm}^\ell(\beta)e^{im\gamma}, \tag{4.31}$$

where $d_{m'm}^\ell(\beta)$ is *Wigner d-function* (also *Wigner small d-function*)

$$d_{m'm}^\ell(\beta) = \sqrt{(\ell + m')!(\ell - m')!(\ell + m)!(\ell - m)!}$$
$$\times \sum_{s=\max(0,m-m')}^{\min(\ell+m,\ell-m')} \frac{(-1)^{m'-m+s}}{(\ell + m - s)!s!(m' - m + s)!(\ell - m' - s)!} \tag{4.32}$$
$$\times \left(\cos\frac{\beta}{2}\right)^{2\ell+m-m'-2s} \left(\sin\frac{\beta}{2}\right)^{m'-m+2s}.$$

The sum goes over such indices, where the arguments of the factorials are nonnegative. If there are no such arguments, then $d_{m'm}^\ell(\beta) = 0$. The Wigner D-functions are sometimes called Wigner D-matrices, because we can express (4.30) as a matrix multiplication

$$Y_\ell(\mathbf{R}^{-1}\mathbf{x}) = (D^\ell(\mathbf{R}))^T Y_\ell(\mathbf{x}), \tag{4.33}$$

where

$$Y_\ell(\mathbf{x}) = \begin{pmatrix} Y_\ell^{-\ell}(\mathbf{x}) \\ Y_\ell^{-\ell+1}(\mathbf{x}) \\ \vdots \\ Y_\ell^{\ell}(\mathbf{x}) \end{pmatrix}$$

is a vector from the spherical harmonics of the same degree ℓ, and

$$D^\ell(\mathbf{R}) = \begin{pmatrix} D_{-\ell,-\ell}^\ell(\mathbf{R}) & D_{-\ell,-\ell+1}^\ell(\mathbf{R}) & \cdots & D_{-\ell,\ell}^\ell(\mathbf{R}) \\ D_{-\ell+1,-\ell}^\ell(\mathbf{R}) & D_{-\ell+1,-\ell+1}^\ell(\mathbf{R}) & \cdots & D_{-\ell+1,\ell}^\ell(\mathbf{R}) \\ \vdots & \vdots & \ddots & \vdots \\ D_{\ell,-\ell}^\ell(\mathbf{R}) & D_{\ell,-\ell+1}^\ell(\mathbf{R}) & \cdots & D_{\ell,\ell}^\ell(\mathbf{R}) \end{pmatrix}$$

is the Wigner D-matrix.

Since the 3D complex moments are based on the spherical harmonics, they are transformed under the rotation similarly as

$$c_{s\ell}^{m(\mathbf{R})} = \sum_{m'=-\ell}^{\ell} D_{m'm}^{\ell}(\mathbf{R}) c_{s\ell}^{m'}.$$ (4.34)

4.5.1 Translation and scaling invariance of 3D complex moments

Translation and scaling invariance can be reached analogously to the geometric moments. If we express the spherical harmonics in Cartesian coordinates, we can use *3D central complex moments* to obtain the translation invariance

$$\bar{c}_{s\ell}^{m} = \int_{-\infty}^{\infty} \int_{-\infty}^{\infty} \int_{-\infty}^{\infty} \varrho^s Y_{\ell}^m(x - x_c, y - y_c, z - z_c) f(x, y, z) \, \mathrm{d}x \, \mathrm{d}y \, \mathrm{d}z.$$ (4.35)

In terms of the complex moments, the centroid can be expressed as

$$x_c = -\sqrt{\frac{2}{3}} \frac{Re\left(c_{11}^1\right)}{c_{00}^0}, \quad y_c = -\sqrt{\frac{2}{3}} \frac{Im\left(c_{11}^1\right)}{c_{00}^0}, \quad z_c = \sqrt{\frac{1}{3}} \frac{c_{11}^0}{c_{00}^0}.$$ (4.36)

The complex moments can be normalized to scaling as

$$\tilde{c}_{s\ell}^{m} = \frac{\bar{c}_{s\ell}^{m}}{(\bar{c}_{00}^0)^{\frac{s}{3}+1}}.$$ (4.37)

The symbols "tilde" and "bar" in $\bar{c}$, $\tilde{c}$ are often dropped for simplicity. In the sequel, the notation $c_{s\ell}^m$ actually means central moments and, if relevant and useful, also the scale-normalized moments (the actual meaning is apparent from the context).

4.5.2 Invariants to rotation by means of the group representation theory

The rotation invariants can be derived by means of the group representation theory. The detailed description of this theory is beyond the scope of this book; it can be be found in some specialized textbook, for instance, in [37]. The first attempt to use the group representation theory for derivation of invariants can be found in [8]. The approach we propose here is slightly different and more general.

Rotations in both 2D and 3D form groups. The 2D rotations are commutative, while in 3D the order of rotations does matter.

A moment after the rotation depends linearly on all original moments of the same order. If we transform a moment by a general linear transformation, the magnitude of the Jacobian of the transformation appears in the integrand [38, 39]. Since the Jacobian of the rotation equals one, the relation between the original and transformed moments is determined solely by the way in which the basic monomials of the same order are transformed. This conclusion is independent of the number of dimensions. In the following, we thus employ an apparatus of the theory of group representations, which can describe related properties of the rotation groups.

Group representations

Group elements G_a (in our context individual rotations) can be represented by corresponding operators $T(G_a)$ such that

$$T(G_a) \circ T(G_b) = T(G_a \circ G_b). \tag{4.38}$$

The set of operators is called *group representation*. Choosing a vector space L (we are interested in monomials of the same order) and its basis $\mathbf{e}_1, \mathbf{e}_2, \ldots, \mathbf{e}_s$ we can also define corresponding matrix representation $T_{ji}(G_a)$ satisfying

$$T(G_a)\mathbf{e}_i = \sum_j T_{ji}(G_a)\mathbf{e}_j. \tag{4.39}$$

Often we can find such a basis that the corresponding matrix representation is block-diagonal

$$\begin{pmatrix} T^{(1)} & 0 & 0 & \cdots \\ 0 & T^{(2)} & 0 & \cdots \\ 0 & 0 & T^{(3)} & \cdots \\ \vdots & \vdots & \vdots & \ddots \end{pmatrix}. \tag{4.40}$$

The $T^{(i)}(G_a)$'s are representations themselves. The space L can be divided into invariant subspaces L_i

$$L = L_1 + L_2 + L_3 + \cdots.$$

$T^{(i)}(G_a)$ acting on an element of L_i gives again a linear combination of elements of L_i. If $T^{(i)}(G_a)$'s cannot be split further, they are called irreducible[10] components of the reducible $T(G_a)$, and we write

$$T(G_a) = T^{(1)}(G_a) \oplus T^{(2)}(G_a) \oplus T^{(3)}(G_a) \oplus \cdots. \tag{4.41}$$

In the context of matrix representations the symbol $\oplus$ denotes block sum of two matrices

$$\mathbf{A} \oplus \mathbf{B} = \begin{pmatrix} \mathbf{A} & 0 \\ 0 & \mathbf{B} \end{pmatrix}.$$

2D rotation representations

The aforementioned theory is well elaborated for both 2D and 3D rotations in [37]. In the case of the 2D rotations, the following set of one-dimensional irreducible representations could appear in (4.41)

$$T^{(m)}(a) = \exp(-ima), \quad m = 0, \pm 1, \pm 2, \ldots. \tag{4.42}$$

They all represent 2D rotation by an angle a. This can be easily verified by substituting into (4.38).

If the space L consists of functions $\psi(r, \theta)$ (including monomials) defined in polar coordinates on $\mathbb{R}^2$, then (4.40) can be diagonalized using (4.42), and the basis functions are

$$\psi^m = \exp(im\theta), \quad m = 0, \pm 1, \pm 2, \ldots. \tag{4.43}$$

---

[10] We have to distinguish between reducible/irreducible invariant and reducible/irreducible group representation.

If we rotate ψ^m by rotation $\mathbf{R}_a$ we obtain

$$
\begin{aligned}
T(\mathbf{R}_a)\exp(im\theta) &= \exp(im(\theta - a)) \\
&= \exp(-ima)\exp(im\theta).
\end{aligned}
\tag{4.44}
$$

In (4.44) we can verify that the function ψ^m is transformed according to the one-dimensional representation $T^{(m)}(a)$ from (4.42). We can see that the basis functions (4.43) are multiplied by $\exp(-ima)$ under the rotation by angle a.

There is an important conclusion motivating the transition to 3D. Returning back to the Cartesian coordinates we get

$$
r^p \exp(ip\theta) = (x + iy)^p,
$$

$$
r^q \exp(-iq\theta) = (x - iy)^q.
$$

Complex polynomials on the right-hand side appear in the definition of the complex moments (3.20). The basis functions thus originate from the basis functions of one-dimensional irreducible representations of the 2D rotation group. Simple transformational properties of (4.43) induce the same simple transformational properties of the complex moments, as was explained in Chapter 3.

3D rotation representations

We want to proceed in the same way in the case of 3D rotations. Their irreducible representations are usually denoted $D^{(j)}, j = 0, 1, 2, \ldots$. Dimension of $D^{(j)}$ is $2j + 1$. If the space L consists of functions $\psi(\varrho, \vartheta, \varphi)$ defined on $\mathbb{R}^3$ and expressed in spherical coordinates $(\varrho, \vartheta, \varphi)$, it can be decomposed into a series with basis functions

$$
R(\varrho)Y_m^\ell(\vartheta, \varphi).
\tag{4.45}
$$

The radial part $R(\varrho)$ can be for instance monomials or spherical Bessel functions. The angular part consists of the spherical harmonics

$$
Y_m^\ell(\vartheta, \varphi), \quad m = -\ell, -\ell + 1, \ldots, \ell - 1, \ell.
\tag{4.46}
$$

For one particular ℓ, they form a basis of the irreducible representation $D^{(\ell)}$. The corresponding matrix representations are the Wigner D-matrices.

We are mainly interested in the invariant one-dimensional irreducible subspace transforming itself according to the irreducible representation $D^{(0)}$. Single element of the corresponding matrix representation equals one. Basis function Y_0^0 is thus absolute invariant to 3D rotation.

Basis functions of the irreducible representations with $\ell \neq 0$ are not invariants. However, we can achieve the invariance by multiplying the basis functions. If we combine basis functions corresponding to the representations $D^{(j_1)}$ and $D^{(j_2)}$, the result is transformed according to their tensor product (in the context of matrix representations also Kronecker product, denoted as $\otimes$). It can be shown [37] that it is reducible into

$$
D^{(j_1)} \otimes D^{(j_2)} = \sum_{j=|j_1-j_2|}^{j=j_1+j_2} D^{(j)}.
\tag{4.47}
$$

Thus, we can construct, for instance, the tensor product of $D^{(2)}$ with itself. According to (4.47), it is reducible, and one of its irreducible components is $D^{(0)}$

$$D^{(2)} \otimes D^{(2)} = D^{(0)} \oplus D^{(1)} \oplus D^{(2)} \oplus D^{(3)} \oplus D^{(4)}.$$

As it was explained above, we systematically create such one-dimensional representations to find invariants.

In the context of matrix representations, the tensor product of two representations $T_{ik}^{(\alpha)}$ and $T_{j\ell}^{(\beta)}$ is

$$T_{ij,k\ell}^{(\alpha\times\beta)} \equiv T_{ik}^{(\alpha)} T_{j\ell}^{(\beta)}.$$

Denoting now generally the basis functions of the irreducible representations from (4.47) $\varphi_{j_1}^i$ and $\varphi_{j_2}^i$, respectively, and the basis functions of one particular new irreducible representation Ψ_j^i (for $D^{(j)}$), we can write

$$\Psi_j^k = \sum_{m=\max(-j_1, k-j_2)}^{\min(j_1, k+j_2)} \langle j_1, j_2, m, k-m | j, k \rangle \, \varphi_{j_1}^m \, \varphi_{j_2}^{k-m}. \qquad (4.48)$$

Functional Ψ_0^0 is an absolute invariant. The coefficients $\langle j_1, j_2, m, k-m | j, k \rangle$ are called *Clebsch-Gordan coefficients* [37]. They are given as

$$\langle j_1, j_2, m_1, m_2 | j, m \rangle$$

$$= \delta_{m, m_1+m_2} \sqrt{\frac{(2j+1)(j+j_1-j_2)!(j-j_1+j_2)!(j_1+j_2-j)!}{(j_1+j_2+j+1)!}}$$

$$\times \sqrt{(j+m)!(j-m)!(j_1+m_1)!(j_1-m_1)!(j_2+m_2)!(j_2-m_2)!}$$

$$\times \sum_k \frac{(-1)^k}{k!(j_1+j_2-j-k)!(j_1-m_1-k)!(j_2+m_2-k)!(j-j_1-m_2+k)!(j-j_2+m_1+k)!}, \qquad (4.49)$$

where the sum goes over such k that the arguments of the factorials are nonnegative. The factor δ_{m, m_1+m_2} is a Kronecker delta symbol.

4.5.3 Construction of the rotation invariants

In this section, we construct absolute invariants to 3D rotations which are based on the complex moments. We prove their invariance by the way in which they have been derived – we are looking for such bases of representations $D^{(0)}$ that are absolute invariants, as it was explained in the previous subsections.

The moment c_{s0}^0 itself is an invariant. For other invariants, we have to construct other forms which transform according to $D^{(0)}$, see also (4.47) and (4.48).

We can now proceed in accordance with Subsection 4.5.2. Based on (4.48), we construct *composite complex moment forms*[11]

$$c_s(\ell, \ell')_j^k = \sum_{m=\max(-\ell, k-\ell')}^{\min(\ell, k+\ell')} \langle \ell, \ell', m, k-m | j, k \rangle \, c_{s\ell}^m \, c_{s\ell'}^{k-m} \tag{4.50}$$

which combine the bases of representations $D^{(\ell)}$ and $D^{(\ell')}$ to get the basis of representation $D^{(j)}$. The form $c_s(\ell, \ell')_0^0$ is then 3D rotation invariant. Substituting the Clebsch-Gordan coefficients $\langle \ell, \ell, m, -m | 0, 0 \rangle = (-1)^{\ell-m} / \sqrt{2\ell + 1}$ we obtain

$$c_s(\ell, \ell')_0^0 = \frac{1}{\sqrt{2\ell + 1}} \sum_{m=-\ell}^{\ell} (-1)^{\ell-m} c_{s\ell}^m c_{s\ell}^{-m}. \tag{4.51}$$

We can further combine the composite complex moment forms and the complex moments. Basis of the corresponding tensor product is then

$$c_s(\ell, \ell')_j c_{s'} = \frac{1}{\sqrt{2j+1}} \sum_{k=-j}^{j} (-1)^{j-k} c_s(\ell, \ell')_j^k c_{s'j}^{-k}. \tag{4.52}$$

Analogously, the basis of a tensor product corresponding to combining two composite complex moment forms is

$$c_s(\ell, \ell')_j c_{s'}(\ell'', \ell''')_j$$
$$= \frac{1}{\sqrt{2j+1}} \sum_{k=-j}^{j} (-1)^{j-k} c_s(\ell, \ell')_j^k c_{s'}(\ell'', \ell''')_j^{-k}. \tag{4.53}$$

If both forms in (4.53) are identical, we obtain

$$c_s^2(\ell, \ell')_j$$
$$= \frac{1}{\sqrt{2j+1}} \sum_{k=-j}^{j} (-1)^{j-k} c_s(\ell, \ell')_j^k c_s(\ell, \ell')_j^{-k}. \tag{4.54}$$

In all the formulas, the parameter j of the Clebsch-Gordan coefficients must be even; see also the indexing in (4.28). Conceivably, we can multiply more forms, but as we will see later, the product of two factors in (4.52) – (4.54) is usually sufficient.

We can denote one particular $D^{(j)}$ on the right-hand side of (4.47) as

$$(D^{(j_1)} \otimes D^{(j_2)})^{(j)}.$$

Then c_{s0}^0 corresponds to $D^{(0)}$, $c_s(\ell, \ell)_0^0$ to $(D^{(\ell)} \otimes D^{(\ell)})^{(0)}$, $c_s(\ell, \ell')_j c_{s'}$ to $((D^{(\ell)} \otimes D^{(\ell')})^{(j)} \otimes D^{(j)})^{(0)}$ and $c_s(\ell, \ell')_j c_{s'}(\ell'', \ell''')_j$ to $((D^{(\ell)} \otimes D^{(\ell')})^{(j)} \otimes (D^{(\ell'')} \otimes D^{(\ell''')})^{(j)})^{(0)}$. We can use the same expressions for various moment orders. For instance, $D^{(0)}$ leads to a sequence of invariants of all even orders: $c_{20}^0, c_{40}^0, \ldots$. The moment c_{00}^0 is also a rotation invariant, but it is usually used for scaling normalization.

The above observations can be summarized in the following Theorem.

---

[11] This term was introduced in [8].

Theorem 4.1 *1) The complex moment c_{s0}^0, 2) the composite complex moment form $c_s(\ell,\ell')_0^0$, 3) the tensor product of a composite complex moment form and a complex moment $c_s(\ell,\ell')_j c_{s'}$, and 4) the tensor product of two composite complex moment forms $c_s(\ell,\ell')_j c_{s'}(\ell'',\ell''')_j$ are 3D rotation invariants.*

The proof of this Theorem was in fact presented in a constructive way before the Theorem has been formulated. In Appendix B, we present an alternative proof by direct substitution.

4.5.4   Automated generation of the invariants

To express the invariants in explicit forms, we implemented the formulas from the previous subsection by means of symbolic calculation. According to Theorem 4.1, we generate the TRS invariants of four kinds.

1. Single complex moments c_{s0}^0,
$$s = 2, 4, 6, \ldots$$

2. Single composite complex moment forms $c_s(\ell,\ell')_0^0$,
$$s = 2, 3, 4, \ldots,$$
$$\ell = s, s-2, s-4, \ldots, 1.$$

   The last value of ℓ in the loop is not always one, but $1 + \text{rem}(s+1, 2)$, where "rem" is a remainder after division. Regardless, the loop breaking condition $\ell \geq 1$ is sufficient. It applies to the following cases, too.

3. Products of a form and a complex moment $c_s(\ell,\ell')_j c_{s'}$,
$$s = 2, 3, 4, \ldots,$$
$$s' = 2, 4, 6, \ldots, s,$$
$$j = 2, 4, 6, \ldots, s',$$
$$\ell = s, s-2, s-4, \ldots, 1,$$
$$\ell' = \ell, \ell-2, \ell-4, \ldots, \max(\ell-j, 1).$$

   where ℓ' must satisfy both $\ell' \geq \ell - j$ and $\ell' \geq 1$.

4. Products of two forms $c_s(\ell,\ell')_j c_{s'}(\ell'',\ell''')_j$,
$$s = 2, 3, 4, \ldots,$$
$$s' = 2, 3, 4, \ldots, s,$$
$$j = 2, 4, 6, \ldots, s',$$
$$\ell = s, s-2, s-4, \ldots, 1,$$
$$\ell' = \ell, \ell-2, \ell-4, \ldots, \max(\ell-j, 1),$$
$$\ell'' = s', s'-2, s'-4, \ldots, 1,$$
$$\ell''' = \ell'', \ell''-2, \ell''-4, \ldots, \max(\ell''-j, 1).$$

4.5.5   Elimination of the reducible invariants

The above algorithm generates a number of reducible and dependent invariants, which must be identified and eliminated afterwards. If we consider, for instance, two invariants which have been obtained as products of different moment forms, they could be dependent. In this sense, the situation is similar to the case of the invariants derived by the tensor method.

The algorithm guarantees the invariants cannot be identically zero, but they can equal products, linear combinations, or polynomials of other invariants. They can even be identical with the others. For example, the following identities have been found among the invariants of the fourth order:

$$
\begin{aligned}
c_4(4,2)_4 c_4 &= c_4(4,4)_2 c_4, \\
c_4(2,2)_4 c_4 &= c_4(4,2)_2 c_4, \\
c_4(4,4)_4 c_4(4,2)_4 &= \frac{9}{\sqrt{143}} \, c_4(4,4)_2 c_4(4,2)_2, \\
c_4(4,2)_4 c_4(2,2)_4 &= c_4(4,2)_2 c_4(2,2)_2, \\
c_4^2(2,2)_4 &= \frac{3}{5}\sqrt{5} \; c_4^2(2,2)_2.
\end{aligned}
$$

The numbers of the generated invariants are listed in Table 4.2. The first row of the table shows the orders of the invariants, the second row contains the cumulative number of all generated invariants up to the given order, the third row contains the number of the irreducible invariants, and the fourth row contains the theoretical maximum number of the independent invariants calculated using (4.17).

4.5.6   The irreducible invariants

The main limitation, which prevents us from using moments of higher orders, is numerical precision of the computers. Even if the calculation is symbolic, we face this limitation because we have to calculate the coefficients of individual terms. These coefficients can theoretically be expressed as a square root of a rational number. If such a coefficient is stored as a floating-point number, then it sometimes cannot be stored precisely, which may affect subsequent calculations. Generation of higher-order invariants brings also other problems. One of them is the insufficient precision of the moments themselves; another one is the high computing complexity and memory demands associated with the elimination of the reducible invariants.

The rotation invariants can be expressed in explicit form in terms of the complex moments. For the second order, we have three independent invariants (for readers' convenience, we list

Table 4.2   The numbers of the 3D irreducible and independent complex moment rotation invariants

Order	2	3	4	5	6	7
All	4	16	49	123	280	573
Irreducible	3	13	37	100	228	486
Independent	3	13	28	49	77	113

them in the form which includes the scaling normalization).

$$\Phi_1 = (c_{20}^0)/(c_{00}^0)^{5/3}$$

$$\Phi_2 = c_2(2,2)_0^0 = \sqrt{\frac{1}{5}}\left((c_{22}^0)^2 - 2c_{22}^{-1}c_{22}^1 + 2c_{22}^{-2}c_{22}^2\right)/(c_{00}^0)^{10/3}$$

$$\Phi_3 = c_2(2,2)_2c_2 = \sqrt{\frac{1}{35}}\left(-\sqrt{2}(c_{22}^0)^3 + 3\sqrt{2}c_{22}^{-1}c_{22}^0c_{22}^1\right.$$
$$\left. - 3\sqrt{3}(c_{22}^{-1})^2c_{22}^2 - 3\sqrt{3}c_{22}^{-2}(c_{22}^1)^2 + 6\sqrt{2}c_{22}^{-2}c_{22}^0c_{22}^2\right)/(c_{00}^0)^5$$

The invariants of the third order are presented as the products of the corresponding composite complex moment forms only

$$\Phi_4 = c_3(3,3)_0^0,$$
$$\Phi_5 = c_3(1,1)_0^0,$$
$$\Phi_6 = c_3(3,3)_2c_2,$$
$$\Phi_7 = c_3(3,1)_2c_2,$$
$$\Phi_8 = c_3(1,1)_2c_2,$$
$$\Phi_9 = c_3(3,3)_2c_2(2,2)_2,$$
$$\Phi_{10} = c_3^2(3,3)_2,$$
$$\Phi_{11} = c_3(3,3)_2c_3(3,1)_2,$$
$$\Phi_{12} = c_3(3,3)_2c_3(1,1)_2,$$
$$\Phi_{13} = c_3(3,1)_2c_3(1,1)_2.$$

In [8], the authors published twelve invariants (we keep the ordering used in [8]): Φ_1, Φ_2 and Φ_3 of the second order and Φ_4, Φ_5, Φ_{10}, $c_3^2(3,1)_2$, Φ_{11}, Φ_{13}, Φ_6, Φ_7 and Φ_8 of the third order. However, their system is dependent because the invariants Φ_{12}, $c_3^2(3,1)_2$ and $\Phi_4\Phi_5$ are linearly dependent. According to (4.17), there should be another independent invariant of the third order, not listed in [8] – it is that one which we denote Φ_9 here.

In practice, we recommend using the independent invariants $\Phi_1,\ldots,\Phi_{13}$ (we prefer Φ_{12} to $c_3^2(3,1)_2$ since it has fewer terms) and, if necessary, some (or all) of the following irreducible invariants of the fourth order: c_4, $c_4(4,4)_0^0$, $c_4(2,2)_0^0$, $c_4(4,4)_2c_2$, $c_4(2,2)_2c_2$, $c_4(4,4)_2c_4$, $c_4(4,2)_2c_4$, $c_4(2,2)_2c_4$, $c_4(4,4)_4c_4$, $c_4(2,2)_2c_3(1,1)_2$, $c_4^2(4,4)_2$, $c_4(4,4)_2c_4(4,2)_2$, $c_4(4,4)_2c_4(2,2)_2$, $c_4^2(4,2)_2$, and $c_4(4,2)_2c_4(2,2)_2$.

In Appendix 4.C the reader may find more detailed expressions of these invariants. Note that the invariants from complex moments derived in this section are formally different from the invariants derived by the tensor method. However, since both systems are equivalent (they generate the same subspace of the invariants) and mutually dependent, we denote the invariants of both systems as Φ_k for simplicity.

4.6 3D translation, rotation, and scale invariants via normalization

As in the 2D case, in 3D we can also define a normalized position of the object by constraining certain moments. The generalization of translation and scale normalization to 3D is straightforward. The object is shifted such that the new centroid is in the origin of the coordinates and is scaled such that its volume equals one. Rotation normalization is more difficult even if we again find an inspiration in the 2D case.

4.6.1   Rotation normalization by geometric moments

Rotation normalization constraints are $\mu'_{110} = \mu'_{101} = \mu'_{011} = 0$. The normalization rotation matrix can be obtained by solving the eigenproblem of the second-order moment matrix

$$
\mathbf{M_C} = \begin{pmatrix} \mu_{200} & \mu_{110} & \mu_{101} \\ \mu_{110} & \mu_{020} & \mu_{011} \\ \mu_{101} & \mu_{011} & \mu_{002} \end{pmatrix} \tag{4.55}
$$

which is an analogy with the 2D case and also with the PCA method. The object must be rotated such that the eigenvectors of $\mathbf{M_C}$ coincide with x, y and z axes. The eigenvalues of $\mathbf{M_C}$ define the extent of the *reference ellipsoid*.

There is, however, another possibility for defining the normalization constraints. Instead of the covariance matrix $\mathbf{M_C}$, we can diagonalize the *inertia matrix* $\mathbf{M_I}$ defined as

$$
\mathbf{M_I} = \begin{pmatrix} \mu_{020} + \mu_{002} & -\mu_{110} & -\mu_{101} \\ -\mu_{110} & \mu_{200} + \mu_{002} & -\mu_{011} \\ -\mu_{101} & -\mu_{011} & \mu_{200} + \mu_{020} \end{pmatrix}. \tag{4.56}
$$

This matrix is known from classical mechanics; the values on its diagonal are moments of inertia with respect to the x, y and z axes, respectively. The eigenvalues of $\mathbf{M_C}$ and $\mathbf{M_I}$ are generally distinct, but their eigenvectors are the same up to a permutation. Hence, the reference ellipsoid obtained from $\mathbf{M_I}$ has the same axes as that one given by $\mathbf{M_C}$, but its extent is different. There is very little difference between these two approaches; both can be found in the literature[12]. Cybenko et al. [40] used the covariance matrix $\mathbf{M_C}$, while Galvez and Canton [10] used the matrix of inertia $\mathbf{M_I}$. The following thoughts are valid for both, so we do not distinguish between these two matrices and we use $\mathbf{M}$ as a joint notation.

Regardless of the particular choice of $\mathbf{M}$, we find the normalized position as follows. Assuming $\mathbf{M}$ has three distinct[13] eigenvalues $\lambda_1 > \lambda_2 > \lambda_3$ (which must be real due to the symmetry of $\mathbf{M}$), its eigenvectors $\mathbf{u_1}, \mathbf{u_2}$ and $\mathbf{u_3}$ are defined unambiguously (up to a multiple). Since the eigenvectors of any symmetric matrix are orthogonal, they form a rotation matrix $\mathbf{R} = (\mathbf{u_1}, \mathbf{u_2}, \mathbf{u_3})$ the inverse of which $\mathbf{R}^{-1} = \mathbf{R}^T$ rotates the object into the normalized position.

To calculate the moments of the normalized object (which is the main purpose of the normalization), it is not necessary to rotate and resample the object itself. The moments in the normalized position can be calculated by means of the moments of the original object.

The general formula for the conversion of the moments under an arbitrary linear transformation is obtained from the polynomial theorem

$$
\mu'_{pqr} = \sum_{i_1=0}^{p} \sum_{j_1=0}^{p-i_1} \sum_{i_2=0}^{q} \sum_{j_2=0}^{q-i_2} \sum_{i_3=0}^{r} \sum_{j_3=0}^{r-i_3}
$$

$$
\times \frac{p!}{i_1! j_1! (p - i_1 - j_1)!} \frac{q!}{i_2! j_2! (q - i_2 - j_2)!} \frac{r!}{i_3! j_3! (r - i_3 - j_3)!}
$$

---

[12] In 2D, the inertia matrix is the same as the covariance matrix.

[13] In practice we check whether or not the differences between them exceed a given threshold.

$$\times a_{11}^{i_1} d_{12}^{j_1} a_{13}^{p-i_1-j_1} a_{21}^{i_2} d_{22}^{j_2} a_{23}^{q-i_2-j_2} a_{31}^{i_3} d_{32}^{j_3} a_{33}^{r-i_3-j_3}$$

$$\times \mu_{i_1+i_2+i_3, j_1+j_2+j_3, p+q+r-i_1-j_1-i_2-j_2-i_3-j_3}, \tag{4.57}$$

where

$$\mathbf{A} = \begin{pmatrix} a_{11} & a_{12} & a_{13} \\ a_{21} & a_{22} & a_{23} \\ a_{31} & a_{32} & a_{33} \end{pmatrix}$$

is the transformation matrix. In the case of rotation, $\mathbf{A} = \mathbf{R}^T$.

The above normalization constraints $\mu'_{110} = \mu'_{101} = \mu'_{011} = 0$ do not define the normalized position unambiguously. They align the object axes but still do not define the unique object orientation because they are fulfilled if we rotate the object by 180° around any of three principal coordinate axes (these rotations are not independent because the coordinate system must stay dextrorotatory; constraining two of them fully determines the object orientation). Some authors have resolved this ambiguity by other techniques than by moments. For instance, Galvez and Canton [10] proposed to find the most distant point from the centroid and orient the object such that this point lies in the first octant. This is, however, an unstable solution because the most distant point could be found inaccurately or there even may exist multiple such points. Additional constraining of higher-order moments provides a more robust solution (let us recall that in 2D this ambiguity is removed by the constraint $\mu'_{30} > 0$). There are many possibilities of how to define these additional constraints. The simplest one is to rotate the object around the z axis such that $\mathcal{Re}(c_{31}^{1\,\prime}) > 0$, which is equivalent to the requirement

$$\mu'_{300} + \mu'_{120} + \mu'_{102} < 0,$$

and around the x axis such that $\mathcal{Im}(c_{31}^{1\,\prime}) > 0$, which means in terms of geometric moments

$$\mu'_{210} + \mu'_{030} + \mu'_{012} < 0.$$

If the above constraints are not fulfilled after the object has been normalized by $\mathbf{R}^T$, the normalization matrix should be multiplied by

$$\begin{pmatrix} -1 & 0 & 0 \\ 0 & -1 & 0 \\ 0 & 0 & 1 \end{pmatrix}$$

to satisfy the first constraint and/or by

$$\begin{pmatrix} 1 & 0 & 0 \\ 0 & -1 & 0 \\ 0 & 0 & -1 \end{pmatrix}$$

to meet the other one. After that, the normalized position is defined unambiguously an we can use all unconstrained moments as the invariants.

The above procedure applies if matrix $\mathbf{M}$ has three distinct eigenvalues. If two (or even all three) eigenvalues are equal, it is an indicator of some type of rotational symmetry of the reference ellipsoid (which does not necessarily mean that the object itself exhibits the same symmetry). Two equal eigenvalues mean that two principal semi-axes are of the same length and the ellipsoid has a rotational symmetry around the third axis. Three identical eigenvalues mean that the ellipsoid is a sphere. Such ellipsoids cannot define the normalized position of the object correctly because the principal axes can be any two (three) perpendicular lines (we recall that any non-zero vector of the respective subspace is an eigenvector of $\mathbf{M}$). If the object itself was also symmetric, then this ambiguity would not cause any problems. If, however, the object is not symmetric because of the presence of some fine details, then we have to avoid this ambiguity by using higher-order moments for normalization. In such a case, we should, for instance, find the non-zero complex moment with the lowest order. More precisely, we search for the non-zero moment $c_{s\ell}^m$ with the lowest latitudinal repetition ℓ and the longitudinal repetition $m \neq 0$. Its order then corresponds to the fold number of the rotation symmetry.

The normalization procedure is illustrated by the following simple experiment, which we carried out on the object from the Princeton Shape Benchmark [1] (the microscope number 589). In Figure 4.5a)-c), you can see the object in three distinct orientations (the first one is the object position in the PSB, the other two were generated randomly). All three objects were first normalized by constraining the covariance matrix $\mathbf{M_C}$. Their normalized positions were visually identical (as one may expect), one instance is depicted in Figure 4.5d). To test the differences between the normalized positions quantitatively, we calculated seventy-seven moments up to the sixth order of each normalized object. The moments should theoretically

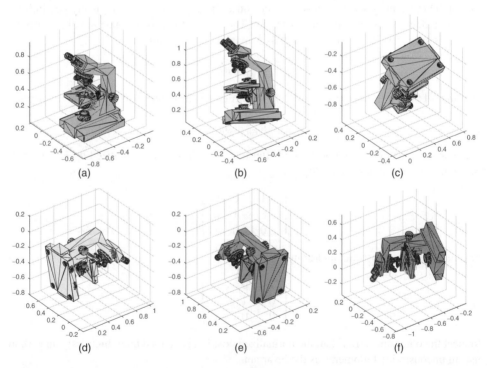

Figure 4.5 The microscope: (a) the original position, (b) the first random rotation, (c) the second random rotation, (d) the standard position by the covariance matrix, (e) the standard position by the matrix of inertia and (f) the standard position by the 3D complex moments

be the same. Actually, their average relative error was only $3 \cdot 10^{-14}$, which illustrates a very precise normalization.

For a comparison, we computed the standard position also by means of the matrix of inertia $\mathbf{M_I}$ (Figure 4.5e). Since the matrices $\mathbf{M_C}$ and $\mathbf{M_I}$ have the same eigenvectors but differently ordered eigenvalues, the axes in Figures 4.5d) and e) are permuted. The stability of the normalized position is the same as before. Finally, we normalized the three objects by constraining the third-order 3D complex moments c_{31}^1, c_{31}^0 and c_{22}^1 (see Figure 4.5f). The standard position is different from the previous cases, but again it is the same (up to an error of order 10^{-14}) for all three objects.

4.6.2   Rotation normalization by complex moments

As in 2D, the normalized position can be alternatively defined by means of the complex moments. In this way we avoid solving the eigenproblem of the moment matrix and looking for the unique orientation. When normalizing by complex moments, the rotation matrix cannot be computed directly. We first find the normalization Euler angles, and then the rotation matrix can be computed by means of (4.7) or (4.8). Burel and Hénocq [41] proposed the following normalization.

1. Calculate the normalizing Euler angle α as the phase of c_{31}^1. Normalize all 3D complex moments by the angle $-\alpha$. After that, c_{31}^1 becomes real and positive.
2. Set the normalizing Euler angle $\beta = \arctan\left(\sqrt{2}Re(c_{31}^1)/c_{31}^0\right)$. It follows from Eqs. (4.31), (4.32), and (4.34) that c_{31}^1 after a rotation by β equals

$$
\begin{aligned}
c_{31}^{1\,\prime} &= \frac{1 + \cos\beta}{2}c_{31}^1 - \frac{\sin\beta}{\sqrt{2}}c_{31}^0 + \frac{1 - \cos\beta}{2}c_{31}^{-1} \\
&= Im(c_{31}^1) + Re(c_{31}^1)\cos\beta - c_{31}^0 \sin\beta/\sqrt{2}.
\end{aligned}
\tag{4.58}
$$

   This means, for this particular choice of β, that $c_{31}^{1\,\prime} = 0$. Normalize all 3D complex moments by the angle $-\beta$.
3. Calculate the normalizing Euler angle γ as the phase of c_{22}^1. Normalize all 3D complex moments by the angle $-\gamma$. After that, c_{22}^1 becomes real and positive.

Since we used only the 3D complex moments with longitudinal repetition $m = 0$ and $m = 1$, the object orientation is unambiguous. A drawback of this method is that the third-order moments employed here are slightly less stable than the second-order moments used for the computation of the characteristic ellipsoid.

When applying this method to normalization of symmetric objects, where some complex moments vanish, we must proceed with care and avoid using vanishing moments (although this may look trivial, in practice the vanishing moments may not exactly equal to zero but rather to some small values). While the first and the third Euler angles α and γ can be found analogously to 2D as the phases of some suitable complex moments

$$
\alpha = \frac{1}{m}\arctan\left(\frac{Im(c_{s\ell}^m)}{Re(c_{s\ell}^m)}\right),
$$

the second Euler angle β must be found from the corresponding Wigner d-matrix. We obtain from Eqs. (4.31), (4.32), and (4.34)

$$c_{s\ell}^{m(\alpha,\beta)} = \sum_{m'=-\ell}^{\ell} d_{m'm}^{\ell}(\beta) c_{s\ell}^{m'(\alpha)}, \tag{4.59}$$

where $c_{s\ell}^{m(\alpha)}$ is a complex moment normalized by the angle α and $c_{s\ell}^{m(\alpha,\beta)}$ is the prescribed value (usually 0) of the complex moment normalized by the angles α and β. It leads to a polynomial equation of the ℓ-th degree for $\cos\beta$ that must be solved numerically for the fold number of the rotation symmetry higher than four. This is another reason why in the case of symmetric bodies we prefer invariants to the normalization (the first reason is, as always in the normalization, the potential instability of the normalization constraints). However, by using invariants we cannot completely avoid the problems induced by vanishing moments. For each type of the object symmetry, a special set of the invariants should be designed, as explained in the next section.

4.7 Invariants of symmetric objects

Similarly as in 2D, the objects having some kind of symmetry (circular, rotational, axial, dihedral, etc.) often cause difficulties because some of their moments are zero. If such a moment is used as a factor in a product, the product always vanishes. Since the invariants are often products or linear combinations of products of the moments, the object symmetry can make certain invariants trivial. This is not a serious problem if there are no two or more classes with the same symmetry; in such a case the vanishing invariants may serve as discriminatory features. However, if more classes have the same symmetry, the vanishing invariants should not be used for recognition, as they decrease the recognition power of the system while increasing the computing complexity. For 2D rotation invariants, this problem was discussed in the previous chapter. Now we are going to analyze the 3D case.

We deal with rotation and reflection symmetry only, the other kinds of symmetry, such as translational, scaling, and skewing symmetry are irrelevant in the context of object recognition. They require infinite support, and the scaling symmetry even requires infinite spatial resolution. It cannot be fulfilled in practice, so in reality we do not meet these two kinds of symmetry.

The problem is analogous both in 2D and in 3D. The main difference is that the class of symmetries in 3D is richer, and the proper treatment is much more complicated. A detailed comprehensive study of this problem was firstly published in [12].

4.7.1 Rotation and reflection symmetry in 3D

In this section, we present a survey of all existing reflection and rotation symmetries (i.e., symmetry groups) in 3D. Let us recall that in 2D, only two kinds of symmetry groups exist – N-fold rotation symmetry groups (denoted as C_N) and N-fold dihedral symmetry D_N. In 3D, the reflection symmetry is always with respect to a reflection plane and rotation symmetry is w.r.t. a symmetry axis. When categorizing the symmetry groups, we assume that the axis with the maximum fold number of the rotation symmetry coincides with the z-axis, and if there is a plane of the reflection symmetry containing the axis, it coincides with the $x - z$ plane. If the reflection plane is perpendicular to the z axis, it coincides with the $x - y$ plane. We can make this assumption without a loss of generality because we study such object properties which do not depend on their particular orientation. A survey of all 3D symmetry groups with the rotation and reflection symmetry is presented below.

1. Asymmetric extension of the 2D groups

$$C_1, \ C_2, \ C_3, \ldots$$

into 3D. We can imagine, e.g., some chiral pyramid constructed on the 2D base with the symmetry group C_n. The group has still n elements. An example of a body with this symmetry is in Figure 4.6a.

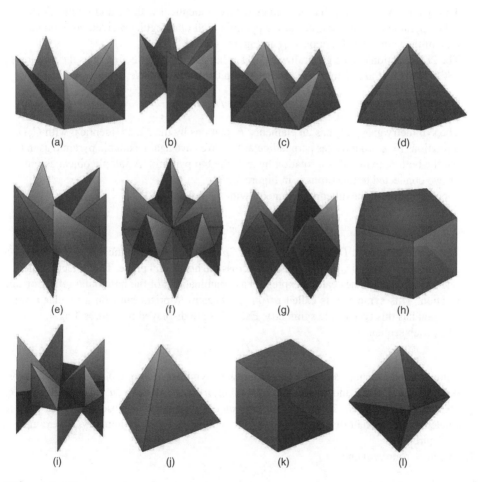

(a)

(b)

(c)

(d)

(e)

(f)

(g)

(h)

(i)

(j)

(k)

(l)

Figure 4.6 The symmetric bodies with the symmetry groups (a) C_5 (the repeated tetrahedrons), (b) C_{5h} (the repeated tetrahedrons), (c) C_{5v} (the repeated tetrahedrons), (d) C_{5v} (the pentagonal pyramid), (e) D_5 (the repeated pyramids), (f) D_{5d} (the repeated tetrahedrons), (g) D_{5h} (the repeated pyramids), (h) D_{5h} (the pentagonal prism), (i) S_{10} (the repeated tetrahedrons), (j) T_d (the regular tetrahedron), (k) O_h (the cube) and (l) O_h (the octahedron)

2. A single horizontal plane of the reflection symmetry. These symmetry groups are denoted as

$$C_{1h}, \; C_{2h}, \; C_{3h}, \ldots$$

The symmetry group C_{nh} has $2n$ elements: n identical with C_n (the identity and $n-1$ rotations) and other n, where the rotations are combined with the reflection. The objects resemble a "double pyramid", where the bottom pyramid is a reflection of the top pyramid. An example of a body with this symmetry is in Figure 4.6b.

3. n vertical planes of the reflection symmetry. In this case we have

$$C_{1v}, \; C_{2v}, \; C_{3v}, \ldots$$

The symmetry group C_{nv} has also $2n$ elements: n identical with C_n and n reflections. It is called pyramidal symmetry, because a pyramid built on a regular n-sided polygon has this type of the symmetry. Examples are in Figures 4.6c–d.

4. The 2D reflection can be generalized to 3D either again as a reflection or as the rotation by $180°$. The dihedral symmetry is generalized in the latter way. We obtain a sequence

$$D_1, \; D_2, \; D_3, \ldots$$

The symmetry group D_n has $2n$ elements: n rotations by the z-axis (identical with C_n) and n rotations by π around axes perpendicular to z. We have again a double pyramid as in C_{nh}, but the bottom pyramid is a rotation by π of the top pyramid. A sample object assembled of two connected tetrahedrons is in Figure 4.6e.

5. A combination of the dihedral symmetry with the reflection

$$D_{1h}, \; D_{2h}, \; D_{3h}, \ldots$$

The symmetry group D_{nh} has $4n$ elements: $2n$ of them being identical with D_n and other $2n$ which are coupled with the reflection across the horizontal plane. The reflections across one of n vertical planes can be expressed as combinations of the horizontal reflection and rotation. This symmetry is called *prismatic*, because a prism built on a regular n-sided polygon has this type of the symmetry. Examples are displayed in Figures 4.6g–h.

6. The group sequence

$$D_{1d}, \; D_{2d}, \; D_{3d}, \ldots$$

without the horizontal reflection plane, but with symmetry to compound reflection across the horizontal plane and rotation by $(2k+1)\pi/n$ for $k = 0, 1, \ldots, n-1$. The number of elements of D_{nd} is $4n$. It is called *anti-prismatic* symmetry, because we would have to rotate the bottom half of the prism by π/n to obtain this type of symmetry. An example is in Figure 4.6f.

7. The last group sequence

$$S_2, \; S_4, \; S_6, \ldots$$

contains the symmetry to the compound transformation from the previous case but without the corresponding simple rotations. S_{2n} has $2n$ elements: n rotations by $4k\pi/n$ and n compound transformations from the rotations by $2(2k+1)\pi/n$ and the reflections. The actual fold number of the rotation symmetry is n. An example is in Figure 4.6i.

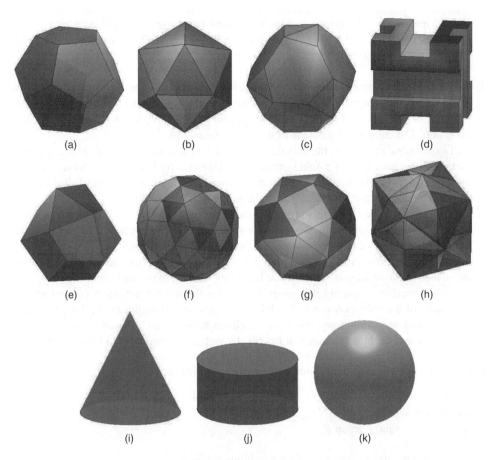

Figure 4.7 The symmetric bodies with the symmetry groups: (a) I_h (the dodecahedron), (b) I_h (the icosahedron), (c) T_h (the simplest canonical polyhedron), (d) T_h (the pyritohedral cube), (e) T (the propello-tetrahedron), (f) I (the snub dodecahedron), (g) O (the snub cube), (h) O (the chiral cube), (i) $C_{\infty v}$ (the conic), (j) $D_{\infty h}$ (the cylinder) and (k) K (the sphere)

The bodies in Figures 4.6 and 4.7 were created by a triangulation, the coplanar triangles were colored by the same color. The color is used for distinction of the different faces of the polyhedrons only, it has no connection with the symmetry.

Note that some of these groups are equivalent. For instance, $C_{1h} = C_{1v}$ (a single reflection), $D_1 = C_2$ (a single rotation by π), $D_{1h} = C_{2v}$ (a reflection in a plane and a rotation by π along a line in that plane), and $D_{1d} = C_{2h}$ (a reflection in a plane and a rotation by π along a line perpendicular to that plane). The group S_2 (sometimes called central symmetry) is a composition of a reflection in a plane and a rotation by π through a line perpendicular to that plane. Unlike D_{1d}, in S_2 both these transformations must be performed together.

All the above symmetry group sequences include at most one rotation axis with the fold number higher then two. However, there exist other symmetry groups, such as regular polyhedra, which do not fit into the above framework.

1. Full *tetrahedral* symmetry: the regular tetrahedron has 4 axes of 3-fold symmetry, 3 axes of 2-fold symmetry, and 6 planes of the reflection symmetry. The symmetry group T_d has 24 elements in total (the letter d means diagonal mirror planes). The regular tetrahedron is in Figure 4.6j.
2. Full *octahedral* symmetry: the cube and the octahedron have the same symmetry. It includes 3 axes of 4-fold symmetry, 4 axes of 3-fold symmetry, 6 axes of 2-fold symmetry, and 9 planes of the reflection symmetry. The symmetry group O_h has 48 elements, T_d is its subgroup. The cube and the octahedron are in Figures 4.6k–l.
3. Full *icosahedral* symmetry: the dodecahedron and the icosahedron have also the same symmetry. It includes 6 axes of 5-fold symmetry, 10 axes of 3-fold symmetry, 15 axes of 2-fold symmetry, and 15 planes of the reflection symmetry. The symmetry group I_h has 120 elements. The dodecahedron and the icosahedron are in Figures 4.7a–b.
4. *Pyritohedral* symmetry T_h: this group includes the same rotations as T_d, but it has only 3 planes of reflection symmetry, the rest are rotoreflections. This group has 24 elements (the letter h means horizontal mirror planes). Examples are in Figure 4.7c–d.
5. *Chiral tetrahedral* symmetry: the group T includes the same rotations as T_d, but without the planes of the reflection symmetry. It has 12 elements. An example is in Figure 4.7e.
6. *Chiral octahedral* symmetry: the group O includes the same rotations as O_h, but without the planes of the reflection symmetry. It has 24 elements. Examples are in Figures 4.7g–h.
7. *Chiral icosahedral* symmetry: the group I includes the same rotations as I_h, but without the planes of the reflection symmetry. It has 60 elements. An example is in Figure 4.7f.

Finally, there exist three groups containing circular symmetry:

1. Conical symmetry group $C_{\infty v}$.
2. Cylindrical symmetry group $D_{\infty h}$.
3. Spherical symmetry group K.

Examples of these three groups are shown in Figures 4.7i–k.
   Summarizing, in 3D we have

- 7 infinite sequences of the symmetry groups, where each group has one main rotation axis and the other axes are at most twofold,
- 7 separate symmetry groups with at least two rotation axes with the fold number higher than two,
- 3 infinite symmetry groups with some kind of circular symmetry.

The presented list of symmetry groups is complete. It was proved that there are no other independent combinations of the rotation and reflection symmetries (see, for instance, [42] for the proof[14]).

4.7.2   The influence of symmetry on 3D complex moments

For each kind of object symmetry, certain complex moments vanish. We can investigate this influence either by successive substitution of all elements of the corresponding symmetry group

---

[14] This proof had been originally published independently by three researches: Y. S. Fyodorov (1891), A. M. Schön-flies (1891), and W. Barlow (1894).

to Equation (4.34) or by the group representation theory as Mamone et al. [43] proposed. The best way is to combine both approaches, as was proposed by Suk in [12]. In that paper, the readers can find a detailed derivation of vanishing moments. The result can be summarized as follows (we list the symmetry groups and the corresponding vanishing moments). For rotation symmetry groups it holds

- C_i, $i = 1, 2, \ldots$: $c_{s\ell}^m = 0$ for $m \neq ki$.
- C_{ih}, $i = 1, 2, \ldots$: $c_{s\ell}^m = 0$ for $m \neq 2ki$ when ℓ is even and for $m \neq (2k + 1)i$ when ℓ is odd.
- C_{iv} (pyramid), $i = 1, 2, \ldots$: $\mathcal{R}e(c_{s\ell}^m) = 0$ for $m \neq ki$. $\mathcal{I}m(c_{s\ell}^m) = 0$ always.
- D_i, $i = 1, 2, \ldots$: $\mathcal{R}e(c_{s\ell}^m) = 0$ for $m \neq 2ki$ and $\mathcal{I}m(c_{s\ell}^m) = 0$ for $m \neq (2k + 1)i$ when ℓ is even. $\mathcal{R}e(c_{s\ell}^m) = 0$ for $m \neq (2k + 1)i$ and $\mathcal{I}m(c_{s\ell}^m) = 0$ for $m \neq 2ki$ when ℓ is odd.
- D_{ih} (prism), $i = 1, 2, \ldots$: $\mathcal{R}e(c_{s\ell}^m) = 0$ for $m \neq 2ki$ when ℓ is even and $\mathcal{R}e(c_{s\ell}^m) = 0$ for $m \neq (2k + 1)i$ when ℓ is odd. $\mathcal{I}m(c_{s\ell}^m) = 0$ always.
- D_{id} (antiprism), $i = 1, 2, \ldots$: $\mathcal{R}e(c_{s\ell}^m) = 0$ for $m \neq 2ki$ and $\mathcal{I}m(c_{s\ell}^m) = 0$ for $m \neq (2k + 1)i$ when ℓ is even. $c_{s\ell}^m = 0$ for any odd ℓ.
- S_{2k}, $i = 1, 2, \ldots$: $c_{s\ell}^m = 0$ for $m \neq ki$ when ℓ is even. $c_{s\ell}^m = 0$ for any odd ℓ.

In all cases, k is an arbitrary non-negative integer. Note that always $\mathcal{I}m(c_{s\ell}^0) = 0$.
For the polyhedral symmetries we have

- T_d (full tetrahedron): $\mathcal{R}e(c_{s\ell}^m) = 0$ when $m \neq 3k$ or $\ell = 1, 2$ or 5. $\mathcal{I}m(c_{s\ell}^m) = 0$ always.
- O_h (full octahedron): $\mathcal{R}e(c_{s\ell}^m) = 0$ when $m \neq 4k$ or ℓ is odd or $\ell = 2$. $\mathcal{I}m(c_{s\ell}^m) = 0$ always.
- I_h (full icosahedron): $\mathcal{R}e(c_{s\ell}^m) = 0$ when $m \neq 5k$ or ℓ is odd or $\ell = 2, 4, 8$ and 14. $\mathcal{I}m(c_{s\ell}^m) = 0$ always.
- T_h (pyritohedron): $\mathcal{R}e(c_{s\ell}^m) = 0$ when $m \neq 2k$ or ℓ is odd or $\ell = 2$. Also $\mathcal{R}e(c_{s4}^2) = 0$, $\mathcal{R}e(c_{s8}^2) = 0$, and $\mathcal{R}e(c_{s8}^6) = 0$. $\mathcal{I}m(c_{s\ell}^m) = 0$ always.
- T (chiral tetrahedron): $\mathcal{R}e(c_{s\ell}^m) = 0$ when $m \neq 2k$ or ℓ is odd or $\ell = 2$. Also $\mathcal{R}e(c_{s4}^2) = 0$, $\mathcal{R}e(c_{s8}^2) = 0$, and $\mathcal{R}e(c_{s8}^6) = 0$. $\mathcal{I}m(c_{s\ell}^m) = 0$ when $m \neq 2k$ or ℓ is even or $\ell = 5$. Also, $\mathcal{I}m(c_{s\ell}^m) = 0$ when $\ell = 4k + 3$ and $m = 4k_1$, where k_1 is arbitrary non-negative integer.
- O (chiral octahedron): $\mathcal{R}e(c_{s\ell}^m) = 0$ when $m \neq 4k$ or $\ell = 1, 2, 3, 5, 7$ and 11. $\mathcal{I}m(c_{s\ell}^m) = 0$ when $m \neq 4k$ or $\ell \neq 4k + 1$ or $\ell = 5$.
- I (chiral icosahedron): $\mathcal{R}e(c_{s\ell}^m) = 0$ when $m \neq 2k$ or $\ell = 1, 2, 3, 4, 5, 7, 8, 9, 11, 13, 14, 17, 19, 23$ or 29. $\mathcal{I}m(c_{s\ell}^m) = 0$ always.

For the groups with circular symmetry, the vanishing moments are

- $C_{\infty v}$ (conic): $c_{s\ell}^m = 0$ when $m \neq 0$.
- $D_{\infty h}$ (cylinder): $c_{s\ell}^m = 0$ when $m \neq 0$ or ℓ is odd.
- K (sphere): $c_{s\ell}^m = 0$ when $m \neq 0$ or $\ell \neq 0$.

As we can see, the number of the vanishing moments may be very high: for instance, spherically symmetric bodies have only c_{s0}^0 non-zero. This is a very important observation – if we want to recognize objects with certain symmetry, we should avoid invariants containing vanishing moments.

4.7.3 Dependencies among the invariants due to symmetry

The moments vanishing due to symmetry induce new dependencies even among the invariants that do not vanish. The invariants, which are generally independent, may be for symmetric objects simplified thanks to some vanishing terms such that their simplified forms become mutually dependent.

Circular symmetry may serve as a typical example. If we take a body with the conic symmetry $C_{\infty v}$, its 3D complex moments $c_{s\ell}^m = 0$ for $m \neq 0$. If we substitute them into the formulas for Φ_1, Φ_2, and Φ_3, we obtain the simplified forms

$$\Phi_1 = c_{20}^0, \quad \Phi_2 = \frac{1}{\sqrt{5}}\left(c_{22}^0\right)^2, \quad \Phi_3 = -\sqrt{\frac{2}{35}}\left(c_{22}^0\right)^3.$$

We can immediately observe the polynomial dependency $\Phi_3^2 = \frac{2\sqrt{5}}{7}\Phi_2^3$. Hence, it is sufficient to use either Φ_2 or Φ_3 but it is useless to use both. Similarly, for the invariants of the third order $\Phi_4, \ldots, \Phi_{13}$ we get

$$\Phi_4 = -\frac{1}{\sqrt{7}}\left(c_{33}^0\right)^2, \qquad \Phi_5 = -\frac{1}{\sqrt{3}}\left(c_{31}^0\right)^2, \qquad \Phi_6 = \frac{2}{\sqrt{105}}c_{22}^0\left(c_{33}^0\right)^2,$$

$$\Phi_7 = -\frac{3}{\sqrt{105}}c_{22}^0 c_{31}^0 c_{33}^0, \qquad \Phi_8 = \sqrt{2}c_{22}^0\left(c_{31}^0\right)^2, \qquad \Phi_9 = -\frac{2}{7}\sqrt{\frac{2}{15}}\left(c_{22}^0\right)^2\left(c_{33}^0\right)^2,$$

$$\Phi_{10} = \frac{4}{21\sqrt{5}}\left(c_{33}^0\right)^4, \qquad \Phi_{11} = -\frac{2}{7\sqrt{5}}c_{31}^0\left(c_{33}^0\right)^3, \qquad \Phi_{12} = \frac{2}{3}\sqrt{\frac{2}{35}}\left(c_{31}^0\right)^2\left(c_{33}^0\right)^2,$$

$$\Phi_{13} = -\sqrt{\frac{2}{35}}\left(c_{31}^0\right)^2 c_{33}^0.$$

Only two of the invariants $\Phi_4 - \Phi_{13}$ are independent. If we simplified in the same way the fourth-order invariants, we would realize that only three of them are independent. A similar situation appears in case of all symmetry groups, although the number of vanishing moments may not be so high, and searching for polynomial dependencies is computationally more demanding. For instance, for the two fold rotation symmetry C_2 we identified the following dependencies among the third-order invariants

$$\Phi_6\Phi_5^2 - 2\sqrt{5}\Phi_{12}\Phi_8 + \frac{\sqrt{70}}{3}\Phi_{13}\Phi_7 = 0,$$

$$\frac{12}{5\sqrt{5}}\Phi_{12}\Phi_{10} - \Phi_{12}\Phi_4^2 + \frac{1}{5}\sqrt{\frac{63}{10}}\Phi_{11}^2 = 0,$$

$$\Phi_{13}^2 - \frac{9}{\sqrt{70}}\Phi_{12}\Phi_5^2 = 0.$$

So, we have two kinds of the polynomial dependencies among the invariants. The dependencies of the first kind are the general polynomial dependencies that equal zero for all bodies. The second kind has been induced by symmetry and applies to the objects with the respective symmetry only. Knowledge of the dependencies of both kinds is important for recognition because dependent invariants only increase the dimensionality but do not contribute to the discrimination power of the system.

4.8   Invariants of 3D vector fields

Rotation invariants of vector fields are in 3D even more desirable than in 2D because the fluid flow data (real as well as simulated) are intrinsically three-dimensional. Surprisingly, only very little work has been done in this area. The reason is probably because the generalization of the 2D vector field invariants is difficult. Langbein and Hagen [44] employed tensor algebra and proposed a general framework for designing moment invariants of 3D vector fields and even of tensor fields of arbitrary dimension. However, they neither provided explicit formulas for the invariants nor demonstrated their usage by experiments, which makes the applicability of their work very limited.

Respectable work was done by Bujack et al. [45, 46], who adopted a normalization approach (similar to that one described earlier in this chapter for scalar 3D images) and formulated the normalization constraints to inner, outer, and total rotations in terms of geometric moments. The weak link of their approach is, similar to all normalization techniques, the potential sensitivity to errors in the constraint calculation and a possible instability of the normalized position. Unfortunately, these stability issues have not been studied and experimentally tested yet.

4.9   Numerical experiments

To demonstrate the performance and numerical properties of the invariants, we present several instructive experiments. Before describing the experiments themselves, let us mention certain important implementation details and tricks.

4.9.1   Implementation details

In the experiments, we work not only with volumetric moments (geometric as well as complex ones), but in some experiments we also used the *surface* moments [47] for comparison. They are defined as

$$m_{pqr}^{(S)} = \iint\limits_{S} x^p y^q z^r \; dS, \tag{4.60}$$

where S is the surface of the object. The surface moments can be interpreted as volume moments of the hollow body, the surface of which is of infinite width and each "surface pixel" has a unit mass. Under noise-free conditions, surface moments could provide better discriminability, but they are more vulnerable to an inaccurate surface detection.

For objects given by volumetric representation, we calculated the moments from the definition (of course after a proper discretization of the formulas which basically consists of using some of the traditional methods of numerical integration)[15]. If the objects are given by their triangulated surfaces, we employed the efficient algorithm [48], which allows us to calculate both surface and volume moments directly from the vertices of the triangulation. The details are in Chapter 8.

---

[15] Several techniques to speed up the evaluation of the definition integral have been proposed, such as the Green-Gauss-Ostrogradski Theorem and various object decomposition schemes. These methods are discussed in Chapter 8.

Unfortunately, there is no formula for direct computation of the 3D complex moments from the triangulation. One way is to convert the triangular representation of the object into the volumetric one. This can be done by ray tracing, but unfortunately many triangulated models in public databases are imperfect, and their surface contains unwanted holes. Standard ray-tracing algorithms fail, and the procedure must detect the holes by checking the parity of the edges and must handle these situations carefully. Anyway, the conversion to volumetric representation, even if implemented correctly, is inefficient in terms of memory usage and computational time. Another possibility is to calculate geometric moments first and then to calculate complex moments as functions of the geometric ones. To do so, we have to express the corresponding spherical harmonics in Cartesian coordinates as a polynomial

$$(x^2 + y^2 + z^2)^{s/2} Y_\ell^m(x, y, z) = \sum_{\substack{k_x, k_y, k_z = 0 \\ k_x + k_y + k_z = s}}^{s} a_{k_x k_y k_z} x^{k_x} y^{k_y} z^{k_z}. \tag{4.61}$$

As soon as we have obtained the coefficients $a_{k_x k_y k_z}$, we can compute the complex moments as

$$c_{s\ell}^m = \sum_{\substack{k_x, k_y, k_z = 0 \\ k_x + k_y + k_z = s}}^{s} a_{k_x k_y k_z} m_{k_x k_y k_z}. \tag{4.62}$$

This algorithm works for both volume as well as surface moments.

The last implementation "trick", which does not follow from the theory of the invariants but may influence the classification rate substantially, is a proper scaling of the moment values.

The values of the moments often rapidly decrease (if the moments have been normalized to image scaling) or increase as their order increases. To keep them in a reasonable interval, we apply so-called *magnitude normalization to the order*. Assuming the moments already have been normalized to image scaling, we use the following magnitude normalization

$$\widehat{\mu_{pqr}} = \frac{(p + q + r + 3)\pi^{\frac{p+q+r}{6}}}{\left(\frac{3}{2}\right)^{\frac{p+q+r}{3}+1}} \mu_{pqr} \qquad \widehat{c_{s\ell}^m} = \frac{(s + 3)\pi^{\frac{s}{6}}}{\left(\frac{3}{2}\right)^{\frac{s}{3}+1}} c_{s\ell}^m \tag{4.63}$$

for volume moments and

$$\widehat{\mu_{pqr}} = \pi^{\frac{p+q+r}{4}} 2^{\frac{p+q+r}{2}} \mu_{pqr} \qquad \widehat{c_{s\ell}^m} = \pi^{\frac{s}{4}} 2^{\frac{s}{2}} c_{s\ell}^m \tag{4.64}$$

for surface moments. After the magnitude normalization, we substitute the normalized moments to the formulas of the invariants as usually. Sometimes the magnitude normalization to the order is not sufficient, and so-called *magnitude normalization to the degree* or *magnitude normalization to the weight* is necessary

$$\hat{\Phi}_i = \text{sign}\Phi_i \sqrt[g]{|\Phi_i|}, \tag{4.65}$$

where g is either the degree or the weight of the invariant Φ_i. The degree equals the number of moments contained in a single term, while the weight equals the sum of the indices in a single term divided by the dimension of the space d.

The magnitude normalization is of particular importance when we use the invariant vector as an input to NN or k-NN classifier with Euclidean distance. Without this normalization, the invariants for which the range of values is low, would actually have less influence on the classification result than the invariants with high dynamic range.

4.9.2   Experiment with archeological findings

The first simple experiment on real data verifies that the irreducible invariants from 3D geometric moments derived by the tensor method are actually invariant to rotation, that is, they were derived correctly. Moreover, it demonstrates that two similar but different objects have different values of (at least some) invariants.

The experiment was carried out on real data. We used two ancient Greek amphoras scanned using a laser rangefinder from various sides[16]. Consequently all measurements were combined to obtain a 3D binary image of the amphora. Since the rangefinder cannot penetrate inside the amphora, it is considered filled up and closed on the top. The surface of the amphora was divided into small triangles (42,400 and 23,738 triangles, respectively). The amphoras were then represented only by their triangulated surfaces. The test data are shown in Figure 4.8. The photographs are for illustration only; no graylevel/color information was used in moment calculation.

For each amphora we generated ten random rotations and translations of its triangular representation. A more realistic scenario would be to rotate the amphora physically in the capturing device and scan it again in each position. We used this data acquisition scheme in one of the following experiments, but here we want to avoid any unpredictable scanning errors. Then we

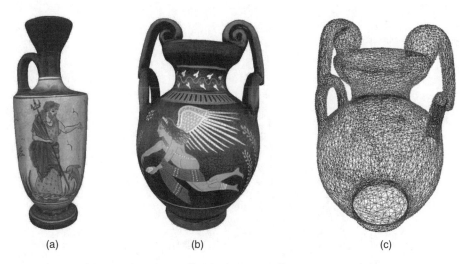

(a)   (b)   (c)

Figure 4.8   Ancient Greek amphoras: (a) photo of A1, (b) photo of A2, (c) wire model of the triangulation of A2

---

[16] Thanks to Clepsydra (ΚΛΕΨΥΔΡΑ), The Digitization Center of Cultural Heritage in Xanthi, Greece, and especially to professor Christodoulos Chamzas for providing the models of the amphoras and other objects used in this section.

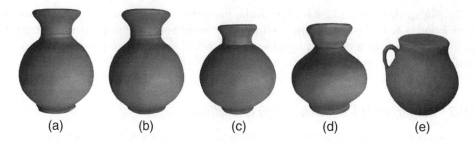

(a)   (b)   (c)   (d)   (e)

Figure 4.9   The archeological findings used for the experiment

calculated the values of the first 242 invariants up to the fourth order. We tested both volume moments and surface moments.

The maximum relative standard deviation was $3.2 \cdot 10^{-13}$ in the case of the invariants computed from the volume moments and $4.5 \cdot 10^{-13}$ for the invariants from the surface moments, which illustrates a perfect invariance in all cases.

The second experiment was performed on similar data and demonstrates the ability of the invariants to distinguish different objects. We chose five real artifacts from ancient Greece (see Figure 4.9) and scanned them as described above, which led to object representation by their triangulated surfaces. We did not introduce any artificial rotations. We used thirty-seven invariants Φ_1 , ... , Φ_{37} in this experiment.

The absolute values of the invariants up to the 4th order of the objects can be seen in Figure 4.10. The objects in Figures 4.9a and 4.9b are two scans of the same vase, which was physically rotated between the scans. Their feature values differ from each other only slightly.

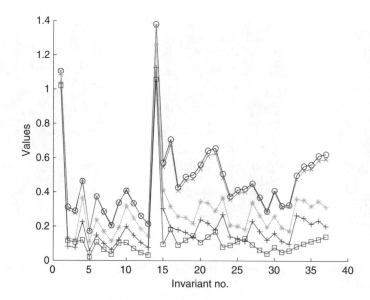

Figure 4.10   The absolute values of the invariants in the experiment with the archeological findings from Figure 4.9. Legend: (a) ○, (b) ×, (c) +, (d) ∗ and (e) □

Theoretically they should be the same, but the range finder produced some errors. The objects in Figures 4.9c and 4.9d are two different objects of the same kind. We can see that the differences between them are bigger than those between Figures 4.9a and 4.9b. The object in Figure 4.9e differs by a handle, but it is still similar. The feature values are further more different. The graphs demonstrate not only the rotation invariance of the features (in the case of the vase which was scanned twice) but also their ability to distinguish similar but different objects.

An interesting question is what are the actual values of those invariants which should theoretically vanish due to the object symmetries. The objects in Figures 4.9a–d have approximate circular symmetry $C_{\infty v}$, and the object in Figure 4.9e has reflection symmetry C_{1v}. The invariants which should be zero due to the circular symmetry are in Table 4.3. We can see that the values of the objects 4.9a–d are really very small; nevertheless it is true for 4.9e, too. This is because the violation of the circular symmetry by the handle is small.

The values of the invariants which should vanish due to the two fold rotation symmetry are in Table 4.4. This experiment shows that also in the case of real objects, where the symmetry is not exact due to measurement errors and object imperfection, the features which should theoretically vanish actually get very low values close to zero, providing no discrimination power. So, one has to analyze the object symmetry and discard the vanishing invariants from the feature vector.

4.9.3   Recognition of generic classes

In this experiment, we used the Princeton Shape Benchmark [1], which was at the time of the preparation of this book probably the most widely used dataset of 3D objects. The database

Table 4.3   The actual values of the invariants that should vanish due to the circular symmetry of the vases

Invariant	(a)	(b)	(c)	(d)	(e)
Φ_3^{∞}	−0.113	−0.097	−0.035	−0.073	−0.057
Φ_6^{∞}	−0.123	−0.129	−0.074	−0.110	−0.063
Φ_7^{∞}	−0.085	−0.084	−0.048	−0.073	−0.039
Φ_{17}^{∞}	−0.103	−0.088	−0.034	−0.092	0.054
Φ_{19}^{∞}	−0.101	−0.085	−0.030	−0.073	−0.044

Table 4.4   The actual values of the invariants that should vanish due to the two fold rotation symmetry of the vases

Invariant	(a)	(b)	(c)	(d)	(e)
Φ_{12}^{C2}	−0.114	−0.146	−0.053	−0.125	−0.046
Φ_{13}^{C2}	0.056	0.074	−0.038	−0.069	−0.016
Φ_6^{C2}	−0.058	0.090	−0.044	0.078	−0.024
Φ_{27}^{C2}	0.080	−0.116	0.059	−0.069	−0.031
Φ_{29}^{C2}	−0.054	−0.102	−0.053	−0.080	−0.030
Φ_{25}^{C2}	−0.063	−0.063	0.040	0.064	−0.035

contains 1814 3D binary objects given by triangulated surfaces. The experiment was performed on generic objects that are easy to recognize by humans, but their big intra-class variability makes the task very difficult for computers. The invariants have not been designed for recognition of generic objects. Classification to generic classes (such as "car", "bike", "dog", etc.) requires handling the intra-class variations properly. It is easy for humans because they incorporate their life-time experience, prior knowledge, and imagination. For automatic systems this task is very difficult and usually requires a high level of abstraction, structural description, and advanced classification techniques. Here we tried to resolve this task for a limited number of classes by the same invariants as in the previous experiment.

We have chosen six generic classes: sword, two-story home, dining chair, handgun, ship, and car. The classes were represented by their training sets consisting of 15 swords, 11 houses, 11 chairs, 10 handguns, 10 ships and 10 cars (see Figure 4.11 for one template per class). The independent test sets contained 16 swords, 10 houses, 11 chairs, 10 handguns, 11 ships, and 10 cars. The test objects were not incorporated into the training set.

We used 49 independent invariants up to the fifth order in this experiment. We compared the invariants from the 3D geometric moments derived by the tensor method, the invariants from the 3D complex moments, and the 3D normalized moments based on the covariance matrix, the matrix of inertia, and the third-order moments. All the sets of the invariants were computed both from surface moments and from the volume moments. The results are summarized

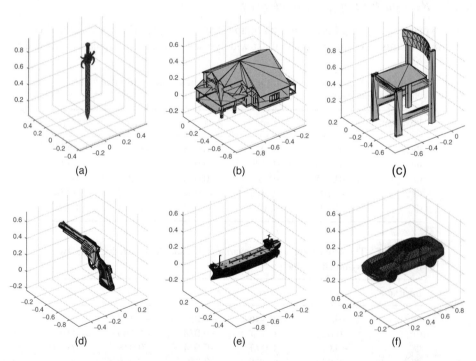

Figure 4.11 Examples of the class representatives from the Princeton Shape Benchmark: (a) sword, (b) two-story home, (c) dining chair, (d) handgun, (e) ship, and (f) car

Table 4.5 The success rates of recognition of the generic classes

	Surface moments	Volume moments
Geometric m. inv.	82%	71%
Complex m. inv.	82%	63%
Normalized m. $\mathbf{M_C}$	74%	75%
Normalized m. $\mathbf{M_I}$	72%	72%
Normalized m. c_{31}^1	74%	63%

in Table 4.5. The success rates are generally higher than one would expect. It shows that the invariants capture certain global characteristics that are common for all representatives of a generic class. This property does not have any theoretical justification – generic classes are defined by means of other properties than those measurable by moments. For other classes and/or other training/test sets, the success rates could be different. The surface moment invariants yielded better results than the volume moments, namely because the objects were noise free and the surface was given without errors.

4.9.4 Submarine recognition – robustness to noise test

The aim of this experiment was to compare behavior of the various types of the invariants on the classes that are similar to one another and that are even difficult to distinguish visually. We worked with six classes of submarines from the Princeton Shape Benchmark. Each class was represented by only one template submarine, see Figure 4.12. To create test submarines, we rotated each template ten times randomly and added a zero-mean Gaussian noise to the coordinates of the triangle vertices of the surface. This resulted in rotated submarines with noisy surfaces, see Figure 4.13 for two examples of those "shaky" submarines.

To measure the impact of the noise, we used signal-to-noise ratio (SNR), defined in this case as

$$\text{SNR} = 10\log(\sigma_{sa}^2/\sigma_{na}^2), \tag{4.66}$$

where σ_{sa} is the standard deviation of the object coordinates in the corresponding axis, that is, σ_{sx} in the x-axis, σ_{sy} in the y-axis, and σ_{sz} in the z-axis, analogously σ_{na} is the standard deviation of the noise in the a-axis, $a \in \{x, y, z\}$. The σ_{na}'s are set to equalize SNR for every axis in the experiment. We used ten SNRs from 3.4 dB to 55 dB. On each level, the noise was generated twice and the average success rate was calculated.

First, we computed the volume moments from the triangulated surfaces of the objects. We compared 100 irreducible invariants up to the fifth order from the 3D complex moments, 49 irreducible (assumed to be independent) invariants up to the fifth order derived by the tensor method and 49 normalized moments up to the fifth order. The normalized moments were obtained by diagonalization of the object covariance matrix.

To compare the results achieved by means of moments to a different method, we also computed a spherical harmonic representation (SHR) from [15]. The method first decomposes

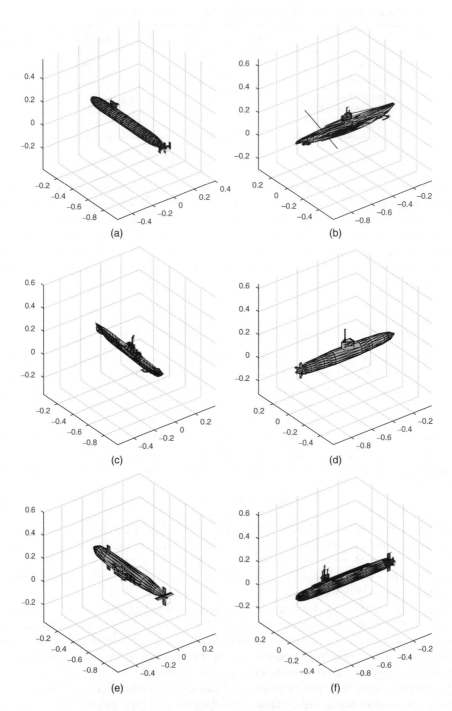

Figure 4.12   Six submarine classes used in the experiment

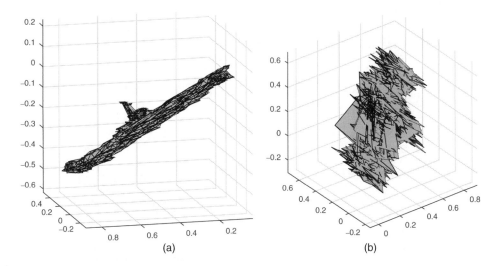

Figure 4.13   Example of the rotated and noisy submarine from Figure 4.12c with SNR (a) 26 dB and (b) 3.4 dB

an object into concentric spherical layers of thickness Δr and then computes 3D complex moments of the individual layers[17]

$$a_{r\ell m} = \int\limits_{0}^{2\pi} \int\limits_{0}^{\pi} \int\limits_{r-\Delta r/2}^{r-\Delta r/2} (Y_\ell^m(\theta, \varphi))^* f(\varrho, \theta, \varphi) \quad \sin\theta \; d\varrho \; d\theta \; d\varphi. \tag{4.67}$$

The rotation invariance is provided by summation over the orders of the spherical harmonics

$$f_{r\ell}(\theta, \varphi) = \sum_{m=-\ell}^{\ell} Y_\ell^m(\theta, \varphi) a_{r\ell m}. \tag{4.68}$$

At the end, the amplitudes of frequency components of all orders form the rotation invariants

$$SHR_{r\ell} = \sqrt{\int\limits_{0}^{2\pi} \int\limits_{0}^{\pi} f_{r\ell}^2(\theta, \varphi) r^2 \sin\theta \; d\theta \; d\varphi}. \tag{4.69}$$

Since this method requires volumetric data, we converted the objects to the volumetric representation with resolution 200 voxels in the directions of their maximum sizes. We used 48 layers approximately 2 voxels thick. Spherical harmonics from the zeroth to the fifth orders were used, i.e. we had 6 features in each layer and 288 features altogether.

We used the Euclidean nearest-neighbor classifier for submarine classification. The obtained success rates are visualized in Figure 4.14. All the methods except the surface geometric

---

[17] The factor ϱ^2 from the Jacobian of the spherical coordinates is canceled by normalization to the layer area.

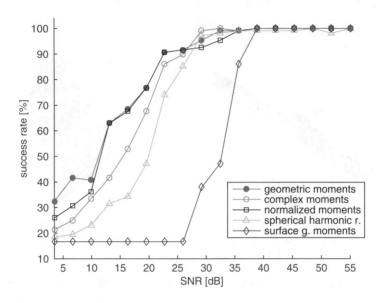

Figure 4.14   The success rate of the invariants from the volume geometric, complex, and normalized moments compared with the spherical harmonic representation and with the surface geometric moment invariants

moments yielded comparable results. The slightly worse results of SHR can be explained by sampling errors when working with such thin spherical layers in a voxel raster.

The results become dramatically worse, when we use surface moments instead of the volume moments. We computed the forty-nine invariants up to the fifth order derived by the tensor method from the surface geometric moments. The surface moments are much less robust to surface noise; see Figure 4.14. It does not mean the surface moments cannot be successfully used in applications with less noise, but it illustrates their high sensitivity to a noise in the coordinates and to inaccurate surface detection.

The heaviest noise used in this experiment really devastated all features of the submarines. An example of the submarine distorted by the noise 26 dB, where the surface moments failed completely but volume moments performed with 90% success, is in Figure 4.13a. The visual recognition is difficult but possible for a non-trained person in this case. An example of the heaviest noise 3.44 dB is in Figure 4.13b. In this case, not only all tested methods, but also the people failed in the attempts to recognize the submarines; the results converge to a random guess.

If we want to compare the computing complexity, we should consider the numbers of triangles of the individual models. Theoretically, the computing complexity is $\mathcal{O}(NS!)$ for moments and $\mathcal{O}(N_1^3 \ell^2)$ for the spherical harmonic representation, where N is the number of triangles, S is the maximum moment order, N_1^3 is the number of voxels in the 3D image, and ℓ is the maximum degree of the spherical harmonics. The computing complexity of the conversion from the triangulation to the volumetric data is $\mathcal{O}(NN_1^2)$; that is, it increases linearly with the number of triangles.

In our experiment, the individual models have 1236, 5534, 10,583, 1076, 684, and 3176 triangles, respectively, that is, 3714.8 triangles on average. We measured the total time and divided it by the number of models. The average times for recognition of one query object are 3.0 s for geometric models, 3.8 s for complex moments, 5.5 s for normalized moments, and 5.5 s for spherical harmonic representation (the last method require additional 18.7 s for the conversion of the object to the volumetric representation). It means the times of all the methods are comparable. The computation was performed in Matlab on a computer with the processor Intel Core i7-2600K CPU @ 3.40GHz, 16 GB memory and 64-bit operating system Windows 7 Professional.

4.9.5 Teddy bears – the experiment on real data

This experiment was done with real 3D objects and their physical rotations. We scanned a teddy bear by means of the Kinect device and created its 3D model. Then we repeated this process five times with different orientations of the teddy bear in the space. Hence, we obtained six scans differing from each other by rotation and also slightly by scale, quality of details, and perhaps by some random errors. For a comparison we also scanned another teddy bear, similar to the first one at the first sight but actually of different shape (see Figure 4.15). When using Kinect, one has to scan the object from several views, and Kinect software then produces a triangulated surface of the object, which is basically the same representation as is employed in the PSB database. In order to demonstrate that the invariants can be calculated not only from the surface representation but also from the volumetric representation, we converted each teddy bear figure into 3D volumetric representation of the size approximately $100 \times 100 \times 100$ voxels. Then we calculated the volumetric geometric moments of each scan, converted them to the complex moments by (4.62) and computed the invariants $\Phi_1, \ldots, \Phi_{100}$ from them. In Figure 4.16 you can see the values of the invariants Φ_2, Φ_{12}, Φ_{15}. The values of of the first teddy bear almost do not depend on the particular rotation (the standard deviation over the six rotations is 0.019 for Φ_2, 0.019 for Φ_{12}, and 0.017 for Φ_{15}). On the other hand, they are significantly different from the values of the second teddy bear, which demonstrates both desirable properties – invariance and discrimination power (note that the second teddy bear was scanned only once).

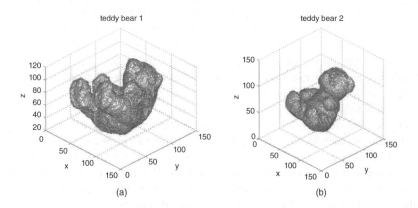

Figure 4.15 3D models of two teddy bears created by Kinect

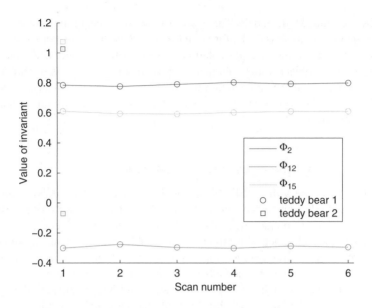

Figure 4.16   The values of the invariants of two teddy bears

4.9.6   Artificial symmetric bodies

The aim of this experiment is to show the behavior of the rotation invariants from 3D complex moments on the objects with a precise symmetry. It compares the ideal computer-generated bodies in a surface representation with the volumetric data corrupted by a sampling error.

We synthetically generated objects of various kinds and degrees of symmetry. The object in Figure 4.17a is asymmetric, and the one in Figure 4.17b has reflection symmetry only. The rectangular pyramid in Figure 4.17c has the symmetry group C_{2v}, the pyramid and the prism in Figures 4.17d–e have 3-fold rotation symmetry. Unlike the regular tetrahedron, only one side of the triangular pyramid is an equilateral triangle; the other three are isosceles triangles. We also used the bodies with the symmetry group O_h: a cube and an octahedron (Figures 4.6k–l) and the bodies with circular symmetry (Figures 4.7i–k). The circular symmetry is in the discrete space only approximative. We used twenty-eight irreducible invariants $\Phi_1, \dots, \Phi_{17}, \Phi_{19}, \dots, \Phi_{23}$ and $\Phi_{32}, \dots, \Phi_{37}$ computed from the triangulated surfaces of the bodies.

For each object we generated 1000 random rotations; then we calculated the values of the twenty-eight irreducible invariants for each object and each rotation. The mean values of the invariants are in Table 4.6.

The only sources of errors are numerical inaccuracies of the computer arithmetic (in case of the cone, the cylinder, and the sphere, we also encounter an error caused by approximation by a finite number of triangles), therefore we can consider the values in Table 4.6 precise. We can see they correspond to the theoretical assumptions. The invariants $\Phi_4, \dots, \Phi_{13}$ and Φ_{32} of the cylinder, the cube and the octahedron are actually zero, exactly as predicted by the theory. The values 1 of Φ_1 and Φ_{14} of the sphere are not accidental; the invariants were normalized such that the values of c_{s0}^0 of the sphere should be one, see (4.63).

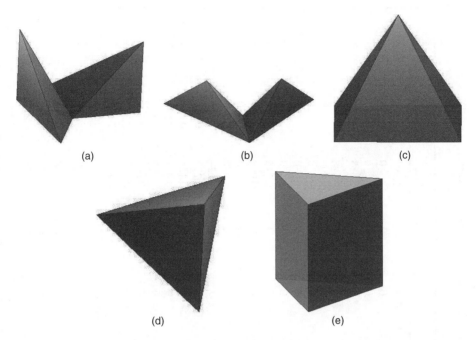

Figure 4.17 The bodies used in the experiment: (a) two connected tetrahedrons with the symmetry group C_1, (b) the body with the reflection symmetry C_{1v}, (c) the rectangular pyramid with the symmetry group C_{2v}, (d) the triangular pyramid with the symmetry group C_{3v} and (e) the triangular prism with the symmetry group D_{3h}

The maximum standard deviations within one class (one body) are in the second row of Table 4.7; they illustrate the perfect invariance of the features. The standard deviation is calculated over the rotations, while the maximum is calculated over the invariants. The slightly higher standard deviation $4 \cdot 10^{-5}$ in the case of O_h is caused by the normalization to the degree, see (4.65). Without this normalization, the standard deviation is $2 \cdot 10^{-14}$.

Then we repeated this experiment with the same bodies in volumetric representation, which is less accurate due to the sampling error. Each body at each attitude was converted into the volumetric form in the cube $100 \times 100 \times 100$ voxels, which is a relatively coarse resolution making it possible to study the impact of discretization. The examples of the volumetric data are in Figure 4.18, the standard deviations caused by the sampling are in the third row of Table 4.7. Although they are by several orders higher than in the case of the ideal bodies, they are still small enough and illustrate the close-to-perfect invariance property achieved on the sampled bodies.

4.9.7 Symmetric objects from the Princeton Shape Benchmark

This experiment was again performed on objects from the Princeton Shape Benchmark. Its main mission is to illustrate that it is really important to use non-vanishing invariants. We demonstrate that the invariants which are for the given objects zero or close to zero are useless for object recognition.

Table 4.6 The mean values of the invariants in the experiment with the artificial objects. In the first row, there is the corresponding symmetry group, "c" denotes the cube and "o" the octahedron

Inv.	C_1	C_{1v}	C_{2v}	C_{3v}	D_{3h}	O_h (c)	O_h (o)	$C_{\infty v}$	$D_{\infty h}$	K
Φ_1	2.882	3.136	1.430	1.377	1.261	1.083	1.073	1.191	1.178	1.000
Φ_2	2.164	2.500	0.707	0.294	0.471	0	0	0	0.503	0
Φ_3	−1.865	−1.593	0.534	−0.273	−0.438	0	0	0	0.467	0
Φ_4	−4.351	−5.041	−1.252	−1.381	−0.479	0	0	−0.813	0	0
Φ_5	−1.240	−1.312	−0.069	−0.159	0	0	0	0	0	0
Φ_6	2.943	3.678	0.935	0.554	−0.471	0	0	0.003	0	0
Φ_7	−1.976	−2.140	0.188	−0.373	0	0	0	0	0	0
Φ_8	1.125	−0.765	0.084	0.192	0	0	0	0	0	0
Φ_9	−1.956	2.014	0.851	−0.447	0.445	0	0	0	0	0
Φ_{10}	3.999	4.803	1.190	0.760	0.470	0	0	0.715	0	0
Φ_{11}	−2.366	−2.478	−0.478	−0.565	0	0	0	−0.009	0	0
Φ_{12}	1.631	1.892	0.260	0.343	0	0	0	0	0	0
Φ_{13}	−1.392	1.585	−0.135	−0.255	0	0	0	0	0	0
Φ_{14}	11.006	13.214	2.556	2.454	1.804	1.247	1.227	1.627	1.470	1.000
Φ_{15}	6.078	8.685	0.681	1.090	0.347	0.521	0.609	0.273	0.074	0
Φ_{16}	8.562	10.815	1.402	0.822	0.749	0	0	0.126	0.683	0
Φ_{17}	−3.365	−2.540	0.490	−0.437	−0.351	0	0	−0.001	0.128	0
Φ_{19}	−4.606	−3.849	0.923	−0.542	−0.596	0	0	0	0.572	0
Φ_{20}	−5.276	−3.871	0.581	−0.616	−0.410	0	0	−0.192	0.142	0
Φ_{21}	6.484	8.348	0.934	0.782	−0.565	0	0	0.159	0.317	0
Φ_{22}	−7.248	−5.896	1.197	−0.763	−0.695	0	0	−0.117	0.634	0
Φ_{23}	−2.635	5.327	0.601	−0.934	−0.308	−0.489	0.572	0.242	0.066	0
Φ_{32}	−2.718	3.504	0.277	−0.337	0	0	0	0	0	0
Φ_{33}	5.060	5.509	0.543	0.533	0.303	0	0	0.238	0.065	0
Φ_{34}	−4.410	5.311	0.413	−0.638	0.386	0	0	−0.206	0.118	0
Φ_{35}	5.211	7.290	−0.539	0.626	0.450	0	0	0.164	0.199	0
Φ_{36}	6.111	8.103	0.784	0.763	0.491	0	0	0.178	0.216	0
Φ_{37}	−6.343	6.979	0.976	−0.749	0.573	0	0	−0.141	0.363	0

Table 4.7 The standard deviations of the invariants in the experiment with the artificial objects. In the first row, there is the corresponding symmetry group, "c" means cube, and "o" means octahedron. In the second row, there are the values of the ideal triangulated bodies, and in the third row the values of the sampled volumetric data

C_1	C_{1v}	C_{2v}	C_{3v}	D_{3h}	O_h (c)	O_h (o)	$C_{\infty v}$	$D_{\infty h}$	K
$1 \cdot 10^{-13}$	$2 \cdot 10^{-13}$	$3 \cdot 10^{-14}$	$3 \cdot 10^{-14}$	$2 \cdot 10^{-8}$	$4 \cdot 10^{-5}$	$4 \cdot 10^{-5}$	$2 \cdot 10^{-10}$	$2 \cdot 10^{-8}$	$7 \cdot 10^{-11}$
0.3	0.3	0.3	0.009	0.02	0.03	0.04	0.1	0.02	0.0003

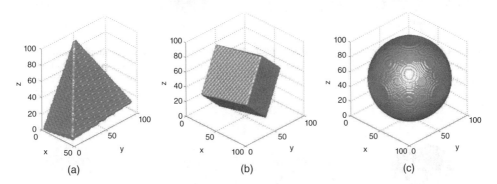

Figure 4.18 Randomly rotated sampled bodies: (a) the rectangular pyramid, (b) the cube and (c) the sphere

We intentionally chose eight objects that have a high degree of symmetry; see Figure 4.19. In the PSB, they are labeled as 1338, 1337, 760, 1343, 1344, 741, 527, and 1342, respectively. In a continuous representation, they should have the $C_{\infty v}$ symmetry. This is, however, not exactly true in the actual triangular representation. The symmetry group of the gear No. 741 is C_{36v}, that of the hot air balloons No. 1338, and 1337 is C_{20v}, the balloon No. 1343 and the vase No. 527 have C_{16v}, and the ice cream No. 760 has only C_{8v}. The symmetry group of the hot air balloon No. 1344 is C_{24v}, but its gondola has only C_{4v}. The balloon No. 1342 is not fully inflated, so it is not symmetric at all.

In our experiment we did not use the texture and worked with the shape only, hence our templates (class representatives) looked as depicted in Figure 4.20. To generate test objects, we added noise to the coordinates of each vertex of the templates. The amount of the noise measured in terms of signal-to-noise ratio (SNR) was 25 dB, where SNR is here defined as

$$\text{SNR} = 10 \log_{10} \frac{\sigma_o^2}{\sigma_n^2}, \tag{4.70}$$

with σ_o being the standard deviation of the object coordinates and σ_n the standard deviation of the noise. In other words, the noise corresponds to about 5% of the object coordinate extent in each direction. We generated nine noisy instances of each template, and each of these noisy objects was randomly rotated (see Figure 4.21 for some examples).

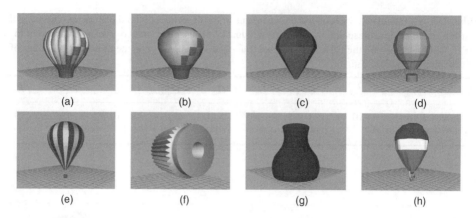

Figure 4.19 The images of the symmetric objects downloaded from the PSB: (a) hot air balloon 1338, (b) hot air balloon 1337, (c) ice cream 760, (d) hot air balloon 1343, (e) hot air balloon 1344, (f) gear 741, (g) vase 527, (h) hot air balloon 1342

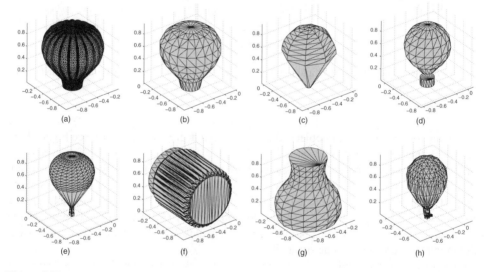

Figure 4.20 The objects without texture and color, used as the database templates: (a) hot air balloon 1338, (b) hot air balloon 1337, (c) ice cream 760, (d) hot air balloon 1343, (e) hot air balloon 1344, (f) gear 741, (g) vase 527, (h) hot air balloon 1342

The test objects were classified by invariants using the minimum-distance rule and majority vote: the minimum distance was found for each individual invariant, and the final class assignment was made by majority vote.

First, we used the five invariants that are identically zero in the case of circular symmetry: Φ_3^∞, Φ_6^∞, Φ_7^∞, Φ_{17}^∞, and Φ_{19}^∞. Since the objects do not have exact circular symmetry, these invariants are able to recognize some test objects. The result was twelve errors out of seventy-two

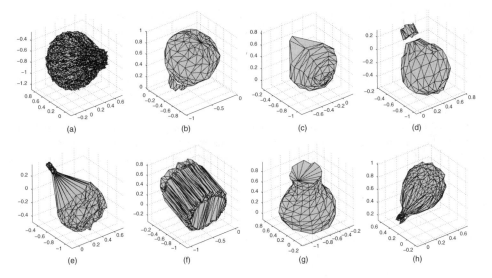

Figure 4.21 Examples of the noisy and rotated query objects: (a) hot air balloon 1338, (b) hot air balloon 1337, (c) ice cream 760, (d) hot air balloon 1343, (e) hot air balloon 1344, (f) gear 741, (g) vase 527, (h) hot air balloon 1342

trials; that is, the success rate was 83%. When we added Φ_{14}, which is non-zero, to the invariant vector, the number of errors decreased to six; that is, the success rate increased to 92%. Then we removed all vanishing invariants and recognized solely by Φ_{14}. The result was the same as before – six misclassifications, but the algorithm ran much faster. On the other hand, when we use the vector of non-vanishing invariants (Φ_{14}, Φ_{10}, Φ_6) or (Φ_{14}, Φ_{10}, Φ_4) we end up with only three misclassifications. This clearly illustrates that the vanishing invariants are useless even if they are not exactly zero and should not be included into the feature vector.

4.10   Conclusion

This chapter was devoted to the invariants of 3D objects to translation, rotation, and scaling. We presented three approaches to this task. Each of them has its pros and cons.

The tensor method generates the invariants from 3D geometric moments. We explicitly derived 1185 irreducible invariants up to the sixteenth order, but the set is complete to the eighth order only. The advantage of this method is that the coefficients of individual terms are integers, which enables us to keep their precision even for high orders. Our method includes the elimination of all reducible invariants, but the subsequent step from irreducible to independent invariants has not been fully resolved yet.

The 3D rotation invariants can also be constructed from 3D complex moments, but the generalization from 2D is rather complicated. We have tensor products of composite complex moment forms here instead of simple products of scalar complex moments. While the dependence/independence in 2D can be seen directly, in 3D, we have to perform complicated tests of reducibility/irreducibility similar to that required in the tensor method. The coefficients are real-valued; they are often given as square roots of fractions, whose numerator is an integer and denominator is a multiple of π. It is difficult to keep their precision. We generated 100 irreducible invariants up to the fifth order with precise coefficients. The other 386 invariants from the sixth to the seventh order have the coefficients expressed as floating-point numbers with less precision. On the other hand, the clear advantage of the complex moment method is that the impact of the object symmetry can be analyzed much easier than in the case of the geometric moments.

The last method is based on image normalization. Its main advantage is the fact that it directly yields an independent set of invariants. In this way, we can avoid the time-consuming elimination of reducible invariants. The main drawback is a potential instability of the normalized position.

Although all these three approaches are theoretically equivalent and yield the same feature space, they differ from one another by numerical properties, as we explained and demonstrated experimentally. The user should make a choice with respect to the particular dataset, time constraints, and other requirements specific for the given task.

Appendix 4.A

Independent invariants of order three derived by the tensor method.

$$
\begin{aligned}
\Phi_4 = \big(&\mu_{300}^2 + \mu_{030}^2 + \mu_{003}^2 + 3\mu_{210}^2 + 3\mu_{201}^2 + 3\mu_{120}^2 + 3\mu_{102}^2 + 3\mu_{021}^2 + 3\mu_{012}^2 \\
&+ 6\mu_{111}^2 \big) / \mu_{000}^5
\end{aligned}
$$

Generating graph:

$$
\begin{aligned}
\Phi_5 = \big(&\mu_{300}^2 + 2\mu_{300}\mu_{120} + 2\mu_{300}\mu_{102} + 2\mu_{210}\mu_{030} + 2\mu_{201}\mu_{003} + \mu_{030}^2 \\
&+ 2\mu_{030}\mu_{012} + 2\mu_{021}\mu_{003} + \mu_{003}^2 + \mu_{210}^2 + 2\mu_{210}\mu_{012} + \mu_{201}^2 + 2\mu_{201}\mu_{021} \\
&+ \mu_{120}^2 + 2\mu_{120}\mu_{102} + \mu_{102}^2 + \mu_{021}^2 + \mu_{012}^2 \big) / \mu_{000}^5
\end{aligned}
$$

Generating graph:

$$
\begin{aligned}
\Phi_6 = \big(&\mu_{300}^4 + 6\mu_{300}^2\mu_{210}^2 + 6\mu_{300}^2\mu_{201}^2 + 2\mu_{300}^2\mu_{120}^2 + 4\mu_{300}^2\mu_{111}^2 + 2\mu_{300}^2\mu_{102}^2 \\
&+ 8\mu_{300}\mu_{210}^2\mu_{120} + 16\mu_{300}\mu_{210}\mu_{201}\mu_{111} + 4\mu_{300}\mu_{210}\mu_{120}\mu_{030} \\
&+ 8\mu_{300}\mu_{210}\mu_{111}\mu_{021} + 4\mu_{300}\mu_{210}\mu_{102}\mu_{012} + 8\mu_{300}\mu_{201}^2\mu_{102} \\
&+ 4\mu_{300}\mu_{201}\mu_{120}\mu_{021} + 8\mu_{300}\mu_{201}\mu_{111}\mu_{012} + 4\mu_{300}\mu_{201}\mu_{102}\mu_{003} \\
&+ 2\mu_{210}^2\mu_{030}^2 + 4\mu_{210}\mu_{201}\mu_{030}\mu_{021} + 4\mu_{210}\mu_{201}\mu_{012}\mu_{003} + 8\mu_{210}\mu_{120}^2\mu_{030} \\
&+ 8\mu_{210}\mu_{111}\mu_{102}\mu_{003} + 2\mu_{201}^2\mu_{003}^2 + 8\mu_{201}\mu_{120}\mu_{111}\mu_{030} + 8\mu_{201}\mu_{102}^2\mu_{003} \\
&+ 6\mu_{120}^2\mu_{030}^2 + 16\mu_{120}\mu_{111}\mu_{030}\mu_{021} + 8\mu_{120}\mu_{111}\mu_{012}\mu_{003} \\
&+ 4\mu_{120}\mu_{102}\mu_{030}\mu_{012} + 4\mu_{120}\mu_{102}\mu_{021}\mu_{003} + 4\mu_{111}^2\mu_{030}^2 + 4\mu_{111}^2\mu_{003}^2 \\
&+ 8\mu_{111}\mu_{102}\mu_{030}\mu_{021} + 16\mu_{111}\mu_{102}\mu_{012}\mu_{003} + 6\mu_{102}^2\mu_{003}^2 + \mu_{030}^4
\end{aligned}
$$

$$+ 6\mu_{030}^2\mu_{021}^2 + 2\mu_{030}^2\mu_{012}^2 + 8\mu_{030}\mu_{021}^2\mu_{012} + 4\mu_{030}\mu_{021}\mu_{012}\mu_{003}$$

$$+ 2\mu_{021}^2\mu_{003}^2 + 8\mu_{021}\mu_{012}^2\mu_{003} + 6\mu_{012}^2\mu_{003}^2 + \mu_{003}^4 + 5\mu_{210}^4 + 10\mu_{210}^2\mu_{201}^2$$

$$+ 16\mu_{210}^2\mu_{120}^2 + 20\mu_{210}^2\mu_{111}^2 + 4\mu_{210}^2\mu_{102}^2 + 4\mu_{210}^2\mu_{021}^2 + 2\mu_{210}^2\mu_{012}^2$$

$$+ 24\mu_{210}\mu_{201}\mu_{120}\mu_{111} + 24\mu_{210}\mu_{201}\mu_{111}\mu_{102} + 8\mu_{210}\mu_{201}\mu_{021}\mu_{012}$$

$$+ 24\mu_{210}\mu_{120}\mu_{111}\mu_{021} + 8\mu_{210}\mu_{120}\mu_{102}\mu_{012} + 16\mu_{210}\mu_{111}^2\mu_{012} + 5\mu_{201}^4$$

$$+ 4\mu_{201}^2\mu_{120}^2 + 20\mu_{201}^2\mu_{111}^2 + 16\mu_{201}^2\mu_{102}^2 + 2\mu_{201}^2\mu_{021}^2 + 4\mu_{201}^2\mu_{012}^2$$

$$+ 8\mu_{201}\mu_{120}\mu_{102}\mu_{021} + 16\mu_{201}\mu_{111}^2\mu_{021} + 24\mu_{201}\mu_{111}\mu_{102}\mu_{012} + 5\mu_{120}^4$$

$$+ 20\mu_{120}^2\mu_{111}^2 + 2\mu_{120}^2\mu_{102}^2 + 10\mu_{120}^2\mu_{021}^2 + 4\mu_{120}^2\mu_{012}^2 + 16\mu_{120}\mu_{111}^2\mu_{102}$$

$$+ 24\mu_{120}\mu_{111}\mu_{021}\mu_{012} + 20\mu_{111}^2\mu_{102}^2 + 20\mu_{111}^2\mu_{021}^2 + 20\mu_{111}^2\mu_{012}^2$$

$$+ 24\mu_{111}\mu_{102}\mu_{021}\mu_{012} + 5\mu_{102}^4 + 4\mu_{102}^2\mu_{021}^2 + 10\mu_{102}^2\mu_{012}^2 + 5\mu_{021}^4$$

$$+ 16\mu_{021}^2\mu_{012}^2 + 5\mu_{012}^4 + 12\mu_{111}^4) / \mu_{000}^{10}$$

Generating graph:

$$\Phi_7 = (\mu_{300}^4 + \mu_{300}^3\mu_{120} + \mu_{300}^3\mu_{102} + 5\mu_{300}^2\mu_{210}^2 + \mu_{300}^2\mu_{210}\mu_{030} + \mu_{300}^2\mu_{210}\mu_{012}$$

$$+ 5\mu_{300}^2\mu_{201}^2 + \mu_{300}^2\mu_{201}\mu_{021} + \mu_{300}^2\mu_{201}\mu_{003} + \mu_{300}^2\mu_{120}^2 + 2\mu_{300}^2\mu_{111}^2$$

$$+ \mu_{300}^2\mu_{102}^2 + 11\mu_{300}\mu_{210}^2\mu_{120} + 4\mu_{300}\mu_{210}^2\mu_{102} + 14\mu_{300}\mu_{210}\mu_{201}\mu_{111}$$

$$+ 4\mu_{300}\mu_{210}\mu_{120}\mu_{030} + 2\mu_{300}\mu_{210}\mu_{120}\mu_{012} + 6\mu_{300}\mu_{210}\mu_{111}\mu_{021}$$

$$+ 2\mu_{300}\mu_{210}\mu_{111}\mu_{003} + 2\mu_{300}\mu_{210}\mu_{102}\mu_{012} + 4\mu_{300}\mu_{201}^2\mu_{120}$$

$$+ 11\mu_{300}\mu_{201}^2\mu_{102} + 2\mu_{300}\mu_{201}\mu_{120}\mu_{021} + 2\mu_{300}\mu_{201}\mu_{111}\mu_{030}$$

$$+ 6\mu_{300}\mu_{201}\mu_{111}\mu_{012} + 2\mu_{300}\mu_{201}\mu_{102}\mu_{021} + 4\mu_{300}\mu_{201}\mu_{102}\mu_{003}$$

$$+ 3\mu_{300}\mu_{120}^3 + \mu_{300}\mu_{120}^2\mu_{102} + 8\mu_{300}\mu_{120}\mu_{111}^2 + \mu_{300}\mu_{120}\mu_{102}^2$$

$$+ \mu_{300}\mu_{120}\mu_{030}^2 + 2\mu_{300}\mu_{120}\mu_{021}^2 + \mu_{300}\mu_{120}\mu_{012}^2 + 8\mu_{300}\mu_{111}^2\mu_{102}$$

$$+ 2\mu_{300}\mu_{111}\mu_{030}\mu_{021} + 4\mu_{300}\mu_{111}\mu_{021}\mu_{012} + 2\mu_{300}\mu_{111}\mu_{012}\mu_{003}$$

$$+ 3\mu_{300}\mu_{102}^3 + \mu_{300}\mu_{102}\mu_{021}^2 + 2\mu_{300}\mu_{102}\mu_{012}^2 + \mu_{300}\mu_{102}\mu_{003}^2 + 3\mu_{210}^3\mu_{030}$$

$$+ 2\mu_{210}^2\mu_{201}\mu_{003} + \mu_{210}^2\mu_{030}^2 + \mu_{210}^2\mu_{030}\mu_{012} + \mu_{210}^2\mu_{021}\mu_{003}$$

$$+ 2\mu_{210}\mu_{201}^2\mu_{030} + 2\mu_{210}\mu_{201}\mu_{030}\mu_{021} + 2\mu_{210}\mu_{201}\mu_{012}\mu_{003}$$

$$+ 11\mu_{210}\mu_{120}^2\mu_{030} + 4\mu_{210}\mu_{120}\mu_{111}\mu_{003} + 2\mu_{210}\mu_{120}\mu_{102}\mu_{030}$$

$$+ 8\mu_{210}\mu_{111}^2\mu_{030} + 6\mu_{210}\mu_{111}\mu_{102}\mu_{003} + \mu_{210}\mu_{102}^2\mu_{030} + \mu_{210}\mu_{030}^3$$

$$+ 4\mu_{210}\mu_{030}\mu_{021}^2 + \mu_{210}\mu_{030}\mu_{012}^2 + 2\mu_{210}\mu_{021}\mu_{012}\mu_{003} + \mu_{210}\mu_{012}\mu_{003}^2$$

$$+ 3\mu_{201}^3\mu_{003} + \mu_{201}^2\mu_{030}\mu_{012} + \mu_{201}^2\mu_{021}\mu_{003} + \mu_{201}^2\mu_{003}^2 + \mu_{201}\mu_{120}^2\mu_{003}$$

$$+ 6\mu_{201}\mu_{120}\mu_{111}\mu_{030} + 2\mu_{201}\mu_{120}\mu_{102}\mu_{003} + 8\mu_{201}\mu_{111}^2\mu_{003}$$

$$+ 4\mu_{201}\mu_{111}\mu_{102}\mu_{030} + 11\mu_{201}\mu_{102}^2\mu_{003} + \mu_{201}\mu_{030}^2\mu_{021}$$

$$+ 2\mu_{201}\mu_{030}\mu_{021}\mu_{012} + \mu_{201}\mu_{021}^2\mu_{003} + 4\mu_{201}\mu_{012}^2\mu_{003} + \mu_{201}\mu_{003}^3$$

$$+ 5\mu_{120}^2\mu_{030}^2 + 4\mu_{120}^2\mu_{030}\mu_{012} + 2\mu_{120}^2\mu_{021}\mu_{003} + 14\mu_{120}\mu_{111}\mu_{030}\mu_{021}$$

$$+ 2\mu_{120}\mu_{111}\mu_{030}\mu_{003} + 6\mu_{120}\mu_{111}\mu_{012}\mu_{003} + \mu_{120}\mu_{102}\mu_{030}^2$$

$$+ 2\mu_{120}\mu_{102}\mu_{030}\mu_{012} + 2\mu_{120}\mu_{102}\mu_{021}\mu_{003} + \mu_{120}\mu_{102}\mu_{003}^2 + 2\mu_{111}^2\mu_{030}^2$$

$$+ 8\mu_{111}^2\mu_{030}\mu_{012} + 8\mu_{111}^2\mu_{021}\mu_{003} + 2\mu_{111}^2\mu_{003}^2 + 6\mu_{111}\mu_{102}\mu_{030}\mu_{021}$$

$$+ 2\mu_{111}\mu_{102}\mu_{030}\mu_{003} + 14\mu_{111}\mu_{102}\mu_{012}\mu_{003} + 2\mu_{102}^2\mu_{030}\mu_{012}$$

$$+ 4\mu_{102}^2\mu_{021}\mu_{003} + 5\mu_{102}^2\mu_{003}^2 + \mu_{030}^4 + \mu_{030}^3\mu_{012} + 5\mu_{030}^2\mu_{021}^2$$

$$+ \mu_{030}^2\mu_{021}\mu_{003} + \mu_{030}^2\mu_{012}^2 + 11\mu_{030}\mu_{021}^2\mu_{012} + 4\mu_{030}\mu_{021}\mu_{012}\mu_{003}$$

$$+ 3\mu_{030}\mu_{012}^3 + \mu_{030}\mu_{012}\mu_{003}^2 + 3\mu_{021}^3\mu_{003} + \mu_{021}^2\mu_{003}^2 + 11\mu_{021}\mu_{012}^2\mu_{003}$$

$$+ \mu_{021}\mu_{003}^3 + 5\mu_{012}^2\mu_{003}^2 + \mu_{003}^4 + 2\mu_{210}^4 + 2\mu_{210}^3\mu_{012} + 4\mu_{210}^2\mu_{201}^2$$

$$+ 5\mu_{210}^2\mu_{201}\mu_{021} + 10\mu_{210}^2\mu_{120}^2 + 5\mu_{210}^2\mu_{120}\mu_{102} + 6\mu_{210}^2\mu_{111}^2 + \mu_{210}^2\mu_{102}^2$$

$$+ \mu_{210}^2\mu_{021}^2 + 5\mu_{210}\mu_{201}^2\mu_{012} + 18\mu_{210}\mu_{201}\mu_{120}\mu_{111} + 18\mu_{210}\mu_{201}\mu_{111}\mu_{102}$$

$$+ 4\mu_{210}\mu_{201}\mu_{021}\mu_{012} + 5\mu_{210}\mu_{120}^2\mu_{012} + 18\mu_{210}\mu_{120}\mu_{111}\mu_{021}$$

$$+ 4\mu_{210}\mu_{120}\mu_{102}\mu_{012} + 12\mu_{210}\mu_{111}^2\mu_{012} + 12\mu_{210}\mu_{111}\mu_{102}\mu_{021}$$

$$+ 5\mu_{210}\mu_{102}^2\mu_{012} + 5\mu_{210}\mu_{021}^2\mu_{012} + 2\mu_{210}\mu_{012}^3 + 2\mu_{201}^4 + 2\mu_{201}^3\mu_{021}$$

$$+ \mu_{201}^2\mu_{120}^2 + 5\mu_{201}^2\mu_{120}\mu_{102} + 6\mu_{201}^2\mu_{111}^2 + 10\mu_{201}^2\mu_{102}^2 + \mu_{201}^2\mu_{012}^2$$

$$+ 5\mu_{201}\mu_{120}^2\mu_{021} + 12\mu_{201}\mu_{120}\mu_{111}\mu_{012} + 4\mu_{201}\mu_{120}\mu_{102}\mu_{021}$$

$$+ 12\mu_{201}\mu_{111}^2\mu_{021} + 18\mu_{201}\mu_{111}\mu_{102}\mu_{012} + 5\mu_{201}\mu_{102}^2\mu_{021} + 2\mu_{201}\mu_{021}^3$$

$$+ 5\mu_{201}\mu_{021}\mu_{012}^2 + 2\mu_{120}^4 + 2\mu_{120}^3\mu_{102} + 6\mu_{120}^2\mu_{111}^2 + 4\mu_{120}^2\mu_{021}^2$$

$$+ \mu_{120}^2\mu_{012}^2 + 12\mu_{120}\mu_{111}^2\mu_{102} + 18\mu_{120}\mu_{111}\mu_{021}\mu_{012} + 2\mu_{120}\mu_{102}^3$$

$$+ 5\mu_{120}\mu_{102}\mu_{021}^2 + 5\mu_{120}\mu_{102}\mu_{012}^2 + 6\mu_{111}^2\mu_{102}^2 + 6\mu_{111}^2\mu_{021}^2 + 6\mu_{111}^2\mu_{012}^2$$

$$+ 18\mu_{111}\mu_{102}\mu_{021}\mu_{012} + 2\mu_{102}^4 + \mu_{102}^2\mu_{021}^2 + 4\mu_{102}^2\mu_{012}^2 + 2\mu_{021}^4$$

$$+ 10\mu_{021}^2\mu_{012}^2 + 2\mu_{012}^4 \Big) / \mu_{000}^{10}$$

Generating graph:

$$
\begin{aligned}
\Phi_8 = \big(& \mu_{300}^4 + 2\mu_{300}^3\mu_{120} + 2\mu_{300}^3\mu_{102} + 4\mu_{300}^2\mu_{210}^2 + 2\mu_{300}^2\mu_{210}\mu_{030} \\
& + 2\mu_{300}^2\mu_{210}\mu_{012} + 4\mu_{300}^2\mu_{201}^2 + 2\mu_{300}^2\mu_{201}\mu_{021} + 2\mu_{300}^2\mu_{201}\mu_{003} \\
& + 2\mu_{300}^2\mu_{120}^2 + 2\mu_{300}^2\mu_{120}\mu_{102} + 2\mu_{300}^2\mu_{111}^2 + 2\mu_{300}^2\mu_{102}^2 \\
& + 10\mu_{300}\mu_{210}^2\mu_{120} + 6\mu_{300}\mu_{210}^2\mu_{102} + 8\mu_{300}\mu_{210}\mu_{201}\mu_{111} \\
& + 8\mu_{300}\mu_{210}\mu_{120}\mu_{030} + 6\mu_{300}\mu_{210}\mu_{120}\mu_{012} + 8\mu_{300}\mu_{210}\mu_{111}\mu_{021} \\
& + 4\mu_{300}\mu_{210}\mu_{111}\mu_{003} + 2\mu_{300}\mu_{210}\mu_{102}\mu_{030} + 4\mu_{300}\mu_{210}\mu_{102}\mu_{012} \\
& + 6\mu_{300}\mu_{201}^2\mu_{120} + 10\mu_{300}\mu_{201}^2\mu_{102} + 4\mu_{300}\mu_{201}\mu_{120}\mu_{021} \\
& + 2\mu_{300}\mu_{201}\mu_{120}\mu_{003} + 4\mu_{300}\mu_{201}\mu_{111}\mu_{030} + 8\mu_{300}\mu_{201}\mu_{111}\mu_{012} \\
& + 6\mu_{300}\mu_{201}\mu_{102}\mu_{021} + 8\mu_{300}\mu_{201}\mu_{102}\mu_{003} + 2\mu_{300}\mu_{120}^3 \\
& + 2\mu_{300}\mu_{120}^2\mu_{102} + 4\mu_{300}\mu_{120}\mu_{111}^2 + 2\mu_{300}\mu_{120}\mu_{102}^2 \\
& + 2\mu_{300}\mu_{120}\mu_{030}^2 + 2\mu_{300}\mu_{120}\mu_{030}\mu_{012} + 2\mu_{300}\mu_{120}\mu_{021}^2 \\
& + 2\mu_{300}\mu_{120}\mu_{021}\mu_{003} + 4\mu_{300}\mu_{111}^2\mu_{102} + 4\mu_{300}\mu_{111}\mu_{030}\mu_{021} \\
& + 8\mu_{300}\mu_{111}\mu_{021}\mu_{012} + 4\mu_{300}\mu_{111}\mu_{012}\mu_{003} + 2\mu_{300}\mu_{102}^3 \\
& + 2\mu_{300}\mu_{102}\mu_{030}\mu_{012} + 2\mu_{300}\mu_{102}\mu_{021}\mu_{003} + 2\mu_{300}\mu_{102}\mu_{012}^2 \\
& + 2\mu_{300}\mu_{102}\mu_{003}^2 + 2\mu_{210}^3\mu_{030} + 2\mu_{210}^2\mu_{201}\mu_{003} + 2\mu_{210}^2\mu_{030}^2 \\
& + 2\mu_{210}^2\mu_{030}\mu_{012} + 2\mu_{210}\mu_{201}^2\mu_{030} + 4\mu_{210}\mu_{201}\mu_{030}\mu_{021} \\
& + 2\mu_{210}\mu_{201}\mu_{030}\mu_{003} + 4\mu_{210}\mu_{201}\mu_{012}\mu_{003} + 10\mu_{210}\mu_{120}^2\mu_{030} \\
& + 8\mu_{210}\mu_{120}\mu_{111}\mu_{003} + 6\mu_{210}\mu_{120}\mu_{102}\mu_{030} + 4\mu_{210}\mu_{111}^2\mu_{030} \\
& + 8\mu_{210}\mu_{111}\mu_{102}\mu_{003} + 2\mu_{210}\mu_{030}^3 + 2\mu_{210}\mu_{030}^2\mu_{012} + 6\mu_{210}\mu_{030}\mu_{021}^2 \\
& + 2\mu_{210}\mu_{030}\mu_{021}\mu_{003} + 2\mu_{210}\mu_{030}\mu_{012}^2 + 6\mu_{210}\mu_{021}\mu_{012}\mu_{003} \\
& + 2\mu_{210}\mu_{012}\mu_{003}^2 + 2\mu_{201}^3\mu_{003} + 2\mu_{201}^2\mu_{021}\mu_{003} + 2\mu_{201}^2\mu_{003}^2 \\
& + 8\mu_{201}\mu_{120}\mu_{111}\mu_{030} + 6\mu_{201}\mu_{120}\mu_{102}\mu_{003} + 4\mu_{201}\mu_{111}^2\mu_{003} \\
& + 8\mu_{201}\mu_{111}\mu_{102}\mu_{030} + 10\mu_{201}\mu_{102}^2\mu_{003} + 2\mu_{201}\mu_{030}^2\mu_{021} \\
& + 6\mu_{201}\mu_{030}\mu_{021}\mu_{012} + 2\mu_{201}\mu_{030}\mu_{012}\mu_{003} + 2\mu_{201}\mu_{021}^2\mu_{003} \\
& + 2\mu_{201}\mu_{021}\mu_{003}^2 + 6\mu_{201}\mu_{012}^2\mu_{003} + 2\mu_{201}\mu_{003}^3 + 4\mu_{120}^2\mu_{030}^2 \\
& + 6\mu_{120}^2\mu_{030}\mu_{012} + 2\mu_{120}^2\mu_{021}\mu_{003} + 8\mu_{120}\mu_{111}\mu_{030}\mu_{021} \\
& + 4\mu_{120}\mu_{111}\mu_{030}\mu_{003} + 8\mu_{120}\mu_{111}\mu_{012}\mu_{003} + 2\mu_{120}\mu_{102}\mu_{030}^2 \\
& + 4\mu_{120}\mu_{102}\mu_{030}\mu_{012} + 4\mu_{120}\mu_{102}\mu_{021}\mu_{003} + 2\mu_{120}\mu_{102}\mu_{003}^2 \\
& + 2\mu_{111}^2\mu_{030}^2 + 4\mu_{111}^2\mu_{030}\mu_{012} + 4\mu_{111}^2\mu_{021}\mu_{003} + 2\mu_{111}^2\mu_{003}^2 \\
& + 8\mu_{111}\mu_{102}\mu_{030}\mu_{021} + 4\mu_{111}\mu_{102}\mu_{030}\mu_{003} + 8\mu_{111}\mu_{102}\mu_{012}\mu_{003}
\end{aligned}
$$

$$+ 2\mu_{102}^2\mu_{030}\mu_{012} + 6\mu_{102}^2\mu_{021}\mu_{003} + 4\mu_{102}^2\mu_{003}^2 + \mu_{030}^4$$

$$+ 2\mu_{030}^3\mu_{012} + 4\mu_{030}^2\mu_{021}^2 + 2\mu_{030}^2\mu_{021}\mu_{003} + 2\mu_{030}^2\mu_{012}^2$$

$$+ 10\mu_{030}\mu_{021}^2\mu_{012} + 8\mu_{030}\mu_{021}\mu_{012}\mu_{003} + 2\mu_{030}\mu_{012}^3 + 2\mu_{030}\mu_{012}\mu_{003}^2$$

$$+ 2\mu_{021}^3\mu_{003} + 2\mu_{021}^2\mu_{003}^2 + 10\mu_{021}\mu_{012}^2\mu_{003} + 2\mu_{021}\mu_{003}^3$$

$$+ 4\mu_{012}^2\mu_{003}^2 + \mu_{003}^4 + \mu_{210}^4 + 2\mu_{210}^3\mu_{012} + 2\mu_{210}^2\mu_{201}^2$$

$$+ 2\mu_{210}^2\mu_{201}\mu_{021} + 8\mu_{210}^2\mu_{120}^2 + 8\mu_{210}^2\mu_{120}\mu_{102} + 2\mu_{210}^2\mu_{111}^2$$

$$+ 2\mu_{210}^2\mu_{102}^2 + 2\mu_{210}^2\mu_{021}^2 + 2\mu_{210}^2\mu_{012}^2 + 2\mu_{210}\mu_{201}^2\mu_{012}$$

$$+ 12\mu_{210}\mu_{201}\mu_{120}\mu_{111} + 12\mu_{210}\mu_{201}\mu_{111}\mu_{102} + 6\mu_{210}\mu_{201}\mu_{021}\mu_{012}$$

$$+ 8\mu_{210}\mu_{120}^2\mu_{012} + 12\mu_{210}\mu_{120}\mu_{111}\mu_{021} + 6\mu_{210}\mu_{120}\mu_{102}\mu_{012}$$

$$+ 4\mu_{210}\mu_{111}^2\mu_{012} + 12\mu_{210}\mu_{111}\mu_{102}\mu_{021} + 2\mu_{210}\mu_{102}^2\mu_{012}$$

$$+ 8\mu_{210}\mu_{021}^2\mu_{012} + 2\mu_{210}\mu_{012}^3 + \mu_{201}^4 + 2\mu_{201}^3\mu_{021} + 2\mu_{201}^2\mu_{120}^2$$

$$+ 8\mu_{201}^2\mu_{120}\mu_{102} + 2\mu_{201}^2\mu_{111}^2 + 8\mu_{201}^2\mu_{102}^2 + 2\mu_{201}^2\mu_{021}^2$$

$$+ 2\mu_{201}^2\mu_{012}^2 + 2\mu_{201}\mu_{120}^2\mu_{021} + 12\mu_{201}\mu_{120}\mu_{111}\mu_{012}$$

$$+ 6\mu_{201}\mu_{120}\mu_{102}\mu_{021} + 4\mu_{201}\mu_{111}^2\mu_{021} + 12\mu_{201}\mu_{111}\mu_{102}\mu_{012}$$

$$+ 8\mu_{201}\mu_{102}^2\mu_{021} + 2\mu_{201}\mu_{021}^3 + 8\mu_{201}\mu_{021}\mu_{012}^2 + \mu_{120}^4 + 2\mu_{120}^3\mu_{102}$$

$$+ 2\mu_{120}^2\mu_{111}^2 + 2\mu_{120}^2\mu_{102}^2 + 2\mu_{120}^2\mu_{021}^2 + 2\mu_{120}^2\mu_{012}^2$$

$$+ 4\mu_{120}\mu_{111}^2\mu_{102} + 12\mu_{120}\mu_{111}\mu_{021}\mu_{012} + 2\mu_{120}\mu_{102}^3 + 2\mu_{120}\mu_{102}\mu_{021}^2$$

$$+ 2\mu_{120}\mu_{102}\mu_{012}^2 + 2\mu_{111}^2\mu_{102}^2 + 2\mu_{111}^2\mu_{021}^2 + 2\mu_{111}^2\mu_{012}^2$$

$$+ 12\mu_{111}\mu_{102}\mu_{021}\mu_{012} + \mu_{102}^4 + 2\mu_{102}^2\mu_{021}^2 + 2\mu_{102}^2\mu_{012}^2 + \mu_{021}^4$$

$$+ 8\mu_{021}^2\mu_{012}^2 + \mu_{012}^4 \big) / \mu_{000}^{10}$$

Generating graph:

$$\Phi_9 = \big(\mu_{200}\mu_{300}^2 + 2\mu_{110}\mu_{300}\mu_{210} + 2\mu_{110}\mu_{120}\mu_{030} + 2\mu_{101}\mu_{300}\mu_{201}$$

$$+ 2\mu_{101}\mu_{102}\mu_{003} + \mu_{020}\mu_{030}^2 + 2\mu_{011}\mu_{030}\mu_{021} + 2\mu_{011}\mu_{012}\mu_{003} + \mu_{002}\mu_{003}^2$$

$$+ 2\mu_{200}\mu_{210}^2 + 2\mu_{200}\mu_{201}^2 + \mu_{200}\mu_{120}^2 + 2\mu_{200}\mu_{111}^2 + \mu_{200}\mu_{102}^2$$

$$+ 4\mu_{110}\mu_{210}\mu_{120} + 4\mu_{110}\mu_{201}\mu_{111} + 4\mu_{110}\mu_{111}\mu_{021} + 2\mu_{110}\mu_{102}\mu_{012}$$

$$+ 4\mu_{101}\mu_{210}\mu_{111} + 4\mu_{101}\mu_{201}\mu_{102} + 2\mu_{101}\mu_{120}\mu_{021} + 4\mu_{101}\mu_{111}\mu_{012}$$

$$+ \mu_{020}\mu_{210}^2 + 2\mu_{020}\mu_{120}^2 + 2\mu_{020}\mu_{111}^2 + 2\mu_{020}\mu_{021}^2 + \mu_{020}\mu_{012}^2$$

$$+ 2\mu_{011}\mu_{210}\mu_{201} + 4\mu_{011}\mu_{120}\mu_{111} + 4\mu_{011}\mu_{111}\mu_{102} + 4\mu_{011}\mu_{021}\mu_{012}$$
$$+ \mu_{002}\mu_{201}^2 + 2\mu_{002}\mu_{111}^2 + 2\mu_{002}\mu_{102}^2 + \mu_{002}\mu_{021}^2 + 2\mu_{002}\mu_{012}^2) / \mu_{000}^7$$

Generating graph:

$$\Phi_{10} = \left(\mu_{200}\mu_{300}^2 + \mu_{200}\mu_{300}\mu_{120} + \mu_{200}\mu_{300}\mu_{102} + \mu_{200}\mu_{210}\mu_{030}\right.$$
$$+ \mu_{200}\mu_{201}\mu_{003} + 2\mu_{110}\mu_{300}\mu_{210} + 2\mu_{110}\mu_{120}\mu_{030} + 2\mu_{110}\mu_{111}\mu_{003}$$
$$+ 2\mu_{101}\mu_{300}\mu_{201} + 2\mu_{101}\mu_{111}\mu_{030} + 2\mu_{101}\mu_{102}\mu_{003} + \mu_{020}\mu_{300}\mu_{120}$$
$$+ \mu_{020}\mu_{210}\mu_{030} + \mu_{020}\mu_{030}^2 + \mu_{020}\mu_{030}\mu_{012} + \mu_{020}\mu_{021}\mu_{003}$$
$$+ 2\mu_{011}\mu_{300}\mu_{111} + 2\mu_{011}\mu_{030}\mu_{021} + 2\mu_{011}\mu_{012}\mu_{003} + \mu_{002}\mu_{300}\mu_{102}$$
$$+ \mu_{002}\mu_{201}\mu_{003} + \mu_{002}\mu_{030}\mu_{012} + \mu_{002}\mu_{021}\mu_{003} + \mu_{002}\mu_{003}^2$$
$$+ \mu_{200}\mu_{210}^2 + \mu_{200}\mu_{210}\mu_{012} + \mu_{200}\mu_{201}^2 + \mu_{200}\mu_{201}\mu_{021}$$
$$+ 4\mu_{110}\mu_{210}\mu_{120} + 2\mu_{110}\mu_{210}\mu_{102} + 2\mu_{110}\mu_{201}\mu_{111} + 2\mu_{110}\mu_{120}\mu_{012}$$
$$+ 2\mu_{110}\mu_{111}\mu_{021} + 2\mu_{101}\mu_{210}\mu_{111} + 2\mu_{101}\mu_{201}\mu_{120} + 4\mu_{101}\mu_{201}\mu_{102}$$
$$+ 2\mu_{101}\mu_{111}\mu_{012} + 2\mu_{101}\mu_{102}\mu_{021} + \mu_{020}\mu_{201}\mu_{021} + \mu_{020}\mu_{120}^2$$
$$+ \mu_{020}\mu_{120}\mu_{102} + \mu_{020}\mu_{021}^2 + 2\mu_{011}\mu_{210}\mu_{021} + 2\mu_{011}\mu_{201}\mu_{012}$$
$$+ 2\mu_{011}\mu_{120}\mu_{111} + 2\mu_{011}\mu_{111}\mu_{102} + 4\mu_{011}\mu_{021}\mu_{012} + \mu_{002}\mu_{210}\mu_{012}$$
$$\left.+ \mu_{002}\mu_{120}\mu_{102} + \mu_{002}\mu_{102}^2 + \mu_{002}\mu_{012}^2\right) / \mu_{000}^7$$

Generating graph:

$$\Phi_{11} = \left(\mu_{200}\mu_{300}^2 + 2\mu_{200}\mu_{300}\mu_{120} + 2\mu_{200}\mu_{300}\mu_{102} + 2\mu_{110}\mu_{300}\mu_{210}\right.$$
$$+ 2\mu_{110}\mu_{300}\mu_{030} + 2\mu_{110}\mu_{300}\mu_{012} + 2\mu_{110}\mu_{120}\mu_{030} + 2\mu_{110}\mu_{102}\mu_{030}$$
$$+ 2\mu_{101}\mu_{300}\mu_{201} + 2\mu_{101}\mu_{300}\mu_{021} + 2\mu_{101}\mu_{300}\mu_{003} + 2\mu_{101}\mu_{120}\mu_{003}$$
$$+ 2\mu_{101}\mu_{102}\mu_{003} + 2\mu_{020}\mu_{210}\mu_{030} + \mu_{020}\mu_{030}^2 + 2\mu_{020}\mu_{030}\mu_{012}$$
$$+ 2\mu_{011}\mu_{210}\mu_{003} + 2\mu_{011}\mu_{201}\mu_{030} + 2\mu_{011}\mu_{030}\mu_{021} + 2\mu_{011}\mu_{030}\mu_{003}$$
$$+ 2\mu_{011}\mu_{012}\mu_{003} + 2\mu_{002}\mu_{201}\mu_{003} + 2\mu_{002}\mu_{021}\mu_{003} + \mu_{002}\mu_{003}^2$$
$$+ \mu_{200}\mu_{120}^2 + 2\mu_{200}\mu_{120}\mu_{102} + \mu_{200}\mu_{102}^2 + 2\mu_{110}\mu_{210}\mu_{120}$$
$$+ 2\mu_{110}\mu_{210}\mu_{102} + 2\mu_{110}\mu_{120}\mu_{012} + 2\mu_{110}\mu_{102}\mu_{012} + 2\mu_{101}\mu_{201}\mu_{120}$$
$$+ 2\mu_{101}\mu_{201}\mu_{102} + 2\mu_{101}\mu_{120}\mu_{021} + 2\mu_{101}\mu_{102}\mu_{021} + \mu_{020}\mu_{210}^2$$

$$+ 2\mu_{020}\mu_{210}\mu_{012} + \mu_{020}\mu_{012}^2 + 2\mu_{011}\mu_{210}\mu_{201} + 2\mu_{011}\mu_{210}\mu_{021}$$

$$+ 2\mu_{011}\mu_{201}\mu_{012} + 2\mu_{011}\mu_{021}\mu_{012} + \mu_{002}\mu_{201}^2 + 2\mu_{002}\mu_{201}\mu_{021}$$

$$+ \mu_{002}\mu_{021}^2 \big) \, / \mu_{000}^7$$

Generating graph:

$$\Phi_{12} = \big(\mu_{200}^2\mu_{300}^2 + 4\mu_{200}\mu_{110}\mu_{300}\mu_{210} + 4\mu_{200}\mu_{101}\mu_{300}\mu_{201}$$

$$+ 2\mu_{200}\mu_{020}\mu_{300}\mu_{120} + 2\mu_{200}\mu_{020}\mu_{210}\mu_{030} + 4\mu_{200}\mu_{011}\mu_{300}\mu_{111}$$

$$+ 2\mu_{200}\mu_{002}\mu_{300}\mu_{102} + 2\mu_{200}\mu_{002}\mu_{201}\mu_{003} + 4\mu_{110}\mu_{020}\mu_{120}\mu_{030}$$

$$+ 4\mu_{110}\mu_{002}\mu_{111}\mu_{003} + 4\mu_{101}\mu_{020}\mu_{111}\mu_{030} + 4\mu_{101}\mu_{002}\mu_{102}\mu_{003}$$

$$+ \mu_{020}^2\mu_{030}^2 + 4\mu_{020}\mu_{011}\mu_{030}\mu_{021} + 2\mu_{020}\mu_{002}\mu_{030}\mu_{012}$$

$$+ 2\mu_{020}\mu_{002}\mu_{021}\mu_{003} + 4\mu_{011}\mu_{002}\mu_{012}\mu_{003} + \mu_{002}^2\mu_{003}^2$$

$$+ \mu_{200}^2\mu_{210}^2 + \mu_{200}^2\mu_{201}^2 + 4\mu_{200}\mu_{110}\mu_{210}\mu_{120}$$

$$+ 4\mu_{200}\mu_{110}\mu_{201}\mu_{111} + 4\mu_{200}\mu_{101}\mu_{210}\mu_{111} + 4\mu_{200}\mu_{101}\mu_{201}\mu_{102}$$

$$+ 2\mu_{200}\mu_{020}\mu_{201}\mu_{021} + 4\mu_{200}\mu_{011}\mu_{210}\mu_{021} + 4\mu_{200}\mu_{011}\mu_{201}\mu_{012}$$

$$+ 2\mu_{200}\mu_{002}\mu_{210}\mu_{012} + 4\mu_{110}^2\mu_{210}^2 + 4\mu_{110}^2\mu_{120}^2$$

$$+ 8\mu_{110}\mu_{101}\mu_{210}\mu_{201} + 8\mu_{110}\mu_{101}\mu_{120}\mu_{111} + 8\mu_{110}\mu_{101}\mu_{111}\mu_{102}$$

$$+ 4\mu_{110}\mu_{020}\mu_{210}\mu_{120} + 4\mu_{110}\mu_{020}\mu_{111}\mu_{021} + 8\mu_{110}\mu_{011}\mu_{210}\mu_{111}$$

$$+ 8\mu_{110}\mu_{011}\mu_{120}\mu_{021} + 8\mu_{110}\mu_{011}\mu_{111}\mu_{012} + 4\mu_{110}\mu_{002}\mu_{210}\mu_{102}$$

$$+ 4\mu_{110}\mu_{002}\mu_{120}\mu_{012} + 4\mu_{101}^2\mu_{201}^2 + 4\mu_{101}^2\mu_{102}^2$$

$$+ 4\mu_{101}\mu_{020}\mu_{201}\mu_{120} + 4\mu_{101}\mu_{020}\mu_{102}\mu_{021} + 8\mu_{101}\mu_{011}\mu_{201}\mu_{111}$$

$$+ 8\mu_{101}\mu_{011}\mu_{111}\mu_{021} + 8\mu_{101}\mu_{011}\mu_{102}\mu_{012} + 4\mu_{101}\mu_{002}\mu_{201}\mu_{102}$$

$$+ 4\mu_{101}\mu_{002}\mu_{111}\mu_{012} + \mu_{020}^2\mu_{120}^2 + \mu_{020}^2\mu_{021}^2$$

$$+ 4\mu_{020}\mu_{011}\mu_{120}\mu_{111} + 4\mu_{020}\mu_{011}\mu_{021}\mu_{012} + 2\mu_{020}\mu_{002}\mu_{120}\mu_{102}$$

$$+ 4\mu_{011}^2\mu_{021}^2 + 4\mu_{011}^2\mu_{012}^2 + 4\mu_{011}\mu_{002}\mu_{111}\mu_{102}$$

$$+ 4\mu_{011}\mu_{002}\mu_{021}\mu_{012} + \mu_{002}^2\mu_{102}^2 + \mu_{002}^2\mu_{012}^2 + 4\mu_{110}^2\mu_{111}^2$$

$$+ 4\mu_{101}^2\mu_{111}^2 + 4\mu_{011}^2\mu_{111}^2 \big) \, / \mu_{000}^9$$

Generating graph:

$$
\begin{aligned}
\Phi_{13} = \big(&\mu_{200}^2\mu_{300}^2 + 2\mu_{200}\mu_{110}\mu_{300}\mu_{210} + 2\mu_{200}\mu_{110}\mu_{120}\mu_{030} \\
&+ 2\mu_{200}\mu_{101}\mu_{300}\mu_{201} + 2\mu_{200}\mu_{101}\mu_{102}\mu_{003} + \mu_{110}^2\mu_{300}^2 \\
&+ \mu_{110}^2\mu_{030}^2 + 2\mu_{110}\mu_{101}\mu_{030}\mu_{021} + 2\mu_{110}\mu_{101}\mu_{012}\mu_{003} \\
&+ 2\mu_{110}\mu_{020}\mu_{300}\mu_{210} + 2\mu_{110}\mu_{020}\mu_{120}\mu_{030} + 2\mu_{110}\mu_{011}\mu_{300}\mu_{201} \\
&+ 2\mu_{110}\mu_{011}\mu_{102}\mu_{003} + \mu_{101}^2\mu_{300}^2 + \mu_{101}^2\mu_{003}^2 \\
&+ 2\mu_{101}\mu_{011}\mu_{300}\mu_{210} + 2\mu_{101}\mu_{011}\mu_{120}\mu_{030} + 2\mu_{101}\mu_{002}\mu_{300}\mu_{201} \\
&+ 2\mu_{101}\mu_{002}\mu_{102}\mu_{003} + \mu_{020}^2\mu_{030}^2 + 2\mu_{020}\mu_{011}\mu_{030}\mu_{021} \\
&+ 2\mu_{020}\mu_{011}\mu_{012}\mu_{003} + \mu_{011}^2\mu_{030}^2 + \mu_{011}^2\mu_{003}^2 \\
&+ 2\mu_{011}\mu_{002}\mu_{030}\mu_{021} + 2\mu_{011}\mu_{002}\mu_{012}\mu_{003} + \mu_{002}^2\mu_{003}^2 \\
&+ 2\mu_{200}^2\mu_{210}^2 + 2\mu_{200}^2\mu_{201}^2 + \mu_{200}^2\mu_{120}^2 + 2\mu_{200}^2\mu_{111}^2 \\
&+ \mu_{200}^2\mu_{102}^2 + 4\mu_{200}\mu_{110}\mu_{210}\mu_{120} + 4\mu_{200}\mu_{110}\mu_{201}\mu_{111} \\
&+ 4\mu_{200}\mu_{110}\mu_{111}\mu_{021} + 2\mu_{200}\mu_{110}\mu_{102}\mu_{012} + 4\mu_{200}\mu_{101}\mu_{210}\mu_{111} \\
&+ 4\mu_{200}\mu_{101}\mu_{201}\mu_{102} + 2\mu_{200}\mu_{101}\mu_{120}\mu_{021} + 4\mu_{200}\mu_{101}\mu_{111}\mu_{012} \\
&+ 3\mu_{110}^2\mu_{210}^2 + 2\mu_{110}^2\mu_{201}^2 + 3\mu_{110}^2\mu_{120}^2 + \mu_{110}^2\mu_{102}^2 \\
&+ 2\mu_{110}^2\mu_{021}^2 + \mu_{110}^2\mu_{012}^2 + 2\mu_{110}\mu_{101}\mu_{210}\mu_{201} \\
&+ 4\mu_{110}\mu_{101}\mu_{120}\mu_{111} + 4\mu_{110}\mu_{101}\mu_{111}\mu_{102} + 4\mu_{110}\mu_{101}\mu_{021}\mu_{012} \\
&+ 4\mu_{110}\mu_{020}\mu_{210}\mu_{120} + 4\mu_{110}\mu_{020}\mu_{201}\mu_{111} + 4\mu_{110}\mu_{020}\mu_{111}\mu_{021} \\
&+ 2\mu_{110}\mu_{020}\mu_{102}\mu_{012} + 4\mu_{110}\mu_{011}\mu_{210}\mu_{111} + 4\mu_{110}\mu_{011}\mu_{201}\mu_{102} \\
&+ 2\mu_{110}\mu_{011}\mu_{120}\mu_{021} + 4\mu_{110}\mu_{011}\mu_{111}\mu_{012} + 2\mu_{101}^2\mu_{210}^2 \\
&+ 3\mu_{101}^2\mu_{201}^2 + \mu_{101}^2\mu_{120}^2 + 3\mu_{101}^2\mu_{102}^2 + \mu_{101}^2\mu_{021}^2 \\
&+ 2\mu_{101}^2\mu_{012}^2 + 4\mu_{101}\mu_{011}\mu_{210}\mu_{120} + 4\mu_{101}\mu_{011}\mu_{201}\mu_{111} \\
&+ 4\mu_{101}\mu_{011}\mu_{111}\mu_{021} + 2\mu_{101}\mu_{011}\mu_{102}\mu_{012} + 4\mu_{101}\mu_{002}\mu_{210}\mu_{111} \\
&+ 4\mu_{101}\mu_{002}\mu_{201}\mu_{102} + 2\mu_{101}\mu_{002}\mu_{120}\mu_{021} + 4\mu_{101}\mu_{002}\mu_{111}\mu_{012} \\
&+ \mu_{020}^2\mu_{210}^2 + 2\mu_{020}^2\mu_{120}^2 + 2\mu_{020}^2\mu_{111}^2 + 2\mu_{020}^2\mu_{021}^2 \\
&+ \mu_{020}^2\mu_{012}^2 + 2\mu_{020}\mu_{011}\mu_{210}\mu_{201} + 4\mu_{020}\mu_{011}\mu_{120}\mu_{111} \\
&+ 4\mu_{020}\mu_{011}\mu_{111}\mu_{102} + 4\mu_{020}\mu_{011}\mu_{021}\mu_{012} + \mu_{011}^2\mu_{210}^2 \\
&+ \mu_{011}^2\mu_{201}^2 + 2\mu_{011}^2\mu_{120}^2 + 2\mu_{011}^2\mu_{102}^2 + 3\mu_{011}^2\mu_{021}^2 \\
&+ 3\mu_{011}^2\mu_{012}^2 + 2\mu_{011}\mu_{002}\mu_{210}\mu_{201} + 4\mu_{011}\mu_{002}\mu_{120}\mu_{111}
\end{aligned}
$$

$$+ 4\mu_{011}\mu_{002}\mu_{111}\mu_{102} + 4\mu_{011}\mu_{002}\mu_{021}\mu_{012} + \mu_{002}^2\mu_{201}^2$$

$$+ 2\mu_{002}^2\mu_{111}^2 + 2\mu_{002}^2\mu_{102}^2 + \mu_{002}^2\mu_{021}^2 + 2\mu_{002}^2\mu_{012}^2$$

$$+ 4\mu_{110}^2\mu_{111}^2 + 4\mu_{101}^2\mu_{111}^2 + 4\mu_{011}^2\mu_{111}^2) / \mu_{000}^9$$

Generating graph:

Appendix 4.B

In this Appendix we present a direct proof of Theorem 4.1.

The behavior of the complex moments under rotation is given by

$$(c_{s\ell}^m)^{(\mathbf{R})} = \sum_{m'=-\ell}^{\ell} D_{m'm}^{\ell}(\mathbf{R}) c_{s\ell}^{m'}. \tag{4.71}$$

1. If we take c_{s0}^0, it becomes $(c_{s0}^0)^{(\mathbf{R})} = D_{00}^0(\mathbf{R})c_{s0}^0$. $D_{00}^0(\mathbf{R}) = 1$ for any rotation $\mathbf{R}$, therefore c_{s0}^0 is a rotation invariant.
2. The composite complex moment form after rotation is

$$(c_s(\ell,\ell')_j^k)^{(\mathbf{R})} = \sum_{m=\max(-\ell,k-\ell')}^{\min(\ell,k+\ell')} \langle \ell,\ell',m,k-m|j,k\rangle \, (c_{s\ell}^m)^{(\mathbf{R})} \, (c_{s\ell'}^{k-m})^{(\mathbf{R})} =$$

$$= \sum_{m=\max(-\ell,k-\ell')}^{\min(\ell,k+\ell')} \langle \ell,\ell',m,k-m|j,k\rangle \sum_{m'=-\ell}^{\ell} D_{m'm}^{\ell} c_{s\ell}^{m'} \sum_{m''=-\ell'}^{\ell'} D_{m'',k-m}^{\ell'} c_{s\ell'}^{m''}. \tag{4.72}$$

We dropped the symbol $(\mathbf{R})$ in $D_{m'm}^{\ell}(\mathbf{R})$ for simplicity. The tensor product of two Wigner D-functions is

$$D_{m'm}^{\ell} D_{n'n}^{\ell'} = \sum_{j=|\ell-\ell'|}^{\ell+\ell'} \langle \ell,\ell',m',n'|j,m'+n'\rangle \, D_{m'+n',m+n}^{j} \langle \ell,\ell',m,n|j,m+n\rangle. \tag{4.73}$$

If we substitute it into (4.72), we obtain

$$\sum_{m=\max(-\ell,k-\ell')}^{\min(\ell,k+\ell')} \langle \ell,\ell',m,k-m|j,k\rangle \sum_{m'=-\ell}^{\ell} \sum_{m''=-\ell'}^{\ell'} c_{s\ell}^{m'} c_{s\ell'}^{m''} \times$$

$$\times \sum_{j'=|\ell-\ell'|}^{\ell+\ell'} \langle \ell,\ell',m',m''|j',m'+m''\rangle D_{m'+m'',k}^{j'} \langle \ell,\ell',m,k-m|j',k\rangle. \tag{4.74}$$

There are two relations of orthogonality of the Clebsch-Gordan coefficients. One of them can be written as

$$\sum_{m=-\ell}^{\ell} \sum_{m'=-\ell'}^{\ell'} \langle \ell,\ell',m,m'|j,m+m'\rangle\langle \ell,\ell',m,m'|j',m+m'\rangle = \delta_{jj'}. \tag{4.75}$$

If we substitute it into (4.74), we obtain

$$\sum_{m'=-\ell}^{\ell} \sum_{m''=-\ell'}^{\ell'} \langle \ell, \ell', m', m'' | j, m' + m'' \rangle \, D^j_{m'+m'',k} c^{m'}_{s\ell} c^{m''}_{s\ell'}. \tag{4.76}$$

Now we introduce the new index $k' = m' + m''$, i.e. $m'' = k' - m'$ and obtain

$$\sum_{k'=-\ell-\ell'}^{\ell+\ell'} D^j_{k',k} \sum_{m'=-\ell}^{\ell} \langle \ell, \ell', m', k' - m' | j, k' \rangle c^{m'}_{s\ell} c^{k'-m'}_{s\ell'} = \sum_{k'=-\ell-\ell'}^{\ell+\ell'} D^j_{k',k} c_s(\ell, \ell')^{k'}_j. \tag{4.77}$$

The Wigner D-function is zero for $k' < -j$ and $k' > j$, therefore we can omit the zero terms and simplify the sum as

$$(c_s(\ell, \ell')^k_j)^{(\mathbf{R})} = \sum_{k'=-j}^{j} D^j_{k',k} c_s(\ell, \ell')^{k'}_j. \tag{4.78}$$

This means the composite complex moment form behaves under rotation much like the moment itself. If $j = k = 0$, then again $D^0_{00} = 1$ and $c_s(\ell, \ell')^0_0$ is rotation invariant.

3. The product of a form and a complex moment after rotation is

$$(c_s(\ell, \ell')_j c_{s'})^{(\mathbf{R})} = \frac{1}{\sqrt{2j+1}} \sum_{k=-j}^{j} (-1)^{j-k} (c_s(\ell, \ell')^k_j)^{(\mathbf{R})} (c^{-k}_{s'j})^{(\mathbf{R})}$$

$$= \frac{1}{\sqrt{2j+1}} \sum_{k=-j}^{j} (-1)^{j-k} \sum_{k'=-j}^{j} D^j_{k'k} c_s(\ell, \ell')^{k'}_j \sum_{m'=-j}^{j} D^j_{m',-k} c^{m'}_{s'j}. \tag{4.79}$$

Substituting the tensor product of the Wigner D-functions we obtain (4.73)

$$\frac{1}{\sqrt{2j+1}} \sum_{k=-j}^{j} (-1)^{j-k} \sum_{k'=-j}^{j} \sum_{m'=-j}^{j} c_s(\ell, \ell')^{k'}_j c^{m'}_{s'j} \times$$

$$\times \sum_{j'=0}^{j} \langle j, j, k', m' | j', k' + m' \rangle D^{j'}_{k'+m',0} \langle j, j, k, -k | j', 0 \rangle. \tag{4.80}$$

The second relation of orthogonality of the Clebsch-Gordan coefficients

$$\sum_{j=|\ell-\ell'|}^{\ell+\ell'} \sum_{m=-j}^{j} \langle \ell, \ell', m, m' | j, m \rangle \langle \ell, \ell', m'', m''' | j, m \rangle = \delta_{mm''} \delta_{m'm'''} \tag{4.81}$$

implies the only relevant terms in the sum are those for $k' = k$ and $m' = -k$. After the substitution we obtain

$$(c_s(\ell, \ell')_j c_{s'})^{(\mathbf{R})} = \frac{1}{\sqrt{2j+1}} \sum_{k=-j}^{j} (-1)^{j-k} c_s(\ell, \ell')^k_j c^{-k}_{s'j} = c_s(\ell, \ell')_j c_{s'}, \tag{4.82}$$

which means that $c_s(\ell, \ell')_j c_{s'}$ is a rotation invariant.

4. Let us consider a product of two forms after the rotation

$$
\begin{aligned}
&(c_s(\ell,\ell')_j c_{s'}(\ell'',\ell''')_j)^{(\mathbf{R})} \\
&= \frac{1}{\sqrt{2j+1}} \sum_{k=-j}^{j} (-1)^{j-k} (c_s(\ell,\ell')_j^k)^{(\mathbf{R})} (c_{s'}(\ell'',\ell''')_j^{-k})^{(\mathbf{R})} \\
&= \frac{1}{\sqrt{2j+1}} \sum_{k=-j}^{j} (-1)^{j-k} \sum_{k'=-j}^{j} D_{k'k}^{j} c_s(\ell,\ell')_j^{k'} \sum_{k''=-j}^{j} D_{k'',-k}^{j} c_{s'}(\ell'',\ell''')_j^{k''}.
\end{aligned}
\tag{4.83}
$$

The use of (4.73) yields

$$
\begin{aligned}
&\frac{1}{\sqrt{2j+1}} \sum_{k=-j}^{j} (-1)^{j-k} \sum_{k'=-j}^{j} \sum_{k''=-j}^{j} c_s(\ell,\ell')_j^{k'} \, c_{s'}(\ell'',\ell''')_j^{k''} \\
&\times \sum_{j'=0}^{j} \langle j,j,k',k''|j',k'+k''\rangle D_{k'+k'',0}^{j'} \langle j,j,k,-k|j',0\rangle.
\end{aligned}
\tag{4.84}
$$

Now we substitute (4.81) again with $k' = k$ and $k'' = -k$

$$
\begin{aligned}
(c_s(\ell,\ell')_j c_{s'}(\ell'',\ell''')_j)^{(\mathbf{R})} &= \frac{1}{\sqrt{2j+1}} \sum_{k=-j}^{j} (-1)^{j-k} c_s(\ell,\ell')_j^{k} c_{s'}(\ell'',\ell''')_j^{-k} \\
&= c_s(\ell,\ell')_j c_{s'}(\ell'',\ell''')_j,
\end{aligned}
\tag{4.85}
$$

which proves that $c_s(\ell,\ell')_j c_{s'}(\ell'',\ell''')_j$ is also a rotation invariant. $\qquad\square$

Appendix 4.C

Irreducible 3D rotation invariants from complex moments of the second and third order.

- Second-order invariants

$$
\begin{aligned}
\phi_1 =&\ \ c_{20}^0 = \frac{1}{2}\sqrt{\frac{1}{\pi}}(\mu_{200} + \mu_{200} + \mu_{200}) \\
\phi_2 =&\ \ c_2(2,2)_0^0 = \frac{1}{\sqrt{5}}\left((c_{22}^0)^2 - 2c_{22}^1 c_{22}^{-1} + 2c_{22}^2 c_{22}^{-2} \right) \\
&= \frac{\sqrt{5}}{4\pi}\big(\mu_{200}^2 + \mu_{020}^2 + \mu_{002}^2 - \mu_{200}\mu_{020} - \mu_{200}\mu_{002} - \mu_{020}\mu_{002} \\
&\qquad\quad + 3\mu_{110}^2 + 3\mu_{101}^2 + 3\mu_{011}^2 \big) \\
\phi_3 =&\ c_2(2,2)_2 c_2 = \frac{1}{\sqrt{35}}(-\sqrt{2}(c_{22}^0)^3 + 3\sqrt{2}c_{22}^1 c_{22}^0 c_{22}^{-1} - 3\sqrt{3}(c_{22}^{-1})^2 c_{22}^2 \\
&\qquad\qquad -3\sqrt{3}c_{22}^{-2}(c_{22}^1)^2 + 6\sqrt{2}c_{22}^2 c_{22}^0 c_{22}^{-2}) \\
&= \frac{5}{8\pi}\frac{1}{\sqrt{14\pi}}(-2\mu_{200}^3 - 2\mu_{020}^3 - 2\mu_{002}^3 + 3\mu_{200}^2\mu_{020} + 3\mu_{200}^2\mu_{002} \\
&\qquad\quad +3\mu_{020}^2\mu_{200} + 3\mu_{020}^2\mu_{002} + 3\mu_{002}^2\mu_{200} + 3\mu_{002}^2\mu_{020} \\
&\qquad\quad -12\mu_{200}\mu_{020}\mu_{002} + 18\mu_{200}\mu_{011}^2 + 18\mu_{020}\mu_{101}^2 + 18\mu_{002}\mu_{110}^2 \\
&\qquad\quad -9\mu_{200}\mu_{110}^2 - 9\mu_{200}\mu_{101}^2 - 9\mu_{020}\mu_{110}^2 - 9\mu_{020}\mu_{011}^2 \\
&\qquad\quad -9\mu_{002}\mu_{101}^2 - 9\mu_{002}\mu_{011}^2 - 54\mu_{110}\mu_{101}\mu_{011}).
\end{aligned}
\tag{4.86}
$$

- Third-order invariants

$$\phi_4 = \quad c_3(3,3)_0^0 = \frac{1}{\sqrt{7}}(2c_{33}^3 c_{33}^{-3} - 2c_{33}^2 c_{33}^{-2} + 2c_{33}^1 c_{33}^{-1} - \left(c_{33}^0\right)^2)$$

$$\phi_5 = \quad c_3(1,1)_0^0 = \frac{1}{\sqrt{3}}(2c_{31}^1 c_{31}^{-1} - \left(c_{31}^0\right)^2)$$

$$\phi_6 = c_3(3,3)_2 c_2 = \frac{1}{\sqrt{210}}((10c_{33}^3 c_{33}^{-3} - 10\sqrt{2}c_{33}^2 c_{33}^0 + 2\sqrt{15}\left(c_{33}^1\right)^2)c_{22}^{-2}$$
$$-(5\sqrt{10}c_{33}^3 c_{33}^{-2} - 5\sqrt{6}c_{33}^2 c_{33}^{-1} + 2\sqrt{5}c_{33}^1 c_{33}^0)c_{22}^{-1}$$
$$+(5\sqrt{10}c_{33}^3 c_{33}^{-3} - 3\sqrt{10}c_{33}^1 c_{33}^{-1} + 2\sqrt{10}\left(c_{33}^0\right)^2)c_{22}^0$$
$$-(5\sqrt{10}c_{33}^{-3} c_{33}^2 - 5\sqrt{6}c_{33}^{-2} c_{33}^1 + 2\sqrt{5}c_{33}^{-1} c_{33}^0)c_{22}^1$$
$$+(10c_{33}^{-3} c_{33}^1 - 10\sqrt{2}c_{33}^{-2} c_{33}^0 + 2\sqrt{15}(c_{33}^{-1})^2)c_{22}^2)$$

$$\phi_7 = c_3(3,1)_2 c_2 = \frac{1}{\sqrt{105}}((-\sqrt{5}c_{33}^2 c_{31}^0 + \sqrt{15}c_{33}^3 c_{31}^{-1} + c_{33}^1 c_{31}^1)c_{22}^{-2}$$
$$-(\sqrt{10}c_{33}^2 c_{31}^{-1} + \sqrt{3}c_{33}^0 c_{31}^1 - 2\sqrt{2}c_{33}^1 c_{31}^0)c_{22}^{-1}$$
$$+(-3c_{33}^0 c_{31}^0 + \sqrt{6}c_{33}^1 c_{31}^{-1} + \sqrt{6}c_{33}^{-1} c_{31}^1)c_{22}^0$$
$$-(\sqrt{10}c_{33}^{-2} c_{31}^1 + \sqrt{3}c_{33}^0 c_{31}^{-1} - 2\sqrt{10}c_{33}^{-1} c_{31}^0)c_{22}^1$$
$$+(-\sqrt{5}c_{33}^{-2} c_{31}^0 + \sqrt{15}c_{33}^{-3} c_{31}^1 + c_{33}^{-1} c_{31}^{-1})c_{22}^2)$$

$$\phi_8 = c_3(1,1)_2 c_2 = \frac{1}{\sqrt{15}}(\sqrt{3}\left(c_{31}^1\right)^2 c_{22}^{-2} - \sqrt{6}c_{31}^0 c_{31}^1 c_{22}^{-1} + \sqrt{2}((c_{31}^0)^2 + c_{31}^1 c_{31}^{-1})c_{22}^0$$
$$-\sqrt{6}c_{31}^0 c_{31}^{-1} c_{22}^1 c_{22}^0 + \sqrt{3}(c_{31}^{-1})^2 c_{22}^2) \ .$$

$$(4.87)$$

Other five third-order invariants can be obtained as products (4.53) of two composite moment forms

$$\phi_9 = c_3(3,3)_2 c_2(2,2)_2,$$
$$\phi_{10} = c_3^2(3,3)_2,$$
$$\phi_{11} = c_3(3,3)_2 c_3(3,1)_2,$$
$$\phi_{12} = c_3(3,3)_2 c_3(1,1)_2,$$
$$\phi_{13} = c_3(3,1)_2 c_3(1,1)_2,$$

$$(4.88)$$

where the composite complex moment forms of the second order are

$$c_2(2,2)_2^2 = 2\sqrt{\frac{2}{7}}\sqrt{2}c_{22}^2 c_{22}^0 - \sqrt{\frac{3}{7}}\left(c_{22}^1\right)^2,$$

$$c_2(2,2)_2^1 = 2\sqrt{\frac{3}{7}}c_{22}^2 c_{22}^{-1} - \frac{2}{\sqrt{14}}c_{22}^1 c_{22}^0,$$

$$c_2(2,2)_2^0 = 2\sqrt{\frac{2}{7}}c_{22}^2 c_{22}^{-2} + 2\frac{1}{\sqrt{14}}c_{22}^1 c_{22}^{-1} - \sqrt{\frac{2}{7}}\left(c_{22}^0\right)^2,$$

$$c_2(2,2)_2^{-1} = 2\sqrt{\frac{3}{7}}c_{22}^1 c_{22}^{-2} - \frac{2}{\sqrt{14}}c_{22}^0 c_{22}^{-1},$$

$$c_2(2,2)_2^{-2} = 2\sqrt{\frac{2}{7}}\sqrt{2}c_{22}^0 c_{22}^{-2} - \sqrt{\frac{3}{7}}(c_{22}^{-1})^2,$$

$$(4.89)$$

and that of the third order are

$$c_3(3,3)_2^2 = 2\sqrt{\frac{5}{42}}c_{33}^3 c_{33}^{-1} - 2\sqrt{\frac{5}{21}}c_{33}^2 c_{33}^0 + \sqrt{\frac{2}{7}}(c_{33}^1)^2,$$

$$c_3(3,3)_2^1 = \frac{5}{\sqrt{21}}c_{33}^3 c_{33}^{-2} - \sqrt{\frac{5}{7}}c_{33}^2 c_{33}^{-1} + \frac{2}{\sqrt{42}}c_{33}^1 c_{33}^0,$$

$$c_3(3,3)_2^0 = \frac{5}{\sqrt{21}}c_{33}^3 c_{33}^{-3} - \sqrt{\frac{3}{7}}c_{33}^1 c_{33}^{-1} + \frac{2}{\sqrt{21}}(c_{33}^0)^2,$$

$$c_3(3,3)_2^{-1} = \frac{5}{\sqrt{21}}c_{33}^2 c_{33}^{-3} - \sqrt{\frac{5}{7}}c_{33}^1 c_{33}^{-2} + \frac{2}{\sqrt{42}}c_{33}^0 c_{33}^{-1},$$

$$c_3(3,3)_2^{-2} = 2\sqrt{\frac{5}{42}}c_{33}^1 c_{33}^{-3} - 2\sqrt{\frac{5}{21}}c_{33}^0 c_{33}^{-2} + \sqrt{\frac{2}{7}}(c_{33}^{-1})^2,$$

$$c_3(3,1)_2^2 = \sqrt{\frac{5}{7}}c_{33}^3 c_{31}^{-1} - \sqrt{\frac{5}{21}}c_{33}^2 c_{31}^0 + \frac{1}{\sqrt{21}}c_{33}^1 c_{31}^1,$$

$$c_3(3,1)_2^1 = \sqrt{\frac{10}{21}}c_{33}^2 c_{31}^{-1} - 2\sqrt{\frac{2}{21}}c_{33}^1 c_{31}^0 + \frac{1}{\sqrt{7}}c_{33}^0 c_{31}^1,$$

$$c_3(3,1)_2^0 = \sqrt{\frac{2}{7}}c_{33}^1 c_{31}^{-1} - \sqrt{\frac{3}{7}}c_{33}^0 c_{31}^0 + \sqrt{\frac{2}{7}}c_{33}^{-1} c_{31}^1,$$

$$c_3(3,1)_2^{-1} = \sqrt{\frac{10}{21}}c_{33}^{-2} c_{31}^1 - 2\sqrt{\frac{2}{21}}c_{33}^{-1} c_{31}^0 + \frac{1}{\sqrt{7}}c_{33}^0 c_{31}^{-1},$$

$$c_3(3,1)_2^{-2} = \sqrt{\frac{5}{7}}c_{33}^{-3} c_{31}^1 - \sqrt{\frac{5}{21}}c_{33}^{-2} c_{31}^0 + \frac{1}{\sqrt{21}}c_{33}^{-1} c_{31}^{-1},$$

$$c_3(1,1)_2^2 = (c_{31}^1)^2,$$

$$c_3(1,1)_2^1 = \sqrt{2}c_{31}^0 c_{31}^1,$$

$$c_3(1,1)_2^0 = \sqrt{\frac{2}{3}}\left((c_{31}^0)^2 + c_{31}^1 c_{31}^{-1}\right),$$

$$c_3(1,1)_2^{-1} = \sqrt{2}c_{31}^0 c_{31}^{-1},$$

$$c_3(1,1)_2^{-2} = (c_{31}^{-1})^2. \tag{4.90}$$

References

[1] P. Shilane, P. Min, M. Kazhdan, and T. Funkhouser, "The Princeton Shape Benchmark," 2004. http://shape.cs.princeton.edu/benchmark/.

[2] F. A. Sadjadi and E. L. Hall, "Three dimensional moment invariants," *IEEE Transactions on Pattern Analysis and Machine Intelligence*, vol. 2, no. 2, pp. 127–136, 1980.

[3] X. Guo, "Three-dimensional moment invariants under rigid transformation," in *Proceedings of the Fifth International Conference on Computer Analysis of Images and Patterns CAIP'93* (D. Chetverikov and W. Kropatsch, eds.), vol. 719 of *Lecture Notes in Computer Science*, (Berlin, Heidelberg, Germany), pp. 518–522, Springer, 1993.

[4] D. Cyganski and J. A. Orr, "Object recognition and orientation determination by tensor methods," in *Advances in Computer Vision and Image Processing* (T. S. Huang, ed.), (Greenwich, Connecticut, USA), pp. 101–144, JAI Press, 1988.

[5] D. Xu and H. Li, "3-D affine moment invariants generated by geometric primitives," in *Proceedings of the 18th International Conference on Pattern Recognition ICPR'06*, pp. 544–547, IEEE, 2006.

[6] D. Xu and H. Li, "Geometric moment invariants," *Pattern Recognition*, vol. 41, no. 1, pp. 240–249, 2008.

[7] T. Suk and J. Flusser, "Tensor method for constructing 3D moment invariants," in *Computer Analysis of Images and Patterns CAIP'11* (P. Real, D. Diaz-Pernil, H. Molina-Abril, A. Berciano, and W. Kropatsch, eds.), vol. 6854–6855 of *Lecture Notes in Computer Science*, (Berlin, Heidelberg, Germany), pp. 212–219, Springer, 2011.

[8] C.-H. Lo and H.-S. Don, "3-D moment forms: Their construction and application to object identification and positioning," *IEEE Transactions on Pattern Analysis and Machine Intelligence*, vol. 11, no. 10, pp. 1053–1064, 1989.

[9] J. Flusser, J. Boldyš, and B. Zitová, "Moment forms invariant to rotation and blur in arbitrary number of dimensions," *IEEE Transactions on Pattern Analysis and Machine Intelligence*, vol. 25, no. 2, pp. 234–246, 2003.

[10] J. M. Galvez and M. Canton, "Normalization and shape recognition of three-dimensional objects by 3D moments," *Pattern Recognition*, vol. 26, no. 5, pp. 667–681, 1993.

[11] T. Suk, J. Flusser, and J. Boldyš, "3D rotation invariants by complex moments," *Pattern Recognition*, vol. 48, no. 11, pp. 3516–3526, 2015.

[12] T. Suk and J. Flusser, "Recognition of symmetric 3D bodies," *Symmetry*, vol. 6, no. 3, pp. 722–757, 2014.

[13] N. Canterakis, "3D Zernike moments and Zernike affine invariants for 3D image analysis and recognition," in *Proceedings of the 11th Scandinavian Conference on Image Analysis SCIA'99* (B. K. Ersbøll and P. Johansen, eds.), DSAGM, 1999.

[14] B. Yang, J. Flusser, and T. Suk, "3D rotation invariants of Gaussian–Hermite moments," *Pattern Recognition Letters*, vol. 54, no. 1, pp. 18–26, 2015.

[15] M. Kazhdan, T. Funkhouser, and S. Rusinkiewicz, "Rotation invariant spherical harmonic representation of 3D shape descriptors," in *Proceedings of the 2003 Eurographics/ACM SIGGRAPH Symposium on Geometry Processing, SGP '03*, (Aire-la-Ville, Switzerland), pp. 156–164, Eurographics Association, 2003.

[16] R. Kakarala, "The bispectrum as a source of phase-sensitive invariants for Fourier descriptors: A group-theoretic approach," *Journal of Mathematical Imaging and Vision*, vol. 44, no. 3, pp. 341–353, 2012.

[17] J. Fehr and H. Burkhardt, "3D rotation invariant local binary patterns," in *19th International Conference on Pattern Recognition ICPR'08*, pp. 1–4, IEEE, 2008.

[18] T. Ojala, M. Pietikäinen, and T. Mäenpää, "Multiresolution gray-scale and rotation invariant texture classification with local binary patterns," *IEEE Transactions on Pattern Analysis and Machine Intelligence*, vol. 24, no. 7, pp. 971–987, 2002.

[19] J.-F. Mangin, F. Poupon, E. Duchesnay, D. Rivière, A. Cachia, D. L. Collins, A. C. Evans, and J. Régis, "Brain morphometry using 3D moment invariants," *Medical Image Analysis*, vol. 8, no. 3, pp. 187–196, 2004.

[20] M. Kazhdan, "An approximate and efficient method for optimal rotation alignment of 3D models," *IEEE Transactions on Pattern Analysis and Machine Intelligence*, vol. 29, no. 7, pp. 1221–1229, 2007.

[21] M. Trummer, H. Suesse, and J. Denzler, "Coarse registration of 3D surface triangulations based on moment invariants with applications to object alignment and identification," in *Proceedings of the 12th International Conference on Computer Vision, ICCV'09*, pp. 1273–1279, IEEE, 2009.

[22] R. D. Millán, M. Hernandez, D. Gallardo, J. R. Cebral, C. Putman, L. Dempere-Marco, and A. F. Frangi, "Characterization of cerebral aneurysms using 3D moment invariants," in *Proceedings of the conference SPIE Medical Imaging 2005: Image Processing*, vol. 5747, pp. 743–754, 2005.

[23] R. D. Millán, L. Dempere-Marco, J. M. Pozo, J. R. Cebral, and A. F. Frangi, "Morphological characterization of intracranial aneurysms using 3-D moment invariants," *IEEE Transactions on Medical Imaging*, vol. 26, no. 9, pp. 1270–1282, 2007.

[24] K. Shoemake, "Animating rotation with quaternion curves," *SIGGRAPH Computer Graphics*, vol. 19, no. 3, pp. 245–254, 1985.

[25] B. Chen, H. Shu, H. Zhang, G. Chen,C. Toumoulin, J. Dillenseger, and L. Luo, "Quaternion Zernike moments and their invariants for color image analysis and object recognition," *Signal Processing*, vol. 92, no. 2, pp. 308–318, 2012.

[26] F. Bouyachcher, B. Bouali, and M. Nasri, "Quaternion affine moment invariants for color texture recognition," *International Journal of Computer Applications*, vol. 81, no. 16, pp. 1–6, 2013.

[27] D. Rizo-Rodríguez, H. Méndez-Vázquez, and E. García-Reyes, "Illumination invariant face image representation using quaternions," in *Progress in Pattern Recognition, Image Analysis, Computer Vision, and Applications* (I. Bloch and R. M. Cesar, Jr., eds.), vol. 6419 of *Lecture Notes in Computer Science*, pp. 434–441, Berlin, Heidelberg, Germany: Springer, 2010.

[28] J. Angulo, "Structure tensor of colour quaternion image representations for invariant feature extraction," in *Computational Color Imaging, Proceedings of the Second International Workshop CCIW'09* (A. Trémeau, R. Schettini, and S. Tominaga, eds.), vol. 5646 of *Lecture Notes in Computer Science* (Berlin, Heidelberg, Germany), pp. 91–100, Springer, 2009.

[29] M. Pedone, E. Bayro-Corrochano, J. Flusser, and J. Heikkilä, "Quaternion Wiener deconvolution for noise robust color image registration," *IEEE Signal Processing Letters*, vol. 22, no. 9, pp. 1278–1282, 2015.

[30] Wikipedia, "Einstein notation."

[31] E. P. L. Van Gool, T. Moons and A. Oosterlinck, "Vision and Lie's approach to invariance," *Image and Vision Computing*, vol. 13, no. 4, pp. 259–277, 1995.

[32] DIP, "3D rotation moment invariants," 2011. http://zoi.utia.cas.cz/3DRotationInvariants.

[33] W. E. Byerly, *An Elementary Treatise on Fourier's Series and Spherical, Cylindrical, and Ellipsoidal Harmonics with Applications to Problems in Mathematical Physics*. New York, USA: Dover Publications, 1893.

[34] E. W. Hobson, *The Theory of Spherical and Ellipsoidal Harmonics*. Cambridge, U.K.: Cambridge University Press, 1931.

[35] N. M. Ferrers, *An Elementary Treatise on Spherical Harmonics and Subjects Connected with Them*. London, U.K.: Macmillan, 1877.

[36] Y.-F. Yen, "3D rotation matching using spherical harmonic transformation of k-space navigator," in *Proceedings of the ISMRM 12th Scientific Meeting*, p. 2154, International Society for Magnetic Resonance in Medicine, 2004.

[37] J. P. Elliot and P. G. Dawber, *Symmetry in Physics*. London, U.K.: MacMillan, 1979.

[38] A. G. Mamistvalov, "*n*-dimensional moment invariants and conceptual mathematical theory of recognition *n*-dimensional solids," *IEEE Transactions on Pattern Analysis and Machine Intelligence*, vol. 20, no. 8, pp. 819–831, 1998.

[39] J. Boldyš and J. Flusser, "Extension of moment features' invariance to blur," *Journal of Mathematical Imaging and Vision*, vol. 32, no. 3, pp. 227–238, 2008.

[40] G. Cybenko, A. Bhasin, and K. D. Cohen, "Pattern recognition of 3D CAD objects: Towards an electronic yellow pages of mechanical parts," *International Journal of Smart Engineering System Design*, vol. 1, no. 1, pp. 1–13, 1997.

[41] G. Burel and H. Hénocq, "Determination of the orientation of 3D objects using spherical harmonics," *Graphical Models and Image Processing*, vol. 57, no. 5, pp. 400–408, 1995.

[42] H. Weyl, *Symmetry*. Princeton, USA: Princeton University Press, 1952.

[43] S. Mamone, G. Pileio, and M. H. Levitt, "Orientational sampling schemes based on four dimensional polytopes," *Symmetry*, vol. 2, no. 3, pp. 1423–1449, 2010.

[44] M. Langbein and H. Hagen, "A generalization of moment invariants on 2D vector fields to tensor fields of arbitrary order and dimension," in *Proceedings of 5th International Symposium Advances in Visual Computing, ISVC'09, Part II*, vol. 5876 of *Lecture Notes in Computer Science*, pp. 1151–1160, Springer, 2009.

[45] R. Bujack, *Orientation Invariant Pattern Detection in Vector Fields with Clifford Algebra and Moment Invariants*. PhD thesis, Fakultät für Mathematik und Informatik, Universität Leipzig, Leipzig, Germany, 2014.

[46] R. Bujack, J. Kasten, I. Hotz, G. Scheuermann, and E. Hitzer, "Moment invariants for 3D flow fields via normalization," in *Pacific Visualization Symposium, PacificVis'15*, pp. 9–16, IEEE, 2015.

[47] D. Xu and H. Li, "3-D surface moment invariants," in *Proceedings of the 18th International Conference on Pattern Recognition ICPR'06*, pp. 173–176, IEEE, 2006.

[48] S. A. Sheynin and A. V. Tuzikov, "Explicit formulae for polyhedra moments," *Pattern Recognition Letters*, vol. 22, no. 10, pp. 1103–1109, 2001.

5

Affine Moment Invariants in 2D and 3D

5.1 Introduction

This chapter is devoted to the *affine moment invariants* (AMIs), which play a very important role in moment-based theories and in invariant object recognition. They are invariant with respect to *affine transformation* (AT) of the spatial coordinates

$$\mathbf{x}' = \mathbf{A}\mathbf{x} + \mathbf{b} \tag{5.1}$$

where $\mathbf{A}$ is a regular $d \times d$ matrix with constant coefficients and $\mathbf{b}$ is a vector of translation.

Since the TRS group is a special case of the AT, one might expect there would be no big difference between these two groups (and also between the respective invariants) because both transformations are linear. However, the opposite is true. The design of affine invariants requires different (and unfortunately more complicated) mathematical tools and approaches. Also the occurrence and the origin of AT and TRS are different from one another. While the TRS appears naturally whenever the object or the camera is rotated and shifted, pure AT which exactly follows the model (5.1) is very rare. An example in 2D is capturing of a flat moving scene by a push-broom camera. We can meet this scenario in practice in satellite remote sensing, where the scanner creates the image of the rotating Earth line by line, which leads to image skewing. So, why are the affine invariants so important? Most of the "traditional" photographs are central projections of the 3D world onto a plane, which is modeled by a perspective projection. As we will see later, the perspective projection is hard to work with, and the affine transformation serves as a reasonably accurate and simple approximation. This is a justification of the need for 2D affine invariants. The demand for 3D affine invariants is much lower because there is no "fourth dimension" which would be projected onto 3D scene and captured by a 3D sensor. In 3D we only face the intrinsic affine transforms, which are very rare.

The substantial part of this chapter is devoted to the 2D AMIs. We start with an explanation of the projective imaging and the relationship between projective and affine transforms.

2D and 3D Image Analysis by Moments, First Edition. Jan Flusser, Tomáš Suk and Barbara Zitová.
© 2017 John Wiley & Sons, Ltd. Published 2017 by John Wiley & Sons, Ltd.
Companion website: www.wiley.com/go/flusser/2d3dimageanalysis

Figure 5.1   The projective deformation of a scene due to a non-perpendicular view. Square tiles appear as quadrilaterals; the transformation preserves straight lines but does not preserve their collinearity

Then we present four different methods that allow a systematic (and automatic) design of the AMIs of any order – the graph method, the normalization method, the method based on the Caley-Aronhold equation, and the transvectant method. We discuss the differences between them and illustrate the numerical properties of the AMIs. At the end of the chapter, we introduce affine invariants of color images and vector fields, 3D AMIs, and the idea of moment matching.

5.1.1   2D projective imaging of 3D world

In traditional photography, 3D scenes, structures, and objects are represented by their projections onto a 2D plane because photography is a 2D medium (here, we deal neither with intrinsically 3D imaging devices, nor with a coupling of camera and rangefinder, which actually yields 3D information). In such cases, we often face object deformations that are beyond the linear model (see Figure 5.1).

An exact model of photographing a planar scene by a pinhole camera whose optical axis is not perpendicular to the scene is a *projective transformation* (sometimes also called *perspective projection*)

$$x' = \frac{a_0 + a_1 x + a_2 y}{1 + c_1 x + c_2 y},$$

$$y' = \frac{b_0 + b_1 x + b_2 y}{1 + c_1 x + c_2 y}.$$

(5.2)

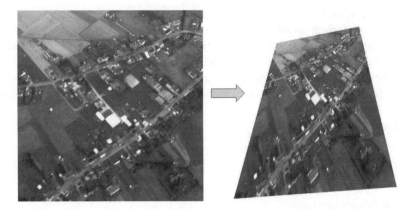

Figure 5.2 The projective transformation maps a square onto a quadrilateral (computer-generated example)

Jacobian J of the projective transformation is

$$J = \frac{(a_1 b_2 - a_2 b_1) + (a_2 b_0 - a_0 b_2)c_1 + (a_0 b_1 - a_1 b_0)c_2}{(1 + c_1 x + c_2 y)^3}$$

$$= \frac{\begin{vmatrix} a_0 & a_1 & a_2 \\ b_0 & b_1 & b_2 \\ 1 & c_1 & c_2 \end{vmatrix}}{(1 + c_1 x + c_2 y)^3}. \tag{5.3}$$

The projective transformation is not a linear transformation but exhibits certain linear properties. It maps a square onto a general quadrilateral (see Figure 5.2), and, importantly, it preserves straight lines. However, it does not preserve their collinearity. The composition of two projective transformations is again a projective transformation. The inverse of a projective transformation is also a projective transformation. Projective transforms form the *projective group*, which indicates it is meaningful to look for projective invariants.

5.1.2   *Projective moment invariants*

Projective invariants are in great demand, especially in computer vision and robotics, where we analyze 2D images of 3D scenes. Unfortunately, construction of moment invariants to projective transformation is practically impossible. Major difficulties with projective moment invariants originate from the fact that projective transformation is not linear; its Jacobian is a function of spatial coordinates, and the transformation does not preserve the center of gravity of the object. The tools that have been successfully applied to derive moment invariants to various linear transformations cannot be exploited in the case of projective invariants.

Moments are of course not the only possibility to construct projective invariants. Differential projective invariants based on local shape properties were described in [1–3], and various invariants defined by means of salient points can be found in [4–8]. These local invariants were already briefly reviewed in Chapter 2. Since they are of a different nature and have different usage from global moment invariants, we will not discuss them further in this chapter.

Concerning projective moment invariants, Van Gool et al. [9] employed the Lie group theory to prove the non-existence of projective invariants from a finite set of moments. Here, we present another proof that provides better insight and does not use the group theory tools.

Let us decompose the projective transformation (5.2) into eight one-parametric transformations:

Horizontal and vertical translations

$$(a)\ \begin{aligned} x' &= x + \alpha \\ y' &= y \end{aligned} \qquad\qquad (b)\ \begin{aligned} x' &= x \\ y' &= y + \beta, \end{aligned}$$

uniform scaling and stretching

$$(c)\ \begin{aligned} x' &= \omega x \\ y' &= \omega y \end{aligned} \qquad\qquad (d)\ \begin{aligned} x' &= rx \\ y' &= \tfrac{1}{r}y, \end{aligned}$$

horizontal and vertical skewing

$$(e)\ \begin{aligned} x' &= x + t_1 y \\ y' &= y \end{aligned} \qquad\qquad (f)\ \begin{aligned} x' &= x \\ y' &= t_2 x + y, \end{aligned}$$

and horizontal and vertical pure projections

$$(g)\ \begin{aligned} x' &= \frac{x}{1 + c_1 x} \\ y' &= \frac{y}{1 + c_1 x} \end{aligned} \qquad\qquad (h)\ \begin{aligned} x' &= \frac{x}{1 + c_2 y} \\ y' &= \frac{y}{1 + c_2 y}. \end{aligned}$$

Any projective invariant F would have to be invariant to all elementary transformations $(a) - (h)$. Each elementary transformation imposes special constraints on F. Particularly, from the invariance to horizontal pure projection (g) it follows that the derivative of F with respect to parameter c_1 must be zero (assuming that all derivatives exist):

$$\frac{dF}{dc_1} = \sum_p \sum_q \frac{\partial F}{\partial m_{pq}} \frac{dm_{pq}}{dc_1} = 0. \tag{5.4}$$

The derivatives of moments can be expressed as

$$\begin{aligned} \frac{dm_{pq}}{dc_1} &= \frac{d}{dc_1} \iint_S \frac{x^p y^q}{(1 + c_1 x)^{p+q}} \frac{1}{|1 + c_1 x|^3} f(x, y)\, dx\, dy \\ &= \iint_S (-p - q - 3) \frac{x^{p+1} y^q}{(1 + c_1 x)^{p+q+1}} \frac{1}{|1 + c_1 x|^3} f(x, y)\, dx\, dy \\ &= -(p + q + 3) m_{p+1,q}. \end{aligned} \tag{5.5}$$

Thus, equation (5.4) becomes

$$-\sum_p \sum_q (p+q+3)m_{p+1,q}\frac{\partial F}{\partial m_{pq}} = 0. \qquad (5.6)$$

The constraint (5.6) must be fulfilled for any image. Assuming that F contains only a finite number of moments, we denote their maximum order as r. However, equation (5.6) contains moments up to the order $r+1$. If equation (5.6) was satisfied for a certain image, we could always construct another image with identical moments up to the order r and different moments of the order $r+1$ such that equation (5.6) does not hold for this new image. Thus, finite-term projective moment invariants cannot exist.

The above result can be extended to prove the non-existence of invariants, that would have a form of infinite series having each term in a form of a finite product of moments.

Suk and Flusser [10] proved the formal existence of projective invariants in a form of infinite series of products of moments with both positive and negative indices. However, practical applicability of these invariants is very limited because of their high dynamic range and slow convergence or divergence. Moreover, these invariants cannot be called moment invariants in the strict sense because their basis functions are not only polynomials but also rational functions.

Some authors proved that finite moment-like projective invariants do exist for certain particular projective transformations (see for instance reference [11] for the case $a_0 = a_2 = b_0 = b_1 = 0$ and reference [12] for the case $a_0 = a_2 = b_0 = 0$). Unfortunately, those special forms of invariants cannot be generalized to the projective transformation (5.2).

5.1.3 Affine transformation

Affine transformation has been thoroughly studied in image analysis because it can – under certain circumstances – approximate the projective transformation. As we already mentioned in the introduction, the affine transformation can be expressed in matrix notation as

$$\mathbf{x}' = \mathbf{A}\mathbf{x} + \mathbf{b} \qquad (5.7)$$

which in 2D leads to the well-known equations

$$\begin{aligned} x' &= a_0 + a_1 x + a_2 y \\ y' &= b_0 + b_1 x + b_2 y. \end{aligned} \qquad (5.8)$$

where

$$\mathbf{A} = \begin{pmatrix} a_1 & a_2 \\ b_1 & b_2 \end{pmatrix} \quad \text{and} \quad \mathbf{b} = \begin{pmatrix} a_0 \\ b_0 \end{pmatrix}. \qquad (5.9)$$

Affine transformation maps a square onto a parallelogram, preserves the straight lines and also preserves the collinearity (see Figure 5.3). Affine transformations form the *affine group*[1] since

---

[1] Affine transformations with $\det(\mathbf{A}) = 1$ form a subgroup, sometimes referred to as the *special affine group*.

Figure 5.3   The affine transformation maps a square to a parallelogram

their composition as well as the inverse are still within the model (5.1). Apparently, TRS is a subgroup of the affine group when $a_1 = b_2$ and $a_2 = -b_1$.

The Jacobian of the affine transformation is $J = a_1 b_2 - a_2 b_1$. Contrary to the projective transformation, the Jacobian does not depend on the spatial coordinates x and y, which makes the design of invariants easier. Affine transformation is a particular case of projective transformation with $c_1 = c_2 = 0$. If the object is small compared to the camera-to-scene distance, then both c_1 and c_2 approach zero, the perspective effect becomes negligible and the affine model is a reasonable approximation of the projective model, as shown in case of the characters in Figure 5.4.

5.1.4   2D Affine moment invariants – the history

For the reasons mentioned above, the AMIs play an important role in view-independent object recognition and have been widely used not only in tasks where image deformation is intrinsically affine but also commonly substitute projective invariants.

The history of the AMIs can surprisingly be traced back to the end of the 19th century, to the times when neither computers nor automatic object recognition existed. Probably the first one who systematically studied invariants to affine transform was the famous German mathematician David Hilbert[2] Hilbert did not work explicitly with moments but studied so called *algebraic invariants*. The algebraic invariants are polynomials of coefficients of a binary form, which are invariant w.r.t. an affine transformation. The algebraic invariants are closely linked with the AMIs through the *Fundamental theorem of the AMIs*, formulated by Hu [13]. This link is so tight that it allows us to consider Hilbert to be the father of the AMIs. Many parts

---

[2] Born on January 23, 1862, Koningsberg; died on February 14, 1943, in Gottingen.

Figure 5.4   The affine transformation approximates the perspective projection if the objects are small

of his book[3] [14], dealing actually with the algebraic invariants, can be adapted to moments in a relatively straightforward way.

Hu [13] tried to introduce the AMIs in 1962 but his version of the Fundamental theorem of affine invariants was incorrect, which slowed the development of the AMIs theory as well as of their applications for a long time. Thirty years later, Reiss [15] and Flusser and Suk [16, 17] independently discovered and corrected his mistake, published new sets of valid AMIs and proved their applicability in simple recognition tasks. Their derivation of the AMIs originated from the traditional theory of algebraic invariants from the late 19th and early 20th century, elaborated after Hilbert by many mathematicians; see for instance [18–22]. The same results achieved by a slightly different approach were later published by Mamistvalov [23][4].

The AMIs can be derived in several ways that differ from each other in the mathematical tools used. Apart from the above-mentioned algebraic invariants, one may use graph theory [25, 26], tensor algebra [27], direct solution of proper partial differential equations [28],

---

[3] The Hilbert's results on algebraic invariants were not published during his life, probably because Hilbert considered his work incomplete. The fundamental book [14] is based on the original notes of the course, that Hilbert delivered in 1897 at the university in Gottingen, Germany. The notes were discovered in 1990, translated into English and first published in 1993, fifty years after the Hilbert's death.

[4] The author probably published the same results already in 1970 [24] in Russian, but that paper is not commonly accessible.

transvectants [29], and derivation via image normalization [30]. These approaches are all equivalent in the following sense. They yield different explicit forms and even different numbers of the invariants, but between any two systems of the invariants up to certain order a polynomial one-to-one mapping exists. Still, each approach has its own pros and cons. That is why all main methods are worth reviewing as we do in the following sections.

A common point of all automatic methods is that they can generate an arbitrary number of the invariants, but only some of them are independent. Since dependent invariants are useless in practice, much effort has been spent on their elimination. The possibility of identifying and eliminating the dependent invariants is one of the criterions for the selection of the proper method for invariant construction.

5.2   AMIs derived from the Fundamental theorem

As we already mentioned above, the theory of the AMIs is closely connected with the theory of algebraic invariants. The Fundamental theorem, describing this connection, was the earliest tool used for the AMIs construction.

The algebraic invariant is a polynomial of coefficients of a binary form, whose terms all have the same structure and whose value remains constant under an affine transformation of the coordinates. In the theory of algebraic invariants, only the transformations without translation (i.e., $\mathbf{b} = 0$) are considered. Moment invariants should handle the translation as well, which is fulfilled automatically when using central moments.

The *binary algebraic form* of order p is a homogeneous polynomial

$$\sum_{k=0}^{p} \binom{p}{k} a_k x^{p-k} y^k \tag{5.10}$$

with coefficients a_k. Under linear transformation

$$\mathbf{x}' = \mathbf{A}\mathbf{x} \tag{5.11}$$

the coefficients are changed, and the binary form becomes

$$\sum_{k=0}^{p} \binom{p}{k} a_k' x'^{\,p-k} y'^{\,k}. \tag{5.12}$$

Hilbert [14] defined an *algebraic invariant* as a polynomial function of the coefficients $a, b, \ldots$ of several binary forms that satisfies the equation

$$I(a_0', a_1', \ldots, a_{p_a}'; b_0', b_1', \ldots, b_{p_b}'; \ldots) = J^w I(a_0, a_1, \ldots, a_{p_a}; b_0, b_1, \ldots, b_{p_b}; \ldots). \tag{5.13}$$

The w is called the *weight* of the invariant. If $w = 0$, the invariant is called *absolute*; otherwise it is called *relative*. If there is only one binary form involved, the invariant is called *homogeneous*; otherwise it is called *simultaneous*. The algebraic invariant is a polynomial, whose terms consist of products of coefficients of the algebraic forms. Their number r multiplied in one term is called the *degree* of the invariant.

The Fundamental theorem of the AMIs (FTAMI) can be formulated as follows.

Theorem 5.1 (Fundamental theorem of the AMIs) If the binary forms of orders $p_a, p_b, \ldots$ have an algebraic invariant of weight w and degree r

$$I(a_0', a_1', \ldots, a_{p_a}'; b_0', b_1', \ldots, b_{p_b}'; \ldots) = J^w I(a_0, a_1, \ldots, a_{p_a}; b_0, b_1, \ldots, b_{p_b}; \ldots),$$

then the moments of the same orders have the same invariant but with the additional multiplicative factor $|J|^r$:

$$I(\mu_{p_a 0}', \mu_{p_a-1,1}', \ldots, \mu_{0 p_a}'; \mu_{p_b 0}', \mu_{p_b-1,1}', \ldots, \mu_{0 p_b}'; \ldots)$$

$$= J^w |J|^r I(\mu_{p_a 0}, \mu_{p_a-1,1}, \ldots, \mu_{0 p_a}; \mu_{p_b 0}, \mu_{p_b-1,1}, \ldots, \mu_{0 p_b}; \ldots).$$

A traditional proof of the FTAMI[5] can be found in reference [15]. A 3D version of this theorem along with its proof was published in reference [23]. In Appendix 5.A, we present another, much simpler proof using the Fourier transform.

The power of the FTAMI depends on our ability to find algebraic invariants of binary forms. The FTAMI does not provide any instructions on how to find them; it only establishes the connection between algebraic and moment invariants. To construct algebraic invariants, one has to use the classical theory from [14, 18–22]. Having an algebraic invariant, we can construct a corresponding AMI easily just by interchanging the coefficients and the central moments. To obtain absolute AMIs, we normalize the invariant by μ_{00}^{w+r} to eliminate the factor $J^w \cdot |J|^r$ (note that $\mu_{00}' = |J|\mu_{00}$). If $J < 0$ and w is odd, then this normalization yields only a pseudoinvariant, i.e. $I' = -I$.

This kind of AMIs derivation was used for instance in [15] and [16] but was later outperformed by more efficient methods.

5.3 AMIs generated by graphs

The simplest and the most transparent way for the AMIs of any orders and weights to be generated systematically is based on representing invariants by graphs. This *graph method* was originally proposed by Suk and Flusser in [25] and further developed in [26]. The reader may notice a certain analogy with an earlier *tensor method* [27]. One of the main advantages of the graph method is that it provides an insight into the structure of the invariants. The main drawback is that the method generates a huge number of dependent invariants that must be identified and removed afterwards, which makes the invariant design slow. On the other hand, the method allows the dependencies to be eliminated automatically and selects such independent invariants, which have a minimum number of terms. So, the final set is optimal in this sense.

---

[5] Probably the first version of this theorem appeared in the original Hu paper [13] but was stated incorrectly without the factor $|J|^r$. That is why the Hu affine invariants are not actually invariant to a general affine transformation; they perform well only if $|J| = 1$ or $r = 0$.

5.3.1 The basic concept

Let us consider image f and two arbitrary points (x_1, y_1), (x_2, y_2) from its support. Let us denote the "cross-product" of these points as C_{12}:

$$C_{12} = x_1 y_2 - x_2 y_1.$$

After an affine transformation has been applied (as in the previous section, we consider no shift) it holds $C'_{12} = J \cdot C_{12}$, which means that C_{12} is a relative affine invariant[6]. We consider various numbers of points (x_i, y_i), and we integrate their cross-products (or some integer powers of their cross-products) over the support of f. These integrals can be expressed in terms of moments, and, after eliminating the Jacobian by a proper normalization, they yield affine invariants.

More precisely, having $r > 1$ points $(x_1, y_1), \cdots, (x_r, y_r)$, we define functional I depending on r and on non-negative integers n_{kj} as

$$I(f) = \int_{-\infty}^{\infty} \cdots \int_{-\infty}^{\infty} \prod_{k,j=1}^{r} C_{kj}^{n_{kj}} \cdot \prod_{i=1}^{r} f(x_i, y_i) \, dx_i \, dy_i. \tag{5.14}$$

Note that it is meaningful to consider only $j > k$, because $C_{kj} = -C_{jk}$ and $C_{kk} = 0$. After an affine transformation, $I(f)$ becomes

$$I(f)' = J^w |J|^r \cdot I(f),$$

where $w = \sum_{k,j} n_{kj}$ is the *weight* of the invariant and r is its *degree*. If $I(f)$ is normalized by μ_{00}^{w+r}, we obtain a desirable absolute affine invariant

$$\left(\frac{I(f)}{\mu_{00}^{w+r}} \right)' = \left(\frac{I(f)}{\mu_{00}^{w+r}} \right)$$

(if w is odd and $J < 0$ then there is an additional factor -1 and I is a pseudoinvariant).

The maximum order of moments of which the invariant is composed is called the *order* of the invariant. The order is always less than or equal to the weight. Another important characteristic of the invariant is its *structure*. The structure of the invariant is defined by an integer vector $\mathbf{s} = (k_2, k_3, \ldots, k_s)$, where s is the invariant order and k_j is the total number of moments of the j-th order contained in each term of the invariant (all terms always have the same structure)[7].

We illustrate the general formula (5.14) on two simple invariants. First, let $r = 2$ and $w = n_{12} = 2$. Then

$$I(f) = \int_{-\infty}^{\infty} \cdots \int_{-\infty}^{\infty} (x_1 y_2 - x_2 y_1)^2 f(x_1, y_1) f(x_2, y_2) \, dx_1 \, dy_1 \, dx_2 \, dy_2 = 2(m_{20} m_{02} - m_{11}^2). \tag{5.15}$$

When dropping the constant factor 2, replacing the geometric moments by corresponding central moments and normalizing the invariant by μ_{00}^4, we obtain an absolute invariant to general affine transformation

$$I_1 = (\mu_{20}\mu_{02} - \mu_{11}^2)/\mu_{00}^4.$$

---

[6] The geometric meaning of C_{12} is the oriented double area of the triangle, whose vertices are (x_1, y_1), (x_2, y_2), and $(0, 0)$.

[7] Since always $k_0 = k_1 = 0$, these two quantities are not included in the structure vector.

This is the simplest affine invariant, uniquely containing the second-order moments only. Its structure is $\mathbf{s} = (2)$. In this form, I_1 is commonly referred to in the literature regardless of the method used for derivation.

Similarly, for $r = 3$ and $n_{12} = 2, n_{13} = 2, n_{23} = 0$ we obtain

$$
I(f) = \int_{-\infty}^{\infty} \cdots \int_{-\infty}^{\infty} (x_1 y_2 - x_2 y_1)^2 (x_1 y_3 - x_3 y_1)^2
$$

$$
\times f(x_1, y_1) f(x_2, y_2) f(x_3, y_3) \, \mathrm{d}x_1 \, \mathrm{d}y_1 \, \mathrm{d}x_2 \, \mathrm{d}y_2 \, \mathrm{d}x_3 \, \mathrm{d}y_3
$$

$$
= m_{20}^2 m_{04} - 4 m_{20} m_{11} m_{13} + 2 m_{20} m_{02} m_{22} + 4 m_{11}^2 m_{22} - 4 m_{11} m_{02} m_{31} + m_{02}^2 m_{40} \quad (5.16)
$$

and the normalizing factor is in this case μ_{00}^7. The weight of this invariant is $w = 4$, the order $s = 4$ and its structure is $\mathbf{s} = (2, 0, 1)$.

5.3.2 Representing the AMIs by graphs

The affine moment invariant generated by the formula (5.14) can be represented by a connected multigraph, where each point (x_k, y_k) corresponds to a node and each cross-product C_{kj} corresponds to an edge of the graph. If $n_{kj} > 1$, the respective term $C_{kj}^{n_{kj}}$ corresponds to n_{kj} edges connecting the kth and jth nodes. Thus, the number of nodes equals the degree r of the invariant, and the total number of the graph edges equals the weight w of the invariant. From the graph, we can also learn about the orders of the moments the invariant is composed of and about its structure. The number of edges originating from each node equals the order of the moments involved. Any invariant of the form (5.14) is in fact a sum where each term is a product of a certain number of moments. This number, which equals the degree of the invariant, is constant for all terms of one particular invariant and is equal to the total number of the graph nodes.

This graph representation of the AMIs is similar to the representation of 3D rotation invariant. However, the graph edges have a different meaning. In the "2D affine" graphs the edges represent the cross-products C_{kj} while the edges of the "3D rotation" graphs stand for the moment tensors. It implies for instance that the affine graphs of a non-trivial invariant cannot contain self-loops because $C_{kk} = 0$ always.

Now one can see that the problem of derivation of the AMIs up to the given weight w is equivalent to generating all connected multigraphs with at least two nodes and at most w edges. Let us denote this set of graphs as G_w. Generating all graphs from G_w is a combinatorial task with exponential complexity but formally easy to implement. Two examples of such graphs are in Figure 5.5.

5.3.3 Automatic generation of the invariants by the graph method

The algorithm for the generation of the set G_w is similar to that one used in Section 4.4.6. More precisely, the algorithm for generation of the next graph from the previous one is the same. The only difference is that the graphs with the self-loops are omitted. In the case of AMIs, we

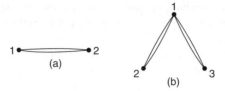

Figure 5.5   The graphs corresponding to the invariants (a) (5.14) and (b) (5.15)

Table 5.1   The numbers of the irreducible and independent affine invariants

Weight	2	3	4	5	6	7	8	9	10	11	12
Irreducible	1	1	4	5	18	31	77	160	362	742	1589
Independent	1	1	4	5	17	25	33	44	55	66	80

initialize the algorithm by the graph

$$
\begin{matrix}
1 & 1 & \cdots & 1 & 1 \\
2 & 2 & \cdots & 2 & 2
\end{matrix}
\tag{5.17}
$$

and terminate it as soon as the last graph

$$
\begin{matrix}
1 & 1 & 3 & 4 & \cdots & w-2 & w-1 & w-1 \\
2 & 3 & 4 & 5 & \cdots & w-1 & w & w
\end{matrix}
\tag{5.18}
$$

has been reached. The numbers of irreducible and independent invariants generated up to the given weight is in Table 5.1.

5.3.4   Independence of the AMIs

We already pointed out that dependent invariants do not increase the discrimination power of the recognition system at all while increasing the dimensionality of the problem, which leads to the growth of complexity and even to misclassifications. Using dependent features in recognition tasks is a serious mistake. Thus, identifying and discarding dependent invariants is highly desirable. In the case of the AMIs generated by the graph method, this is extremely important because, as we will see later, most of them are dependent.

To estimate the number of independent invariants, we can apply the same consideration as in the case of the TRS invariants (4.17). The number of independent invariants up to a certain order equals the number of independent moments minus the number of independent constraints that must be satisfied. However, the number of independent constraints is sometimes hard to determine because the dependencies may be hidden.

In most cases we estimate the number of independent constraints as the number of transformation parameters. This works perfectly in the case of rotation, but for affine transformation this estimate is not true in general due to the hidden dependency. Affine transformation has six parameters, so we would not expect any second-order affine invariant (6 moments – 6 parameters = 0). However, in the previous section we proved the existence of the non-trivial second-order invariant I_1. On the other hand, for invariants up to the third order we have 10 moments – 6 parameters = 4, which is actually the correct number of the independent invariants. The same estimation works well for the fourth order ($15 - 6 = 9$ invariants) and for all higher orders where the actual number of independent invariants is known. It is a common belief (although not exactly proven) that the second order is the only exception to this rule.

A detailed analysis of the number of independent AMIs regarding their weights, degrees, and structures can be found in [31]. It is based on classic results of the theory of algebraic invariants, particularly on the Cayley-Sylvester theorem [21, 14].

Possible dependencies among the AMIs

The dependencies among the AMIs may be of the same five types as the dependencies among the TRS invariants. The invariant may be be zero, identical with some other invariant, product and/or linear combination of other invariants, and some invariants may exhibit a polynomial dependency.

Most automatically generated AMIs are identically zero regardless of the image they have been calculated from. Some of these vanishing invariants can be recognized directly from their graphs. If there are nodes with one adjacent edge only, then all terms of the invariants contain first-order moment(s). When using central moments, they are zero by definition, and, consequently, such an invariant is zero, too. However, this is not the only case of vanishing invariants. Some invariants are zero because of the term cancelation, which is not apparent from the graph at the first sight. Consider, for instance, the graph in Figure 5.6, which leads to the invariant

$$I(f) = \int\limits_{-\infty}^{\infty} \cdots \int\limits_{-\infty}^{\infty} (x_1 y_2 - x_2 y_1)^3 f(x_1, y_1) f(x_2, y_2) \, dx_1 \, dy_1 \, dx_2 \, dy_2 \tag{5.19}$$

$$= m_{30} m_{03} - 3 m_{21} m_{21} + 3 m_{21} m_{21} - m_{30} m_{03} = 0.$$

Elimination of identical invariants is accomplished by comparing the invariants termwise. All isomorphic graphs obviously generate the same invariants, but also some non-isomorphic graphs can generate identical invariants due to a possible term cancelation. This is why the test of identity cannot be reduced to the test of graph isomorphism.

For identification of linear and product dependencies, we use essentially the same procedure as we proposed in Sec. 4.4.5. We perform an exhaustive search over all possible linear combinations and products. We sort the invariants according to their structure first. For each group of the same structure, we construct a matrix of coefficients of all invariants. If this matrix has a full rank, all corresponding invariants are linearly independent. On the other hand, each zero singular value corresponds to a linearly dependent invariant that can be removed from the set.

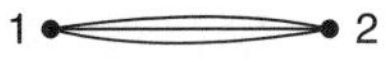

Figure 5.6   The graph leading to a vanishing invariant (5.18)

After the above procedure has been applied, we end up with a set of all irreducible invariants up to the given weight or order. It should be emphasized that the vast majority of the AMIs generated at the beginning are reducible. For example, for the weight $w \leq 12$ we generated 2,533,942,752 graphs altogether, 2,532,349,394 of them leading to vanishing invariants. 1,575,126 invariants were equal to some other invariants, 2105 invariants were products of other invariants, and 14,538 invariants were linearly dependent. After eliminating all these reducible invariants, we obtained 1589 irreducible invariants. This very high ratio between the number of all generated invariants and of the irreducible invariants might give a feeling of inefficiency of this method. That is partially true – there are methods which generate irreducible or even independent invariants in a more straightforward way without the necessity of such time-consuming elimination (the method of normalization is a typical example). On the other hand, the invariants produced by these methods have more complicated forms, which not only make their theoretical studies harder but also slow numerical calculations with them. The graph method enables us to select such irreducible invariants which have minimum number of terms.

Below we present the first ten examples of irreducible invariants in explicit forms along with their generating graphs and structures. Four other important invariants are listed in Appendix 5.B. The complete list of all 1589 irreducible invariants along with their MATLAB implementation is available on the accompanying website [32].

$$I_1 = (\mu_{20}\mu_{02} - \mu_{11}^2)/\mu_{00}^4$$

$w = 2, \mathbf{s} = (2)$

$$I_2 = (-\mu_{30}^2\mu_{03}^2 + 6\mu_{30}\mu_{21}\mu_{12}\mu_{03} - 4\mu_{30}\mu_{12}^3 - 4\mu_{21}^3\mu_{03} + 3\mu_{21}^2\mu_{12}^2)/\mu_{00}^{10}$$

$w = 6, \mathbf{s} = (0, 4)$

$$I_3 = (\mu_{20}\mu_{21}\mu_{03} - \mu_{20}\mu_{12}^2 - \mu_{11}\mu_{30}\mu_{03} + \mu_{11}\mu_{21}\mu_{12} + \mu_{02}\mu_{30}\mu_{12}$$
$$- \mu_{02}\mu_{21}^2)/\mu_{00}^7$$

$w = 4, \mathbf{s} = (1, 2)$

$$I_4 = (-\mu_{20}^3\mu_{03}^2 + 6\mu_{20}^2\mu_{11}\mu_{12}\mu_{03} - 3\mu_{20}^2\mu_{02}\mu_{12}^2 - 6\mu_{20}\mu_{11}^2\mu_{21}\mu_{03}$$

$$-6\mu_{20}\mu_{11}^2\mu_{12}^2 + 12\mu_{20}\mu_{11}\mu_{02}\mu_{21}\mu_{12} - 3\mu_{20}\mu_{02}^2\mu_{21}^2 + 2\mu_{11}^3\mu_{30}\mu_{03}$$

$$+6\mu_{11}^3\mu_{21}\mu_{12} - 6\mu_{11}^2\mu_{02}\mu_{30}\mu_{12} - 6\mu_{11}^2\mu_{02}\mu_{21}^2 + 6\mu_{11}\mu_{02}^2\mu_{30}\mu_{21}$$

$$-\mu_{02}^3\mu_{30}^2)/\mu_{00}^{11}$$

$w = 6, \mathbf{s} = (3, 2)$

$$I_5 = (\mu_{20}^3\mu_{30}\mu_{03}^3 - 3\mu_{20}^3\mu_{21}\mu_{12}\mu_{03}^2 + 2\mu_{20}^3\mu_{12}^3\mu_{03} - 6\mu_{20}^2\mu_{11}\mu_{30}\mu_{12}\mu_{03}^2$$

$$+6\mu_{20}^2\mu_{11}\mu_{21}^2\mu_{03}^2 + 6\mu_{20}^2\mu_{11}\mu_{21}\mu_{12}^2\mu_{03} - 6\mu_{20}^2\mu_{11}\mu_{12}^4$$

$$+3\mu_{20}^2\mu_{02}\mu_{30}\mu_{12}^2\mu_{03} - 6\mu_{20}^2\mu_{02}\mu_{21}^2\mu_{12}\mu_{03} + 3\mu_{20}^2\mu_{02}\mu_{21}\mu_{12}^3$$

$$+12\mu_{20}\mu_{11}^2\mu_{30}\mu_{12}^2\mu_{03} - 24\mu_{20}\mu_{11}^2\mu_{21}^2\mu_{12}\mu_{03} + 12\mu_{20}\mu_{11}^2\mu_{21}\mu_{12}^3$$

$$-12\mu_{20}\mu_{11}\mu_{02}\mu_{30}\mu_{12}^3 + 12\mu_{20}\mu_{11}\mu_{02}\mu_{21}^3\mu_{03} - 3\mu_{20}\mu_{02}^2\mu_{30}\mu_{21}^2\mu_{03}$$

$$+6\mu_{20}\mu_{02}^2\mu_{30}\mu_{21}\mu_{12}^2 - 3\mu_{20}\mu_{02}^2\mu_{21}^3\mu_{12} - 8\mu_{11}^3\mu_{30}\mu_{12}^3 + 8\mu_{11}^3\mu_{21}^3\mu_{03}$$

$$-12\mu_{11}^2\mu_{02}\mu_{30}\mu_{21}^2\mu_{03} + 24\mu_{11}^2\mu_{02}\mu_{30}\mu_{21}\mu_{12}^2 - 12\mu_{11}^2\mu_{02}\mu_{21}^3\mu_{12}$$

$$+6\mu_{11}\mu_{02}^2\mu_{30}^2\mu_{21}\mu_{03} - 6\mu_{11}\mu_{02}^2\mu_{30}^2\mu_{12}^2 - 6\mu_{11}\mu_{02}^2\mu_{30}\mu_{21}^2\mu_{12}$$

$$+6\mu_{11}\mu_{02}^2\mu_{21}^4 - \mu_{02}^3\mu_{30}^3\mu_{03} + 3\mu_{02}^3\mu_{30}^2\mu_{21}\mu_{12} - 2\mu_{02}^3\mu_{30}\mu_{21}^3)/\mu_{00}^{16}$$

$w = 9, \mathbf{s} = (3, 4)$

$$I_6 = (\mu_{40}\mu_{04} - 4\mu_{31}\mu_{13} + 3\mu_{22}^2)/\mu_{00}^6$$

$w = 4, \mathbf{s} = (0, 0, 2)$

$$I_7 = (\mu_{40}\mu_{22}\mu_{04} - \mu_{40}\mu_{13}^2 - \mu_{31}^2\mu_{04} + 2\mu_{31}\mu_{22}\mu_{13} - \mu_{22}^3)/\mu_{00}^9$$

$w = 6, \mathbf{s} = (0,0,3)$

$$I_8 = (\mu_{20}^2 \mu_{04} - 4\mu_{20}\mu_{11}\mu_{13} + 2\mu_{20}\mu_{02}\mu_{22} + 4\mu_{11}^2\mu_{22} - 4\mu_{11}\mu_{02}\mu_{31}$$
$$+\mu_{02}^2\mu_{40})/\mu_{00}^7$$

$w = 4, \mathbf{s} = (2,0,1)$

$$I_9 = (\mu_{20}^2 \mu_{22}\mu_{04} - \mu_{20}^2\mu_{13}^2 - 2\mu_{20}\mu_{11}\mu_{31}\mu_{04} + 2\mu_{20}\mu_{11}\mu_{22}\mu_{13}$$
$$+\mu_{20}\mu_{02}\mu_{40}\mu_{04} - 2\mu_{20}\mu_{02}\mu_{31}\mu_{13} + \mu_{20}\mu_{02}\mu_{22}^2 + 4\mu_{11}^2\mu_{31}\mu_{13} - 4\mu_{11}^2\mu_{22}^2$$
$$-2\mu_{11}\mu_{02}\mu_{40}\mu_{13} + 2\mu_{11}\mu_{02}\mu_{31}\mu_{22} + \mu_{02}^2\mu_{40}\mu_{22} - \mu_{02}^2\mu_{31}^2)/\mu_{00}^{10}$$

$w = 6, \mathbf{s} = (2,0,2)$

$$I_{10} = (\mu_{20}^3 \mu_{31}\mu_{04}^2 - 3\mu_{20}^3\mu_{22}\mu_{13}\mu_{04} + 2\mu_{20}^3\mu_{13}^3 - \mu_{20}^2\mu_{11}\mu_{40}\mu_{04}^2$$
$$-2\mu_{20}^2\mu_{11}\mu_{31}\mu_{13}\mu_{04} + 9\mu_{20}^2\mu_{11}\mu_{22}^2\mu_{04} - 6\mu_{20}^2\mu_{11}\mu_{22}\mu_{13}^2$$
$$+\mu_{20}^2\mu_{02}\mu_{40}\mu_{13}\mu_{04} - 3\mu_{20}^2\mu_{02}\mu_{31}\mu_{22}\mu_{04} + 2\mu_{20}^2\mu_{02}\mu_{31}\mu_{13}^2$$
$$+4\mu_{20}\mu_{11}^2\mu_{40}\mu_{13}\mu_{04} - 12\mu_{20}\mu_{11}^2\mu_{31}\mu_{22}\mu_{04} + 8\mu_{20}\mu_{11}^2\mu_{31}\mu_{13}^2$$
$$-6\mu_{20}\mu_{11}\mu_{02}\mu_{40}\mu_{13}^2 + 6\mu_{20}\mu_{11}\mu_{02}\mu_{31}^2\mu_{04} - \mu_{20}\mu_{02}^2\mu_{40}\mu_{31}\mu_{04}$$
$$+3\mu_{20}\mu_{02}^2\mu_{40}\mu_{22}\mu_{13} - 2\mu_{20}\mu_{02}^2\mu_{31}^2\mu_{13} - 4\mu_{11}^3\mu_{40}\mu_{13}^2 + 4\mu_{11}^3\mu_{31}^2\mu_{04}$$
$$-4\mu_{11}^2\mu_{02}\mu_{40}\mu_{31}\mu_{04} + 12\mu_{11}^2\mu_{02}\mu_{40}\mu_{22}\mu_{13} - 8\mu_{11}^2\mu_{02}\mu_{31}^2\mu_{13}$$
$$+\mu_{11}\mu_{02}^2\mu_{40}^2\mu_{04} + 2\mu_{11}\mu_{02}^2\mu_{40}\mu_{31}\mu_{13} - 9\mu_{11}\mu_{02}^2\mu_{40}\mu_{22}^2$$
$$+6\mu_{11}\mu_{02}^2\mu_{31}^2\mu_{22} - \mu_{02}^3\mu_{40}^2\mu_{13} + 3\mu_{02}^3\mu_{40}\mu_{31}\mu_{22} - 2\mu_{02}^3\mu_{31}^3)/\mu_{00}^{15}$$

$w = 9, \mathbf{s} = (3, 0, 3)$

Removing the polynomial dependencies

Polynomial dependencies of orders higher than one among the irreducible invariants pose a serious problem because the dependent invariants cannot be easily identified and removed. This is not a borderline problem which could be ignored. Out of 1589 irreducible invariants of the weight of 12 and less, only 85 invariants at most can be independent, which means that at least 1504 of them must be polynomially dependent. Although an algorithm for a complete removal of dependent invariants is known in principle (it is nothing but an exhaustive search over all possible polynomials), it would be so difficult to implement and so computationally demanding that even for low weights it lies beyond the capacity of current computers. Nevertheless, many dependencies among the irreducible invariants have been discovered. For instance, among the ten invariants from the previous section, there are two polynomial dependencies

$$-4I_1^3I_2^2 + 12I_1^2I_2I_3^2 - 12I_1I_3^4 - I_2I_4^2 + 4I_3^3I_4 - I_5^2 = 0, \tag{5.20}$$

and

$$-16I_1^3I_7^2 - 8I_1^2I_6I_7I_8 - I_1I_6^2I_8^2 + 4I_1I_6I_9^2 + 12I_1I_7I_8I_9 + I_6I_8^2I_9 - I_7I_8^3 - 4I_9^3 - I_{10}^2 = 0. \tag{5.21}$$

This proves that the invariants I_5 and I_{10} are dependent. To obtain a complete and independent set of AMIs up to the fourth order, I_5 and I_{10} are omitted, and one of I_{11}, I_{19} or I_{25} is added to $\{I_1, I_2, I_3, I_4, I_6, I_7, I_8, I_9\}$ (for explicit forms of I_{11}, I_{19} and I_{25}, see Appendix 5.B).

Searching for polynomial dependencies is extremely expensive on its own. However, even if we found (all or some) polynomial dependencies among the irreducible invariants, selecting independent invariants would not be straightforward. In the previous case of linear dependencies, an invariant that has been proven to be a linear combination of other invariants was simply omitted. This cannot be done in the case of polynomial dependencies because the identified dependencies among the invariants may not be independent. Let us illustrate this on a hypothetical example. Assume I_a, I_b, I_c and I_d to be irreducible invariants with these three dependencies:

$$S_1 : \qquad I_a^2 + I_b I_c = 0,$$

$$S_2 : \qquad I_d^2 - I_b I_c^2 = 0,$$

$$S_3 : \; I_a^4 + 2I_a^2 I_b I_c + I_d^2 I_b = 0.$$

If we claimed three invariants to be dependent and we omitted them, it would be a mistake because the third dependency is a combination of the first and the second one

$$S_3 = S_1^2 + I_b S_2$$

and does not bring any new information. Among these four invariants, only two of them are dependent, and two are independent. This is an example of a *second-order dependency*, that is, "dependency among dependencies", while S_1, S_2, and S_3 are so-called first-order dependencies. The second-order dependencies may be of the same kind as the first-order dependencies – we may have identities, products, linear combinations, and polynomials. They can be found in the same way; the algorithm from the previous section requires only minor modifications.

This consideration can be further extended, and we can define kth order dependencies. The number n of independent invariants is then

$$n = n_0 - n_1 + n_2 - n_3 + \ldots , \tag{5.22}$$

where n_0 is the number of the irreducible invariants, n_1 is the number of the first-order dependencies, n_2 is the number of the second-order dependencies, etc. If we consider only the invariants of a certain finite order, this chain is always finite (the proof of finiteness for algebraic invariants originates from Hilbert [14], the proof for moment invariants is essentially the same).

5.3.5 The AMIs and tensors

There is an analogy between the graph method and the derivation of the AMIs by means of tensors. The tensors were employed in the context of the AMIs: for instance, by Reiss [27]. In this section, we briefly show the link between the graph method and the tensor method.

For an explanation of the method of tensors, it is necessary to introduce a concept of unit polyvectors. Tensor $\epsilon_{i_1 i_2 \cdots i_n}$ is a *covariant unit polyvector*, if it is a *skew-symmetric* tensor over all indices and $\epsilon_{12\cdots n} = 1$. The term "skew-symmetric" means that the tensor component changes its sign and preserves its magnitude when interchanging two indices.

In two dimensions, it means that $\varepsilon_{12} = 1$, $\varepsilon_{21} = -1$, $\varepsilon_{11} = 0$ and $\varepsilon_{22} = 0$. Under an affine transformation, a covariant unit polyvector is changed as

$$\hat{\varepsilon}_{\alpha_1 \alpha_2 \cdots \alpha_n} = J^{-1} p_{\alpha_1}^{i_1} p_{\alpha_2}^{i_2} \cdots p_{\alpha_n}^{i_n} \varepsilon_{i_1 i_2 \cdots i_n}, \tag{5.23}$$

that is, it is a relative affine invariant of weight $w = -1$. Then, if we multiply a proper number of moment tensors and unit polyvectors such that the number of upper indices at the moment tensors equals the number of lower indices at the polyvectors, we obtain a real-valued relative affine invariant. For instance

$$M^{ij} M^{klm} M^{nop} \varepsilon_{ik} \varepsilon_{jn} \varepsilon_{lo} \varepsilon_{mp}$$

$$= 2(m_{20}(m_{21}m_{03} - m_{12}^2) - m_{11}(m_{30}m_{03} - m_{21}m_{12}) + m_{02}(m_{30}m_{12} - m_{21}^2)). \tag{5.24}$$

This method is analogous to the method of graphs. Each moment tensor corresponds to a node of the graph, and each unit polyvector corresponds to an edge. The indices show what edge connects which nodes. The graph corresponding to the invariant (5.23) is in Figure 5.7.

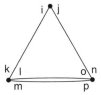

Figure 5.7   The graph corresponding to the invariant from (5.23)

5.4   AMIs via image normalization

In Chapter 3, we demonstrated that the rotation moment invariants can be derived not only in the "direct" way but also by means of image normalization. The normalization approach can also be applied in the case of affine transformation, with similar advantages and drawbacks as in the case of rotation. Recall that in image normalization the object is brought into a certain "normalized" (also "standard" or "canonical") position, which is independent of the actual position, rotation, scale, and skewing of the original object. In this way, the influence of the affine transformation is eliminated. Since the normalized position is the same for all objects differing from each other only by an affine transformation, the moments of the normalized object are in fact affine invariants of the original. The normalized objects can be viewed as representatives of "shape classes".

We emphasize that no actual spatial transformation of the original object is necessary. Such a transformation would slow down the process and would introduce resampling errors. Instead, the moments of the normalized object can be calculated directly from the original object using the normalization constraints. These constraints are formulated by means of low-order moments.

Image normalization to affine transformation is always based on a proper decomposition of the affine transformation into several simpler transformations (usually into six elementary one-parametric transformations). The normalization methods differ from each other by the type of decomposition used and by elementary normalization constraints.

The idea of affine normalization via decomposition was introduced by Rothe et al. [30], where two different decompositions were used – XSR decomposition into skewing, non-uniform scaling and rotation, and XYS decomposition into two skews and non-uniform scaling. However, this approach led to some ambiguities, which were later studied in reference [33] in detail. Pei and Lin [34] presented a method based on the decomposition into two rotations and a non-uniform scaling between them. Their method uses eigenvalues of the covariance matrix and tensors for the computation of the normalization coefficients. However, their method fails on symmetric objects. This is a serious weakness because in many applications we have to classify man-made or specific natural objects that are commonly symmetrical. Since many moments of symmetric objects vanish, the normalization constraints may not be well defined. A similar decomposition scheme with normalization constraints adapted also for symmetric objects was proposed by Suk and Flusser [35]. Shen and Ip [36] used the so-called generalized complex moments for the normalization constraints. Their approach combines moments with the Fourier transformation in polar coordinates and is relatively stable also in the case of symmetric objects, yet exceedingly complicated. Sprinzak

[37] and Heikkilä [38] used Cholesky factorization of the second-order moment matrix for derivation of the normalization constraints.

In this section, we introduce a proper affine decomposition and show how the normalization constraints can be formulated. We also demonstrate that one can consider the moments of the normalized image as another type of AMI, that are, however, theoretically equivalent to the AMIs derived by the graph method. We call them *normalized moments* because of the distinction from the AMIs obtained by other methods.

The normalized moments have certain advantages in comparison with the AMIs, namely the easy creation of a complete and independent set. This set is directly formed by the moments of the normalized image. There is no need of time-consuming algorithms for identification and removal of dependent invariants. On the other hand, the normalized moments are less numerically stable than the AMIs. If some of the constrained moments are calculated with errors, these errors propagate into all normalized moments. Even worse, the normalization may be "discontinuous" on nearly symmetric objects. A small, almost invisible change of the input object may cause a significant change in the normalized position and may, consequently, change all normalized moments in a substantial way. This is the main negative property of the normalization. Another one is that the explicit forms of the normalized moments are more complicated and more computationally demanding than those of the AMIs.

5.4.1 Decomposition of the affine transformation

Decomposition of the affine transformation is not unique. Each decomposition has its own pros and cons. For the purpose of normalization, we prefer the decomposition containing two rotations and a non-uniform scaling. This decomposition avoids skewing, and the normalization constraints to the elementary transformations are formulated in a simple way by geometric and complex moments of low orders. The method is well defined also for objects having an N-FRS. In this method, the affine transformation is decomposed into the following elementary transformations.

Horizontal translation:
$$x' = x + \alpha,$$
$$y' = y. \tag{5.25}$$

Vertical translation:
$$x' = x,$$
$$y' = y + \beta. \tag{5.26}$$

Uniform scaling:
$$x' = \omega x,$$
$$y' = \omega y. \tag{5.27}$$

First rotation:
$$x' = x \cos \alpha - y \sin \alpha,$$
$$y' = x \sin \alpha + y \cos \alpha. \tag{5.28}$$

Stretching:
$$x' = \delta x,$$
$$y' = \frac{1}{\delta} y. \tag{5.29}$$

Second rotation:

$$x' = x \cos \varrho - y \sin \varrho,$$
$$y' = x \sin \varrho + y \cos \varrho. \tag{5.30}$$

Possible mirror reflection:

$$x' = x,$$
$$y' = \pm y. \tag{5.31}$$

These elementary transformations are not commutative; the stretching must always be performed between the two rotations.

In the following sections, we show how to normalize an image with respect to individual elementary transformations.

Normalization to translation

Normalization to translation is analogous to that in the previous methods; we can easily shift the image in such a way that its centroid coincides with the origin. This is ensured by the constraint $m'_{10} = m'_{01} = 0$, which is equivalent to using central moments instead of geometric ones.

Normalization to uniform scaling

Normalization to scaling is expressed by the constraint $\mu'_{00} = 1$, which can be assured by using the scale-normalized moments in exactly the same way as in Chapter 3:

$$\nu_{pq} = \frac{\mu_{pq}}{\mu_{00}^{(p+q)/2+1}}. \tag{5.32}$$

Normalization to the first rotation

Normalization to rotation was already described in Chapter 3. To eliminate the first rotation, we use traditional "principal axis" normalization constrained by $\mu'_{11} = 0$ (i.e., $c'_{20} > 0$). As shown in Chapter 3, this constraint leads to the normalizing angle

$$\alpha = \frac{1}{2} \arctan \left(\frac{2\mu_{11}}{\mu_{20} - \mu_{02}} \right). \tag{5.33}$$

If $\mu_{11} = 0$ and $\mu_{20} = \mu_{02}$, then we consider the image already normalized to rotation and set $\alpha = 0$.

After this normalization to the first rotation, we have for the geometric moments of the normalized image

$$\mu'_{pq} = \sum_{k=0}^{p} \sum_{j=0}^{q} \binom{p}{k} \binom{q}{j} (-1)^{p+j} \sin^{p-k+j}\alpha \cos^{q+k-j}\alpha \, \nu_{k+j,p+q-k-j} \quad . \tag{5.34}$$

Normalization to stretching

Under stretching, the moments are transformed as

$$\mu''_{pq} = \delta^{p-q} \mu'_{pq}$$

(note that the Jacobian of a stretching equals one). Normalization to stretching can be done by imposing a constraint $\mu''_{20} = \mu''_{02}$, which can be fulfilled by choosing

$$\delta = \sqrt[4]{\frac{\mu'_{02}}{\mu'_{20}}}.$$

In terms of the moments of the original image, we have

$$\delta = \sqrt{\frac{\mu_{20} + \mu_{02} - \sqrt{(\mu_{20} - \mu_{02})^2 + 4\mu_{11}^2}}{2\sqrt{\mu_{20}\mu_{02} - \mu_{11}^2}}} \qquad (5.35)$$

(this is well defined because $\mu_{20}\mu_{02} \geq \mu_{11}^2$ always and, moreover, for nondegenerated 2D objects it holds $\mu_{20}\mu_{02} > \mu_{11}^2$) and, for the moments normalized both to the first rotation and stretching,

$$\mu''_{pq} = \delta^{p-q} \sum_{k=0}^{p} \sum_{j=0}^{q} \binom{p}{k} \binom{q}{j} (-1)^{p+j} \ \sin^{p-k+j}\alpha \ \cos^{q+k-j}\alpha \ v_{k+j,p+q-k-j} \qquad (5.36)$$

After this normalization, c''_{20} becomes zero and cannot be further used for another normalization.

Normalization to the second rotation

For normalization to the second rotation, we adopt the Abu-Mostafa and Psaltis [39] normalization scheme by means of complex moments that was used already in Chapter 3. The normalized position is defined in such a way that a certain nonzero moment c_{st} is required to be real and positive. As proven in Chapter 3, it is always possible, but for $s - t > 1$ the normalized position is ambiguous.

Normalization to the second rotation is a critical step. While small errors and inaccuracies in the previous stages can be compensated in the later steps, there is no chance to fix the errors in normalization to the second rotation. The selection of the normalizing moment must be done very carefully. If c''_{21} was nonzero, it would be an optimal choice. However, as we saw in Chapter 3, many complex moments including c''_{21} are zero for images having a certain degree of symmetry. Note that we are speaking about symmetry after normalization of the image with respect to stretching, which may be different from the symmetry of the original. For instance, a rectangle has a fourfold dihedral symmetry after the normalization to stretching, and an ellipse has a circular symmetry after this normalization.

The normalizing moment can be found in the same way as described in Chapter 3 – we sort the moments and look for the first moment whose magnitude exceeds the given threshold. The only difference here is that we cannot consider c''_{20} because, due to the previous normalization, $c''_{20} = 0$.

As soon as the normalizing moment has been determined, the normalizing angle ϱ is given as

$$\varrho = \frac{1}{s-t} \arctan \left(\frac{\mathcal{I}m(c''_{st})}{\mathcal{R}e(c''_{st})} \right). \tag{5.37}$$

Finally, the moments of the object normalized to the second rotation are calculated by means of a similar formula to that used in the first rotation

$$\tau_{pq} = \sum_{k=0}^{p} \sum_{j=0}^{q} \binom{p}{k} \binom{q}{j} (-1)^{p+j} \ \sin^{p-k+j} \varrho \ \cos^{q+k-j} \varrho \ \mu''_{k+j,p+q-k-j} \quad . \tag{5.38}$$

The normalized moments τ_{pq} are new AMIs of the original object. Some of them have "prescribed" values due to the normalization, regardless of image:

$$\tau_{00} = 1, \ \tau_{10} = 0, \ \tau_{01} = 0, \ \tau_{02} = \tau_{20}, \ \tau_{11} = 0, \ \tau_{03} = -\tau_{21}. \tag{5.39}$$

The last equality follows from the normalization to the second rotation and holds only if the normalizing moment was c''_{21} or if $c''_{21} = 0$ because of symmetry; otherwise it is replaced by an analogous condition for other moments.

Normalization to mirror reflection

Although the general affine transformation may contain a mirror reflection, normalization to the mirror reflection should be done separately from the other partial transformations in equations (5.25)–(5.30) for practical reasons. In most affine deformations occurring in practice, no mirror reflection can in principle be present, and it may be necessary to classify mirrored images into different classes (in character recognition we certainly want to distinguish capital S from a question mark, for instance). Normalization to mirror reflection is not desirable in those cases.

If we still want to normalize objects to mirror reflection, we can do that, after all normalizations mentioned above, as follows. We find the first nonzero moment τ_{pq} with an odd second index. If $\tau_{pq} < 0$, then we change the signs of all moments with odd second indices. If $\tau_{pq} > 0$ or if all normalized moments up to the chosen order with odd second indices are zero, no action is necessary.

5.4.2 *Relation between the normalized moments and the AMIs*

In this section we show the close link between the normalized moments and the AMIs derived by the graph method.

The normalization is nothing but a special affine transformation, so the AMIs must be invariant to it. If we substitute τ_{pq} for μ_{pq} in the AMIs, considering the constraints (5.39), we obtain for the second and third orders

$$I_1 = \tau_{20}^2,$$
$$I_2 = -\tau_{30}^2\tau_{21}^2 - 6\tau_{30}\tau_{21}^2\tau_{12} - 4\tau_{30}\tau_{12}^3 + 4\tau_{21}^4 + 3\tau_{21}^2\tau_{12}^2,$$
$$I_3 = \tau_{20}(-2\tau_{21}^2 - \tau_{12}^2 + \tau_{30}\tau_{12}),$$ (5.40)
$$I_4 = -\tau_{20}^3(4\tau_{21}^2 + 3\tau_{12}^2 + \tau_{30}^2).$$

For the fourth order, we obtain

$$I_6 = \tau_{40}\tau_{04} - 4\tau_{31}\tau_{13} + 3\tau_{22}^2,$$
$$I_7 = \tau_{40}\tau_{22}\tau_{04} - \tau_{40}\tau_{13}^2 - \tau_{31}^2\tau_{04} + 2\tau_{31}\tau_{22}\tau_{13} - \tau_{22}^3,$$
$$I_8 = \tau_{20}^2(\tau_{04} + 2\tau_{22} + \tau_{40}),$$ (5.41)
$$I_9 = \tau_{20}^2(\tau_{40}\tau_{04} + \tau_{40}\tau_{22} + \tau_{22}\tau_{04} + \tau_{22}^2 - \tau_{13}^2 - 2\tau_{31}\tau_{13} - \tau_{31}^2),$$
$$I_{25} = \tau_{20}^3(\tau_{12} + \tau_{30})^2(\tau_{04} + \tau_{22}).$$

Similar relations can also be obtained for higher-order invariants.

An inverse problem – calculating the normalized moments from the AMIs – can be resolved by solving the above equations for τ_{pq}. The system is not linear and does not have an analytical solution for orders higher than four. It must be solved numerically in those cases. Doing so, we can find a one-to-one mapping between the two sets of invariants. It is naturally impractical to calculate the normalized moments in this way, but this is a suitable illustration of the theoretical equivalence of the both sets.

5.4.3   Violation of stability

In this simulated experiment, we illustrate the potential vulnerability of the normalized moments. Observe that under certain circumstances the normalization to the second rotation might become unstable and "discontinuous". Even if it is probably a rare case in real applications, one should be aware of this danger.

Let us consider a binary fourfold cross inscribed in a square with a side a. Such a cross has only one parameter – the thickness b, for which $0 < b \leq a$ (see Figure 5.8a). Since the cross has a fourfold symmetry, one would expect c''_{40} to be chosen for the normalization to the second rotation. It holds for the cross

$$c''_{40} = (3a^5b + 3ab^5 + 2b^6 - 10a^3b^3)/120$$ (5.42)

which is generally non-zero, but if $b \to 0.593a$ then $c''_{40} \to 0$. If we take two very similar crosses – one with b slightly less than $0.593a$ and the other one with b slightly greater – then c''_{40} changes its sign and, consequently, the standard position changes from a horizontal-vertical "+" shape to a diagonal "×" shape (see Figure 5.8 for illustration). This leads to discontinuities in the normalized moments and, finally, to false results of shape recognition. The AMI's are not plagued by these problems because they are continuous in the neighborhood of the critical point.

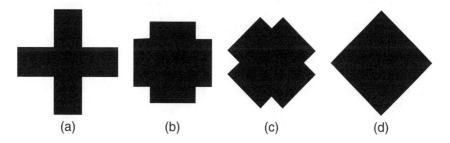

Figure 5.8 The standard positions of the cross with varying thickness b: (a) Thin cross, $b \ll a$, (b) b slightly less than $0.593a$, (c) b slightly greater than $0.593a$, and (d) $b = a$. Note the difference between the two middle positions

5.4.4 Affine invariants via half normalization

Half normalization is a "hybrid" approach to constructing invariants, trying to keep the positive properties of normalization as well as direct construction while avoiding the negative ones. This idea was first introduced by Heikkilä [38].

The first stage of the method is the same as the normalization described previously. The image is normalized to translation, uniform scaling, the first rotation, and stretching, which ends up with the normalized moments μ_{pq}'' and c_{pq}''. The normalization to the second rotation is not performed. Instead, the partially normalized image is described by proper rotation invariants as described in Chapter 3 (potential symmetry of the normalized image must be taken into account). Thus, we obtain affine invariants for instance in the form $c_{pq}'' c_{12}''^{p-q}$.

The half normalization method is, of course, theoretically equivalent to both AMIs and normalized moments but may have certain favorable numerical properties. Avoiding normalization to the second rotation, we are rid of the danger of instability of the normalized position.

5.4.5 Affine invariants from complex moments

In affine normalization (including the half normalization), some constraints are formulated by means of complex moments, and the resulting invariants can be considered both in terms of geometric as well as complex moments. In Chapter 3, we saw that the complex moments provide an excellent insight into the behavior of invariants on symmetric images. For images having an N-FRS, we know precisely which complex moments are zero and can select only nontrivial invariants.

In contrast, the AMIs have been expressed as functions of geometric moments only, thus not providing any hints as to which invariants should be avoided when recognizing symmetric objects. To overcome this hindrance, we show how AMIs can be expressed by means of the complex moments. Such relations also show a link between the AMIs and rotation invariants from Chapter 3.

Theorem 5.2 Let us denote the value of invariant I computed from the geometric moments as $I(\mu_{pq})$ and the value obtained when replacing μ_{pq} by c_{pq} as $I(c_{pq})$. Then it holds

$$I(c_{pq}) = (-2i)^w I(\mu_{pq}),$$ (5.43)

where w is the weight of invariant I.

Proof. Consider the affine transformation

$$\begin{aligned} x' &= x + iy \\ y' &= x - iy. \end{aligned}$$ (5.44)

Under this transformation, the geometric moments are changed as

$$\mu'_{pq} = \int\limits_{-\infty}^{\infty} \int\limits_{-\infty}^{\infty} (x+iy)^p (x-iy)^q |J| f(x,y) \, dx \, dy = |J| c_{pq},$$ (5.45)

where the Jacobian $J = -2i$. It follows from Theorem 5.1 that

$$I(\mu'_{pq}) = J^w |J|^r I(\mu_{pq}),$$ (5.46)

where r is a degree of I. On the other hand, from (5.45) we have

$$I(\mu'_{pq}) = I(|J| c_{pq}) = |J|^r I(c_{pq}).$$ (5.47)

Comparing (5.46) and (5.47), we immediately obtain (5.43).   □

In particular, we obtain for I_1 and I_2

$$(-2i)^2 I_1 = c_{20} c_{02} - c_{11}^2$$ (5.48)

and

$$(-2i)^6 I_2 = -c_{30}^2 c_{03}^2 + 6 c_{30} c_{21} c_{12} c_{03} - 4 c_{30} c_{12}^3 - 4 c_{21}^3 c_{03} + 3 c_{21}^2 c_{12}^2.$$ (5.49)

Similar relations can be derived for all other AMIs.

These formulas are useful not only for choosing proper invariants in case of certain symmetry but also demonstrate the connection between the AMIs and the basic rotation invariants $\Phi(p,q)$.

This link can be advantageously used to prove the independency and completeness of a set of AMIs. Since the basis of rotation invariants is complete and independent, it is sufficient to find a one-to-one mapping between the AMIs and the rotation invariants.

Let us assume that the image has undergone a half normalization. The half normalization reduces an affine transformation to a rotation and imposes $c_{00} = 1$, $c_{10} = c_{01} = 0$ and $c_{20} = c_{02} = 0$.[8] Under these constraints we obtain from Theorem 5.2

$$(-2i)^2 I_1 = -c_{11}^2,$$

$$(-2i)^6 I_2 = -c_{30}^2 c_{03}^2 + 6 c_{30} c_{21} c_{12} c_{03} - 4 c_{30} c_{12}^3 - 4 c_{21}^3 c_{03} + 3 c_{21}^2 c_{12}^2,$$

---

[8] To be precise, we should write $c_{00}'' = 1$, $c_{10}'' = c_{01}'' = 0$, etc. For the sake of similarity, we omit the double-primes in this section.

$$(-2i)^4 I_3 = -c_{11}c_{30}c_{03} + c_{11}c_{21}c_{12},$$
$$(-2i)^6 I_4 = 2c_{11}^3 c_{30}c_{03} + 6c_{11}^3 c_{21}c_{12},$$
$$(-2i)^4 I_6 = c_{40}c_{04} - 4c_{31}c_{13} + 3c_{22}^2,$$
$$(-2i)^6 I_7 = c_{40}c_{22}c_{04} - c_{40}c_{13}^2 - c_{31}^2 c_{04} + 2c_{31}c_{22}c_{13} - c_{22}^3,$$
$$(-2i)^4 I_8 = 4c_{11}^2 c_{22},$$
$$(-2i)^6 I_9 = 4c_{11}^2 c_{31}c_{13} - 4c_{11}^2 c_{22}^2,$$
$$(-2i)^8 I_{25} = -8c_{11}^3 c_{21}^2 c_{13} + 16c_{11}^3 c_{21}c_{12}c_{22} - 8c_{11}^3 c_{12}^2 c_{31}.$$

$$(5.50)$$

These equations can be inverted and resolved for rotation invariants in a much easier way than similar equations for normalized moments (5.41). We obtain

$$c_{11} = 2\sqrt{I_1},$$

$$c_{21}c_{12} = \frac{1}{\sqrt{I_1}}\left(2I_3 - \frac{I_4}{I_1}\right),$$

$$c_{30}c_{03} = -\frac{1}{\sqrt{I_1}}\left(6I_3 + \frac{I_4}{I_1}\right),$$

$$\mathcal{R}e(c_{30}c_{12}^3) = 8I_2 - 12\frac{I_3^2}{I_1} + \frac{I_4^2}{I_1^3},$$

$$c_{22} = \frac{I_8}{I_1},$$

$$c_{31}c_{13} = \frac{I_8^2}{I_1^2} - 4\frac{I_9}{I_1},$$

$$(5.51)$$

$$c_{40}c_{04} = 16I_6 + \frac{I_8^2}{I_1^2} - \frac{I_9}{I_1},$$

$$\mathcal{R}e(c_{40}c_{13}^2) = 32I_7 + 8\frac{I_6 I_8}{I_1} - 12\frac{I_8 I_9}{I_1^2} + \frac{I_8^3}{I_1^3},$$

$$\mathcal{R}e(c_{31}c_{12}^2) = \frac{1}{\sqrt{I_1}}\left(2\frac{I_3 I_8}{I_1} - \frac{I_4 I_8}{I_1^2} - 2\frac{I_{25}}{I_1}\right).$$

The set of rotation invariants $\{c_{11}, c_{21}c_{12}, c_{30}c_{03}, \mathcal{R}e(c_{30}c_{12}^3), c_{22}, c_{31}c_{13}, c_{40}c_{04}, \mathcal{R}e(c_{40}c_{13}^2),$ $\mathcal{R}e(c_{31}c_{12}^2)\}$ is independent, which implies that the set of AMIs $\{I_1, I_2, I_3, I_4, I_6, I_7, I_8, I_9, I_{25}\}$ must be independent, too.

For higher-order invariants, it is difficult to resolve the equations with respect to the rotation invariants in general. As proposed in references [31] and [40], for the independence test it is sufficient to analyze one particular point in the space of the invariants, in which most rotation invariants are ± 1 and many AMIs equal zero. A detailed description of this method is beyond the scope of the book, but it provides the most successful current method for independence tests of the AMIs. Thereby, the independence of the following set of the AMIs of the weight $w \le 12$ has been proven: $\{I_1, I_2, I_3, I_4, I_6, I_7, I_8, I_9, I_{19}, I_{47}, I_{48}, I_{49}, I_{50}, I_{64}, I_{65}, I_{249}, I_{251}, I_{252}, I_{253}, I_{254}, I_{256}, I_{270}, I_{587}, I_{588}, I_{599}, I_{600}, I_{601}, I_{602}, I_{603}, I_{604}, I_{960}, I_{962}, I_{963}, I_{965}, I_{973}, I_{974}, I_{975}, I_{976}, I_{990}, I_{1241}, I_{1243}, I_{1249}, I_{1250}, I_{1251}, I_{1267}, I_{1268}, I_{1284}, I_{1285}, I_{1297}, I_{1423}, I_{1424}, I_{1425}, I_{1427}, I_{1429}, I_{1430}, I_{1434}, I_{1436}, I_{1437}, I_{1438}, I_{1456}, I_{1524}, I_{1525}, I_{1526}, I_{1527}, I_{1528}, I_{1531}, I_{1532}, I_{1536}, I_{1537}, I_{1538}, I_{1542}, I_{1545}, I_{1569}, I_{1570}, I_{1571}, I_{1574}, I_{1575}, I_{1576}, I_{1578}, I_{1579}\}$. This is the largest set of independent AMIs ever published; it contains eighty invariants. The explicit formulas of these AMIs can be found on the accompanying website [32].

5.5   The method of the transvectants

This method for the design of the AMIs has been proposed very recently by Hickman [29], who employed the theory of algebraic invariants described 100 years ago in [18]. The term *transvectant* denotes an invariant formed from M invariants of M variables as a result of so-called Cayley's Omega process [41]. Comparing to the graph method, the method of transvectants is much more complicated and less transparent but directly generates an independent set of invariants.

Let us consider the Kronecker (block or direct)[9] product of two matrices

$$\mathbf{A} \otimes \mathbf{B} = \begin{pmatrix} a_{11}\mathbf{B} & a_{12}\mathbf{B} & \cdots & a_{1n}\mathbf{B} \\ a_{21}\mathbf{B} & a_{22}\mathbf{B} & \cdots & a_{2n}\mathbf{B} \\ \vdots & \vdots & \ddots & \vdots \\ a_{m1}\mathbf{B} & a_{m2}\mathbf{B} & \cdots & a_{mn}\mathbf{B} \end{pmatrix}, \tag{5.52}$$

where a_{ij} are elements of $m \times n$ matrix $\mathbf{A}$. We can replace the matrices by vectors. If $\mathbf{x}_{(1)} = (x, y)^T$ then

$$\mathbf{x}_{(2)} = \mathbf{x}_{(1)} \otimes \mathbf{x}_{(1)} = (x^2, xy, yx, y^2)^T,$$
$$\mathbf{x}_{(3)} = \mathbf{x}_{(1)} \otimes \mathbf{x}_{(2)} = (x^3, x^2y, xyx, xy^2, yx^2, yxy, y^2x, y^3)^T, \tag{5.53}$$
$$\vdots$$

generally

$$\mathbf{x}_{(d)} = \overset{d}{\underset{i=1}{\otimes}} \mathbf{x}_{(1)}. \tag{5.54}$$

If we multiply each element of the vector $\mathbf{x}_{(d)}$ by an image function and integrate, we obtain a vector of moments

$$\mathbf{m}_{(1)} = (m_{10}, m_{01})^T,$$
$$\mathbf{m}_{(2)} = (m_{20}, m_{11}, m_{11}, m_{12})^T,$$
$$\mathbf{m}_{(3)} = (m_{30}, m_{21}, m_{21}, m_{12}, m_{21}, m_{12}, m_{12}, m_{03})^T, \tag{5.55}$$
$$\vdots$$

---

[9] The term "tensor product" is not precise, because the tensor product is stored in four-dimensional array.

For convenience we define

$$\mathbf{x}_{(0)} = 1, \quad \mathbf{m}_{(0)} = (m_{00}).$$

Hickman started with designing invariants to 2D rotation. The dot product of the coordinate vector and the moment vector is a rotation covariant. The term "covariant" in this context means an invariant computed from two types of measurements, in this case from the coordinates and from the moments

$$\eta_d = \mathbf{x}_{(d)}^T \mathbf{m}_{(d)} = \sum_{i=0}^{d} \binom{d}{i} x^{d-i} y^i m_{d-i,i}. \tag{5.56}$$

The above definition is used for $d > 0$. The case $d = 0$ is defined separately as

$$\eta_0 = x^2 + y^2.$$

The rth transvectant of two covariants is defined in the case of rotation as

$$\Lambda^r(P, Q) = (\partial_{(r)} P)^T \partial_{(r)} Q, \tag{5.57}$$

where P and Q are some covariants. The rth derivative $\partial_{(r)}$ is again meant as a formal Kronecker product of the first derivatives

$$\partial_{(1)} = \left(\frac{\partial}{\partial x}, \frac{\partial}{\partial y} \right)^T, \quad \partial_{(r)} = \bigotimes_{i=1}^{d} \partial_{(1)}.$$

The derivatives in the transvectant decrease the degree of the coordinate polynomial in the covariants. When we use the transvectant of such degree r that the coordinates vanish and the moments remain, we obtain an invariant. For even order $d = 2q$ we use the formulas

$$\sigma_{0d} = \Lambda^d(\eta_0^q, \eta_d),$$
$$\sigma_{1d} = \Lambda^d(\eta_0^{q-2}\eta_2^2, \eta_d), \qquad d > 2$$
$$\sigma_{id} = \begin{cases} \Lambda^d(\eta_0^{q-i/2}\eta_i, \eta_d), & i = 2, 4, \ldots, d, \\ \Lambda^d(\eta_0^{q-i}\eta_i^2, \eta_d), & i = 3, 5, \ldots, s, \\ \Lambda^d(\Lambda^{i-q}(\eta_i, \eta_i), \eta_d), & i = s+2, s+4, \ldots, d-1, \end{cases} \tag{5.58}$$

where $s = 2[(q+1)/2] - 1$ and $\sigma_{12} = \Lambda^2(\eta_1^2, \eta_2)$ is defined separately. For odd order $d = 2q + 1$ the formulas are different

$$\sigma_{0d} = \Lambda^{2d}(\eta_0^d, \eta_d^2),$$
$$\sigma_{1d} = \Lambda^d(\eta_0^{(d-3)/2}\eta_3, \Lambda^1(\eta_2, \eta_d)), \qquad d > 3$$
$$\sigma_{id} = \begin{cases} \Lambda^d(\eta_0^{(d-3-i)/2}\eta_3\eta_i, \eta_d), & i = 2, 4, \ldots, d-3, \\ \Lambda^d(\eta_0^{(d-i)/2}\eta_i, \eta_d), & i = 3, 5, \ldots, d, \\ \Lambda^{d+2}(\eta_3\eta_{d-1}, \eta_0\eta_d), & i = d-1 \end{cases} \tag{5.59}$$

and $\sigma_{13} = \Lambda^2(\eta_2, \Lambda^2(\eta_0, \eta_3^2))$ is defined separately.

In [29], there is a proof that the set of σ_{id}, where $d > 1$ is the order and $i = 0, 1, \cdots d$ is the serial number, is complete and independent; that is, it is a basis of rotation invariants. This way of the basis construction is more complicated than that based on the complex moments presented in Chapter 3, but it is a good starting point for derivation of affine invariants.

We slightly modify equation (5.56) by incorporating so-called *simplectic matrix* $\mathbf{J}_d$

$$v_d = \mathbf{x}_{(d)}^T \mathbf{J}_d \mathbf{m}_{(d)} = \sum_{i=0}^{d} (-1)^i \binom{d}{i} x^{d-i} y^i m_{i,d-i}. \tag{5.60}$$

The simplectic matrix is defined as

$$\mathbf{J}_d = \overset{d}{\underset{i=1}{\otimes}} \mathbf{J},$$

where

$$\mathbf{J} = \begin{pmatrix} 0 & 1 \\ -1 & 0 \end{pmatrix}.$$

After this modification, v_d is an affine covariant. The definition of the transvectant is analogous

$$\Lambda^r(P, Q) = (\partial_{(r)}P)^T \mathbf{J}_r \partial_{(r)}Q. \tag{5.61}$$

The transvectant of two equal covariants has a special notation

$$\mathcal{D}^r(P) = \Lambda^r(P, P).$$

The formulas for AMIs of even order $d = 2q \geq 4$ are then

$$\theta_{id} = \begin{cases} \Lambda^4(v_4, \mathcal{D}^{d-2}(v_d)), & i = 0, \\ \Lambda^4(v_2^2, \mathcal{D}^{d-2}(v_d)), & i = 1, \\ \Lambda^d(v_i v_{d-i}, v_d), & i = 2, 3, \ldots, q, \\ \Lambda^d(v_2 \mathcal{D}^2(v_i), v_d), & i = q+1, \\ \Lambda^d(\mathcal{D}^{i-q}(v_i), v_d), & i = q+2, q+4, \ldots, d-1, \\ \Lambda^d(\Lambda^1(v_i, v_{d-i+2}), v_d), & i = q+3, q+5, \ldots, d-1, \\ \mathcal{D}^d(v_d), & i = d. \end{cases} \tag{5.62}$$

For odd order $d = 2q + 1 \geq 5$, we have

$$\theta_{id} = \begin{cases} \Lambda^2(v_2 \mathcal{D}^{d-1}(v_d)), & i = 0, \\ \Lambda^2(\mathcal{D}^2(v_3), \mathcal{D}^{d-1}(v_d)), & i = 1, \\ \Lambda^d(v_i v_{d-i}, v_d), & i = 2, 3, \ldots, q, \\ \Lambda^d(\Lambda^1(v_2 v_q, v_{q+1}), v_d), & i = q+1^{10}, \\ \Lambda^d(\Lambda^1(v_i, v_{d-i+2}), v_d), & i = q+2, q+3, \ldots, d-1 \\ \mathcal{D}^2(\mathcal{D}^{d-1}(v_d)), & i = d. \end{cases} \tag{5.63}$$

The invariants

$$\theta_{12} = \Lambda^2(v_1^2, v_2),$$
$$\theta_{22} = \mathcal{D}^2(v_2),$$
$$\theta_{03} = \Lambda^2(v_2, \mathcal{D}^2(v_3)),$$
$$\theta_{13} = \Lambda^3(v_1^3, v_3),$$
$$\theta_{23} = \Lambda^6(v_2^3, v_3^2),$$
$$\theta_{33} = \mathcal{D}^2(\mathcal{D}^2(v_3))$$

are defined separately.

We can write down the relative invariants up to the fifth order in explicit form. To obtain absolute invariants, the normalization by an appropriate power of μ_{00} is required.

$$\theta_{12} = m_{10}^2 m_{02} - 2m_{10}m_{01}m_{11} + m_{01}^2 m_{20}$$

$$\theta_{22} = m_{20}m_{02} - m_{11}^2$$

$$\theta_{03} = m_{20}m_{21}m_{03} - m_{20}m_{12}^2 - m_{11}m_{30}m_{03} + m_{11}m_{21}m_{12} + m_{02}m_{30}m_{12} - m_{02}m_{21}^2$$

$$\theta_{13} = m_{10}^3 m_{03} - 3m_{10}^2 m_{01}m_{12} + 3m_{10}m_{01}^2 m_{21} - m_{01}^3 m_{30}$$

$$\theta_{23} = 5m_{20}^3 m_{03}^2 - 30m_{20}^2 m_{11}m_{12}m_{03} + 6m_{20}^2 m_{02}m_{21}m_{03} + 9m_{20}^2 m_{02}m_{12}^2$$
$$+ 24m_{20}m_{11}^2 m_{21}m_{03} + 36m_{20}m_{11}^2 m_{12}^2 - 6m_{20}m_{11}m_{02}m_{30}m_{03}$$
$$- 54m_{20}m_{11}m_{02}m_{21}m_{12} + 6m_{20}m_{02}^2 m_{30}m_{12} + 9m_{20}m_{02}^2 m_{21}^2 - 4m_{11}^3 m_{30}m_{03}$$
$$- 36m_{11}^3 m_{21}m_{12} + 24m_{11}^2 m_{02}m_{30}m_{12} + 36m_{11}^2 m_{02}m_{21}^2 - 30m_{11}m_{02}^2 m_{30}m_{21}$$
$$+ 5m_{02}^3 m_{30}^2$$

$$\theta_{33} = -m_{30}^2 m_{03}^2 + 6m_{30}m_{21}m_{12}m_{03} - 4m_{30}m_{12}^3 - 4m_{21}^3 m_{03} + 3m_{21}^2 m_{12}^2$$

$$\theta_{04} = m_{40}m_{22}m_{04} - m_{40}m_{13}^2 - m_{31}^2 m_{04} + 2m_{31}m_{22}m_{13} - m_{22}^3$$

$$\theta_{14} = 3m_{20}^2 m_{22}m_{04} - 3m_{20}^2 m_{13}^2 - 6m_{20}m_{11}m_{31}m_{04} + 6m_{20}m_{11}m_{22}m_{13}$$
$$+ m_{20}m_{02}m_{40}m_{04} + 2m_{20}m_{02}m_{31}m_{13} - 3m_{20}m_{02}m_{22}^2 + 2m_{11}^2 m_{40}m_{04}$$
$$+ 4m_{11}^2 m_{31}m_{13} - 6m_{11}^2 m_{22}^2 - 6m_{11}m_{02}m_{40}m_{13} + 6m_{11}m_{02}m_{31}m_{22}$$
$$+ 3m_{02}^2 m_{40}m_{22} - 3m_{02}^2 m_{31}^2$$

$$\theta_{24} = m_{20}^2 m_{04} - 4m_{20}m_{11}m_{13} + 2m_{20}m_{02}m_{22} + 4m_{11}^2 m_{22} - 4m_{11}m_{02}m_{31} + m_{02}^2 m_{40}$$

$$\theta_{34} = m_{20}m_{30}m_{12}m_{04} - m_{20}m_{30}m_{03}m_{13} - m_{20}m_{21}^2 m_{04} + m_{20}m_{21}m_{12}m_{13}$$
$$+ m_{20}m_{21}m_{03}m_{22} - m_{20}m_{12}^2 m_{22} - 2m_{11}m_{30}m_{12}m_{13} + 2m_{11}m_{30}m_{03}m_{22}$$
$$+ 2m_{11}m_{21}^2 m_{13} - 2m_{11}m_{21}m_{12}m_{22} - 2m_{11}m_{21}m_{03}m_{31} + 2m_{11}m_{12}^2 m_{31}$$
$$+ m_{02}m_{30}m_{12}m_{22} - m_{02}m_{30}m_{03}m_{31} - m_{02}m_{21}^2 m_{22} + m_{02}m_{21}m_{12}m_{31}$$

---

[10] This formula was stated incorrectly in [29] as $\Lambda^d(\Lambda^2(v_2 v_q, v_{q+1}), v_d)$.

$$+ m_{02}m_{21}m_{03}m_{40} - m_{02}m_{12}^2m_{40}$$

$$\theta_{44} = m_{40}m_{04} - 4m_{31}m_{13} + 3m_{22}^2$$

$$\theta_{05} = m_{20}m_{41}m_{05} - 4m_{20}m_{32}m_{14} + 3m_{20}m_{23}^2 - m_{11}m_{50}m_{05} + 3m_{11}m_{41}m_{14}$$
$$- 2m_{11}m_{32}m_{23} + m_{02}m_{50}m_{14} - 4m_{02}m_{41}m_{23} + 3m_{02}m_{32}^2$$

$$\theta_{15} = 2m_{30}m_{12}m_{41}m_{05} - 8m_{30}m_{12}m_{32}m_{14} + 6m_{30}m_{12}m_{23}^2 - m_{30}m_{03}m_{50}m_{05}$$
$$+ 3m_{30}m_{03}m_{41}m_{14} - 2m_{30}m_{03}m_{32}m_{23} - 2m_{21}^2m_{41}m_{05} + 8m_{21}^2m_{32}m_{14}$$
$$- 6m_{21}^2m_{23}^2 + m_{21}m_{12}m_{50}m_{05} - 3m_{21}m_{12}m_{41}m_{14} + 2m_{21}m_{12}m_{32}m_{23}$$
$$+ 2m_{21}m_{03}m_{50}m_{14} - 8m_{21}m_{03}m_{41}m_{23} + 6m_{21}m_{03}m_{32}^2 - 2m_{12}^2m_{50}m_{14}$$
$$+ 8m_{12}^2m_{41}m_{23} - 6m_{12}^2m_{32}^2$$

$$\theta_{25} = m_{20}m_{30}m_{05} - 3m_{20}m_{21}m_{14} + 3m_{20}m_{12}m_{23} - m_{20}m_{03}m_{32} - 2m_{11}m_{30}m_{14}$$
$$+ 6m_{11}m_{21}m_{23} - 6m_{11}m_{12}m_{32} + 2m_{11}m_{03}m_{41} + m_{02}m_{30}m_{23} - 3m_{02}m_{21}m_{32}$$
$$+ 3m_{02}m_{12}m_{41} - m_{02}m_{03}m_{50}$$

$$\theta_{35} = m_{20}^2m_{21}m_{05} - 2m_{20}^2m_{12}m_{14} + m_{20}^2m_{03}m_{23} - m_{20}m_{11}m_{30}m_{05}$$
$$- m_{20}m_{11}m_{21}m_{14} + 5m_{20}m_{11}m_{12}m_{23} - 3m_{20}m_{11}m_{03}m_{32} + m_{20}m_{02}m_{30}m_{14}$$
$$- m_{20}m_{02}m_{21}m_{23} - m_{20}m_{02}m_{12}m_{32} + m_{20}m_{02}m_{03}m_{41} + 2m_{11}^2m_{30}m_{14}$$
$$- 2m_{11}^2m_{21}m_{23} - 2m_{11}^2m_{12}m_{32} + 2m_{11}^2m_{03}m_{41} - 3m_{11}m_{02}m_{30}m_{23}$$
$$+ 5m_{11}m_{02}m_{21}m_{32} - m_{11}m_{02}m_{12}m_{41} - m_{11}m_{02}m_{03}m_{50} + m_{02}^2m_{30}m_{32}$$
$$- 2m_{02}^2m_{21}m_{41} + m_{02}^2m_{12}m_{50}$$

$$\theta_{45} = -m_{30}m_{31}m_{05} + 3m_{30}m_{22}m_{14} - 3m_{30}m_{13}m_{23} + m_{30}m_{04}m_{32} + m_{21}m_{40}m_{05}$$
$$- m_{21}m_{31}m_{14} - 3m_{21}m_{22}m_{23} + 5m_{21}m_{13}m_{32} - 2m_{21}m_{04}m_{41} - 2m_{12}m_{40}m_{14}$$
$$+ 5m_{12}m_{31}m_{23} - 3m_{12}m_{22}m_{32} - m_{12}m_{13}m_{41} + m_{12}m_{04}m_{50} + m_{03}m_{40}m_{23}$$
$$- 3m_{03}m_{31}m_{32} + 3m_{03}m_{22}m_{41} - m_{03}m_{13}m_{50}$$

$$\theta_{55} = -m_{50}^2m_{05}^2 + 10m_{50}m_{41}m_{14}m_{05} - 4m_{50}m_{32}m_{23}m_{05} - 16m_{50}m_{32}m_{14}^2$$
$$+ 12m_{50}m_{23}^2m_{14} - 16m_{41}^2m_{23}m_{05} - 9m_{41}^2m_{14}^2 + 12m_{41}m_{32}^2m_{05}$$
$$+ 76m_{41}m_{32}m_{23}m_{14} - 48m_{41}m_{23}^3 - 48m_{32}^3m_{14} + 32m_{32}^2m_{23}^2$$

If we use central moments to achieve translation invariance, the invariants θ_{12} and θ_{13} vanish.

Comparing the Hickman's method to the graph method, the major difference is that the Hickman's set is independent. We can see that most of the Hickman's invariants are equal to some AMI while the others are linear combinations of the AMIs or of their products, such as $\theta_{23} = -5I_4 + 6I_1I_3$ and $\theta_{14} = 3I_9 - 2I_1I_6$. This leads to slightly more complicated explicit

Table 5.2   The independent graph-generated AMIs up to the sixth order, which were selected thanks to their correspondence with the Hickman's set

Order	Independent invariants
2	I_1
3	I_2, I_3, I_4
4	$I_6, I_7, I_8, I_9, I_{19}$
5	$I_{47}, I_{48}, I_{54}, I_{64}, I_{65}, I_{120}$
6	$I_{249}, I_{252}, I_{256}, I_{263}, I_{269}, I_{291}, I_{294}$

formulas. To avoid this and to preserve independence, we can use the Hickman's set to identify independent AMIs. In Table 5.2, one can see such selection up to the order six.

Hickman intentionally excluded the pseudoinvariants (i.e., the invariants that change the sign under mirror reflection) from his set. If we need to distinguish between mirrored images, we have to include at least one AMI of an odd weight.

5.6   Derivation of the AMIs from the Cayley-Aronhold equation

In this section, we present yet another method for deriving AMIs. This method uses neither the theory of the algebraic invariants nor any normalization. It also does not generate all possible invariants, as does the graph method. The basic idea is simple, and its "manual" implementation for finding a few low-order AMIs is easy (this approach has been used already in references [16] and [42]). The solution can be automated [28], which leads to a method of polynomial complexity as compared to the exponential complexity of the graph method. On the other hand, the implementation and programming of this method is much more difficult than that of the graph method.

5.6.1   Manual solution

This method relies on a decomposition of the affine transformation into elementary transformations. This is a similar idea as in the normalization approach, but here we do not perform any normalization, and the decomposition is different from the one we used for the normalization purposes. We employ the following partial transformations.
Horizontal translation:

$$
\begin{aligned}
x' &= x + \alpha, \\
y' &= y;
\end{aligned}
\tag{5.64}
$$

vertical translation:

$$
\begin{aligned}
x' &= x, \\
y' &= y + \beta;
\end{aligned}
\tag{5.65}
$$

scaling:

$$
\begin{aligned}
x' &= \omega x, \\
y' &= \omega y;
\end{aligned}
\tag{5.66}
$$

stretching:

$$x' = \delta x,$$
$$y' = \frac{1}{\delta} y;$$

(5.67)

horizontal skewing:

$$x' = x + ty,$$
$$y' = y;$$

(5.68)

vertical skewing:

$$x' = x,$$
$$y' = y + sx;$$

(5.69)

possible mirror reflection:

$$x' = x,$$
$$y' = \pm y.$$

(5.70)

Unlike the decomposition for image normalization, no rotation is included since it has been replaced here by horizontal and vertical skewing.

Any function I of moments is invariant under these seven transformations if and only if it is invariant under the general affine transformation (5.8). Each of these transformations imposes one constraint on the invariants.

The invariance to translation and scaling can be achieved in exactly the same way as in the image normalization method; we do not repeat it here. The invariance to stretching is provided by the constraint that the sum of the first indices in each term equals the sum of the second indices.

If a function I should be invariant to the horizontal skew, then its derivative with respect to the skew parameter t must be zero

$$\frac{dI}{dt} = \sum_p \sum_q \frac{\partial I}{\partial \mu_{pq}} \frac{d\mu_{pq}}{dt} = 0.$$

(5.71)

The derivative of the moment is

$$\frac{d\mu_{pq}}{dt} = \frac{d}{dt} \int_{-\infty}^{\infty} \int_{-\infty}^{\infty} (x + ty - x_c - ty_c)^p (y - y_c)^q f(x, y) \; dx \; dy$$

$$= \int_{-\infty}^{\infty} \int_{-\infty}^{\infty} p(x - x_c + t(y - y_c))^{p-1} (y - y_c)^{q+1} f(x, y) \; dx \; dy$$

$$= p\mu_{p-1,q+1}.$$

(5.72)

If we substitute this into (5.71), we obtain the *Cayley-Aronhold differential equation*

$$\sum_p \sum_q p\mu_{p-1,q+1} \frac{\partial I}{\partial \mu_{pq}} = 0.$$

(5.73)

We can derive a similar equation for the vertical skew

$$\sum_p \sum_q q\mu_{p+1,q-1} \frac{\partial I}{\partial \mu_{pq}} = 0. \tag{5.74}$$

From the mirror reflection we can derive the condition of symmetry. If we combine it with interchanging of the indices

$$\begin{aligned} x' &= y, \\ y' &= -x \end{aligned} \tag{5.75}$$

then the moment after this transformation is

$$\mu'_{pq} = (-1)^q \mu_{qp}. \tag{5.76}$$

If there are two symmetric terms

$$c_1 \prod_{\ell=1}^{r} \mu_{p_{j\ell},q_{j\ell}} \quad \text{and} \quad c_2 \prod_{\ell=1}^{r} \mu_{q_{j\ell},p_{j\ell}} \tag{5.77}$$

in the invariant, then they become after the transformation (5.75)

$$c_1(-1)^w \prod_{\ell=1}^{r} \mu_{q_{j\ell},p_{j\ell}} \quad \text{and} \quad c_2(-1)^w \prod_{\ell=1}^{r} \mu_{p_{j\ell},q_{j\ell}}, \tag{5.78}$$

therefore $c_1 = (-1)^w c_2$. We can write a symbolic formula

$$I(\mu'_{pq}) = (-1)^w I(\mu_{qp}). \tag{5.79}$$

This means that the coefficients of the symmetric terms can be computed together.

The Cayley-Aronhold differential equation leads to the solution of a system of linear equations, and the AMIs can be derived in this way.

The invariants (i.e., the solutions of the Cayley-Aronhold equation) are usually supposed to have a form of a linear combination of moment products

$$I = \left(\sum_{j=1}^{n_t} c_j \prod_{\ell=1}^{r} \mu_{p_{j\ell},q_{j\ell}} \right) / \mu_{00}^{r+w} \quad . \tag{5.80}$$

In reference [16], four solutions representing four low-order AMIs were found manually. According to the notation introduced for the graph-generated AMIs, the invariants proposed in [16] are I_1, I_2, I_3 and $-(I_4 + 6I_1I_3)$.[11]

It is natural to ask the question whether or not all the AMIs generated by the graph method are combinations of the AMIs found by means of the Cayley-Aronhold equation, and vice versa. Although a positive answer seems to be intuitive (and is actually correct), to prove this assertion is rather difficult. We refer to Gurevich's proof of a similar theorem [22] dealing with tensors, which was later adapted for moments [43].

---

[11] In reference [16], historical names of the invariants, such as "Apolar" and "Hankel determinant" are used.

5.6.2   Automatic solution

The need for higher-order AMIs has led to development of an automatic method for solving the Cayley-Aronhold equation. The input parameters of the algorithm are the order s and the structure $\mathbf{s} = (k_2, k_3, \ldots, k_s)$ of the desired invariant. Then, the degree of the invariant is $r = k_2 + k_3 + \cdots + k_s$, and its weight is $w = (2k_2 + 3k_3 + \cdots + sk_s)/2$.

First, we generate moment indices of all possible terms of the invariant. This means we generate all possible partitions of the number w into the sums of k_2 integers from 0 to 2, k_3 integers from 0 to 3 up to k_s integers from 0 to s. The generation of the partitions must be divided into two stages. In the first stage, we partition w among orders. In the second stage, we partition the quotas of specific moment orders among individual moments. The elements of the partitions are used as the first indices of the moments. When we complete the computation of the second indices from the moment orders, we have nearly generated the invariant; only the coefficients of the terms remain unknown. The number of all possible terms of the invariant is labeled n_t.

We sort both moments in a term and all terms in an invariant, because we need to find a term in an invariant or to detect identical invariants quickly.

The next step of the algorithm is to look for the symmetric terms. The first and second indices of the moments in each term are interchanged, and the counterpart is sought. If it is found, then we compute coefficients of both terms together. The number of the terms differing in another way is labeled n_s. Then we compute the derivatives of all terms with respect to all moments (except μ_{0q}) and the terms of the Cayley-Aronhold equation from them.

Now we can construct the system of linear equations for unknown coefficients. The Cayley-Aronhold equation must hold for all values of the moments; therefore the sum of coefficients of identical terms must equal zero. The horizontal size of the matrix of the system equals the number of unknown coefficients, the vertical size equals the number of different derivative terms.

The right-hand side of the equation is zero, which means that the equation can have one zero solution or an infinite number of solutions. If it has the zero solution only, it means that only invariants identically equaling zero can exist within the given structure. If it has infinite solutions, then the dimension of the space of the solutions equals the number of linearly independent invariants. If the dimension is n, and we have found n linearly independent solutions $I_1, I_2, \ldots, I_n$ of the equation, then each solution can be expressed as $k_1 I_1 + k_2 I_2 + \cdots + k_n I_n$, where $k_1, k_2, \ldots, k_n$ are arbitrary real numbers.

The basis $I_1, I_2, \ldots, I_n$ of the solution can be found by SVD of the system matrix, which is decomposed to the product $\mathbf{U} \cdot \mathbf{W} \cdot \mathbf{V}^T$, where $\mathbf{U}$ and $\mathbf{V}$ are orthonormal matrices and $\mathbf{W}$ is a diagonal matrix of the singular values. The solution of the system of linear equations is in each row of the matrix $\mathbf{V}$ that corresponds to a zero singular value in the matrix $\mathbf{W}$.

The solution from the matrix $\mathbf{V}$ does not have integer coefficients. Since it is always possible (and desirable) to obtain invariants with integer coefficients, we impose this requirement as an additional constraint on the solution.

An example with the structure s $=$ (2, 0, 2)

An illustrative explanation of the above method can be seen from the following example. Let us consider the order $s = 4$ and the structure $\mathbf{s} = (2, 0, 2)$ as input parameters. Then the weight $w = 6$, and the degree $r = 4$.

Table 5.3 All possible terms of the invariant

1st stage	2nd stage	coefficient	term
6+0	4+2+0+0	c_1	$\mu_{02}^2\mu_{40}\mu_{22}$
	3+3+0+0	c_2	$\mu_{02}^2\mu_{31}^2$
5+1	4+1+1+0	c_3	$\mu_{11}\mu_{02}\mu_{40}\mu_{13}$
	3+2+1+0	c_4	$\mu_{11}\mu_{02}\mu_{31}\mu_{22}$
4+2	4+0+2+0	c_5	$\mu_{20}\mu_{02}\mu_{40}\mu_{04}$
	4+0+1+1	c_8	$\mu_{11}^2\mu_{40}\mu_{04}$
	3+1+2+0	c_6	$\mu_{20}\mu_{02}\mu_{31}\mu_{13}$
	3+1+1+1	c_9	$\mu_{11}^2\mu_{31}\mu_{13}$
	2+2+2+0	c_7	$\mu_{20}\mu_{02}\mu_{22}^2$
	2+2+1+1	c_{10}	$\mu_{11}^2\mu_{22}^2$
3+3	3+0+2+1	c_3	$\mu_{20}\mu_{11}\mu_{31}\mu_{04}$
	2+1+2+1	c_4	$\mu_{20}\mu_{11}\mu_{22}\mu_{13}$
2+4	2+0+2+2	c_1	$\mu_{20}^2\mu_{22}\mu_{04}$
	1+1+2+2	c_2	$\mu_{20}^2\mu_{13}^2$

Now we need to generate all partitions of number 6 into a sum of two integers from 0 to 2 and two integers from 0 to 4; see Table 5.3. The first column contains partitions at the first stage among different orders, the second column shows partitions at the second stage inside the orders. One can see that the partitions 4+2 and 2+4 in the first stage are different; hence we must consider both cases. If we have partitions 3+2+1+0, 3+2+0+1, 2+3+1+0 and 2+3+0+1 in the second stage, we can see that it would be only interchanging of the factors in the term, so dealing with only one of these partitions is sufficient.

The moments in the terms are then sorted, so there are moments of the second order first in the fourth column of the table. The terms are also sorted, so the unknown coefficients in the third column are numbered according to this sorted order of terms. The coefficients are unknown at this phase of the computation, but their correct labeling is important. We can see that the theoretical number of terms $n_t = 14$. Then, the symmetric counterparts of each term are searched. The coefficients at the symmetric terms are labeled by the same number, as can be seen in the third column of the table; the number of different coefficients is $n_s = 10$.

Now, we need to compose the Cayley-Aronhold equation. The derivatives with respect to the particular moments are

$$D(\mu_{13}) = 2c_2\mu_{20}^2\mu_{13}\mu_{04} + c_4\mu_{20}\mu_{11}\mu_{22}\mu_{04} + c_6\mu_{20}\mu_{02}\mu_{31}\mu_{04} + c_9\mu_{11}^2\mu_{31}\mu_{04}$$

$$+ c_3\mu_{11}\mu_{02}\mu_{40}\mu_{04}$$

$$D(\mu_{22}) = 2c_1\mu_{20}^2\mu_{13}\mu_{04} + 2c_4\mu_{20}\mu_{11}\mu_{13}^2 + 4c_7\mu_{20}\mu_{02}\mu_{22}\mu_{13} + 4c_{10}\mu_{11}^2\mu_{22}\mu_{13}$$

$$+ 2c_1\mu_{02}^2\mu_{40}\mu_{13} + 2c_4\mu_{11}\mu_{02}\mu_{31}\mu_{13}$$

$$D(\mu_{31}) = 3c_3\mu_{20}\mu_{11}\mu_{22}\mu_{04} + 3c_6\mu_{20}\mu_{02}\mu_{22}\mu_{13} + 3c_9\mu_{11}^2\mu_{22}\mu_{13} + 6c_2\mu_{02}^2\mu_{31}\mu_{22}$$

$$+ 3c_4\mu_{11}\mu_{02}\mu_{22}^2$$

$$D(\mu_{40}) = 4c_5\mu_{20}\mu_{02}\mu_{31}\mu_{04} + 4c_8\mu_{11}^2\mu_{31}\mu_{04} + 4c_1\mu_{02}^2\mu_{31}\mu_{22} + 4c_3\mu_{11}\mu_{02}\mu_{31}\mu_{13}$$

$$D(\mu_{11}) = c_3\mu_{20}\mu_{02}\mu_{31}\mu_{04} + c_4\mu_{20}\mu_{02}\mu_{22}\mu_{13} + 2c_8\mu_{11}\mu_{02}\mu_{40}\mu_{04}$$

$$+ 2c_9\mu_{11}\mu_{02}\mu_{31}\mu_{13} + 2c_{10}\mu_{11}\mu_{02}\mu_{22}^2 + c_3\mu_{02}^2\mu_{40}\mu_{13} + c_4\mu_{02}^2\mu_{31}\mu_{22}$$

$$D(\mu_{20}) = 4c_1\mu_{20}\mu_{11}\mu_{22}\mu_{04} + 4c_2\mu_{20}\mu_{11}\mu_{13}^2 + 2c_3\mu_{11}^2\mu_{31}\mu_{04} + 2c_4\mu_{11}^2\mu_{22}\mu_{13}$$

$$+ 2c_5\mu_{11}\mu_{02}\mu_{40}\mu_{04} + 2c_6\mu_{11}\mu_{02}\mu_{31}\mu_{13} + 2c_7\mu_{11}\mu_{02}\mu_{22}^2.$$

The equation

$$D(\mu_{13}) + D(\mu_{22}) + D(\mu_{31}) + D(\mu_{40}) + D(\mu_{11}) + D(\mu_{20}) = 0 \tag{5.81}$$

should hold for arbitrary values of the moments. From this constraint we can put together the matrix of the system of linear equations for the coefficients as can be seen in Table 5.4.

After the SVD, the diagonal of the **W** matrix contains the values 9.25607, 7.34914, 6.16761, $3.03278 \cdot 10^{-16}$, 2.41505, 3.51931, 4.02041, 4.65777, 5.11852 and $4.49424 \cdot 10^{-16}$. We can see two values, $3.03278 \cdot 10^{-16}$ and $4.49424 \cdot 10^{-16}$, are less than the threshold, so the system of equations has two independent solutions. Integer coefficients of the solutions are in the last two rows of Table 5.4. We obtain two solutions:

$$I_a = (\mu_{20}^2\mu_{22}\mu_{04} - \mu_{20}^2\mu_{13}^2 - 2\mu_{20}\mu_{11}\mu_{31}\mu_{04} + 2\mu_{20}\mu_{11}\mu_{22}\mu_{13} + 2\mu_{20}\mu_{02}\mu_{31}\mu_{13}$$
$$-2\mu_{20}\mu_{02}\mu_{22}^2 + \mu_{11}^2\mu_{40}\mu_{04} - \mu_{11}^2\mu_{22}^2 - 2\mu_{11}\mu_{02}\mu_{40}\mu_{13} + 2\mu_{11}\mu_{02}\mu_{31}\mu_{22}$$
$$+\mu_{02}^2\mu_{40}\mu_{22} - \mu_{02}^2\mu_{31}^2)/\mu_{00}^{10}$$

and

$$I_b = (\mu_{20}\mu_{02}\mu_{40}\mu_{04} - 4\mu_{20}\mu_{02}\mu_{31}\mu_{13} + 3\mu_{20}\mu_{02}\mu_{22}^2 - \mu_{11}^2\mu_{40}\mu_{04} + 4\mu_{11}^2\mu_{31}\mu_{13}$$
$$-3\mu_{11}^2\mu_{22}^2)/\mu_{00}^{10}.$$

Table 5.4 The matrix of the system of linear equations for the coefficients. The empty elements are zero. The solution is in the last two rows

term	c_1	c_2	c_3	c_4	c_5	c_6	c_7	c_8	c_9	c_{10}
$\mu_{20}^2\mu_{13}\mu_{04}$	2	2								
$\mu_{20}\mu_{11}\mu_{22}\mu_{04}$	4		3	1						
$\mu_{20}\mu_{02}\mu_{31}\mu_{04}$			1		4	1				
$\mu_{11}^2\mu_{31}\mu_{04}$			2					4	1	
$\mu_{11}\mu_{02}\mu_{40}\mu_{04}$			1		2			2		
$\mu_{20}\mu_{11}\mu_{13}^2$		4			2					
$\mu_{20}\mu_{02}\mu_{22}\mu_{13}$					1	3	4			
$\mu_{11}^2\mu_{22}\mu_{13}$					2				3	4
$\mu_{11}\mu_{02}\mu_{31}\mu_{13}$				4	2	2			2	
$\mu_{02}^2\mu_{40}\mu_{13}$	2			1						
$\mu_{11}\mu_{02}\mu_{22}^2$					3		2			2
$\mu_{02}^2\mu_{31}\mu_{22}$	4	6			1					
I_a	1	-1	-2	2		2	-2	1		-1
I_b					1	-4	3	-1	4	-3

We can easily recognize the link to the AMIs yielded by the graph method:

$$I_a = I_9, \quad I_b = I_1 I_6.$$

5.7 Numerical experiments

In this section, we demonstrate the performance of both the AMIs and the normalized moments on artificial data in simulated recognition experiments and in a controlled experiment on real data, where the known ground truth allows the success rate to be evaluated. Practical experiments on real data can be found in Chapter 9.

5.7.1 Invariance and robustness of the AMIs

In the first simple experiment, we illustrate the invariance property of I_1 in the case when the affine model is completely valid. We generated a range of affine transformations from identity to severe deformations and calculated the relative error $(I_1' - I_1)/I_1$, where I_1' is the invariant value of the deformed image. The plot in Figure 5.9 demonstrates a perfect invariance (small errors are caused by the image resampling). The behavior of other AMIs is quite similar.

In the second experiment, we studied the stability/vulnerability of the AMIs if the object has undergone a perspective projection. We used real images and real projective transformations. We took five pictures of a comb lying on a black background by a digital camera (see Figure 5.10). The images differ from each other by viewing angles. We used five viewing angles ranging from a perpendicular view (0°) to approximately 75°, that yielded perspective deformations of the image to various extents. To eliminate the influence of varying lighting conditions, the images were normalized to the same contrast, and the background was thresholded.

The values of the thirty independent AMIs up to the seventh order were computed for each image (see Table 5.5 for eight independent invariants). Observe that they change a little when the viewing angle is changed. This is because the invariants are invariant to affine transformation but not to perspective projection, which occurs in this experiment. Observe also the loss of invariance property when taking the picture from sharp angles. On the other hand, notice the reasonable stability in case of weak perspective projections.

5.7.2 Digit recognition

This experiment illustrates recognition of digits $1, 2, \ldots, 9, 0$ (see Figure 5.11) deformed by various affine transformations. We deliberately chose a font where the digit 9 is not exactly a rotated version of 6.

The original binary image of each digit of the size 48×32 was deformed by 100 affine transformations, whose parameters were generated as random values with Gaussian distribution. The "mean value" was the identity, the standard deviation of the translation was 8 pixels, and the standard deviation σ of the other parameters was set to 0.25. Examples of the deformed digits can be seen in Figure 5.12.

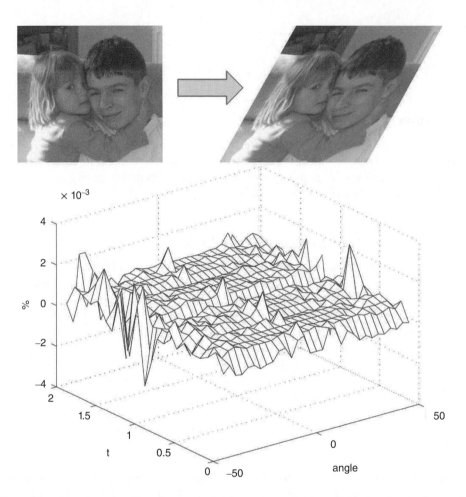

Figure 5.9 Numerical test of invariance of I_1. Horizontal axes: horizontal skewing and rotation angle, respectively. Vertical axis – relative deviation (in %) between I_1 of the original and that of the transformed image. The test proves the invariance of I_1

All deformed digits were classified independently by a minimum-distance classifier acting in the space of 30 AMIs. The values of the invariants were normalized by their standard deviations to have a comparable dynamic range. The results are summarized in Table 5.6a. It is clearly visible that the AMIs yielded an excellent 100% success rate. A part of the results is graphically visualized in Figure 5.13, where one can see the distribution of the digits in the space of two AMIs I_1 and I_6. All digits form compact clusters, well separated from each other (the clusters of "6" and "9" are close to one another but still separable). The same situation can be observed in other feature subspaces.

To test the robustness with respect to noise, we ran this experiment again. Gaussian white noise inducing SNR 26 dB was added to each deformed digit. To eliminate the role of the background, the noise was added to the interior of the bounding box of the digit only. The

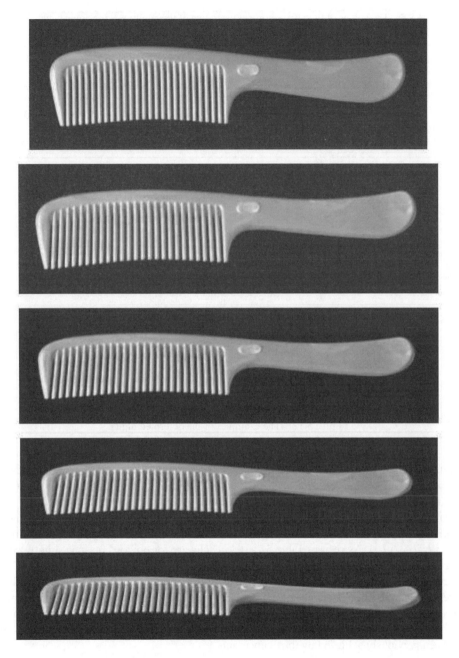

Figure 5.10 The comb. The viewing angle increases from 0° (top) to 75° (bottom)

Table 5.5 The values of the affine moment invariants of the comb; γ is the approximate viewing angle

$\gamma[°]$	$I_1[10^{-2}]$	$I_2[10^{-6}]$	$I_3[10^{-4}]$	$I_4[10^{-5}]$	$I_6[10^{-3}]$	$I_7[10^{-5}]$	$I_8[10^{-3}]$	$I_9[10^{-4}]$
0	2.802	1.538	−2.136	−1.261	4.903	6.498	4.526	1.826
30	2.679	1.292	−1.927	−1.109	4.466	5.619	4.130	1.589
45	2.689	1.328	−1.957	−1.129	4.494	5.684	4.158	1.605
60	2.701	1.398	−2.011	−1.162	4.528	5.769	4.193	1.626
75	2.816	1.155	−1.938	−1.292	5.033	6.484	4.597	1.868

1 2 3 4 5 6 7 8 9 0

Figure 5.11   The original digits used in the recognition experiment

1 2 3 4 5 6 7 8 9 0

Figure 5.12   The examples of the deformed digits

results are summarized in Table 5.6b. We may still observe an excellent recognition rate with the exception of certain "9"s misclassified as "6". When repeating this experiment with a heavier noise of 14 dB, the success rate dropped (see Table 5.6c).

In the next part of this experiment, we demonstrated the importance of choosing independent invariants. We classified the noisy digits (SNR = 14 dB) again but only by eight invariants $(I_1,\ldots,I_4,I_6,\ldots,I_9,)$. As seen in Table 5.7a, the results are for obvious reasons slightly worse than those achieved by 30 AMIs.

Then, the irreducible but dependent invariant I_{10} was added to the feature set, and the digits were reclassified. The results are presented in Table 5.7b. The overall percentage of correct classification increased only slightly. Then, the invariant I_{10} was replaced by the independent invariant I_{25}. As seen in Table 5.7c, the recognition rate increased significantly to the level that had been previously achieved by thirty invariants.

5.7.3   Recognition of symmetric patterns

This experiment demonstrates the power of the AMIs and of the normalized moments to recognize affinely distorted symmetric objects. The simple binary patterns used in the test are

Table 5.6   The recognition rate (in %) of the AMIs

	1	2	3	4	5	6	7	8	9	0	overall
(a) Noise-free case	100	100	100	100	100	100	100	100	100	100	100
(b) Additive noise, SNR = 26 dB	100	100	100	100	100	100	100	100	71	100	97.1
(c) Additive noise, SNR = 14 dB	99	82	78	36	100	52	95	100	27	70	73.9

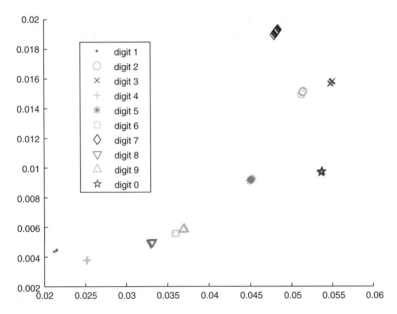

Figure 5.13   The digits in the feature space of affine moment invariants I_1 and I_6 (ten various transformations of each digit, a noise-free case)

Table 5.7   The recognition rate (in %) of the limited number of the AMIs

	1	2	3	4	5	6	7	8	9	0	overall
(a) Independent set $I_1, I_2, I_3, I_4, I_6, I_7, I_8, I_9$	76	80	31	34	100	46	76	100	68	51	66.2
(b) Dependent set $I_1, I_2, I_3, I_4, I_6, I_7, I_8, I_9, I_{10}$	51	100	46	73	53	56	74	100	72	70	69.5
(c) Independent set $I_1, I_2, I_3, I_4, I_6, I_7, I_8, I_9, I_{25}$	89	77	73	63	94	66	79	100	54	50	74.5

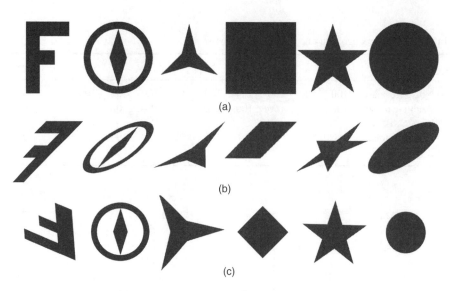

Figure 5.14 The test patterns. (a) originals, (b) examples of distorted patterns, and (c) the normalized positions

shown in Figure 5.14a. They have N-FRS with $N = 1, 2, 3, 4, 5$ and ∞, respectively. Each pattern was deformed successively by ten randomly generated affine transformations (for one instance, see Figure 5.14b). The affine deformation matrices were created as matrix products

$$\begin{pmatrix} a_1 & a_2 \\ b_1 & b_2 \end{pmatrix} = \begin{pmatrix} 1 & 0 \\ 0 & z \end{pmatrix} \begin{pmatrix} \cos \varrho & -\sin \varrho \\ \sin \varrho & \cos \varrho \end{pmatrix} \begin{pmatrix} \omega & 0 \\ 0 & \omega \end{pmatrix} \begin{pmatrix} \delta & 0 \\ 0 & \frac{1}{\delta} \end{pmatrix} \begin{pmatrix} \cos \alpha & -\sin \alpha \\ \sin \alpha & \cos \alpha \end{pmatrix},$$

(5.82)

where α and ϱ were uniformly distributed between 0 and 2π, ω and δ were normally distributed with mean value 1 and standard deviation 0.1. The value of z was set randomly to ± 1. We added white Gaussian noise of standard deviation 0.25 to the deformed objects.

First, we used the AMIs up to the fifth order. All patterns in all instances were recognized correctly. An example of the feature space of the AMIs I_6 and I_{47} is in Figure 5.15.

Then, the normalized moments were used as the features for recognition. In Figure 5.14c the normalized positions of the test patterns are shown. Recall that this figure is for illustration purposes only; transforming the objects is not required for calculation of the normalized moments. In the experiment with noisy patterns (as well as in reality), the choice of threshold used for the decision whether or not a moment is zero is very important and may greatly affect the results. In this case, the threshold 0.175 provided an error-free result, while the threshold 0.17 as well as the threshold 0.18 caused one misclassification. For an example of the feature space of the normalized moments, see Figure 5.16.

The clusters of both types of feature are a little spread, but, because they do not overlap each other, the recognition power of the features is still preserved. Comparing Figure 5.15 and Figure 5.16, one can see that the discrimination power of both feature types is roughly the same.

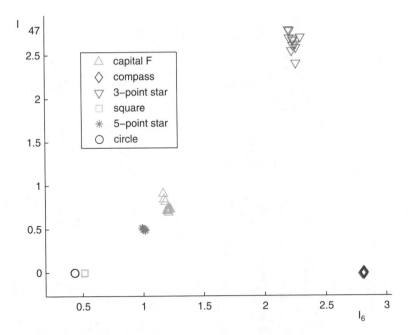

Figure 5.15   The space of two AMIs I_6 and I_{47}

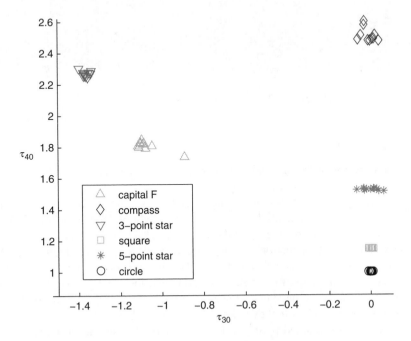

Figure 5.16   The space of two normalized moments τ_{30} and τ_{40}

Figure 5.17   The part of the mosaic used in the experiment

5.7.4   The children's mosaic

To show the behavior of the invariants on real images, we have chosen an experiment with a mosaic carpet (Figure 5.17) consisting of various tiles. Each tile contains a simple pattern – geometric figure (a polygon or a conic section) or an animal silhouette. Each tile was snapped by a camera perpendicularly and then with a slight projective distortion (5.2) from eight sides, so we have nine photographs of each tile. The slant angle of the camera was about 20° from the perpendicular direction. For an example of a tile picture, see Figure 5.18. We can see a noticeable trapezoidal distortion of the tile. The tiles are colored, but in this experiment we used the color information for the segmentation only; the moments were then computed from binary images. The perpendicular snaps were used as class representatives (templates), and the other snaps were classified by a minimum distance classifier.

The tiles were first classified by the AMIs and then by the normalized moments. In both cases, the magnitudes of the individual features were scaled to the same range. The recognition rate depends significantly on the order (and the number) of the invariants used. In Table 5.8,

Figure 5.18   The image of a tile with a slight projective distortion

Table 5.8   The numbers of errors in recognition of 200 tile snaps

Maximum order of invariants	6	7	8	9
Affine moment invariants	6	3	0	7
Normalized moments	2	4	1	4

we can see the total number of errors that occurred for various maximum orders of the invariants. Clearly, the maximum order 8 is an optimal choice both for the AMIs as well as for the
normalized moments. For the detailed contingency table of the AMIs, see Table 5.9. Note that
the geometric patterns include objects that can be affinely transformed one to another. The first
such group consists of an equilateral and an isosceles triangles; the second group includes a
square, a rhombus, small and big rectangles, and a rhomboid; the third group includes a regular
and an irregular hexagon; and the last group of the affine-equivalent objects contains a circle
and an ellipse. If a pattern in a group is classified as another pattern within the same group, it
cannot be considered a mistake.

For an example of the feature subspace of the AMIs, see Figures 5.19 and 5.20. The patterns
were divided into two graphs, for lucidity. In the first graph, the affinely equivalent geometric
patterns actually create one mixed cluster. In the second graph, the clusters of animal silhouettes are partially overlapping, the correct recognition by means of two features is not possible.
Fortunately, by using thirty-nine features up to the eighth order, a correct recognition was
achieved. For analogous graphs for the normalized moments, see Figures 5.21 and 5.22.

Table 5.9 The contingency table of the AMIs up to eighth order

equilateral triangle	4 4 0 0 0 0 0 0 0 0 0 0 0 0 0 0 0 0 0 0 0 0 0 0 0 0 0
isosceles triangle	4 4 0 0 0 0 0 0 0 0 0 0 0 0 0 0 0 0 0 0 0 0 0 0 0 0 0
square	0 0 0 0 0 0 0 0 0 0 0 0 0 0 0 0 0 0 0 0 0 0 0 0 0 0 0
rhombus	0 0 1 4 0 0 1 0 0 0 0 0 0 0 0 0 0 0 0 0 0 0 0 0 0 0 0
big rectangle	0 0 7 4 8 7 7 0 0 0 0 0 0 0 0 0 0 0 0 0 0 0 0 0 0 0 0
small rectangle	0 0 0 0 0 1 0 0 0 0 0 0 0 0 0 0 0 0 0 0 0 0 0 0 0 0 0
rhomboid	0 0 0 0 0 0 0 0 0 0 0 0 0 0 0 0 0 0 0 0 0 0 0 0 0 0 0
trapezoid	0 0 0 0 0 0 0 8 0 0 0 0 0 0 0 0 0 0 0 0 0 0 0 0 0 0 0
pentagon	0 0 0 0 0 0 0 8 0 0 0 0 0 0 0 0 0 0 0 0 0 0 0 0 0 0 0
regular hexagon	0 0 0 0 0 0 0 0 0 2 0 0 0 0 0 0 0 0 0 0 0 0 0 0 0 0 0
irregular hexagon	0 0 0 0 0 0 0 0 8 6 0 0 0 0 0 0 0 0 0 0 0 0 0 0 0 0 0
circle	0 0 0 0 0 0 0 0 0 8 6 0 0 0 0 0 0 0 0 0 0 0 0 0 0 0 0
ellipse	0 0 0 0 0 0 0 0 0 0 2 0 0 0 0 0 0 0 0 0 0 0 0 0 0 0 0
bear	0 0 0 0 0 0 0 0 0 0 0 8 0 0 0 0 0 0 0 0 0 0 0 0 0 0 0
squirrel	0 0 0 0 0 0 0 0 0 0 0 0 8 0 0 0 0 0 0 0 0 0 0 0 0 0 0
pig	0 0 0 0 0 0 0 0 0 0 0 0 0 8 0 0 0 0 0 0 0 0 0 0 0 0 0
cat	0 0 0 0 0 0 0 0 0 0 0 0 0 0 8 0 0 0 0 0 0 0 0 0 0 0 0
bird	0 0 0 0 0 0 0 0 0 0 0 0 0 0 0 8 0 0 0 0 0 0 0 0 0 0 0
dog	0 0 0 0 0 0 0 0 0 0 0 0 0 0 0 0 8 0 0 0 0 0 0 0 0 0 0
cock	0 0 0 0 0 0 0 0 0 0 0 0 0 0 0 0 0 8 0 0 0 0 0 0 0 0 0
hedgehog	0 0 0 0 0 0 0 0 0 0 0 0 0 0 0 0 0 0 8 0 0 0 0 0 0 0 0
rabbit	0 0 0 0 0 0 0 0 0 0 0 0 0 0 0 0 0 0 0 8 0 0 0 0 0 0 0
duck	0 0 0 0 0 0 0 0 0 0 0 0 0 0 0 0 0 0 0 0 8 0 0 0 0 0 0
dolphin	0 0 0 0 0 0 0 0 0 0 0 0 0 0 0 0 0 0 0 0 0 8 0 0 0 0 0
cow	0 0 0 0 0 0 0 0 0 0 0 0 0 0 0 0 0 0 0 0 0 0 0 8

For a comparison, we tried to recognize the tiles also by the basic TRS invariants. The best achievable result was 18 errors out of 200 trials, which clearly illustrates the necessity of having affine-invariant features in this case.

To investigate the influence of the perspective projection, we repeated this experiment with a different setup. The tiles were captured by a camera with a slant angle of approximately 45°. As seen in Figure 5.23, the trapezoidal distortion is now significant. The best possible result of the AMIs up to the eighth order yielded 56 errors out of 200 trials, the optimal result of the normalized moments produced 53 errors. It illustrates that a heavy projective distortion cannot be well approximated by an affine transformation, and hence affine invariants, regardless of their particular type, cannot yield acceptable recognition rates.

5.7.5   Scrabble tiles recognition

This experiment, similar to the previous one with the digits, not only illustrates the recognition power of the AMIs in case of weak perspective distortion, but also shows the importance of using independent sets of invariants. We tested the AMIs' ability to recognize letters on the scrabble tiles[12]. First, we photographed the tiles side by side from an almost perpendicular

---

[12] We used the tiles from the Scrabble Upwords by Hasbro, see www.hasbro.com for details.

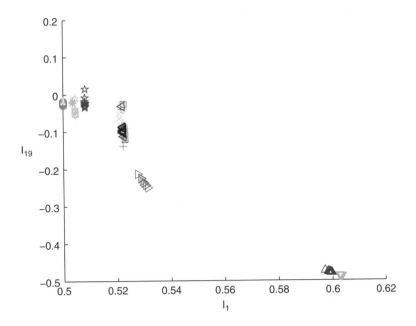

Figure 5.19 The geometric patterns. The feature space of I_1 and I_{19}: ▽ – equilateral triangle, △ – isosceles triangle, □ – square, ◇ – rhombus, + – big rectangle, × – small rectangle, ◁ – rhomboid, ▷ – trapezoid, ☆ – pentagon, ∗ – regular hexagon, ✿ – irregular hexagon, ○ – circle, ● – ellipse

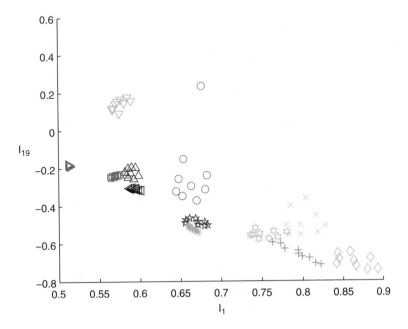

Figure 5.20 The animal silhouettes. The feature space of I_1 and I_{19}: ▽ – bear, △ – squirrel, □ – pig, ◇ – cat, + – bird, × – dog, ◁ – cock, ▷ – hedgehog, ☆ – rabbit, ∗ – duck, ✿ – dolphin, ○ – cow

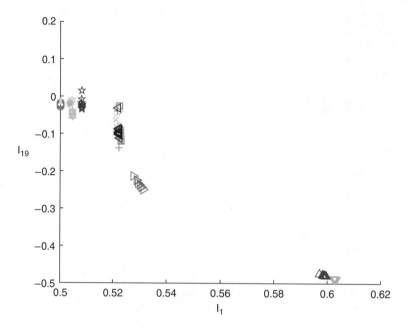

Figure 5.21 The geometric patterns. The feature space of τ_{20} and τ_{50}: $\triangledown$ – equilateral triangle, $\triangle$ – isosceles triangle, $\square$ – square, $\diamondsuit$ – rhombus, $+$ – big rectangle, $\times$ – small rectangle, $\triangleleft$ – rhomboid, $\triangleright$ – trapezoid, $\star$ – pentagon, $*$ – regular hexagon, ✿ – irregular hexagon, $\bigcirc$ – circle, $\bullet$ – ellipse

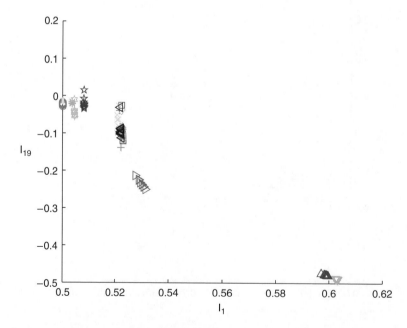

Figure 5.22 The animal silhouettes. The feature space of τ_{20} and τ_{50}: $\triangledown$ – bear, $\triangle$ – squirrel, $\square$ – pig, $\diamondsuit$ – cat, $+$ – bird, $\times$ – dog, $\triangleleft$ – cock, $\triangleright$ – hedgehog, $\star$ – rabbit, $*$ – duck, ✿ – dolphin, $\bigcirc$ – cow

Figure 5.23 The image of a tile with a heavy projective distortion

Figure 5.24 Scrabble tiles – the templates

view (see Figure 5.24). They were automatically segmented and separately binarized. The threshold for conversion to binary images was found by Otsu's method [44]. Then the individual characters were found and labeled. The result is a "gallery" of twenty-one binary letters that served as our template set.

Figure 5.25   Scrabble tiles to be recognized – a sample scene

Then we placed the tiles almost randomly on the table; we only ensured that no tiles are upside down and that all letters are completely visible. We repeated that eight times (see Figure 5.25 for a sample scene). On each photograph, the letters were automatically located and segmented by the same method as in case of the template set.

We classified all the 168 letters by minimum distance with respect to the templates in the space of invariants I_1, I_2, I_3, I_4, I_5, I_6, I_7, I_8, I_9, I_{10}, I_{19}, I_{47}, I_{48}, I_{49}, I_{50}, I_{64} and I_{65} (see Figure 5.26 for a 2D subspace of I_2 and I_6). There were nine misclassifications. However, as shown in the previous section, the AMIs I_5 and I_{10} are dependent. We omitted them and ran the classification again. Now the number of misclassifications decreased to five. Finally, for a comparison, we omitted from the initial set two independent invariants I_{64} and I_{65}. We received sixteen misclassifications which is much worse than before.

This clearly illustrates that dependent features do not contribute to between-class discriminability and may even decrease the performance of the classifier. Thanks to the algorithm introduced in Section 5.3.4, one can identify the dependent AMIs on theoretical basis before they are applied to particular data.

5.8   Affine invariants of color images

Color and multispectral/multichannel images are, according to the terminology introduced in Chapter 3, vector fields that underlie inner spatial transformations; that is, their color values

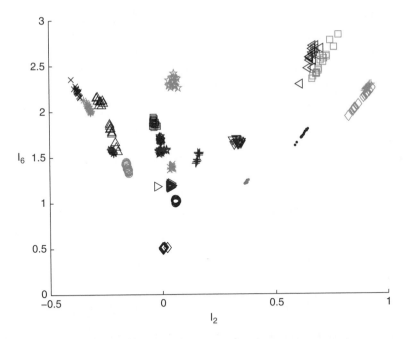

Figure 5.26   The space of invariants I_2 and I_6. Even in two dimensions the clustering tendency is evident. Legend: • A, ○ B, × C, + D, * E, □ H, ◇ I, ▽ J, △ K, ◁ L, ▷ M, ☆ N, ✿ O, • P, ○ R, × S, + T, * U, □ V, ◇ Y, ☆ Z

are not changed if the image has been transformed[13]. In case of affine transformation we have

$$\mathbf{f}'(\mathbf{x}) = \mathbf{f}(\mathbf{A}^{-1} \cdot \mathbf{x})$$

where $\mathbf{f}(x) = (f_1(x,y),\dots,f_m(x,y))^T$ and m is the number of spectral channels. Color images are traditionally encoded as the RGB channels, so $m = 3$ in that case.

Without introducing any new theory, we could calculate "scalar image" moment invariants from each channel separately and use them as a three-component vector. In that way we should obtain $3(P - 6)$ affine invariants, where P is the number of the moments used. However, there is an unexploited link which connects the channels – the affine transformation is supposed to be the same for all of them. This assumption decreases the degree of freedom, and, if properly incorporated, it offers the possibility of designing 12 new independent invariants ($6(m - 1)$ new invariants in general) such that the total number of invariants increases to $3P - 6$, possibly $mP - 6$. This can be accomplished by introducing so-called *joint affine invariants* [54].

Joint invariants are created from simultaneous "scalar" AMIs. By a *simultaneous invariant* we understand an AMI containing moments of at least two different orders[14]. If we take an arbitrary simultaneous invariant and calculate the moments of the same order (let us say of

---

[13] A color transform, if present, is usually independent of the spatial transform and performs mostly a channel-wise contrast stretching.

[14] An invariant containing the moments of one order only is called homogeneous.

order 2) from the first channel, the moments of the next order (order 3) from the second channel, etc., we get a joint invariant, which is independent of all "scalar" invariants.

Let us illustrate this process on two channels labeled as a and b. We take I_3, which is a simultaneous invariant and calculate the second-order moments from channel a and the third-order moments from channel b. We obtain a joint invariant

$$J_{2,3} = (\mu_{20}^{(a)}\mu_{21}^{(b)}\mu_{03}^{(b)} - \mu_{20}^{(a)}(\mu_{12}^{(b)})^2 - \mu_{11}^{(a)}\mu_{30}^{(b)}\mu_{03}^{(b)} + \mu_{11}^{(a)}\mu_{21}^{(b)}\mu_{12}^{(b)} + \mu_{02}^{(a)}\mu_{30}^{(b)}\mu_{12}^{(b)}$$
$$- \mu_{02}^{(a)}(\mu_{21}^{(b)})^2)/\mu_{00}^7.$$

(the normalizing moment μ_{00} is computed from the whole color image, $\mu_{00} = \mu_{00}^{(a)} + \mu_{00}^{(b)} + \mu_{00}^{(c)}$).

To create joint invariants, we can also utilize algebraic invariants of two or more binary forms of the same orders. An example of such an invariant of the second order is

$$J_2 = (\mu_{20}^{(a)}\mu_{02}^{(b)} + \mu_{20}^{(b)}\mu_{02}^{(a)} - 2\mu_{11}^{(a)}\mu_{11}^{(b)})/\mu_{00}^4.$$

If we use moments of one channel only (i.e., $a = b$), we obtain essentially I_1. Another joint invariant of the third order is

$$J_3 = (\mu_{30}^{(a)}\mu_{03}^{(b)} - 3\mu_{21}^{(a)}\mu_{12}^{(b)} + 3\mu_{21}^{(b)}\mu_{12}^{(a)} - \mu_{30}^{(b)}\mu_{03}^{(a)})/\mu_{00}^5.$$

If we use moments of one channel only then $J_3 = 0$. A third-order nontrivial invariant of degree two from one channel does not exist, while that from two channels does.

In color invariants, we can also employ the zero and the first-order moments. The central moments are defined with respect to the joint centroid of the whole image

$$m_{00} = m_{00}^{(a)} + m_{00}^{(b)} + m_{00}^{(c)}, \tag{5.83}$$

$$x_c = (m_{10}^{(a)} + m_{10}^{(b)} + m_{10}^{(c)})/m_{00}, \qquad y_c = (m_{01}^{(a)} + m_{01}^{(b)} + m_{01}^{(c)})/m_{00}.$$

$$\mu_{pq}^{(a)} = \int_{-\infty}^{\infty}\int_{-\infty}^{\infty}(x - x_c)^p(y - y_c)^q a(x,y) \; \mathrm{d}x \; \mathrm{d}y \qquad p,q = 0,1,2,\ldots. \tag{5.84}$$

Since the centroids of the individual channels are generally different from the joint centroid, the first-order moments need not be zero. We can use them for construction of additional joint invariants, such as

$$J_1 = (\mu_{10}^{(a)}\mu_{01}^{(b)} - \mu_{10}^{(b)}\mu_{01}^{(a)})/\mu_{00}^3$$

and

$$J_{1,2} = (\mu_{20}^{(a)}(\mu_{01}^{(b)})^2 + \mu_{02}^{(a)}(\mu_{10}^{(b)})^2 - 2\mu_{11}^{(a)}\mu_{10}^{(b)}\mu_{01}^{(b)})/\mu_{00}^5.$$

There exists even a zero-order joint affine invariant

$$J_0 = \mu_{00}^{(a)}/\mu_{00}^{(b)}.$$

Incorporating the joint invariants into the feature set increases the number of the features and the recognition power when using invariants up to the given moment order. In other words, creating the same number of "scalar" invariants requires the usage of moments of higher orders, which has a negative impact on the robustness of the system. The joint invariants may be particularly useful in satellite and aerial image recognition because contemporary multi- and hyper-spectral sensors produce up to several hundreds of channels (spectral bands).

The bond among the channels can be exploited also in other ways than by means of the joint invariants. Mindru et al. [45, 46] used moments computed from certain products and powers of the channels. They defined the *generalized color moment* of degree $\alpha + \beta + \gamma$ as

$$M_{pq}^{\alpha\beta\gamma} = \iint\limits_{D} x^{p}y^{q}a(x,y)^{\alpha}b(x,y)^{\beta}c(x,y)^{\gamma} \ dx \ dy, \qquad (5.85)$$

where a, b, and c are the image functions of the RGB channels of the image and α, β, γ are non-negative integers. They use these moments for the construction of affine invariants and also of combined invariants to the affine transformation and brightness/contrast linear changes. These features unfortunately exhibit very high redundancy. In the case of an infinite set of moments of all orders, only the moments where $\alpha + \beta + \gamma = 1$ are independent. This redundancy decreases as the maximum order of moments decreases; for low-order moments this method may yield meaningful results. However, even for low orders, using higher powers of the brightness in individual channels makes the method more sensitive to noise and to the model violations. Such moments are more prone to numerical errors, which may lead to misclassifications.

In the most general case, Mindru et al. [46] consider between-channel "cross-talks", when the output intensity of one band is a linear combination of all three input bands. This is actually close to the model of an independent total affine transformation[15], used in the theory of vector fields.

5.8.1 *Recognition of color pictures*

The goal of this experiment is to show how to use the joint invariants for recognition of color images under an affine distortion.

We photographed a series of cards used in a mastercard (pexeso) game. Ten sample cards are shown in Figure 5.27. Note that in this game each picture appears on two cards of the set. Each card was captured eight times, rotated by approximately 45° between consecutive snaps. An example of the rotation of a pair of cards is shown in Figure 5.28. Small deviations from the perpendicular viewing direction led to affine distortions of the cards. The first snap of each card was used as a representative of its class, and the following seven snaps were recognized by the minimum-distance classifier. Our feature set included I_1, I_2, I_3, I_4 and $J_{1,2}^{(a,a)}$ of each channel, $J_0^{(R,G)}, J_0^{(B,G)}, J_1^{(R,B)}, J_{1,2}^{(R,B)}, J_2^{(R,G)}$ and $J_2^{(B,G)}$; that is, twenty-one invariants. The invariants were normalized to magnitude, both to the order and to the degree. All the cards were classified correctly, without any misclassifications. We can see the feature space of the zeroth-order invariants in Figure 5.29.

We repeated the experiment with single-channel invariants $I_1, I_2, I_3,$ and I_4 only; that is, we had twelve features altogether. The cards were classified with three errors, which means the joint invariants bring better discriminability and improve the recognition.

For a comparison, the same experiment was repeated again with the invariants by Mindru et al. called the "GPD invariants" in [46], which should handle also linear changes of

---

[15] However, here the dimensions of the inner and outer transformation matrices are different from one another.

Figure 5.27   The mastercards. Top row from the left: Girl, Carnival, Snowtubing, Room-bell, and Fireplace. Bottom row: Winter cottage, Spring cottage, Summer cottage, Bell, and Star

illumination. The number of the invariants was twenty-one, the same as in the first experiment for a fair comparison. We obtained the card classification with six errors. This can still be considered a good result, but apparently the joint invariants improve the recognition rate.

5.9   Affine invariants of 2D vector fields

Unlike the color images, the "true" vector fields (i.e., the velocity fields, optical flow fields, etc.) are mostly transformed according to the total affine transformation. The change of the spatial coordinates is coupled with the inverse change of the field orientation and magnitudes as

$$\mathbf{f}'(\mathbf{x}) = \mathbf{A}\mathbf{f}(\mathbf{A}^{-1}\mathbf{x})$$

This model called *total affine transformation* is an analogy with Section 3.9, where we studied total rotation of a field and introduced respective rotation invariants. That theory was based on expressing the vector field as a complex number field, and the invariants were derived easily thanks to the rotation property of complex moments. Unfortunately, if **A** is not a rotation matrix, complex moments lose their simple transformation property, and the field representation by complex numbers does not provide any advantage.

To derive *vector field affine moment invariants* (VFAMIs), we keep working with geometric moments. Let us first show how to construct invariants w.r.t. the outer affine transformation

$$\mathbf{f}'(\mathbf{x}) = \mathbf{A}\mathbf{f}(\mathbf{x}).$$

There are various possible choices, but the simplest relative invariants are of the form

$$I_{out} = m_{pq}^{(f)} m_{st}^{(g)} - m_{st}^{(f)} m_{pq}^{(g)},$$

for arbitrary combinations of indexes except $(p, q) = (s, t)$, where $m^{(f)}$ and $m^{(g)}$ are scalar moments of the first and the second component of $\mathbf{f}(\mathbf{x}) = (f(x, y), g(x, y))^T$, respectively. We can easily prove by substitution that

$$I'_{out} = J I_{out}.$$

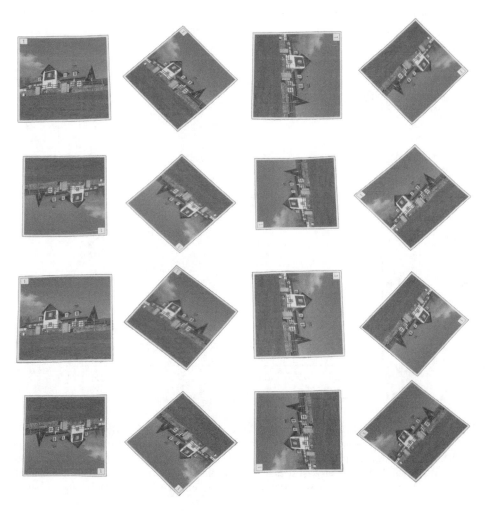

Figure 5.28 The card "Summer cottage" including all its rotations. The rotations are real due to the rotations of the hand-held camera. This acquisition scheme also introduces mild perspective deformations. The third and fourth rows contain the other card from the pair

To couple inner and outer invariance properties, we modified the graph method for scalar images by incorporating the above-mentioned knowledge. We got inspired by (5.14) but modified it into the form

$$V(\mathbf{f}) = \int\limits_{-\infty}^{\infty} \cdots \int\limits_{-\infty}^{\infty} \prod_{k,j=1}^{r} C_{kj}^{n_{kj}} \, F_{kj}^{v_{kj}} \prod_{i=1}^{r} \mathrm{d}x_i \, \mathrm{d}y_i \,, \tag{5.86}$$

where

$$F_{kj} = f(x_k, y_k)g(x_j, y_j) - f(x_j, y_j)g(x_k, y_k)$$

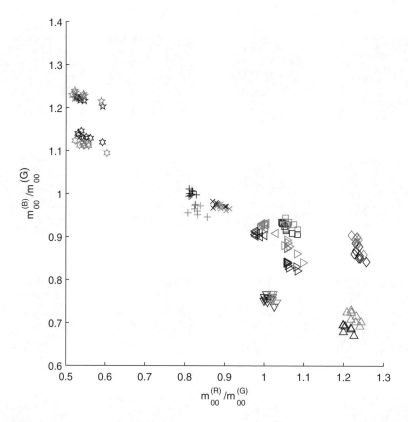

Figure 5.29 The mastercards. Legend: $\triangledown$ – Girl, $\triangle$ – Carnival, $\square$ – Snowtubing, $\diamondsuit$ – Room-bell and $\triangleright$ – Fireplace, $\times$ – Winter cottage, $\triangleleft$ – Spring cottage, $+$ – Summer cottage, $\star$ – Bell and $\hexstar$ – Star. A card from each pair is expressed by the black symbol, while the other card is expressed by the gray symbol. Note that some clusters have been split into two sub-clusters. This is because the two cards of the pair may be slightly different. This minor effect does not influence the recognition rate

and all other symbols and variables have the same meaning as in the case of the "scalar" AMIs. To obtain moments after the integration, we impose the constraint on the "intensity cross products" F_{kj} such that each of the points $(x_1, y_1),\ldots,(x_r, y_r)$ is involved just once in all F_{kj}'s used. If a point was not involved or involved more than once, the integration would not yield moments (note that the moments are obtained only if $f(x_k, y_k)$ and $g(x_k, y_k)$ appear in the integrand in the power of one). It follows immediately from this constraint that r must be even, v_{kj} can only be 0 or 1, and $\sum_{k,j} v_{kj} = r/2$. Actually, v_{kj}'s form an upper triangular matrix which contains just one value 1 in each row and column.. Since $F_{kj} = -F_{jk}$ and $F_{kk} = 0$, we consider only the indexes $j > k$.

Under a total affine transformation, V is changed as

$$V(\mathbf{f}') = J^{w+r/2}|J|^r V(\mathbf{f}),$$

which means it is a relative invariant. We show three simplest invariants in explicit forms below; several thousands of the invariants can be found on the webpage [32].

If we choose $r = 4$, $w = 4$, and $n_{12} = n_{13} = n_{24} = n_{34} = 1$ we obtain

$$
\begin{aligned}
V_1 = & -(m_{20}^{(f)})^2(m_{02}^{(g)})^2 + 4m_{20}^{(f)}m_{11}^{(f)}m_{11}^{(g)}m_{02}^{(g)} + 2m_{20}^{(f)}m_{02}^{(f)}m_{20}^{(g)}m_{02}^{(g)} \\
& -4m_{20}^{(f)}m_{02}^{(f)}(m_{11}^{(g)})^2 - 4(m_{11}^{(f)})^2m_{20}^{(g)}m_{02}^{(g)} + 4m_{11}^{(f)}m_{02}^{(f)}m_{20}^{(g)}m_{11}^{(g)} \\
& -(m_{02}^{(f)})^2(m_{20}^{(g)})^2.
\end{aligned}
$$

The choice of $r = 2$, $w = 3$, and $n_{12} = 3$ yields

$$
V_2 = m_{30}^{(f)}m_{03}^{(g)} - 3m_{21}^{(f)}m_{12}^{(g)} + 3m_{12}^{(f)}m_{21}^{(g)} - m_{03}^{(f)}m_{30}^{(g)}.
$$

The parameters $r = 2$, $w = 5$, and $n_{12} = 5$ lead to the invariant

$$
V_3 = m_{50}^{(f)}m_{05}^{(g)} - 5m_{41}^{(f)}m_{14}^{(g)} + 10m_{32}^{(f)}m_{23}^{(g)} - 10m_{23}^{(f)}m_{32}^{(g)} + 5m_{14}^{(f)}m_{41}^{(g)} - m_{05}^{(f)}m_{50}^{(g)}.
$$

To eliminate J, we cannot normalize by a power of m_{00} because m_{00} is not a relative invariant w.r.t. the total affine transformation. Instead, we have to take ratios of relative invariants V of proper powers such that the Jacobian is canceled.

Similarly, neither the centroids of the components $m_{10}^{(f)}/m_{00}^{(f)}$, $m_{10}^{(g)}/m_{00}^{(g)}$ nor any their combinations are mapped on themselves during the affine transformation. We can suppose there is no translation in many applications, but if it is not true, we must take two VFAMIs, express the translation by means of them, and normalize the remaining invariants to the translation by them. One of the invariants can be $m_{10}^{(f)}m_{01}^{(g)} - m_{10}^{(g)}m_{01}^{(f)}$; the other must also include the second-order moments.

As in the case of the AMIs, the VFAMIs can also be represented by graphs. To do so, we need multigraphs with edges of two kinds. The graph is constructed similarly to the AMIs case. Each point (x_k, y_k) corresponds to a node, and each cross-product C_{kj} corresponds to an edge of the first kind. If $n_{kj} > 1$, the respective term $C_{kj}^{n_{kj}}$ generates n_{kj} edges connecting the kth and jth nodes. The second-kind edges represent the factors F_{kj} in the same way. Note that the second-kind edges are constrained – each node has just one adjacent edge of this type. Although this might look like a big restriction, there are still exponentially many possibilities how to place the second-kind edges for a fixed particular configuration of the first-kind edges. Identification of reducible and dependent invariants is even more computationally demanding than in the case of scalar AMIs but in principle it can be done in an analogous way.

The VFAMIs are a very recent topic. Only a little work has been done on this field, and many challenges and problems still remain open. At the time of writing this book, no research publications on the VFAMIs were available. The method we briefly described in this section is not the only possibility for designing the VFAMIs. The normalization approach can be used as well, and probably also some other approaches, developed originally for scalar AMIs, can be adapted to vector fields. Since the vector field data has become more and more frequent and important, we can expect an increasing interest of researchers in this topic.

5.10   3D affine moment invariants

As already mentioned in the introduction to this chapter, affine transformation in 3D is much less common than in 2D because there is no "perspective projection from 4D". This is why

3D affine invariants have attracted significantly less attention (measured by the number of research publications) than their 2D counterparts and also than 3D rotation invariants. Affine transformation in 3D occurs mostly as a result of non-uniform spatial resolution of the sensor coupled with object rotations.

In principle, any method of deriving affine invariants in 2D can be extended to 3D. In this section, we present a derivation of 3D AMIs using the method of geometric primitives, which is a generalization of the graph method, and the method of affine normalization. The tensor method and the method using the Cayley-Aronhold equation are mentioned only briefly.

5.10.1   The method of geometric primitives

The method of geometric primitives was introduced by Xu and Li [47] and performs a 3D version of the graph method. It uses tetrahedrons with one vertex in the centroid of the object and three vertices moving through the object. The volume of such a body is

$$V(O, 1, 2, 3) = \frac{1}{6} \begin{vmatrix} x_1 & x_2 & x_3 \\ y_1 & y_2 & y_3 \\ z_1 & z_2 & z_3 \end{vmatrix}. \tag{5.87}$$

When we denote

$$C_{123} = 6V(O, 1, 2, 3),$$

then, having $r \geq 3$ points, a relative invariant can be obtained by integration of the product

$$I(f) = \int\limits_{-\infty}^{\infty} \cdots \int\limits_{-\infty}^{\infty} \prod_{j,k,\ell=1}^{r} C_{jk\ell}^{n_{jk\ell}} \cdot \prod_{i=1}^{r} f(x_i, y_i, z_i) \; dx_i \; dy_i \; dz_i, \tag{5.88}$$

where $n_{jk\ell}$ are non-negative integers,

$$w = \sum_{j,k,\ell} n_{jk\ell}$$

is the weight of the invariant, and r is its degree. When undergoing an affine transformation, $I(f)$ changes as

$$I(f)' = J^w |J|^r I(f).$$

The normalization of $I(f)$ by μ_{000}^{w+r} yields an absolute affine invariant (up to the sign if w is odd).

For example, if $r = 3$ and $n_{123} = 2$, then we obtain

$$I_1^{3D} = \frac{1}{6} \int\limits_{-\infty}^{\infty} \cdots \int\limits_{-\infty}^{\infty} C_{123}^2 f(x_1, y_1, z_1) f(x_2, y_2, z_2)$$

$$\times f(x_3, y_3, z_3) \; dx_1 \; dy_1 \; dz_1 \; dx_2 \; dy_2 \; dz_2 \; dx_3 \; dy_3 \; dz_3 / \mu_{000}^5 \tag{5.89}$$

$$= (\mu_{200}\mu_{020}\mu_{002} + 2\mu_{110}\mu_{101}\mu_{011} - \mu_{200}\mu_{011}^2 - \mu_{020}\mu_{101}^2 - \mu_{002}\mu_{110}^2)/\mu_{000}^5.$$

Another example we obtain for $r = 4$, $n_{123} = 1$, $n_{124} = 1$, $n_{134} = 1$ and $n_{234} = 1$:

$$I_2^{3D} = \frac{1}{36} \int\limits_{-\infty}^{\infty} \cdots \int\limits_{-\infty}^{\infty} C_{123}C_{124}C_{134}C_{234}\, f(x_1,y_1,z_1)\, f(x_2,y_2,z_2)\, f(x_3,y_3,z_3)\, f(x_4,y_4,z_4)$$

$$\mathrm{d}x_1\ \mathrm{d}y_1\ \mathrm{d}z_1\ \mathrm{d}x_2\ \mathrm{d}y_2\ \mathrm{d}z_2\ \mathrm{d}x_3\ \mathrm{d}y_3\ \mathrm{d}z_3\ \mathrm{d}x_4\ \mathrm{d}y_4\ \mathrm{d}z_4/\mu_{000}^8$$

$$= (\mu_{300}\mu_{003}\mu_{120}\mu_{021} + \mu_{300}\mu_{030}\mu_{102}\mu_{012} + \mu_{030}\mu_{003}\mu_{210}\mu_{201} - \mu_{300}\mu_{120}\mu_{012}^2$$

$$-\mu_{300}\mu_{102}\mu_{021}^2 - \mu_{030}\mu_{210}\mu_{102}^2 - \mu_{030}\mu_{201}^2\mu_{012} - \mu_{003}\mu_{210}^2\mu_{021} - \mu_{003}\mu_{201}\mu_{120}^2$$

$$-\mu_{300}\mu_{030}\mu_{003}\mu_{111} + \mu_{300}\mu_{021}\mu_{012}\mu_{111} + \mu_{030}\mu_{201}\mu_{102}\mu_{111} + \mu_{003}\mu_{210}\mu_{120}\mu_{111}$$

$$+\mu_{210}^2\mu_{012}^2 + \mu_{201}^2\mu_{021}^2 + \mu_{120}^2\mu_{102}^2 - \mu_{210}\mu_{120}\mu_{102}\mu_{012} - \mu_{210}\mu_{201}\mu_{021}\mu_{012}$$

$$-\mu_{201}\mu_{120}\mu_{102}\mu_{021} - 2\mu_{210}\mu_{012}\mu_{111}^2 - 2\mu_{201}\mu_{021}\mu_{111}^2 - 2\mu_{120}\mu_{102}\mu_{111}^2$$

$$+3\mu_{210}\mu_{102}\mu_{021}\mu_{111} + 3\mu_{201}\mu_{120}\mu_{012}\mu_{111} + \mu_{111}^4)/\mu_{000}^8.$$

$$(5.90)$$

To describe the 3D AMIs by means of graph theory, using ordinary graphs / multigraphs, where each edge connects two nodes, is not sufficient. In 3D, we need *hyperedges* that connect three nodes. This concept is known in the literature as a *hypergraph*. The hypergraphs were used, for instance, in reference [48] for estimation of the parameters of an affine transformation.

The *k-uniform hypergraph* is a pair $G = (V, E)$, where $V = \{1, 2, \ldots, v\}$ is a finite set of nodes and E is a set of hyperedges. Each hyperedge connects up to k nodes. In this context, ordinary graphs are two-uniform hypergraphs. We need three-uniform hypergraphs for description of the 3D AMIs. The hypergraphs corresponding to I_1^{3D} and I_2^{3D} are shown in Figure 5.30, where the hyperedges are drawn as sets containing the connected nodes.

In tensor notation, the 3D moment tensor is

$$M^{i_1 i_2 \cdots i_k} = \int\limits_{-\infty}^{\infty} \int\limits_{-\infty}^{\infty} \int\limits_{-\infty}^{\infty} x^{i_1} x^{i_2} \cdots x^{i_k} f(x^1, x^2, x^3)\ \mathrm{d}x^1\ \mathrm{d}x^2\ \mathrm{d}x^3, \qquad (5.91)$$

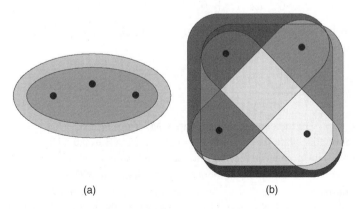

(a)   (b)

Figure 5.30   The hypergraphs corresponding to invariants (a) I_1^{3D} and (b) I_2^{3D}

where $x^1 = x$, $x^2 = y$ and $x^3 = z$. $M^{i_1 i_2 \cdots i_k} = m_{pqr}$ if p indices equal 1, q indices equal 2, and r indices equal 3. The unit polyvector in 3D is $\varepsilon_{123} = \varepsilon_{312} = \varepsilon_{231} = 1$, $\varepsilon_{321} = \varepsilon_{132} = \varepsilon_{213} = -1$, and the remaining twenty-one components are zero. In this notation, I_1^{3D} can be obtained from

$$M^{ij} M^{kl} M^{mn} \varepsilon_{ikm} \varepsilon_{jln} =$$
$$6(m_{200} m_{020} m_{002} + 2m_{110} m_{101} m_{011} - m_{200} m_{011}^2 - m_{020} m_{101}^2 - m_{002} m_{110}^2). \tag{5.92}$$

5.10.2   Normalized moments in 3D

Similarly as in 2D, one possibility of designing affine invariants in 3D is the method of object normalization. Since the 3D affine transformation has twelve parameters, the normalization requires fulfilling of twelve constraints.

Working with central coordinates assures the normalization to shift. To normalize w.r.t. the affine matrix $\mathbf{A}$, we decompose it into two rotations and a non-uniform scaling between them

$$\mathbf{A} = \mathbf{R}_2 \, \mathbf{S} \, \mathbf{R}_1$$
$$= \begin{pmatrix} \cos\beta_{xx} & \cos\beta_{xy} & \cos\beta_{xz} \\ \cos\beta_{yx} & \cos\beta_{yy} & \cos\beta_{yz} \\ \cos\beta_{zx} & \cos\beta_{zy} & \cos\beta_{zz} \end{pmatrix} \begin{pmatrix} \omega_x & 0 & 0 \\ 0 & \omega_y & 0 \\ 0 & 0 & \omega_z \end{pmatrix} \begin{pmatrix} \cos\alpha_{xx} & \cos\alpha_{xy} & \cos\alpha_{xz} \\ \cos\alpha_{yx} & \cos\alpha_{yy} & \cos\alpha_{yz} \\ \cos\alpha_{zx} & \cos\alpha_{zy} & \cos\alpha_{zz} \end{pmatrix}.$$
$$\tag{5.93}$$

It is natural to follow the 2D procedure and to normalize to the first rotation and to the non-uniform scaling by the second-order complex moments and then to normalize w.r.t. the second rotation by the third-order moments. To accomplish the first normalization, we diagonalize the covariance matrix (4.55) (the inertia matrix (4.56) cannot be applied here). The eigenvectors define the principal axes, and the eigenvalues define the size of the reference ellipsoid. We rotate the object such that its reference ellipsoid is in axial position, and then we scale it such that the ellipsoid becomes a unit sphere. The scaling factors are easy to calculate if we decompose the non-uniform scaling into a uniform scaling and stretching

$$\begin{pmatrix} \omega_x & 0 & 0 \\ 0 & \omega_y & 0 \\ 0 & 0 & \omega_z \end{pmatrix} = \begin{pmatrix} \omega & 0 & 0 \\ 0 & \omega & 0 \\ 0 & 0 & \omega \end{pmatrix} \begin{pmatrix} \delta_x & 0 & 0 \\ 0 & \delta_y & 0 \\ 0 & 0 & \delta_z \end{pmatrix}, \tag{5.94}$$

where $\delta_x \delta_y \delta_z = 1$. Then we normalize to the uniform scaling by m_{000} and to the stretching by δ_x, δ_y and δ_z. We have

$$\delta_x = \sqrt{\frac{\sqrt[3]{\lambda_1 \lambda_2 \lambda_3}}{\lambda_1}}, \quad \delta_y = \sqrt{\frac{\sqrt[3]{\lambda_1 \lambda_2 \lambda_3}}{\lambda_2}}, \quad \delta_z = \sqrt{\frac{\sqrt[3]{\lambda_1 \lambda_2 \lambda_3}}{\lambda_3}}, \tag{5.95}$$

where $\lambda_1 \geq \lambda_2 \geq \lambda_3$ are the eigenvalues of matrix $\mathbf{M_C}$ from (4.55).

In order to normalize w.r.t. the second rotation, we use the Burel and Hénocq algorithm [49].

First, we convert the geometric moments to the complex ones by (4.61) and (4.62). The Euler angle α is the phase of 3D complex moment c_{31}^1, β is calculated as $\beta = \arctan(\sqrt{2}\mathcal{R}(c_{31}^1)/c_{31}^0)$, and γ is the phase of c_{33}^1 (note that $c_{22}^1 = 0$ after the previous normalization and therefore it cannot be used at this step). We normalize the object by the angles $-\alpha$, $-\beta$ and $-\gamma$. If the magnitude of c_{31}^1 is close to zero because of symmetry, we must search for some higher-order normalizing moment. The index m of the complex moments plays the same role as the index difference in 2D case. If the object has N-FRS with respect to the current axis, we should use a moment with $m = N$ or generally with $m = kN$, where k is a positive integer as small as possible.

After this normalization, twelve moments have prescribed values

$$\mu_{000} = 1, \quad \mu_{100} = 0, \quad \mu_{010} = 0, \quad \mu_{001} = 0, \quad \mu_{110} = 0, \quad \mu_{101} = 0, \quad \mu_{011} = 0,$$
$$\mu_{020} = \mu_{200}, \quad \mu_{002} = \mu_{200}, \quad \mu_{012} = 0, \quad \mu_{210} = -\mu_{030}, \quad \mu_{102} = -\mu_{300} - \mu_{120}$$

and cannot be used for the recognition. The other moments of the normalized object are affine invariants of the original.

The above moments are not normalized to mirror reflection, because the mirror reflection is a borderline issue in 3D. If we still wanted to do so, we would take the normalized geometric moments and find the first nonzero moment μ_{pqr} with an odd second index q. If it is negative, then we change the signs of all moments the second index of which is odd. If it is positive or if all normalized moments with odd second indices vanish, no action is necessary.

5.10.3 Cayley-Aronhold equation in 3D

The extension of this method to 3D is relatively straightforward; however, the number of terms of the invariants is higher than in 2D.

A 3D affine transformation can be decomposed into three non-uniform scalings, six skews and three translations. The solution is analogous to the 2D case. There are six Cayley-Aronhold differential equations, such as

$$\sum_{p}\sum_{q}\sum_{r} p\mu_{p-1,q+1,r}\frac{\partial I}{\partial \mu_{pqr}} = 0. \qquad (5.96)$$

Solving the other five equations can be substituted by using the symmetry condition. Since we have six permutations of three indices, we obtain up to six symmetric terms with the same coefficient.

We must first enumerate all possible terms of the invariant. We can imagine this task as a table that must be filled in such a way that all row sums equal the weight of the invariants and the column sums equal the orders of the moments. Creating and solving the system of linear equations based on (5.96) is then analogous to the 2D case.

5.11 Beyond invariants

5.11.1 Invariant distance measure between images

The affine invariants introduced in the previous sections cannot be extended beyond the framework of linear spatial deformations of the image. The linear model is, however, insufficient in numerous practical cases, where elastic and/or local deformations may appear. Let us imagine, for instance, recognition of objects printed on the surface of a ball or a bottle, recognition of images taken by a fish-eye-lens or omni-directional camera (see Figure 5.31), and recognition of characters printed on a flexible time-variant surface (see Figure 5.32).

Figure 5.31 Nonlinear deformation of the text captured by a fish-eye-lens camera

It is hopeless to try finding invariants under "general elastic transformation" without assuming anything about its parametric model. In this section, we rely on the general assumption that the spatial deformation of the image can be described by a polynomial transformation[16]. However, even if we restrict to polynomials, moment invariants in the traditional sense cannot exist. The reason is that polynomial transformations of any finite degree greater than one do not preserve the order of the moments and do not form a group. A composition of two transforms of degree n is a transform of degree n^2; an inverse transform to a polynomial is not a polynomial transform at all. This "non-group" property makes the existence of invariants w.r.t. a finite-degree polynomial transformation impossible. If such an invariant existed, it would be invariant also to any composition of such transforms that implies it must be invariant to any polynomial transform of an arbitrary high degree.

We may overcome this by means of a *moment matching* technique. We do not look for invariants in the traditional sense but construct a distance measure d between two objects such that this measure is invariants to polynomial transformations of the objects up to the given degree. More formally, let $\mathbf{r}(\mathbf{x}) = (r_1(\mathbf{x}), r_2(\mathbf{x}))^T$ be a finite-degree polynomial transformation

---

[16] From a practical point of view, this is not a serious limitation because polynomials are dense in the set of all continuous functions, but from a theoretical point of view this assumption simplifies the problem.

Figure 5.32   Character deformations due to the print on a flexible surface

in 2D. The distance measure we are looking for should in an ideal case have the property of invariance

$$d(f(\mathbf{x}),\ f(\mathbf{r}(\mathbf{x}))) = 0$$

for any f and $\mathbf{r}$, and that of discriminability

$$d(f(\mathbf{x}),\ g(\mathbf{x})) > 0$$

for any f, g such that g is not a deformed version of f and vice versa.

This invariant distance performs a more general concept than traditional invariants. If the invariants existed, then it would be possible to define the distance easily as $d(f, g) = \|I(f) - I(g)\|$. The reverse implication, of course, does not hold. The invariant distance does not provide a description of a single image because it is always defined for a pair of images. This is why it cannot be used for shape encoding and reconstruction but can be applied in object recognition and matching. We can, for each database entry g_i, calculate the value of $d(f, g_i)$ and then classify f according to the minimum distance. It should be emphasized that the idea of invariant distance is much more general; it is not restricted to moments and to spatial transformations.

5.11.2   Moment matching

We construct the invariant distance $d(f, g)$ by means of moments of both f and g. To ensure the desired invariance property, we have to study how the moments are transformed if the image has undergone a polynomial deformation. Let us show this in 1D for simplicity; the transition to 2D and 3D is a bit laborious but in principle straightforward.

Let $\boldsymbol{\pi}(x)$ be polynomial basis functions and $\mathbf{m}$ be a vector of moments w.r.t. this basis; that is,

$$\boldsymbol{\pi}(x) = \begin{pmatrix} \pi_0(x) \\ \pi_1(x) \\ \vdots \\ \pi_{n-1}(x) \end{pmatrix} \qquad \text{and} \qquad \mathbf{m} = \begin{pmatrix} m_0 \\ m_1 \\ \vdots \\ m_{n-1} \end{pmatrix}. \tag{5.97}$$

Let the deformed image $\tilde{f}$ be

$$\tilde{f}(r(x)) = f(x). \tag{5.98}$$

We are interested in the relation between the moments $\mathbf{m}$ and the moments

$$\tilde{\mathbf{m}} = \int \tilde{f}(\tilde{x}) \tilde{\boldsymbol{\pi}}(\tilde{x}) \, d\tilde{x}$$

of the transformed image with respect to some other $\tilde{n}$ basis polynomials

$$\tilde{\boldsymbol{\pi}}(\tilde{x}) = \begin{pmatrix} \tilde{\pi}_0(\tilde{x}) \\ \tilde{\pi}_1(\tilde{x}) \\ \vdots \\ \tilde{\pi}_{\tilde{n}-1}(\tilde{x}) \end{pmatrix}. \tag{5.99}$$

By substituting $\tilde{x} = r(x)$ into the definition of $\tilde{\mathbf{m}}$ and by using the composite function integration we obtain the following result:

Theorem 5.3 Denote by $J_r(x)$ the Jacobian of the transform function $r(x)$. If

$$\tilde{\boldsymbol{\pi}}(r(x)) |J_r(x)| = \mathbf{A}\boldsymbol{\pi}(x) \tag{5.100}$$

for some $\tilde{n} \times n$ matrix $\mathbf{A}$ then

$$\tilde{\mathbf{m}} = \mathbf{A}\mathbf{m} \ . \tag{5.101}$$

This is a very strong theorem, which says that there is a linear relationship between the moments of the deformed image and those of the original, even if the deformation is not linear. Matrix $\mathbf{A}$ is, of course, not a square matrix, which means that polynomial deformations do not preserve the moment orders. Matrix $\mathbf{A}$ is a function of the parameters of r. Finding its particular form for the given family of deformations r is the key step in using this theorem but it is always possible for any polynomial transformation r the parametric form of which is known.

For simple transformations r and small number of moments, finding $\mathbf{A}$ is straightforward even manually, using the standard basis $\pi_k(x) = x^k$. However, due to numerical instability, this intuitive approach cannot be used for higher-order polynomial transforms and/or for more moments. To obtain a numerically stable method, it is important to use suitable polynomial bases, such as orthogonal polynomials, without using their expansions into standard (monomial) powers (see Chapter 7 for a review of orthogonal polynomials and moments). A proper implementation is based on the representation of polynomial bases by matrices with special structure [50, 51]. Such a representation makes it possible to evaluate the polynomials efficiently by means of recurrent relations. For constructing the matrix $\mathbf{A}$ and for other implementation details see [52], where Chebyshev polynomials were used as $\boldsymbol{\pi}(x)$.

5.11.3   Object recognition as a minimization problem

The definition of the invariant distance measure d can be formulated as a minimization problem. We calculate the "uniform best fit" from all equations in (5.101). For a given set of values of the moments $\mathbf{m}$ and $\tilde{\mathbf{m}}$, we find the values of the parameters of $\mathbf{A}$ to satisfy (5.101) as best possible in ℓ_2 norm; the error of this fit then becomes the invariant distance. The implementation of the object recognition by invariant distance can be described as follows.

1. Given is a library (database) of images $g_j(x, y)$, $j = 1, \ldots, L$, and a deformed image $\tilde{f}(\tilde{x}, y)$ that is assumed to have been obtained by a transform of a known polynomial form $\mathbf{r}(x, y, \mathbf{a})$ with unknown parameters $\mathbf{a}$.
2. Choose the appropriate domains and polynomial bases $\boldsymbol{\pi}$ and $\tilde{\boldsymbol{\pi}}$.
3. Derive a program to evaluate the matrix $\mathbf{A}(\mathbf{a})$. This critical error-prone step is performed by a symbolic algorithmic procedure that produces the program used then in numerical calculations. This step is performed only once for the given task. (It has to be repeated only if we change the polynomial bases or the form of transform $\mathbf{r}(x, y, \mathbf{a})$, which basically means only if we move to another application).
4. Calculate the moments $\mathbf{m}^{(g_j)}$ of all library images $g_j(x, y)$.
5. Calculate the moments $\tilde{\mathbf{m}}^{(\tilde{f})}$ of the deformed image $\tilde{f}(\tilde{x}, y)$.
6. For all $j = 1, \ldots, L$ calculate the distance measure d using a proper minimization technique

$$d(\tilde{f}, g_j) = \min_{\mathbf{a}} \| \tilde{\mathbf{m}}^{(\tilde{f})} - \mathbf{A}(\mathbf{a})\mathbf{m}^{(g_j)} \| \tag{5.102}$$

and denote

$$M = \min_j d(\tilde{f}, g_j).$$

The norm used here should be weighted, for example, relative to the components corresponding to the same degree.
7. The identified image is such g_k for which $d(\tilde{f}, g_k) = M$. The ratio

$$\gamma = \frac{d(\tilde{f}, g_\ell)}{d(\tilde{f}, g_k)},$$

where $d(\tilde{f}, g_\ell)$ is the second minimum, may be used as a confidence measure of the identification. Note that $1 \leq \gamma < \infty$, the higher γ the more reliable classification. We may accept only decisions the confidence of which exceeds a given threshold.

5.11.4   Numerical experiments

The experiment illustrates the applicability of the invariant distance measure to classify real images in situations when the image deformation is not known and, moreover, is not exactly polynomial. Such a situation occurs in practice when the intention is to recognize for instance letters or digits which are painted/printed on curved surfaces such as on balls, cans, bottles, etc.

We took six photos of capital letters "M", "N", "E" and "K" printed on a label and glued to a bottle. The label was intentionally tilted by a random degree of rotation; see Figure 5.33.

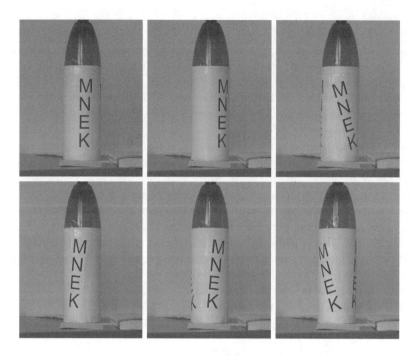

Figure 5.33   The six bottle images used in the experiment

This resulted in six differently deformed instances of each letter. The task was to recognize these letters against a database containing the full English alphabet of identical font.

One can see that, apart from the deformation induced by the cylindrical shape of the bottle, rotation and slight perspective is present. Scaling must be also considered, since the size of the captured letters and database templates may not match.

The letter deformation apparently is not a polynomial, but we assumed it can be approximated by a quadratic polynomial in horizontal direction that approximates the cylinder-to-plane projection. The deformed letters were segmented and binarized first. After the segmentation, we used the distance measure based on six moments to classify them against the undistorted alphabet of twenty-six letters. The recognition was quite successful – only the most deformed letter "N" was misclassified as "H", all other letters were recognized correctly.

For a comparison, we also tried to recognize the letters by two other methods. First, we employed the affine moment invariants. They failed in many cases because they are not de-signed to handle nonlinear deformations. Then we employed spline-based elastic matching as proposed by Kybic in [53]. This method searches the space of parameters of cubic B-spline deformation of the templates, calculates the mean square error as a measure of dissimilarity between the images, and finds the minimum of this error function using hierarchical optimiza-tion. This method was designed in medical imaging for recognition of templates undergoing elastic deformations and has been proved to be very powerful. To make the comparison fair, we used the original author's implementation of the method. The recognition rate was 100%, which is not surprising – the method in fact performs an exhaustive search in the space of deformations while working with full uncompressed representation of each template.

Table 5.10 Classification confidence coefficients of four letters (each having six different instances of distortion) using the invariant distance (top) and the spline-based elastic matching (bottom). MC means a misclassified letter

	Invariant distance confidence					
M	9.4	1.3	2.3	6.4	1.1	1.6
N	28.5	11.9	1.3	9.7	10.9	(MC)
E	5.8	2.2	10.4	5.5	3.3	3.0
K	12.1	10.0	2.1	5.4	6.4	2.6
	Elastic matching confidence					
M	11.5	1.7	1.9	9.8	10.5	1.4
N	8.3	3.9	2.5	9.3	6.8	2.5
E	6.4	3.1	2.2	5.3	3.8	2.0
K	10.8	3.0	1.9	5.0	3.4	2.2

It is interesting to compare the methods also in terms of their speed and confidence. The computational cost of both methods is of vastly different magnitudes. On a computer with the Intel Pentium Core Duo T2400 processor, the time required to classify all twenty-four deformed letters was in the case of elastic matching over one hour. On the other hand, it took less then one minute when using the moment-based distance measure. Such a significant difference was caused mainly by the fact that elastic matching works with full image representation while the distance measure uses a significant compression into six moments. The efficiency of the invariant distance could be further improved by using fast algorithms for moment calculation because we worked with binary images only.

Surprisingly, the use of different methods and different object representations did not lead to different confidences. The confidences γ of both methods are not very high but similar; see Table 5.10. It is interesting to notice that extremely low confidence (close to 1) is frequent for the letter "M", often having the second-closest match "N". The explanation is that the quadratic deformation of "M" can merge the right diagonal line with the right vertical line thus creating an impression of letter "N".

5.12   Conclusion

In this chapter, we presented four different methods of automatic generation of the affine moment invariants. Each of them has its positive and negative properties.

The graph method is transparent and easy to implement, but its drawback is an exponential computing complexity of generating invariants that builds a wall preventing it from being used for higher weights. Until now, the invariants of weight 12 and less have been found explicitly. This is not, however, a limitation for practical use – the explicit forms of the invariants are generated only once, and the weight 12 is sufficient for all applications we have encountered.

The second method was based on image normalization. Its main advantage is that it directly yields an independent set of invariants. In this way, we are rid of the time-consuming elimination of reducible and dependent invariants. On the other hand, the need to search nonzero moments (and the need for the choice of proper thresholds), less numerical stability, propagation of potential errors into all invariants, possible discontinuity in rotation normalization, and also more complicated explicit forms of the invariants are clear drawbacks of this approach. Half normalization provides a compromise between the normalized moments and the AMIs. This approach retains the advantage of producing independent invariants and removes the problems with normalization to the second rotation, but still suffers from lower numerical stability than the AMIs and from the possibility of error propagation.

The transvectants are relatively difficult to understand, but this method generates an independent set of invariants, which is a big advantage. The invariants are similar (and many are even equivalent) to the invariants derived by graphs. However, some of their explicit forms contain more terms.

The last method we presented is based on an automatic direct solution of the Cayley-Aronhold differential equation. Its advantage is the relatively low computing complexity of invariant generation. The disadvantage of the method is a complicated implementation and, similarly to the graph method, producing reducible and dependent invariants that should be identified and removed afterwards.

Although the background, the complexity, and the computational aspects of the above methods are different, we proved that they are theoretically equivalent because they end up with the sets of invariants between which a one-to-one mapping exists.

Finally, we briefly explained the generalization of the AMIs to color images, to vector fields, and to 3D images. At the end of the chapter we mentioned the idea of moment matching in situations, when invariants cannot exist.

Appendix 5.A

Proof of the Fundamental theorem of AMIs (Theorem 5.1) The proof is based on the Taylor expansion of the Fourier transform. Let us start with the identity

$$\mathcal{F}(f)(\mathbf{u}) = \frac{1}{|J|}\mathcal{F}(f')(\mathbf{u}') \tag{5.103}$$

where

$$f'(\mathbf{x}) = f(\mathbf{A}^{-1}\mathbf{x}), \tag{5.104}$$

$$\mathbf{x}' = \mathbf{A}\mathbf{x}, \quad \mathbf{u}' = \mathbf{A}^{-T}\mathbf{u}, \tag{5.105}$$

and $J \equiv \det(\mathbf{A}) \neq 0$. The identity expresses the fact that the Fourier transformation of the image stays the same (up to a multiplicative factor) if the frequency coordinate systems is transformed inversely-transposed to the transformation of spatial coordinates. To see that, we rewrite (5.103) as

$$\int\limits_{-\infty}^{\infty}\int\limits_{-\infty}^{\infty} e^{-2\pi i \mathbf{u}^T \mathbf{x}} f(x,y) \; \mathrm{d}x \; \mathrm{d}y = \frac{1}{|J|} \int\limits_{-\infty}^{\infty}\int\limits_{-\infty}^{\infty} e^{-2\pi i \mathbf{u}'^T \mathbf{x}'} f(x,y) \; \mathrm{d}x' \; \mathrm{d}y' \quad . \tag{5.106}$$

Since the inverse and direct linear transformations are canceled $\mathbf{u}'^T\mathbf{x}' = \mathbf{u}^T\mathbf{A}^{-1}\mathbf{A}\mathbf{x} = \mathbf{u}^T\mathbf{x}$ and $\mathbf{dx}' = |J|\mathbf{dx}$, the identity is apparent (note that the values of f have not been changed under the coordinate transformation).

Both sides of (5.106) can be expanded into Taylor series

$$\sum_{j=0}^{\infty}\sum_{k=0}^{\infty}\frac{(-2\pi i)^{j+k}}{j!k!}m_{jk}u^j v^k = \frac{1}{|J|}\sum_{j=0}^{\infty}\sum_{k=0}^{\infty}\frac{(-2\pi i)^{j+k}}{j!k!}m'_{jk}u'^{\,j}v'^{\,k} \tag{5.107}$$

where m'_{jk} are the geometric moments of image $f'(x, y)$. The monomials $u^j v^k$ of the same order are under an affine transformation transformed among themselves (for instance, from a quadratic monomial we obtain a linear combination of quadratic monomials). Therefore, it is possible to decompose the equality (5.107) and compare the terms with the same monomial order. For one particular order $p = j + k$ we obtain

$$\sum_{j=0}^{p}\frac{p!}{j!(p-j)!}m_{j,p-j}u^j v^{p-j} = \frac{1}{|J|}\sum_{j=0}^{p}\frac{p!}{j!(p-j)!}m'_{j,p-j}u'^{\,j}v'^{\,p-j}, \tag{5.108}$$

which is the expression of the transformation of a binary algebraic form of order p; see (5.10) and (5.12). Thus, we can substitute the coefficients of the binary forms for moments

$$a_j = m_{j,p-j} \tag{5.109}$$

and

$$a'_j = \frac{1}{|J|}m'_{j,p-j} \tag{5.110}$$

in (5.13). Since the invariant is in the form of a polynomial with the same structure of all terms, as is implied by the definition of the algebraic invariant, we can multiply the equality by the factor $|J|^r$ and obtain the proposition of the theorem.

Appendix 5.B

In this appendix, we present explicit forms of the affine invariants that were used in the experiments and have not been listed in the chapter.

$$
\begin{aligned}
I_{11} = (&\mu_{30}^2\mu_{12}^2\mu_{04} - 2\mu_{30}^2\mu_{12}\mu_{03}\mu_{13} + \mu_{30}^2\mu_{03}^2\mu_{22} - 2\mu_{30}\mu_{21}^2\mu_{12}\mu_{04} \\
&+ 2\mu_{30}\mu_{21}^2\mu_{03}\mu_{13} + 2\mu_{30}\mu_{21}\mu_{12}^2\mu_{13} - 2\mu_{30}\mu_{21}\mu_{03}^2\mu_{31} - 2\mu_{30}\mu_{12}^3\mu_{22} \\
&+ 2\mu_{30}\mu_{12}^2\mu_{03}\mu_{31} + \mu_{21}^4\mu_{04} - 2\mu_{21}^3\mu_{12}\mu_{13} - 2\mu_{21}^3\mu_{03}\mu_{22} + 3\mu_{21}^2\mu_{12}^2\mu_{22} \\
&+ 2\mu_{21}^2\mu_{12}\mu_{03}\mu_{31} + \mu_{21}^2\mu_{03}^2\mu_{40} - 2\mu_{21}\mu_{12}^3\mu_{31} - 2\mu_{21}\mu_{12}^2\mu_{03}\mu_{40} + \mu_{12}^4\mu_{40})/\mu_{00}^{13} \\[4pt]
I_{19} = (&\mu_{20}\mu_{30}\mu_{12}\mu_{04} - \mu_{20}\mu_{30}\mu_{03}\mu_{13} - \mu_{20}\mu_{21}^2\mu_{04} + \mu_{20}\mu_{21}\mu_{12}\mu_{13} \\
&+ \mu_{20}\mu_{21}\mu_{03}\mu_{22} - \mu_{20}\mu_{12}^2\mu_{22} - 2\mu_{11}\mu_{30}\mu_{12}\mu_{13} + 2\mu_{11}\mu_{30}\mu_{03}\mu_{22} \\
&+ 2\mu_{11}\mu_{21}^2\mu_{13} - 2\mu_{11}\mu_{21}\mu_{12}\mu_{22} - 2\mu_{11}\mu_{21}\mu_{03}\mu_{31} + 2\mu_{11}\mu_{12}^2\mu_{31} \\
&+ \mu_{02}\mu_{30}\mu_{12}\mu_{22} - \mu_{02}\mu_{30}\mu_{03}\mu_{31} - \mu_{02}\mu_{21}^2\mu_{22} + \mu_{02}\mu_{21}\mu_{12}\mu_{31}
\end{aligned}
$$

$$+ \mu_{02}\mu_{21}\mu_{03}\mu_{40} - \mu_{02}\mu_{12}^2\mu_{40})/\mu_{00}^{10}$$

$$\begin{aligned}
I_{25} = (&\mu_{20}^3\mu_{12}^2\mu_{04} - 2\mu_{20}^3\mu_{12}\mu_{03}\mu_{13} + \mu_{20}^3\mu_{03}^2\mu_{22} - 4\mu_{20}^2\mu_{11}\mu_{21}\mu_{12}\mu_{04} \\
&+ 4\mu_{20}^2\mu_{11}\mu_{21}\mu_{03}\mu_{13} + 2\mu_{20}^2\mu_{11}\mu_{12}^2\mu_{13} - 2\mu_{20}^2\mu_{11}\mu_{03}^2\mu_{31} \\
&+ 2\mu_{20}^2\mu_{02}\mu_{30}\mu_{12}\mu_{04} - 2\mu_{20}^2\mu_{02}\mu_{30}\mu_{03}\mu_{13} - 2\mu_{20}^2\mu_{02}\mu_{21}\mu_{12}\mu_{13} \\
&+ 2\mu_{20}^2\mu_{02}\mu_{21}\mu_{03}\mu_{22} + \mu_{20}^2\mu_{02}\mu_{12}^2\mu_{22} - 2\mu_{20}^2\mu_{02}\mu_{12}\mu_{03}\mu_{31} \\
&+ \mu_{20}^2\mu_{02}\mu_{03}^2\mu_{40} + 4\mu_{20}\mu_{11}^2\mu_{21}^2\mu_{04} - 8\mu_{20}\mu_{11}^2\mu_{21}\mu_{03}\mu_{22} \\
&- 4\mu_{20}\mu_{11}^2\mu_{12}^2\mu_{22} + 8\mu_{20}\mu_{11}^2\mu_{12}\mu_{03}\mu_{31} - 4\mu_{20}\mu_{11}\mu_{02}\mu_{30}\mu_{21}\mu_{04} \\
&+ 4\mu_{20}\mu_{11}\mu_{02}\mu_{30}\mu_{03}\mu_{22} + 4\mu_{20}\mu_{11}\mu_{02}\mu_{21}^2\mu_{13} \\
&- 4\mu_{20}\mu_{11}\mu_{02}\mu_{21}\mu_{12}\mu_{22} + 4\mu_{20}\mu_{11}\mu_{02}\mu_{12}^2\mu_{31} \\
&- 4\mu_{20}\mu_{11}\mu_{02}\mu_{12}\mu_{03}\mu_{40} + \mu_{20}\mu_{02}^2\mu_{30}^2\mu_{04} - 2\mu_{20}\mu_{02}^2\mu_{30}\mu_{21}\mu_{13} \\
&+ 2\mu_{20}\mu_{02}^2\mu_{30}\mu_{12}\mu_{22} - 2\mu_{20}\mu_{02}^2\mu_{30}\mu_{03}\mu_{31} + \mu_{20}\mu_{02}^2\mu_{21}^2\mu_{22} \\
&- 2\mu_{20}\mu_{02}^2\mu_{21}\mu_{12}\mu_{31} + 2\mu_{20}\mu_{02}^2\mu_{21}\mu_{03}\mu_{40} - 8\mu_{11}^3\mu_{21}^2\mu_{13} \\
&+ 16\mu_{11}^3\mu_{21}\mu_{12}\mu_{22} - 8\mu_{11}^3\mu_{12}^2\mu_{31} + 8\mu_{11}^2\mu_{02}\mu_{30}\mu_{21}\mu_{13} \\
&- 8\mu_{11}^2\mu_{02}\mu_{30}\mu_{12}\mu_{22} - 4\mu_{11}^2\mu_{02}\mu_{21}^2\mu_{22} + 4\mu_{11}^2\mu_{02}\mu_{12}^2\mu_{40} \\
&- 2\mu_{11}\mu_{02}^2\mu_{30}^2\mu_{13} + 4\mu_{11}\mu_{02}^2\mu_{30}\mu_{12}\mu_{31} + 2\mu_{11}\mu_{02}^2\mu_{21}^2\mu_{31} \\
&- 4\mu_{11}\mu_{02}^2\mu_{21}\mu_{12}\mu_{40} + \mu_{02}^3\mu_{30}^2\mu_{22} - 2\mu_{02}^3\mu_{30}\mu_{21}\mu_{31} + \mu_{02}^3\mu_{21}^2\mu_{40})/\mu_{00}^{14}
\end{aligned}$$

$$\begin{aligned}
I_{47} = (&-\mu_{50}^2\mu_{05}^2 + 10\mu_{50}\mu_{41}\mu_{14}\mu_{05} - 4\mu_{50}\mu_{32}\mu_{23}\mu_{05} - 16\mu_{50}\mu_{32}\mu_{14}^2 + 12\mu_{50}\mu_{23}^2\mu_{14} \\
&- 16\mu_{41}^2\mu_{23}\mu_{05} - 9\mu_{41}^2\mu_{14}^2 + 12\mu_{41}\mu_{32}^2\mu_{05} + 76\mu_{41}\mu_{32}\mu_{23}\mu_{14} - 48\mu_{41}\mu_{23}^3 \\
&- 48\mu_{32}^3\mu_{14} + 32\mu_{32}^2\mu_{23}^2)/\mu_{00}^{14}.
\end{aligned}$$

The set $\{I_1, I_2, I_3, I_4, I_6, I_7, I_8, I_9, I_{19}\}$ is a complete and independent set of features up to the fourth order. Alternatively, I_{11} or I_{25} can substitute I_{19}. I_{47} is the simplest homogeneous invariant of the fifth order.

References

[1] I. Weiss, "Projective invariants of shapes," in *Proceedings of the Computer Vision and Pattern Recognition CVPR '88*, pp. 1125–1134, IEEE, 1988.

[2] C. A. Rothwell, A. Zisserman, D. A. Forsyth, and J. L. Mundy, "Canonical frames for planar object recognition," in *Proceedings of the Second European Conference on Computer Vision ECCV'92*, vol. LNCS 588, pp. 757–772, Springer, 1992.

[3] I. Weiss, "Differential invariants without derivatives," in *Proceedings of the 11th International Conference on Pattern Recognition ICPR'92*, pp. 394–398, IEEE, 1992.

[4] T. Suk and J. Flusser, "Vertex-based features for recognition of projectively deformed polygons," *Pattern Recognition*, vol. 29, no. 3, pp. 361–367, 1996.

[5] R. Lenz and P. Meer, "Point configuration invariants under simultaneous projective and permutation transformations," *Pattern Recognition*, vol. 27, no. 11, pp. 1523–1532, 1994.

[6] N. S. V. Rao, W. Wu, and C. W. Glover, "Algorithms for recognizing planar polygonal configurations using perspective images," *IEEE Transactions on Robotics and Automation*, vol. 8, no. 4, pp. 480–486, 1992.

[7] J. L. Mundy and A. Zisserman, *Geometric Invariance in Computer Vision*. Cambridge, Massachusetts, USA: MIT Press, 1992.

[8] C. A. Rothwell, D. A. Forsyth, A. Zisserman, and J. L. Mundy, "Extracting projective structure from single perspective views of 3D point sets," in *Proceedings of the Fourth International Conference on Computer Vision ICCV'94*, pp. 573–582, IEEE, 1993.

[9] E. P. L. Van Gool, T. Moons, and A. Oosterlinck, "Vision and Lie's approach to invariance," *Image and Vision Computing*, vol. 13, no. 4, pp. 259–277, 1995.

[10] T. Suk and J. Flusser, "Projective moment invariants," *IEEE Transactions on Pattern Analysis and Machine Intelligence*, vol. 26, no. 10, pp. 1364–1367, 2004.

[11] K. Voss and H. Süße, *Adaptive Modelle und Invarianten für zweidimensionale Bilder*. Aachen, Germany: Shaker, 1995.

[12] W. Yuanbin, Z. Bin, and Y. Tianshun, "Moment invariants of restricted projective transformations," in *Proceedings of the International Symposium on Information Science and Engineering ISISE'08*, vol. 1, pp. 249–253, IEEE, 2008.

[13] M.-K. Hu, "Visual pattern recognition by moment invariants," *IRE Transactions on Information Theory*, vol. 8, no. 2, pp. 179–187, 1962.

[14] D. Hilbert, *Theory of Algebraic Invariants*. Cambridge, U.K.: Cambridge University Press, 1993.

[15] T. H. Reiss, "The revised fundamental theorem of moment invariants," *IEEE Transactions on Pattern Analysis and Machine Intelligence*, vol. 13, no. 8, pp. 830–834, 1991.

[16] J. Flusser and T. Suk, "Pattern recognition by affine moment invariants," *Pattern Recognition*, vol. 26, no. 1, pp. 167–174, 1993.

[17] J. Flusser and T. Suk, "Pattern recognition by means of affine moment invariants," Research Report 1726, Institute of Information Theory and Automation, Prague, 1991.

[18] J. H. Grace and A. Young, *The Algebra of Invariants*. Cambridge, U.K.: Cambridge University Press, 1903.

[19] J. J. Sylvester assisted by F. Franklin, "Tables of the generating functions and groundforms for the binary quantics of the first ten orders," *American Journal of Mathematics*, vol. 2, pp. 223–251, 1879.

[20] J. J. Sylvester assisted by F. Franklin, "Tables of the generating functions and groundforms for simultaneous binary quantics of the first four orders taken two and two together," *American Journal of Mathematics*, vol. 2, pp. 293–306, 324–329, 1879.

[21] I. Schur, *Vorlesungen über Invariantentheorie*. Berlin, Germany: Springer, 1968. (in German).

[22] G. B. Gurevich, *Foundations of the Theory of Algebraic Invariants*. Groningen, The Netherlands: Nordhoff, 1964.

[23] A. G. Mamistvalov, "*n*-dimensional moment invariants and conceptual mathematical theory of recognition *n*-dimensional solids," *IEEE Transactions on Pattern Analysis and Machine Intelligence*, vol. 20, no. 8, pp. 819–831, 1998.

[24] A. G. Mamistvalov, "On the fundamental theorem of moment invariants," *Bulletin of the Academy of Sciences of the Georgian SSR*, vol. 59, pp. 297–300, 1970. In Russian.

[25] T. Suk and J. Flusser, "Graph method for generating affine moment invariants," in *Proceedings of the 17th International Conference on Pattern Recognition ICPR'04*, pp. 192–195, IEEE, 2004.

[26] T. Suk and J. Flusser, "Affine moment invariants generated by graph method," *Pattern Recognition*, vol. 44, no. 9, pp. 2047–2056, 2011.

[27] T. H. Reiss, *Recognizing Planar Objects Using Invariant Image Features*, vol. 676 of *LNCS*. Berlin, Germany: Springer, 1993.

[28] T. Suk and J. Flusser, "Affine moment invariants generated by automated solution of the equations," in *Proceedings of the 19th International Conference on Pattern Recognition ICPR'08*, IEEE, 2008.

[29] M. S. Hickman, "Geometric moments and their invariants," *Journal of Mathematical Imaging and Vision*, vol. 44, no. 3, pp. 223–235, 2012.

[30] I. Rothe, H. Süsse, and K. Voss, "The method of normalization to determine invariants," *IEEE Transactions on Pattern Analysis and Machine Intelligence*, vol. 18, no. 4, pp. 366–376, 1996.

[31] T. Suk and J. Flusser, "Tables of affine moment invariants generated by the graph method," Research Report 2156, Institute of Information Theory and Automation, 2005.

[32] J. Flusser, T. Suk, and B. Zitová. http://zoi.utia.cas.cz/moment_invariants, *the accompanying web pages of the book: "Moments and Moment Invariants in Pattern Recognition"*, Chichester: Wiley, 2009.

[33] Y. Zhang, C. Wen, Y. Zhang, and Y. C. Soh, "On the choice of consistent canonical form during moment normalization," *Pattern Recognition Letters*, vol. 24, no. 16, pp. 3205–3215, 2003.

[34] S.-C. Pei and C.-N. Lin, "Image normalization for pattern recognition," *Image and Vision Computing*, vol. 13, no. 10, pp. 711–723, 1995.

[35] T. Suk and J. Flusser, "Affine normalization of symmetric objects," in *Proceedings of the Seventh International Conference on Advanced Concepts for Intelligent Vision Systems Acivs'05*, vol. LNCS 3708, pp. 100–107, Springer, 2005.

[36] D. Shen and H. H.-S. Ip, "Generalized affine invariant image normalization," *IEEE Transactions on Pattern Analysis and Machine Intelligence*, vol. 19, no. 5, pp. 431–440, 1997.

[37] J. Sprinzak and M. Werman, "Affine point matching," *Pattern Recognition Letters*, vol. 15, no. 4, pp. 337–339, 1994.

[38] J. Heikkilä, "Pattern matching with affine moment descriptors," *Pattern Recognition*, vol. 37, pp. 1825–1834, 2004.

[39] Y. S. Abu-Mostafa and D. Psaltis, "Image normalization by complex moments," *IEEE Transactions on Pattern Analysis and Machine Intelligence*, vol. 7, no. 1, pp. 46–55, 1985.

[40] T. Suk and J. Flusser, "The independence of the affine moment invariants," in *Proceedings of the Fifth International Workshop on Information Optics WIO'06*, pp. 387–396, American Institute of Physics, 2006.

[41] A. Cayley, "On linear transformations," *Cambridge and Dublin Mathematical Journal*, vol. 1, pp. 104–122, 1846.

[42] J. Flusser and T. Suk, "A moment-based approach to registration of images with affine geometric distortion," *IEEE Transactions on Geoscience and Remote Sensing*, vol. 32, no. 2, pp. 382–387, 1994.

[43] T. Suk, "The proof of completeness of the graph method for generation of affine moment invariants," in *7th International Conference on Image Analysis and Recognition ICIAR'10* (A. Campilho and M. Kamel, eds.), vol. LNCS 6111, part 1, (Berlin, Heidelberg, Germany), pp. 157–166, Springer, 2010.

[44] N. Otsu, "A threshold selection method from gray-level histograms," *IEEE Transactions on Systems, Man, and Cybernetics*, vol. 9, no. 1, pp. 62–66, 1979.

[45] F. Mindru, T. Moons, and L. V. Gool, "Color-based moment invariants for viewpoint and illumination independent recognition of planar color patterns," in *Proceedings of the International Conference on Advances in Pattern Recognition ICAPR'98*, pp. 113–122, Springer, 1998.

[46] F. Mindru, T. Tuytelaars, L. V. Gool, and T. Moons, "Moment invariants for recognition under changing viewpoint and illumination," *Computer Vision and Image Understanding*, vol. 94, no. 1–3, pp. 3–27, 2004.

[47] D. Xu and H. Li, "3-D affine moment invariants generated by geometric primitives," in *Proceedings of the 18th International Conference on Pattern Recognition ICPR'06*, pp. 544–547, IEEE, 2006.

[48] S. Rota Bulò, A. Albarelli, A. Torsello, and M. Pelillo, "A hypergraph-based approach to affine parameters estimation," in *Proceedings of the 19th International Conference on Pattern Recognition ICPR'08*, IEEE, 2008.

[49] G. Burel and H. Hénocq, "Determination of the orientation of 3D objects using spherical harmonics," *Graphical Models and Image Processing*, vol. 57, no. 5, pp. 400–408, 1995.

[50] G. H. Golub and J. Kautsky, "Calculation of Gauss quadratures with multiple free and fixed knots," *Numerische Mathematik*, vol. 41, no. 2, pp. 147–163, 1983.

[51] G. H. Golub and J. Kautsky, "On the calculation of Jacobi matrices," *Linear Algebra and Its Applications*, vol. 52/53, pp. 439–455, 1983.

[52] J. Flusser, J. Kautsky, and F. Šroubek, "Implicit moment invariants," *International Journal of Computer Vision*, vol. 86, no. 1, pp. 72–86, 2010.

[53] J. Kybic and M. Unser, "Fast parametric elastic image registration," *IEEE Transactions on Image Processing*, vol. 12, no. 11, pp. 1427–1442, 2003.

[54] T. Suk and J. Flusser, Affine moment invariants of color images, *Computer Analysis of Images and Patterns CAIP'09*, (Xiaoyi Jiang and Nikolai Petkov, eds.), vol. LNCS 5702, (Münster, Germany), pp. 334–341, Springer, 2009

6

Invariants to Image Blurring

6.1 Introduction

In the previous chapters we described moment invariants with respect to various transformations of the spatial coordinates – translation, rotation, scaling, and affine transformation. According to the terminology introduced in Chapter 1, all these deformations fall into the category of *geometric degradations* of an image. However, geometric degradations are not the only source of image imperfection. As pointed out already in Chapter 1, heavy image corruption may be caused by *graylevel/color degradations* that influence the intensity values. The linear contrast changes discussed in Chapter 3 may be considered very simple and easy-to-handle graylevel degradations, but most of the degradations of this category are much more complicated and deriving invariants with respect to them is a challenging problem.

6.1.1 Image blurring – the sources and modeling

The degradations which we consider in this chapter are caused by such factors as a wrong focus, camera and/or scene motion, camera vibrations, and by taking images through a turbulent medium such as atmosphere or water. Along with various second-order factors such as lens aberrations and intrinsic sensor blur caused by a finite-size sampling pulse, they result in the phenomenon called *image blurring* (see Figure 6.1).

Blurring is inevitably present in any image. If it is small compared to other degradations or below the image resolution, we can ignore it. However, in many cases blurring is a primary degradation factor to an extent which cannot be overlooked.

Assuming the image acquisition time is so short that the blurring factors do not change during the image formation and also assuming that the blurring is of the same kind for all colors/graylevels, we can model the blurring process of image f by a *superposition integral*

$$g(x,y) = \int\limits_{-\infty}^{\infty} \int\limits_{-\infty}^{\infty} f(s,t)h(x,y,s,t)dsdt, \tag{6.1}$$

where g is the acquired image and $h(x,y,s,t)$ is the *point-spread function* (PSF) of the system (see Figure 6.2 for a visualization of the PSF). Many other degradations might be present such as geometric deformations and noise, but let us ignore them for the moment.

2D and 3D Image Analysis by Moments, First Edition. Jan Flusser, Tomáš Suk and Barbara Zitová.
© 2017 John Wiley & Sons, Ltd. Published 2017 by John Wiley & Sons, Ltd.
Companion website: www.wiley.com/go/flusser/2d3dimageanalysis

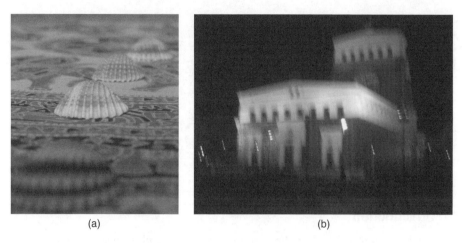

(a)   (b)

Figure 6.1 Two examples of image blurring. Space-variant out-of-focus blur caused by a narrow depth of field of the camera (a) and camera-shake blur at a long exposure (b)

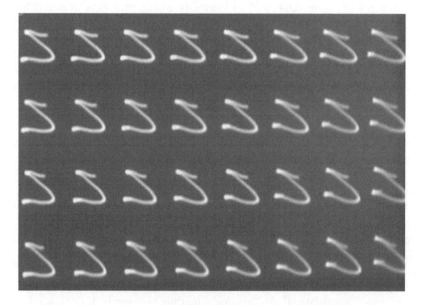

Figure 6.2 Visualization of the space variant PSF. A photograph of a point grid by a hand-held camera blurred by camera shake. The curves are images of bright points equivalent to the PSF at the particular places

If the PSF is *space-invariant*, that is, if

$$h(x, y, s, t) = h(x - s, y - t),$$

then Eq. (6.1) is simplified to the form of a standard convolution

$$g(x, y) = (f * h)(x, y). \tag{6.2}$$

Such an imaging system is called a *linear space-invariant system* and is fully determined by its PSF. The model (6.2) is a frequently used compromise between universality and simplicity –it is general enough to approximate many practical situations such as out-of-focus blur of a flat scene, motion blur of a flat scene in case of translational motion, and media turbulence blur. At the same time, its simplicity allows reasonable mathematical treatment.

6.1.2   The need for blur invariants

As in the case of geometric distortions, it is desirable to find a representation of the image $g(x,y)$ that does not depend on the particular blurring PSF, the actual shape and parameters of which are mostly unknown. As usual, there are two different approaches to this problem: image normalization and direct image description by invariants.

Image normalization in this case consists of *blind deconvolution* that removes or suppresses the blurring. Blind deconvolution has been extensively studied in image processing literature, see [1, 2, 3], for a survey. It is an ill-posed problem of high computational complexity and the quality of the results is questionable[1]. This is why we do not follow the normalization approach in this chapter and rather look for invariant descriptors of the image that do not depend on the PSF.

An invariant approach is much more effective here than the normalization because we avoid a difficult and ambiguous inversion of Eq. (6.2). Unlike deconvolution, invariant descriptors do not restore the original image, which may be understood as a weakness in some applications. On the other hand, in many cases we do not need to know the whole original image; we only need for instance to localize or recognize some objects on it (a typical examples are matching of a blurred query image against a database of clear templates and registration of blurred frames; see Figure 6.3). In such situations, the knowledge of a certain incomplete but robust representation of the image is sufficient (see Figure 6.4 for illustration of the difference between deconvolution and the invariant approach). Such a representation should be independent of the imaging system and should really describe the original image, not the degraded one. In other words, we are looking for a functional I such that

$$I(f) = I(f * h) \tag{6.3}$$

holds for any admissible PSF h. The invariants satisfying condition (6.3) are called *blur invariants* or *convolution invariants*.

6.1.3   State of the art of blur invariants

Unlike the affine invariants, which can be traced over two centuries back to Hilbert, the blur invariants are relatively new topic. The problem formulation along with the roots of the theory of blur invariants appeared originally in the 1990s in the series of papers by Flusser et al. [4–6]. They proposed a system of blur invariants which are recursive functions of geometric moments of the image and proved their invariance under a convolution with arbitrary centrosymmetric PSF. These invariants, along with the centrosymmetry assumption, have been adopted by many researchers and have found a number of practical applications.

---

[1] Blind deconvolution of multiple images and blind deconvolution of a single image supported by another under-exposed image or by certain assumptions about the PSF were proven to be solvable and numerically stable, but these methods are not applicable here.

Figure 6.3 Blurred image (top) to be matched against a database (bottom). A typical situation where the convolution invariants may be employed

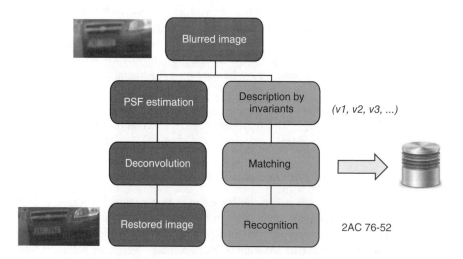

Figure 6.4 The flowchart of image deconvolution (left) and of the recognition by blur invariants (right)

Categorization of blur invariants is possible from various viewpoints. We may categorize according to the PSF they are invariant to, according to the image dimension, according to the domain they are defined in, according to the image degradation (pure blur or blur combined with some spatial transformation), and in case of application papers the sorting according to the application area is also relevant.

Categorization by the admissible PSFs

Sorting the invariants according to the PSFs they are invariant to is the most natural kind of categorization. This is the main criterion, because the definition of the set of the admissible PSFs is crucial. The choice of the admissible PSFs influences not only the existence/non-existence of the invariants but also their number and the discrimination power. An essential assumption about the PSF is that the convolution with them must be *closed* with respect to composition. In other words, convolution of two admissible PSFs must belong to the set of admissible PSFs. We already explained this necessary assumption in a general form in Chapter 2. Here, the convolution plays the role of the operator of intra-class variations $\mathcal{D}$, which is always required to be closed; otherwise the respective classes are not well defined.

The fact that the set of the admissible PSFs is closed under convolution does not necessary imply the existence of non-trivial invariants. Let us assume that the PSF could be an arbitrary image function in the sense of Definition 3.1. Such a set is obviously closed under convolution, but no non-trivial invariants may exist[2].

Now let us assume that the blur preserves the overall brightness of the image, that is,

$$\int\limits_{-\infty}^{\infty} \int\limits_{-\infty}^{\infty} h(x,y)\mathrm{d}x\mathrm{d}y = 1.$$

This is a very natural assumption fulfilled by most real PSFs. Under this assumption, there exists just one non-trivial blur invariant $I(f)$ – the total brightness of the image (in moment terminology $I(f) = m_{00}$). The respective classes of equivalence are composed of images of the same total brightness. The discriminability of this invariant is very limited and would be insufficient in most practical tasks. On the other hand, if we assume that the only admissible PSF is δ-function, then any pixel value $f(x,y)$ (and also any moment of f and any other feature) is "blur" invariant. From these simple examples we can deduce that the narrower the class of admissible PSFs, the more invariants exist.

About 90% of blur invariants reported in the literature have assumed a two-fold rotation symmetry also called the *central symmetry* of the PSF, that is, $h(x,y) = h(-x,-y)$ (to be precise, it is sufficient to assume centrosymmetry with respect to the centroid of $h(x,y)$ that may be generally different from $(0,0)$). Historically, this assumption was introduced in the fundamental paper [6] where *blur moment invariants* were presented, and most authors have adopted it in further theoretical studies [7–31] and in many application papers [32–52].

Several authors have been aware of the limitations of the centrosymmetry assumption. Numerous cameras, scanners, and other image acquisition systems have PSFs with some higher symmetry. Although these symmetries are often particular cases of the central symmetry, incorporating this assumption into the theory allows us to derive specific invariants providing a better discrimination power.

Invariants to blur, the PSF of which is *circularly symmetric* $h(x,y) = h(r)$ were proposed in [53], where complex moments were employed. These invariants were later equivalently expressed and in terms of Zernike moments in [54–56], pseudo-Zernike [11], radial Chebyshev [12], and Fourier-Mellin moments [57]. Practical applications of the circular blur invariants can be found in [58, 59].

---

[2] Let us assume that such invariant I exists and let f_1 and f_2 be two images discriminable by I, i.e. $I(f_1) \neq I(f_2)$. Now consider f_1 and f_2 to be two PSFs acting on δ-function. Since I is an invariant, it must hold $I(\delta) = I(f_i * \delta) = I(f_i)$ and hence $I(f_1) = I(f_2)$, which is a contradiction.

Invariants to linear *motion blur* were introduced in [60] and later proposed in a similar form in [61]. Assuming a linear horizontal motion, the PSF model used in [60] had the following form:

$$h(x, y) = \begin{cases} \dfrac{1}{vt}\delta(y) & \Longleftrightarrow 0 \le x \le vt \\ 0 & \text{otherwise,} \end{cases} \tag{6.4}$$

where v is the motion velocity and t is the exposure time.

Since the set of all motion-blur PSFs (6.4) is not closed under convolution, the "motion blur" invariants proposed in the literature are in fact invariant to a broader class of the *symmetric directional blur*[3]. The respective PSFs have the form

$$h(x, y) = \beta(x)\delta(y)$$

where $\beta(x)$ is an arbitrary 1D finitely-supported even function. Applications of the motion blur invariants can be found in [62–68].

Invariants to *Gaussian blur* are the only ones which actually have used the parametric form of the PSF. Liu and Zhang [69] realized that the complex moments of the image, one index of which is zero, are invariant to a Gaussian blur. Xiao [70] seemingly derived other invariants to Gaussian blur, but he did not employ the parametric Gaussian form explicitly. He only used the circular symmetry property, which led to an incomplete invariant system. An interesting approach was proposed by Zhang et al. [71, 72]. Instead of deriving the invariants explicitly, they proposed a distance measure between two images which is independent of a Gaussian blur. Most recently, Flusser et al. [74] introduced a complete set of moment-based Gaussian blur invariants, which outperform the Zhang distance significantly in terms of noise robustness and speed. Höschl proposed similar invariants for 1D Gaussian blur [75].

Significant progress of the theory of blur invariants was made by Flusser et al. in [76], where the invariants to arbitrary N-fold rotation symmetric blur were proposed. These invariants are based on complex moments and perform a substantial generalization of previous work. They include invariants to centrosymmetric and circular blur as special cases. Another set of invariants to N-fold symmetric blur was found in the Fourier domain by Pedone et al. [77]. Both sets of the invariants were later even further generalized to *N-fold dihedral blur* in [78, 79].

Gopalan et al. [80] derived yet another, conceptually different set of blur invariants –they did not impose any limitations on the shape of the PSF but assumed that the PSF has been sampled and is of small support. This assumption allowed them to work with PSFs as finite matrices. All previously mentioned invariants were designed in a continuous domain for PSFs being real functions.

Categorization by the image dimension

The vast majority of blur invariants have been designed for 2D images and also almost all application papers have described their usage in 2D tasks. In a few papers, 1D versions of blur invariants were used. In 1D it does not make sense to distinguish analogies with various N-fold symmetries of the PSF because all of them are equivalent with centrosymmetric (i.e., even) PSF.

---

[3] Although it has not been commonly mentioned in the papers on motion blur invariants.

1D centrosymmetric blur invariants were firstly described in [81]. Kautsky [14] showed how to construct such invariants in terms of arbitrary moments. Galigekere [28] studied the blur invariants of 2D images in Radon domain which inherently led to 1D blur invariants. Höschl [75] proposed 1D Gaussian blur invariants.

Some papers have been devoted to blur invariants for 3D images. Boldyš et al. first extended centrosymmetric blur invariants into d-D (d being finite) [19] and then, for the 3D case, they also added invariance to rotation [20]. Their latest paper [21] presents a general approach to constructing centrosymmetric blur and affine invariants in an arbitrary number of dimensions. Their invariants were further studied by Candocia [22], who analyzed in detail the impact of discretization, and by Bentoutou et al. [36], who applied them to 3D template matching. A generalization of Gaussian blur invariants to d-D was proposed in [74].

Categorization by the domain of the invariants

Blur invariants may act in the spatial domain, in the Fourier domain, and rarely also in some other transform domains. Spatial domain blur invariants are the most frequently used (measured by the number of relevant papers) and were historically the first ones ever proposed. Fourier domain invariants are formally more simple and more intuitive, because the comvolution which describes the blur in image domain is transformed to the multiplication in the frequency domain.

All spatial domain invariants (except [80] and [72]) are based on image moments. The type of the moments is conditioned by the assumed type of the blurring PSF. Geometric moments are convenient for centrosymmetric PSFs, so they were used in the pioneer papers [4–6] and in most follow-up papers adopting the assumption of centrosymmetry. Geometric moments were also used in all papers on the invariants w.r.t. the directional (motion) blur and 1D blur. In order to achieve better numerical properties of the blur invariants, several researchers replaced the geometric moments by Legendre moments and derived "orthogonal" blur invariants [7–10]. However, as was proved by Kautsky [14], blur invariants in any two distinct polynomial bases are mutually dependent. When knowing invariants from geometric moments, one can easily derive blur invariants in an arbitrary polynomial basis using a simple matrix operations.

For the PSFs exhibiting some kind of rotational symmetry (circular, N-fold rotational, N-fold dihedral, Gaussian), various radial-circular moments are naturally convenient. The use of complex moments [53, 69, 73, 74, 76, 82], Zernike moments [54–56], pseudo-Zernike [11], radial Chebyshev moments [12] and Fourier-Mellin moments [57] have been proposed. Again, we recall the theoretical equivalence of all these moment sets.

As we already pointed out, the Fourier domain seems to be attractive for designing blur invariants because the convolution turns to a product. Blur invariants can be defined as a ratio of $F(\mathbf{u})$ and some other properly chosen function. In a limited form, this idea was proposed as early as in [6] where the tangent of the phase of $F(\mathbf{u})$ was used as a blur invariant for a centrosymmetric blur. Essentially the same idea was proposed later by Ojansivu et al. [23, 24] under the name *blur-invariant phase correlation*. Later on, the quest to find a complete set of N-fold blur invariants in Fourier domain led to introducing various *projection operators* [74, 76–79]. Interestingly, most of these frequency-domain invariants have equivalent moment-based counterparts in the spatial domain. The choice of the working domain in a particular application is then a user's decision, which depends on specific conditions, noise intensity, image size, and other factors.

Local phase quantization (LPQ) [29–31] performs a slightly different approach to designing blur invariants in the Fourier domain. The authors utilized the fact that any centrosymmetric PSF does not change the phase of the image within a certain low-frequency band. This idea is applied locally on the image, and the phase creates (after certain post-processing) the invariant vector. Since the respective bandwidth is unknown, the authors heuristically use a very narrow band. LPQ does not have a counterpart in the spatial domain. Comparing to moment invariants, LPQ produces redundant dependent features of high dimensionality but exhibits good performance (mostly comparable to moment invariants) in experiments.

In addition to Fourier domain, the use of some other transform domains was described. Makaremi [26, 27] observed that the moment blur invariants retain their properties even if a wavelet transform has been applied on the image because the blur and wavelet transform are commutative. Applying wavelet transform before the invariants are calculated may increase their discrimination power [27]. Yet another suitable domain is the Radon domain, because it transforms 2D blur into a 1D blur while preserving certain symmetric properties of the PSF, such as centrosymmetry or Gaussian shape [28, 70] (on the other hand, some other properties such as an N-fold symmetry are inevitably lost).

Categorization by the image distortions

Many authors have tried to combine invariance w.r.t. blur with invariance to certain spatial transformations of the image such as rotation, scaling, and affine transform. This is very important namely for practical applications in object recognition, because the spatial transformations are present very often and it is difficult to suppress them beforehand.

Design of the *combined invariants* to blur and translation is straightforward. In the moment domain, we just replace the basic moments with the central ones, whatever polynomial basis has been used. In the Fourier domain, using the magnitudes ensures the invariance to translation while keeping the invariance to blur. Similarly, the invariance to uniform scaling can be achieved readily by using normalized moments. Deriving combined invariants to blur and rotation is more challenging.

Although a few heuristic attempts to construct 2D blur and rotation invariants from geometric moments can be found already in [6], only the introduction of radial-circular moments into the blur invariants and a consequent understanding of their behavior under geometric transformations made this possible in a systematic way. Combined invariants to convolution and rotation were introduced by Zitová and Flusser [82]. They discovered that the rotation property of complex moments propagates into blur invariants without being corrupted such that the combined invariants can be created analogously to pure rotation invariants. This property does not depend on the PSF type, so the above approach can be generalized for arbitrary N-fold symmetric [76], circular symmetric [53] as well as Gaussian [74] PSF. A further extension to additional invariance to scaling and contrast changes can be found in [15]. The same rule of the rotation property propagation applies to Zernike and other radial moments and makes possible an easy creation of the combined invariants from them [11, 12, 54–57]. Rotation/scale and blur invariants can also be constructed in the Fourier domain by means of polar/log-polar mapping of the Fourier magnitudes [23]. The LPQ method can be extended to handle the image rotation, too [30]. Blur and rotation invariants can be constructed also in the Radon domain [70], where the image rotation turns to a cyclic shift of the rows.

Rotation and blur combined invariants in 3D were proposed in [20] and later generalized to an arbitrary number of dimensions in [21]. They require a centrosymmetric PSF. Designing the combined invariants in 3D is much more difficult than in 2D because the rotation properties of 3D moments are more complicated than in case of the 2D versions. This is why so few invariants have been proposed so far and why there have not been introduced any 3D combined invariants for other than centrosymmetric and Gaussian PSFs.

The first attempt to find the combined invariants both to convolution and 2D full affine transformation was published by Zhang et al. [17], who employed an indirect approach using image normalization. Later on, Suk derived combined affine invariants in explicit forms [18]. In [21], a general approach to constructing blur and affine invariants in arbitrary dimension for a centrosymmetric PSF was proposed. It should be noted that the possibility of combining affine transformation and blur invariants is limited in principle. For other than centrosymmetric PSFs the affine transform violates the assumption about the PSF shape under which the blur invariants have been derived. For instance, neither circular nor N-fold symmetries are preserved, so specific combined invariants cannot exist.

Categorization with respect to the application area[4]

In the moment-based recognition in general, many applications have used only binary data, such as the silhouettes of objects. This is of course irrelevant for application of blur invariants. Hence, the blur invariants have been applied only in the tasks where we work with blurred graylevel or color data without any preprocessing which would violate the convolution model of the acquisition.

The main task where the blur invariants have been applied is *image registration*, both landmark-free (global) as well as landmark-based (local) ones.

Landmark-based blur-invariant registration methods appeared right after the first papers on blur invariants [5, 6] had been published. The basic algorithm is as follows. Significant corners and other dominant points are detected in both frames and considered to be the control point (landmark) candidates. To establish the correspondence between them, a vector of blur invariants is computed for each candidate point over its neighborhood, and then the candidates are matched in the space of the invariants. Various authors have employed this general scenario in several modifications, differing from one another namely by the particular invariants used as the features. The invariants based on geometric moments can only be used for registration of blurred and mutually shifted images [6, 27, 35] and for rough registration if a small rotation is present [32, 34, 36, 37, 83]. Nevertheless, they found successful applications in matching and registration of satellite and aerial images [32–34, 83] and in medical imaging [35–37]. The discovery of combined blur and rotation invariants from complex moments [82] and Zernike moments [55] led to registration algorithms that are able to handle blurred, shifted, and rotated images, as was demonstrated on satellite images [15, 58], indoor images [20, 84], and outdoor scenes [10, 55]. Zuo [13] even combined moment blur invariants and SIFT features [85] into a single vector with weighted components but without a convincing improvement.

Landmark-free registration methods estimate directly the between-image distortion. In basic versions they are mostly limited to translation, but extension to rotation and scaling is mostly possible. The blur invariant phase-correlation [23, 24] was used for registering X-ray images by Tang [25]. More advanced N-fold phase correlation was used to register consecutive blurred

---

[4] This kind of sorting is relevant for application-oriented papers only.

snapshots for a consequent multichannel blind deconvolution [77]. There also exist global methods which are based on image normalization. Zhang et al. [16, 17] proposed to estimate the registration parameters by bringing the input images into a normalized (canonical) form. Since blur-invariant moments were used to define the normalization constraints, neither the type nor the level of the blur influences the parameter estimation.

For the sake of completeness, there have also been other blur-invariant or blur-robust global registration methods which do not use the blur invariants explicitly but rather parameterize the blur space and search it along with the space of the transformation parameters to minimize certain between-image distance [86–88]. Some registration methods were designed particularly for a certain type of blur, such us motion blur [89], out-of-focus blur [90], and camera shake blur [91]. They assume the knowledge of either the parametric form of the blurring function or of its statistical characteristics. Several authors, being motivated namely by super-resolution[5] applications [92], concentrated primarily on blur caused by insufficient resolution of the sensor. Vandewalle [93] proposed a simple translation-rotation registration method working in a low-frequency part of the Fourier domain in a similar way as the traditional phase correlation [94]. For heavily sub-sampled and aliased images he recommended a variable projection method [95] which originates from [96]. However, these methods are robust but not exactly invariant to image blurring.

In addition to image registration, blur invariants (both pure and combined ones) have been used for recognition of objects on blurred images. The following applications of centrosymmetric blur invariants have been reported: face recognition on out-of-focus photographs [80], blurred digit and character recognition [38], robot control [39], image forgery detection [40, 41], traffic sign recognition [42, 43], fish shape-based classification [44], cell recognition [45], aircraft silhouette recognition [46], sign language recognition [47], classification of winged insect [48], and robust digital watermarking [49]. Peng et al. used the motion blur invariants for weed recognition from a camera moving quickly above the field [62] and for classification of wood slices on a moving conveyor belt [63] (these applications were later enhanced by Flusser et al. [64, 65]). Zhong used the motion blur invariants for recognition of reflections on a waved water surface [66], although the motion-blur model used there is questionable. Höschl used Gaussian blur invariants of the image histogram for noise-resistant image retrieval [75, 97].

6.1.4   The chapter outline

The main goal of this chapter is to present the derivation of blur invariants by means of projection operators in the Fourier domain and introduce projection operators as a general framework of blur invariants of any kind, regardless of the PSF type and even regardless of the image dimension. We will see that all existing blur invariants, both in Fourier as well as in the moment domain, are special cases of the invariants obtained by projection operators. We start our exposition with a simple illustrative 1D case, then the general theory of blur invariants is introduced, and, finally, the particular cases of centrosymmetric, circular, N-fold, dihedral, directional, and Gaussian PSF are studied. We also explain how to create the combined invariants to rotation and affine transform whenever it is possible.

---

[5] The term *super-resolution* denotes an operation the input of which are multiple misregistered low-resolution images and the output is a single high-resolution frame.

6.2   An intuitive approach to blur invariants

The first blur invariants that appeared in [4–6] were composed of geometric moments and were derived in an intuitive way that does not require any deep theory. This derivation provides an insight into the mechanism of how the blur invariants are constructed, but it is tedious and must be carried out for any particular type of the PSF separately. Although it has been recently replaced by the use of projection operators (which is, as we will see in the next sections, a formally easy unifying framework of blur invariants), the intuitive approach is still worth studying because of its transparency. We present it here for a 1D case.

First, we explain how the geometric moments change under convolution. It holds for arbitrary image functions f, g and h such that $g = f * h$

$$m_p^{(g)} = \sum_{k=0}^{p} \binom{p}{k} m_k^{(h)} m_{p-k}^{(f)}. \tag{6.5}$$

Expressing few low-order moments explicitly, we have

$$m_0^{(g)} = m_0^{(f)} m_0^{(h)}, \tag{6.6}$$

$$m_1^{(g)} = m_1^{(f)} m_0^{(h)} + m_0^{(f)} m_1^{(h)}, \tag{6.7}$$

$$m_2^{(g)} = m_2^{(f)} m_0^{(h)} + 2 m_1^{(f)} m_1^{(h)} + m_0^{(f)} m_2^{(h)}, \tag{6.8}$$

$$m_3^{(g)} = m_3^{(f)} m_0^{(h)} + 3 m_2^{(f)} m_1^{(h)} + 3 m_1^{(f)} m_2^{(h)} + m_0^{(f)} m_3^{(h)}, \tag{6.9}$$

$$\vdots \tag{6.10}$$

Now we have to accept some assumptions about the PSF which restrict the set of admissible blurs. We will assume that the PSF is even and brightness preserving, that is, $h(x) = h(-x)$ and the integral of h equals one. Since all odd-order moments of even function vanish, the above expressions simplify to the form

$$m_0^{(g)} = m_0^{(f)}, \tag{6.11}$$

$$m_1^{(g)} = m_1^{(f)}, \tag{6.12}$$

$$m_2^{(g)} = m_2^{(f)} + m_0^{(f)} m_2^{(h)}, \tag{6.13}$$

$$m_3^{(g)} = m_3^{(f)} + 3 m_1^{(f)} m_2^{(h)}, \tag{6.14}$$

$$\vdots \tag{6.15}$$

If p is odd, it is possible to eliminate all moments of h in the expression of $m_p^{(g)}$ and then to group all moments of g on the left-hand side and all moments of f on the right-hand side of the equality. For instance for $p = 3$ we obtain

$$m_3^{(g)} - 3 m_1^{(g)} m_2^{(g)} / m_0^{(g)} = m_3^{(f)} - 3 m_1^{(f)} m_2^{(f)} / m_0^{(f)}.$$

Since the expressions on the both sides are the same, we just have found a blur invariant.

If we continue with this process, we derive an invariant for each odd p. By means of induction, these invariants can be expressed as the recursive formula

$$S(p) = m_p - \frac{1}{m_0} \sum_{n=1}^{(p-1)/2} \binom{p}{2n} S(p - 2n) \cdot m_{2n} \qquad (6.16)$$

which is very well known in the theory of blur invariants. Note that for even p no invariants exist because there is no way to eliminate $m_p^{(h)}$. This has a profound reason, which will become apparent later in connection with the projection operators.

Although it is not apparent at first sight, the invariants (6.16) are closely linked to invariants, which can be found in the frequency domain. Due to the convolution theorem, Eq. (6.2) becomes, after the Fourier transform has been applied,

$$G(u) = F(u) \cdot H(u). \qquad (6.17)$$

Considering the amplitude and phase separately, we obtain

$$|G(u)| = |F(u)| \cdot |H(u)| \qquad (6.18)$$

and

$$\mathrm{ph}G(u) = \mathrm{ph}\, F(u) + \mathrm{ph}H(u) \qquad (6.19)$$

(note that the last equation is correct only for those points where $G(u) \neq 0$; $\mathrm{ph}G(u)$ is not defined otherwise). Since h is supposed to be even, its Fourier transform H is real, which means the phase of $H(u)$ is only a two-valued function:

$$\mathrm{ph}H(u) \in \{0; \pi\}.$$

It follows immediately from the periodicity of tangent that

$$\tan(\mathrm{ph}G(u)) = \tan(\mathrm{ph}\, F(u) + \mathrm{ph}H(u)) = \tan(\mathrm{ph}\, F(u)). \qquad (6.20)$$

Thus, a tangent of the phase of the signal is an invariant with respect to convolution with any even PSF[6].

Now we show the close connection between the moment-based and the Fourier-based blur invariants. The tangent of the Fourier transform phase of any image function f can be expanded into a power series (except the points where $F(u) = 0$ or $\mathrm{ph}\, F(u) = \pm\pi/2$)

$$\tan(\mathrm{ph}\, F(u)) = \sum_{k=0}^{\infty} a_k u^k, \qquad (6.21)$$

where

$$a_k = \frac{(-1)^{(k-1)/2} \cdot (-2\pi)^k}{k! \cdot m_0} S(k). \qquad (6.22)$$

if k is odd and $a_k = 0$ otherwise (see Appendix 6.A for the proof).

---

[6] The phase itself is not an invariant. Using a tangent is the simplest but not the only way to make it invariant to blur. Alternatively, as proposed in [24], we reach the same effect by taking the double of the phase.

The same intuitive derivation, both in moment domain and frequency domain, can be repeated in arbitrary dimensions under the assumption of the centrosymmetry $h(\mathbf{x}) = h(-\mathbf{x})$. As the dimension increases, the algebraic manipulations are more laborious, but the idea stays the same. In 2D, Eq. (6.5) turns to

$$\mu_{pq}^{(g)} = \sum_{k=0}^{p} \sum_{j=0}^{q} \binom{p}{k} \binom{q}{j} \mu_{kj}^{(h)} \mu_{p-k,q-j}^{(f)} \tag{6.23}$$

(we use the version with central moments because it is more common in practice; see Appendix 6.B for the proof) and (6.16) becomes

$$C(p,q) = \mu_{pq} - \frac{1}{\mu_{00}} \sum_{\substack{k=0 \\ 0<k+j<p+q}}^{p} \sum_{j=0}^{q} \binom{p}{k} \binom{q}{j} C(p-k,q-j) \cdot \mu_{kj} \tag{6.24}$$

for $p + q$ being odd and $C(p, q) = 0$ otherwise (see [6] for more details and Appendix 6.C for a direct proof independent of the projection operator). A few invariants in explicit form are listed in Appendix 6.D.

Everything we have shown until now can easily be extended into d dimensions ($d > 2$), as was proposed in [19]. Let us define the d-dimensional binomial coefficients as

$$\binom{\mathbf{p}}{\mathbf{k}} \equiv \prod_{i=1}^{d} \binom{p_i}{k_i}.$$

The d-dimensional blur invariants are then defined as

$$C(\mathbf{p}) = \mu_{\mathbf{p}} - \frac{1}{\mu_{\mathbf{0}}} \sum_{\substack{0 \le \mathbf{k} \le \mathbf{p} \\ 0<|\mathbf{k}|<|\mathbf{p}|}} \binom{\mathbf{p}}{\mathbf{k}} C(\mathbf{p} - \mathbf{k}) \cdot \mu_{\mathbf{k}} \tag{6.25}$$

for $|\mathbf{p}|$ being odd and as
$$C(\mathbf{p}) = 0$$

for $|\mathbf{p}|$ even. The blur invariants in d-D Fourier domain are defined analogously to (6.20) as the tangent of the phase.

The main drawback of this intuitive approach is the limited possibility of its generalization to other types of the PSF and a difficult construction of combined blur-rotation invariants. Both are important in practice because the out-of-focus blurring PSF often has higher symmetry than the central one, as we will see later.

6.3   Projection operators and blur invariants in Fourier domain

In this section, we introduce the *projection operators* which create a general theoretical frameworks of blur invariants. Projection operators are classical tools in linear algebra and functional analysis.

Let $\mathcal{I}$ be a vector space and P be a linear idempotent (i.e., $P^2 = P$) operator from $\mathcal{I}$ to itself. Operator P is called *projection operator* or *projector* for short of $\mathcal{I}$ onto $P(\mathcal{I})$. Let us denote

$S = P(I)$. The set S is also a vector space, and I can be expressed as a direct sum $I = S \oplus A$, where A is called the *complement* of S. This means that any $f \in I$ can be unambiguously written as a sum $f = Pf + f_A$, where Pf is a projection of f onto S and $f_A \in A$ is simply defined as $f - Pf$.

Obviously, $Pf_A = \mathbf{0}$ for any f_A, which means that A is a nullspace of P. This follows from the idempotence and linearity of P as $Pf_A = P(f - Pf)$. If $f \in S$ then $f = Pf$ and vice versa.

If P has been given, then S is defined explicitly as $P(I)$. In some cases (which also include, as we will see later, the designing of blur invariants) the construction goes the other way round – for a given S we look for a projector P. If I is a space with a scalar product $(\cdot, \cdot)$ and S has an orthogonal basis $\{\phi_n\}$, then P can be found easily as a partial Fourier-like reconstruction of f

$$Pf = \sum_n (f, \phi_n)\phi_n$$

(the sum may be finite or infinite, depending on the dimension of S). Such P is called an *orthogonal projector* on S. Since these assumptions are too restrictive for our purposes, we will not limit ourselves to orthogonal projectors and will try to construct P without using a scalar product.

When speaking about blur invariants, the space I must be a space where convolution and Fourier transform are well defined. The space of image functions[7] of course fulfills these requirements (see Chapter 3), but we can also employ more general spaces as our I (for instance a space of fast-decayed functions on unlimited support or a space of absolutely integrable functions). If not specified otherwise, I will be the space of the image functions. Many conclusions are, however, valid also in more general spaces.

Let us assume that S is a subspace of I which is closed under convolution[8]. For arbitrary $f \in I$ and $h \in S$ we have

$$f * h = (Pf + f_A) * h = Pf * h + f_A * h. \qquad (6.26)$$

Since the decomposition of any element of I is unique and $Pf * h \in S$ due to the closure of S, $f_A * h$ must lie in A. Applying P on both sides of eq. (6.26) we obtain

$$P(f * h) = Pf * h. \qquad (6.27)$$

This is a key property of how an arbitrary projector acts on a convolution. It allows us to formulate the *Fundamental theorem of blur invariants* (FTBI).

Theorem 6.1 Let S be a linear subspace of I which is closed under convolution. Let P be a projector of I on S. Then

$$I(f) \equiv \frac{\mathcal{F}(f)}{\mathcal{F}(Pf)}$$

is an invariant w.r.t. blur by arbitrary PSF $h \in S$ at all frequencies where $I(f)$ is well defined.

---

[7] Image functions in the sense of Definition 3.1 do not form a vector space, so we drop the assumption that the integral is nonzero to obtain a vector space.

[8] In other words, S is a commutative ring w.r.t. point-wise addition and convolution.

Proof. The proof follows directly from the above mentioned propositions:

$$I(f * h) \equiv \frac{\mathcal{F}(f * h)}{\mathcal{F}(P(f * h))} = \frac{\mathcal{F}(f) \cdot \mathcal{F}(h)}{\mathcal{F}(Pf * h)} = \frac{\mathcal{F}(f) \cdot \mathcal{F}(h)}{\mathcal{F}(h) \cdot \mathcal{F}(Pf)} = I(f). \qquad \square$$

The FTBI is a very strong theorem because it constructs the blur invariants in a unified form regardless of the particular class of the blurring PSFs and regardless of the image dimension d. Its practical usefulness depends on our ability to find, for a given subspace S of the admissible PSFs, the projector P. Although a general algorithm does not exist, it is fortunately not difficult to find the projector for most practically important PSF classes. Before we proceed to individual PSF classes, we discuss a few important general consequences of the FTBI.

The invariant $I(f)$ is not defined whenever $\mathcal{F}(Pf)(\mathbf{u}) = 0$. It depends on S and f whether or not (and on which frequencies) this may happen. If S contains compactly supported functions only, $\mathcal{F}(Pf)(\mathbf{u})$ is non-zero almost everywhere[9], and therefore the frequencies where $I(f)$ is not defined cannot cause serious problems when working with the invariant. This is, however, not true in general. There exist such spaces S that $\mathcal{F}(Pf)(\mathbf{u})$ may vanish on a nonzero-measure set which requires special care when using them. On the other hand, one can construct special spaces such that $\mathcal{F}(Pf)(\mathbf{u}) \neq 0$ everywhere [98], but these spaces do not actually represent real-life PSFs.

Understanding what properties of f are reflected by $I(f)$ is important both for theoretical considerations as well as for practical application of the invariant. $I(f)$ is a ratio of two Fourier transforms which may be interpreted as a deconvolution. Having an image f, we seemingly "deconvolve" it by the kernel Pf. This "deconvolution" exactly eliminates that part of f belonging to S (more precisely, it transfers Pf to δ-function) and acts on the f_A only:

$$I(f) = \frac{\mathcal{F}(f)}{\mathcal{F}(Pf)} = \frac{\mathcal{F}(Pf) + \mathcal{F}(f_A)}{\mathcal{F}(Pf)} = 1 + \frac{\mathcal{F}(f_A)}{\mathcal{F}(Pf)} \equiv 1 + \Psi(f).$$

Hence, $I(f)$, if it exists almost everywhere, can be viewed as a Fourier transform of so-called *primordial image*. This term is one of the central terms of blur invariants. The primordial image is a representative of the equivalence classes. Two images f and g are blur-equivalent w.r.t. S if they have the same primordial image. In other words, all blur-equivalent images lie in a subset of $\mathcal{I}$ called a *semi-orbit* which originates from the primordial image. Note that the existence of the primordial image is not guaranteed – $\mathcal{F}^{-1}(I(f))$ may not exist in $\mathcal{I}$. The primordial image can also be viewed as a kind of normalization. It plays the role of a canonical form of f, obtained as the "maximally deconvolved" f.

Investigation of the discrimination power is crucial for any features. The images within the same equivalence class, that is, those having the same primordial image, cannot be distinguished. We need to know if the equivalence classes are formed just by functions with convolution relations between them or if there might be also some other functions, which would decrease the discriminability of the invariants The answer is provided by the following simple proposition, which is sometimes called the *completeness theorem*.

Theorem 6.2 Let $f, g \in \mathcal{I}$ such that $I(f) = I(g)$ almost everywhere. Then there exist functions $h_1, h_2 \in S$ such that

$$f * h_1 = g * h_2.$$

---

[9] More precisely, $\mathcal{F}(Pf)(\mathbf{u})$ cannot vanish on any open set.

The proof follows immediately from the equality of the invariants when setting $h_1 = Pg$ and $h_2 = Pf$. Note that under the above assumptions the theorem does not guarantee the existence of $h \in S$ such that either $g = f * h$ or $f = g * h$[10].

The completeness theorem shows that two images f and g are distinguishable by the invariant I if and only if there is no "convolution link" between them. In particular, we can never distinguish two elements of S (for any two such functions h_1, h_2 we have $I(h_1) = I(h_2) = 1$ almost everywhere, so their joint primordial image is the δ-function). Although this is a natural consequence of the invariance property, it may sometimes cause problems in applications. As we will see later, S may consist of prominent functions which appear not only as the PSFs but also as the objects themselves. This is why we must pay attention to the choice of S in the particular application (in an ideal case, it should be as small as possible and should describe just the blurs which actually appear in the given task) and why we need specific invariants for each S. Let us imagine that the space of actual blurs is S_1 but the user has chosen its superset S_2, either because of incorrect problem analysis or just because he wants to be on the safe side. This choice does not violate the invariance but decreases the discrimination power of the features as we are not able to distinguish any two elements of S_2. This situation may arise for instance if S_1 is a set of circular symmetric functions while S_2 is a set of centrosymmetric functions.

6.4   Blur invariants from image moments

The blur invariants defined in the frequency domain may suffer from several drawbacks when we use them in practical object recognition tasks. Since $I(f)$ is a ratio, we possibly divide by very small numbers which requires a careful numerical treatment. Moreover, if the input image is noisy, the high-frequency components of $I(f)$ may be significantly corrupted. This can be overcome by suppressing them by a low-pass filter, but this procedure introduces a user-defined parameter (the cut-off frequency) which should be set up with respect to the particular noise level. That is why in most cases we prefer to work directly in the image domain, where moment-based invariants equivalent to $I(f)$ can be constructed. An additional advantage of working with moments is that it is easier to design combined blur-rotation and blur-affine invariants.

Moment blur invariants exist, for the given spaces $\mathcal{I}$, S, and projector P, in principle for any polynomial basis of $\mathcal{I}$ (i.e. any kind of moments), but their direct construction is easy only if the moments fulfill the *separation condition*.

Let $\mathcal{B} = \{\pi_{\mathbf{p}}(\mathbf{x})\}$ be a d-variable polynomial basis of space $\mathcal{I}$, and let $M_{\mathbf{p}}^{(f)}$ be the respective moments of function f (assuming they all exist). Considering the decomposition $f = Pf + f_A$, we have for the moments

$$M_{\mathbf{p}}^{(f)} = M_{\mathbf{p}}^{(Pf)} + M_{\mathbf{p}}^{(f_A)}.$$

---

[10] Certain inconveniences which we encounter when using this theorem and when characterizing the equivalence classes originate from our general conditions imposed on $\mathcal{I}$ and S. If we put more restrictive constraints, we could have a space with an inverse convolution [98]. Then $(S, *)$ would be a group instead of a monoid, the primordial image would always exist, and the equivalence classes would be orbits instead of semi-orbits. The completeness theorem would state that if $I(f) = I(g)$ then $g = f * h$ and $f = g * h^{-1}$. This simplification would be achieved unfortunately at the expense of usability, because real images and PSFs do not fulfil these constraints. This is why we have not followed this direction.

We say that B separates the moments if there exist a non-empty set of multi-indices D such that it holds for any $f \in \mathcal{I}$

$$M_{\mathbf{p}}^{(Pf)} = M_{\mathbf{p}}^{(f)}$$

if $\mathbf{p} \in D$ and

$$M_{\mathbf{p}}^{(Pf)} = 0$$

if $\mathbf{p} \notin D$. In other words, this condition says that the moments are either preserved or vanish under the action of P. If fulfilled, the condition also says that the value of $M_{\mathbf{p}}^{(f_A)}$ is complementary to $M_{\mathbf{p}}^{(Pf)}$. Another consequence is that the moments $M_{\mathbf{p}}^{(h)}$ of any $h \in S$ vanish if $\mathbf{p} \notin D$.

Whether or not the moment separation is fulfilled depends on both S and B, which may not be linked in a transparent way. If P is an orthogonal projector defined through a scalar product, B is an OG basis of $\mathcal{I}$, and $B_s \subset B$ is a basis of S, then

$$Pf = \sum_{\mathbf{p} \in D} M_{\mathbf{p}} \pi_{\mathbf{p}}$$

and the moments $M_{\mathbf{p}}$ are always separated thanks to the orthogonality of B. If the construction of P has been done in a different way, we have to verify the moment separability in each particular case because it is not fulfilled automatically.

As we will see later, a weaker separation condition may be required. It is sufficient if the moments $M_{\mathbf{p}}^{(Pf)}$ can be expressed in terms of $M_{\mathbf{p}}^{(f)}$ if $\mathbf{p} \in D$ and some functions of $M_{\mathbf{p}}^{(f)}$ equal zero for $\mathbf{p} \notin D$. This makes the derivation more laborious and the formulas more complicated but does not perform a principle difference. Anyway, to keep the things simple, we try for any particular S to find such B that provides the strong moment separability.

To get the link between $I(f)$ and the moments $M_{\mathbf{p}}^{(f)}$, we use the Taylor expansion of Fourier transform introduced already in Chapter 3. We demonstrate this approach in the general form for geometric moments only, provided the strong moment separability is satisfied. It can be used for other polynomial bases and other moments as well, but the formulas would be more complicated and inconvenient to work with. To design blur invariants with respect to other than the power basis, the indirect approach proposed in [14] is recommended.

The expansion of the Fourier transform is

$$\mathcal{F}(f)(\mathbf{u}) = \sum_{\mathbf{p}} \frac{(-2\pi i)^{|\mathbf{p}|}}{\mathbf{p}!} m_{\mathbf{p}}^{(f)} \mathbf{u}^{\mathbf{p}}. \tag{6.28}$$

The FTBI can be rewritten as

$$\mathcal{F}(Pf)(\mathbf{u}) \cdot I(f)(\mathbf{u}) = \mathcal{F}(f)(\mathbf{u}).$$

All these three Fourier transforms can be expanded similarly to (6.28) into absolutely convergent Taylor series

$$\sum_{\mathbf{p} \in D} \frac{(-2\pi i)^{|\mathbf{p}|}}{\mathbf{p}!} m_{\mathbf{p}}^{(f)} \mathbf{u}^{\mathbf{p}} \cdot \sum_{\mathbf{p}} \frac{(-2\pi i)^{|\mathbf{p}|}}{\mathbf{p}!} A_{\mathbf{p}} \mathbf{u}^{\mathbf{p}} = \sum_{\mathbf{p}} \frac{(-2\pi i)^{|\mathbf{p}|}}{\mathbf{p}!} m_{\mathbf{p}}^{(f)} \mathbf{u}^{\mathbf{p}}. \tag{6.29}$$

Comparing the coefficients of the same powers of $\mathbf{u}$ we obtain, for any $\mathbf{p}$

$$\sum_{\mathbf{k} \in D}^{\mathbf{p}} \frac{(-2\pi i)^{|\mathbf{k}|}}{\mathbf{k}!} \frac{(-2\pi i)^{|\mathbf{p}-\mathbf{k}|}}{(\mathbf{p}-\mathbf{k})!} m_{\mathbf{k}} A_{\mathbf{p}-\mathbf{k}} = \frac{(-2\pi i)^{|\mathbf{p}|}}{\mathbf{p}!} m_{\mathbf{p}}, \tag{6.30}$$

which can be read as

$$\sum_{\substack{\mathbf{k} \in D}}^{\mathbf{p}} \binom{\mathbf{p}}{\mathbf{k}} m_{\mathbf{k}} A_{\mathbf{p}-\mathbf{k}} = m_{\mathbf{p}}. \tag{6.31}$$

The summation goes over those $\mathbf{k} \in D$ for which $k_i \leq p_i, i = 1, \ldots, d$. After isolating $A_{\mathbf{p}}$ on the left-hand side, we obtain the final recurrence

$$m_0 A_{\mathbf{p}} = m_{\mathbf{p}} - \sum_{\substack{\mathbf{k} \in D \\ \mathbf{k} \neq \mathbf{0}}}^{\mathbf{p}} \binom{\mathbf{p}}{\mathbf{k}} m_{\mathbf{k}} A_{\mathbf{p}-\mathbf{k}}. \tag{6.32}$$

This recurrence formula is a general definition of blur invariants in the image domain. Since $I(f)$ has been proven to be invariant to blur belonging to S, all coefficients $A_{\mathbf{p}}$ must also be blur invariants. The $A_{\mathbf{p}}$'s can be understood as the moments of the primordial image of f (if it exists). The beauty of Eq. (6.32) lies in the fact that we can calculate them from the moments of f, without constructing the primordial image explicitly either in frequency or in the spatial domain.

Thanks to the uniqueness of the Fourier transform, the set $\{A_{\mathbf{p}} | \mathbf{p} \in \mathbb{N}_0^d\}$ carries the same information about the function f as $I(f)$ itself, so the cumulative discrimination power of all $A_{\mathbf{p}}$'s equals to that of $I(f)$.

Some of the invariants $A_{\mathbf{p}}$ are trivial. $A_{\mathbf{0}} = 1$ for any f, and some other invariants may vanish. The vanishing invariants depend on the index set D. Let us explain that for $d = 1$ first, where $D = \{k_n\}$ is a sequence of integers. Assuming that $k_1 \neq 0$, then Eq. (6.32) implies that $A_{k_1} = 0$ because the only term of the sum is m_{k_1} which cancels the first term. If S is a set of even functions, then $k_1 = 2$ and $k_n = 2n$ and all A_{k_n}'s vanish. The same is true whenever D has a regular structure such that $k_n = nk_1$.

If $d > 1$, we observe the same behavior, but its formal description is more complicated. In all particular cases of S which we study in the rest of this chapter (and probably in all meaningful cases ever), D exhibits a kind of regularity in all components of the multi-index. The first invariants always vanish because of the term cancellation, and the others vanish because this cancellation propagates thanks to the regularity into the higher orders. Hence, for any particular space S described below, any $\mathbf{p} \in D$ and any f it holds $A_{\mathbf{p}} = 0$, but this is not a general rule and might be violated in certain (rather obscure) choices of S and of the polynomial basis of the moments. A special case is if $D = \mathbb{N}_0^d$, which means $S = \mathcal{I}$ and all $A_{\mathbf{p}}$'s vanish.

Now let us assume $f \in S$. Then *all* invariants $A_{\mathbf{p}}$ (except $\mathbf{p} = \mathbf{0}$) vanish whatever D is. We already have learned that in such a case $I(f) = 1$ and the primordial image is a δ-function. All its moments (except the zeroth one) vanish, so all $A_{\mathbf{p}}$'s must vanish, too. We can see this property also directly from Eq. (6.32) without using arguments based on $I(f)$.

In the next sections, we study several particular choices of S that are of practical importance. We show how to construct the projector P and how to find the set D. Thanks to Theorem 6.1 and Eq. (6.32), the respective blur invariants are completely determined. There is no need to derive them from scratch in each particular case as it has been attempted so far in the papers cited in the survey part of this chapter.

6.5 Invariants to centrosymmetric blur

As we already pointed out in Section 6.2, the invariants w.r.t. centrosymmetric PSF attracted the researchers' attention from the beginning. Now we illustrate that the intuitive blur invariants introduced in Section 6.2 are nothing but a particular case of eq. (6.32).

Let S be a set of centrosymmetric functions C_2

$$S \equiv C_2 = \{h \in \mathcal{I} \,|\, h(-\mathbf{x}) = h(\mathbf{x})\}$$

of arbitrary dimension d. We define operator $P_2 : \mathcal{I} \to C_2$ as follows

$$(P_2 f)(\mathbf{x}) = (f(\mathbf{x}) + f(-\mathbf{x}))/2.$$

This operator fulfills the conditions of being a projector on C_2 – it is linear, $P_2 f \in C_2$ for any f, and $P_2 P_2 = P_2$.

The FTBI yields the blur invariant in Fourier domain. Note that P_2 commutes with the Fourier transform, which allows to write

$$I_2(f) = \frac{F}{P_2 F}.$$

Since $P_2 F = \mathcal{R}e\, F$, we obtain the invariant in the form

$$I_2(f) = 1 + i \tan(\mathrm{ph}\, F),$$

which is equivalent to the result achieved by the intuitive approach.

The standard power basis separates the geometric moments with $D = \{\mathbf{p}\,|$ such that $|\mathbf{p}|$ is even$\}$. If we assume the PSF has a unit integral, then m_0 itself is an invariant and (6.32) turns to the form

$$C(\mathbf{p}) = \mu_{\mathbf{p}} - \frac{1}{\mu_0} \sum_{\substack{\mathbf{k} \neq 0 \\ |\mathbf{k}|\ \text{even}}}^{\mathbf{p}} \binom{\mathbf{p}}{\mathbf{k}} C(\mathbf{p} - \mathbf{k}) \cdot \mu_{\mathbf{k}} \qquad (6.33)$$

where $C(\mathbf{p}) = m_0 A_{\mathbf{p}}$, which almost exactly matches Eq. (6.25) (we use central moments to also ensure the translation invariance). The only difference is that the recurrence in (6.33) automatically yields $C(\mathbf{p}) = 0$ for any even $|\mathbf{p}|$ thanks to the summation over D, so it is not necessary to define them separately as in Eq. (6.25). An example of a centrosymmetric PSF can be seen in Figure 6.5.

Figure 6.5 The PSF of non-linear motion blur which exhibits an approximate central symmetry but no axial symmetry. Visualization of the PSF was done by photographing a single bright point

6.6   Invariants to circular blur

Circularly symmetric PSF's appear in imaging namely as an out-of-focus blur on a circular aperture; see Figure 6.6 for an example. They include the well-known ideal "pillbox" out-of-focus model, but they are not limited to it – some objectives may exhibit circular PSF's which are far from being constant on a circle and rather resemble a ring (see Figure 6.7).

(a)

(b)

Figure 6.6   Two examples of an out-of-focus blur on a circular aperture: (a) the books, (b) the harbor in the night

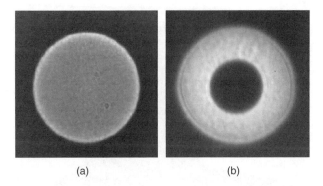

(a)   (b)

Figure 6.7   Real examples of a circularly symmetric out-of-focus PSF: (a) a disk-like PSF on a circular aperture, (b) a ring-shaped PSF of a catadioptric objective

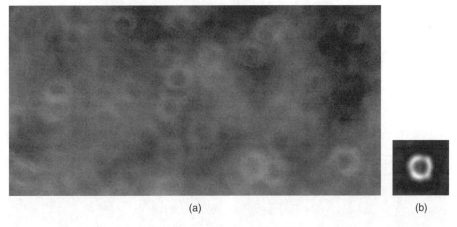

(a)   (b)

Figure 6.8   A bokeh example. The photo was taken with a mirror-lens camera. The picture (a) is a close-up of a larger scene and shows the out-of-focus background. The estimated PSF of an annular shape, which is characteristic for this kind of lenses, is shown in (b)

As in all other cases where the convolution model holds, the PSF shape can be observed in the image as a response to bright points on an out-of-focus background/foreground (in photography, this effect is called *bokeh*). An example of the background bokeh along with the estimated PSF can be seen in Figure 6.8. Diffraction blur, if present, is also given by the aperture shape. For circular aperture it takes a form of a convolution with the Airy function (see Figure 6.9). In common consumers' photography, diffraction blur is usually negligible compared to out-of-focus and camera shake blurs.

The circular symmetry is sometimes called the ∞-fold rotation symmetry, and it is best described in polar coordinates, so we set

$$S \equiv C_\infty = \{h \in \mathcal{I} \,|\, h(r, \theta) = h(r)\}$$

(we consider only a 2D case in this section).

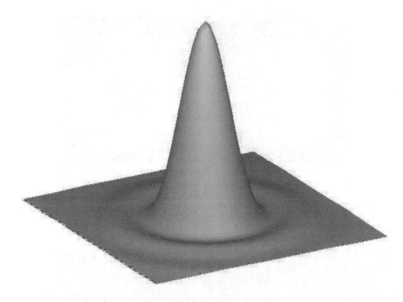

Figure 6.9 Airy function, a diffraction PSF on a circular aperture

| (a) | (b) |

Figure 6.10 (a) The original image (the Rainer cottage in High Tatra mountains, Slovakia) and (b) its projection $P_\infty f$

The projector P_∞ is defined as

$$(P_\infty f)(r) = \frac{1}{2\pi} \int\limits_{0}^{2\pi} f(r, \theta)\, d\theta.$$

We can see the performance of the projector in Figure 6.10.

The standard power basis does not separate the moments. The moments $m_{pq}^{(P_\infty f)}$ with at least one odd index vanish, but the others are generally not preserved (to see this, consider that $m_{2k,2j}^{(f)}$ changes if f is rotated but $m_{2k,2j}^{(P_\infty f)}$ stays constant as $P_\infty f$ is circularly symmetric). On the other hand, complex moments are separated. As we know from Chapter 3, $c_{pq}^{(P_\infty f)} = 0$ for any $p \neq q$. For any $c_{pp}^{(P_\infty f)}$ it holds

$$c_{pp}^{(P_\infty f)} = \int\limits_0^\infty \int\limits_0^{2\pi} r^{2p+1}(P_\infty f)(r)\mathrm{d}\theta\mathrm{d}r$$

$$= \int\limits_0^\infty r^{2p+1} \int\limits_0^{2\pi} \frac{1}{2\pi} \int\limits_0^{2\pi} f(r,\theta)\,\mathrm{d}\theta\mathrm{d}\theta\mathrm{d}r$$

$$= \int\limits_0^\infty r^{2p+1} \int\limits_0^{2\pi} f(r,\theta)\,\mathrm{d}\theta\mathrm{d}r = c_{pp}^{(f)}.$$

Hence, $D = \{(p,p)|p \geq 0\}$.

In order to expand $I(f)$ into a series of complex moments, we cannot use directly (u,v) coordinates, but we have to make a substitution $U = u + v$ and $V = i(u - v)$. Then we can expand $I(f)(U,V)$ in the same way as before (see Chapter 3 for the expansion of $F(U,V)$ in terms of complex moments). The blur invariance has not been violated by the substitution. The expansion leads again to Eq. (6.32) with geometric moments being replaced by the complex moments

$$c_{00}A_{pq} = c_{pq} - \sum_{k=1}^q \binom{p}{k}\binom{q}{k}A_{p-k,q-k} \cdot c_{kk} \tag{6.34}$$

provided that $p \geq q$. For brightness-preserving PSFs, the above expression can be simplified to the final form that defines the circular blur invariants in moment domain

$$K_\infty(p,q) = c_{pq} - \frac{1}{c_{00}}\sum_{k=1}^q \binom{p}{k}\binom{q}{k}K_\infty(p-k,q-k)\cdot c_{kk}.$$

Note that $K_\infty(p,q) = K_\infty(q,p)^*$ and $K_\infty(p,p) = 0$ for any $p > 0$ and any f, as can simply be proved by induction on p. See Appendix 6.E for a few $K_\infty(p,q)$ invariants in explicit form.

6.7 Invariants to N-FRS blur

Invariants to N-fold rotational symmetric blur[11] are very important, both from theoretical and practical points of view. This kind of blur appears in images very often as an out-of-focus blur on a polygonal aperture. Most cameras have an aperture the size of which is controlled by physical diaphragm blades. It is almost impossible to manufacture the diaphragm such that the aperture is always circular regardless of its size. Most producers use straight or slightly

---

[11] We refer to Chapter 3 for the definition and the basic properties of the N-FRS functions.

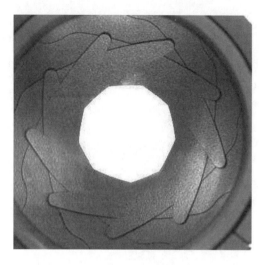

Figure 6.11 Partially open diaphragm with 9 blades forms a polygonal aperture

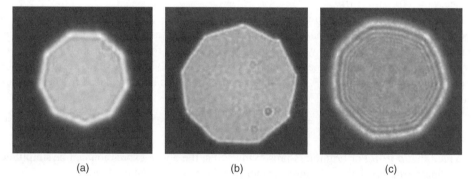

(a)   (b)   (c)

Figure 6.12 Out-of-focus PSFs on polygonal apertures of three different cameras obtained as photographs of a bright point. A nine-blade diaphragm was used in (a) and (b); a seven-blade diaphragm in (c)

curved blades that lead to polygonal or close-to-polygonal aperture shape if the diaphragm is not fully open, as can be seen in Figure 6.11. The rotational symmetry follows immediately from the diaphragm construction, regardless of the blade shape. N is the number of blades and is known and fixed for each objective. Objectives used in consumer cameras are equipped by four to nine blades. In Figure 6.12 we can see three examples of real out-of-focus polygonal PSF, obtained as photographs of a bright point.

Centrosymmetric and circular symmetric PSFs can be viewed as particular cases for $N = 2$ and $N = \infty$, respectively.

As in the previous section, we consider only the case of $d = 2$. We define

$$S \equiv C_N = \{h \in \mathcal{I} | h(r, \theta) = h(r, \theta + 2\pi/N)\}.$$

C_N is a vector space closed under convolution. We can construct projector P_N as

$$(P_N f)(r, \theta) = \frac{1}{N} \sum_{j=1}^{N} f(r, \theta + \alpha_j),$$

where $\alpha_j = 2\pi j/N$. Operator P_N rotates f repeatedly by $2\pi/N$ and calculates an average. P_N is actually a projector on C_N. Since P_N commutes with Fourier transform, the invariant $I_N(f)$ can be equivalently expressed as

$$I_N(f) = \frac{F}{P_N F}.$$

Examples of the projection operators for $N = 2, 3, 4$, and 5 are in Figure 6.13.

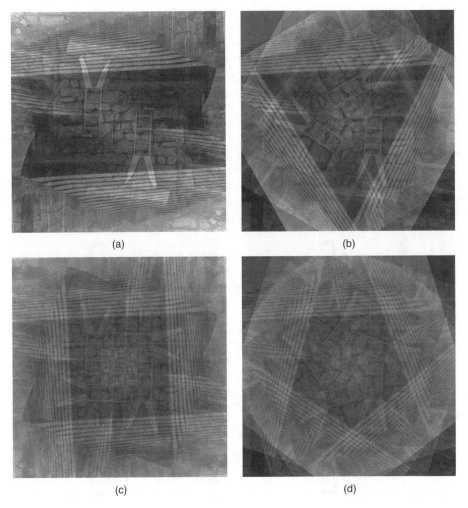

(a)

(b)

(c)

(d)

Figure 6.13 Various projections of the original image from Figure 6.10: (a) $P_2 f$, (b) $P_3 f$, (c) $P_4 f$ and (d) $P_5 f$

As in the case of the circular blur, we again employ the complex moments. They are separated and $D_N = \{(p,q)|(p-q)/N \text{ is integer}\}$, as we know from Chapter 3. Using the substitution $U = u + v$ and $V = i(u - v)$, we can expand $I_N(f)(U, V)$ similarly as in the previous case and end up with the recurrent relation for the invariants w.r.t. N-FRS brightness preserving blur

$$K_N(p,q) = c_{pq} - \frac{1}{c_{00}} \sum_{\substack{j=0 \\ 0<j+k \\ (j-k)/N \text{ is integer}}}^{p} \sum_{k=0}^{q} \binom{p}{j}\binom{q}{k} K_N(p-j,q-k) \cdot c_{jk}. \qquad (6.35)$$

Let us discuss some important properties of the blur invariants $K_N(p,q)$.

First of all, note that $K_N(p,q) = 0$ for any N, p, q such that $(p-q)/N$ is an integer (except $p = q = 0$). A formal proof can easily be done by induction; we present only the core idea here. If $(p-q)/N$ is integer and the summation in (6.35) goes over $(j-k)/N$ integer only, then also $((p-j) - (q-k))/N$ is always an integer. In other words, if $(p-q)/N$ is an integer, then on the right-hand side of (6.35) we have only K_N's of the same property. Since the highest-order term of the sum always cancels with the c_{pq} standing outside the sum, all other terms also vanish because they contain the K_N of orders less than $p + q$ and of the same nature.

The number of non-trivial invariants up to the given order r increases as N increases, reaching its maximum for all $N > r$. This is in accordance with our intuitive expectation – the more we know about the PSF (i.e., the fewer "degrees of freedom" the PSF has), the more invariants are available. The structure of invariant matrix $\mathbf{K}_{ij} = K_N(i-1, j-1)$ is shown in Figure 6.14. There is always a zero-stripe on the main diagonal regardless of N, and other zero-stripes are on all minor diagonals for $(p-q)/N$ being an integer. In other words, the

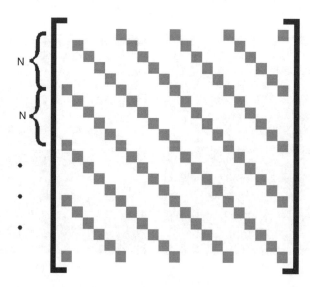

Figure 6.14 The structure of the matrix of N-FRS blur invariants. The gray elements on the diagonals are identically zero regardless of f. The white elements stand for non-trivial invariants

invariant matrix of the image is complementary to the moment matrix of the PSF in terms of zero and non-zero elements.

The invariants (6.35) are generally complex-valued. If we prefer working with real-valued invariants, we should use their real and imaginary parts separately. On the other hand, for any invariant (6.35) $K_N(p,q) = K_N(q,p)^*$, so it is sufficient to consider the cases $p \geq q$ only.

For any N, invariant $K_N(1,0) = c_{10}$ is non-trivial. However, since in practical applications c_{10} is usually employed for image normalization to shift, $K_N(1,0)$ becomes useless for image recognition.

The invariants (6.35) formally exist also for $N = 1$ (i.e., PSF without any symmetry) but all invariants except $K_1(0,0) = c_{00}$ are trivial, which is in accordance with our intuition. The existence of the "singular" invariant $K_1(0,0)$ is not because of the symmetry but because the PSF has been supposed to be brightness-preserving.

The sets C_N create interesting structures inside $\mathcal{I}$. Assuming N can be factorized as

$$N = \prod_{i=1}^{L} N_i,$$

where $N_i = k_i^{n_i}$ and k_i are distinct primes. Then

$$C_N = \bigcap_{i=1}^{L} C_{N_i}.$$

For each N_i we have a nested sequence of the sets

$$C_{k_i} \underset{\neq}{\supseteq} C_{k_i^2} \underset{\neq}{\supseteq} \cdots \underset{\neq}{\supseteq} C_{k_i^{n_i}} \equiv C_{N_i}.$$

Particularly,

$$C_\infty = \bigcap_{k=1}^{\infty} C_k.$$

If the blur is from C_N, then it has also k_n-fold symmetry for any $n = 1, 2, \ldots, L$. Thus, any $K_{k_n}(p,q)$ is also a blur invariant. However, using these invariants together with $K_N(p,q)$'s is useless since they are all dependent on the $K_N(p,q)$'s. (Note that the dependence here does not mean equality, generally $K_N(p,q) \neq K_{k_n}(p,q)$).

Let us conclude this section by a remark concerning the proper choice of N. It is a user-defined parameter of the method, and its correct estimation may be a tricky problem. Ideally, it should be deduced from the number of the diaphragm blades of the objective. If the camera is of an unknown model and physically unavailable, we may try to estimate N directly from the blurred image by analyzing its spectral patterns or the response to an ideal bright point in the scene(see Figure 6.15).

If none of the above is applicable and we overestimate N in order to have more invariants available, then we lose the invariance property of some (or even all) of them. On the other hand, some users might rather underestimate it to be on the safe side (they might choose for instance $N = 2$ instead of the ground-truth $N = 16$). This would always lead to the loss of discriminability and sometimes could also violate the invariance, depending on the actual and chosen N. Obviously, the worst combination occurs if the true and estimated fold numbers are

(a)

(b)

Figure 6.15 The shape of the out-of-focus PSF can be observed in the background. This bokeh effect may serve for estimation of N. In this case, $N = 7$

small coprime integers such as 2 and 3 or 3 and 4. If h is not exactly N-fold symmetric, then the more the ratio $H/P_N H$ differs from a constant 1, the more violated the invariance of the K_N's.

In some cases, the unknown objects and the database objects were acquired by different cameras, one having the number of blades N_1 and the other one N_2. If both of them are blurred, how can we choose proper invariants K_N? To ensure the invariance to both and to get the best possible discriminability, we may choose N as the greatest common divisor of N_1 and N_2. If they are coprime, the task is not correctly solvable. In such a case, we choose either N_1 or N_2 depending on which blur is more severe.

6.8   Invariants to dihedral blur

As we have seen in Chapter 3, the N-FRS may be coupled with the axial symmetry. In such a case, the number of the axes equals N, and we speak about the N-fold dihedral symmetry. Many out-of-focus blur PSFs with N-FRS are actually dihedral, particularly if the diaphragm blades are straight, see Figure 6.16.

Before we construct the projector on a set of dihedral functions, we need to formalize the reflection of function f across a line a. The *reflection matrix S* is

$$S = \begin{pmatrix} \cos 2\alpha & \sin 2\alpha \\ \sin 2\alpha & -\cos 2\alpha \end{pmatrix} \tag{6.36}$$

where $\alpha \in \langle 0, \pi/2 \rangle$ is the angle between the reflection line a and the x-axis[12]. Then the reflected function f^α is given as

$$f^\alpha(\mathbf{x}) = f(S\mathbf{x}).$$

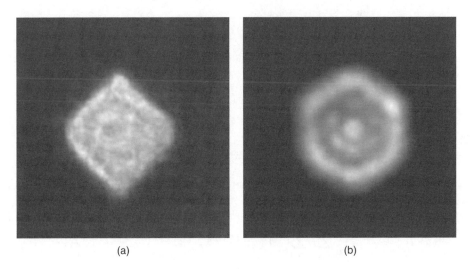

(a)   (b)

Figure 6.16   Two examples of real out-of-focus PSF on apertures with dihedral symmetry, obtained by photographing a bright point. (a) $N = 4$, (b) $N = 6$

---

[12] Note that S is a symmetric, orthogonal, and *involutory* matrix, which means $S = S^{-1} = S^T$ and $S^2 = I$.

The special case when $\alpha = 0$ leads to our intuitive notion of the reflection as $f^0(x, y) = f(x, -y)$. We can copy the previous approach and define

$$S \equiv D_N = \{h \in C_N | \exists \alpha \text{ such that } h(x, y) = h^\alpha(x, y)\}.$$

There is, however, a substantial difference from the N-FRS functions. For any finite N, the set D_N is *not* closed under convolution. We cannot define a projector and specific blur invariants do not exist[13]. The closure property is preserved only if the angle α is *known and fixed* for the whole S.

If we define S as

$$S \equiv D_N^\alpha = \{h \in C_N | h(x, y) = h^\alpha(x, y)\},$$

which is a set of all N-fold dihedral functions with the same symmetry axes oriented into direction α, then D_N^α is a vector space closed under convolution. Projection operator Q_N^α from $\mathcal{I}$ onto D_N^α is defined as

$$Q_N^\alpha f = P_N((f + f^\alpha)/2),$$

where P_N is the projector on the set C_N defined in the previous section. It is easy to prove that Q_N^α exhibits all required properties of a projector for arbitrary fixed α. Examples of the projection operators are in Figure 6.17.

Neither geometric nor complex moments are separated, but complex moments are separated in a weak sense. For any N and α,

$$c_{pq}^{(Q_N^\alpha f)} = 0$$

whenever $(p - q)/N$ is not an integer. If $(p - q)/N$ equals an integer, we have

$$2c_{pq}^{(Q_N^\alpha f)} = c_{pq}^{(f)} + c_{pq}^{(f^\alpha)}.$$

To express $c_{pq}^{(f^\alpha)}$, note that f^α is nothing but f rotated by α such that a coincides with x, reflected across x and rotated backwards. Hence,

$$c_{pq}^{(f^\alpha)} = c_{qp}^{(f)} \cdot e^{2i\alpha(p-q)}$$

and

$$2c_{pq}^{(Q_N^\alpha f)} = c_{pq}^{(f)} + c_{qp}^{(f)} \cdot e^{2i\alpha(p-q)}.$$

Using the same expansion of $I(f)(U, V)$ as before we obtain the recurrence for the dihedral invariants

$$L_N^\alpha(p, q) = c_{pq} - \frac{1}{2c_{00}} \sum_{\substack{j=0 \\ 0 < j+k \\ (j-k)/N \text{ is integer}}}^{p} \sum_{k=0}^{q} \binom{p}{j} \binom{q}{k} L_N^\alpha(p-j, q-k) \cdot (c_{jk} + c_{kj} e^{2i\alpha(p-q)}). \tag{6.37}$$

---

[13] In other words, the knowledge that $h \in C_N$ is dihedral, without knowing α, does not reduce the number of the degrees of freedom of h.

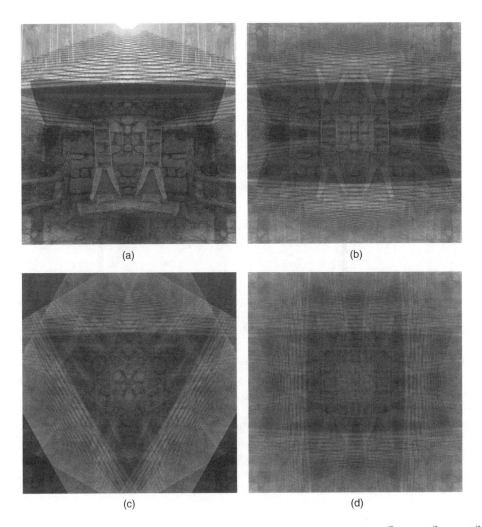

Figure 6.17   Dihedral projections of the original image from Figure 6.10. (a) $Q_1^{\pi/2}$, (b) $Q_2^{\pi/2}$, (c) $Q_3^{\pi/2}$ and (d) $Q_4^{\pi/2}$

In addition to the symmetry axis a, the functions from D_N^α also have other symmetry axes of angles $\beta_k = \alpha + k\pi/N$, $k = 1, \ldots, N - 1$. However, their usage in eq. (6.37) does not create any new invariants because

$$L_N^{\beta_k}(p,q) = L_N^\alpha(p,q)$$

for any $N > 1$ and $k = 1, \ldots, N - 1$.

The blur invariants $L_N^\alpha(p,q)$ share some of their properties with $K_N(p,q)$, but some others are different. General properties do not depend on α.

We may expect that the number of non-trivial invariants $L_N^\alpha(p,q)$ up to the given order r is greater than the number of $K_N(p,q)$, because $D_N^\alpha \subset C_N$.

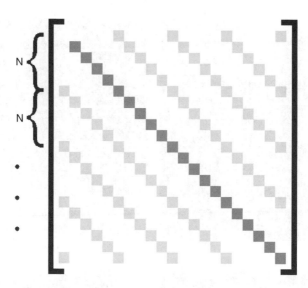

Figure 6.18 The structure of the dihedral invariant matrix. The dark gray elements on the main diagonal vanish for any f. The light gray elements on the minor diagonals are non-trivial, but their real and imaginary parts are constrained. The white elements stand for non-trivial complex-valued invariants

It holds that $L_N^\alpha(p,p) = 0$ for any N and $p \neq 0$. The invariant matrix $\mathbf{L}_{ij} = L_N^\alpha(i-1,j-1)$ has zeros on the main diagonal. The minor diagonals with $(p-q)/N$ being an integer are, however, not zero. They contain non-trivial invariants which are complex-valued, but there is a band between their real and imaginary parts constrained by the value of α. These invariants provide only one independent real-valued feature. The existence of non-trivial invariants on the minor diagonals is the main advantage of the dihedral invariants (see Figure 6.18 for the structure of the invariant matrix). If the object has an N-fold rotation symmetry but not an axial symmetry, these minor-diagonal invariants are the only ones which do not vanish. The invariants $L_N^\alpha(p,q)$ with a non-integer $(p-q)/N$ are dependent on the invariants $K_N(p,q)$. It again holds $L_N^\alpha(q,p) = L_N^\alpha(p,q)^*$.

Comparing to the N-FRS blur, dihedral blur reduces the number of trivial invariants to approximately one half. This is in accordance with our intuitive expectation because, for the given N, the dihedral functions have "a half" of the degrees of freedom of the N-FRS functions.

In case of dihedral blur, it is meaningful to consider also the case of $N = 1$. In that case, we obtain a full matrix $\mathbf{L}$ with the main diagonal being zero and with all other elements being constrained. On the other hand, it is useless to study the case of ∞-fold dihedral blur separately because $D_\infty^\alpha = D_\infty = C_\infty$.

The practical importance of the dihedral blur invariants is less than that of the invariants to N-FRS blur. The major limitation comes from the fact that the orientation of the symmetry axis must be known. When working with multiple images, it must be even the same for all of them. This is far from being realistic. Unlike N, the angle α can hardly be deduced from the hardware construction. Even if the camera was on a tripod, α would change as we change the aperture size. In case of a hand-held camera, α depends on the actual hand rotation. The only possibility is to estimate α from the blurred image itself. Pedone et al. [79] proposed a fast heuristic algorithm working in the Fourier domain, which can be used for that purpose.

Another restriction is that the dihedral blur invariants cannot be combined with an unknown object rotation since it changes the angle α. Summarizing, dihedral blur invariants should be used only in a controlled environment when α is known and constant and we aim to distinguish among various N-fold symmetric objects.

6.9   Invariants to directional blur

Directional blur (sometimes inaccurately called motion blur) is a 2D blur of a 1D nature which acts in a single constant direction only. In the perpendicular direction, the image is not blurred at all. Directional blur may be caused by camera shake, scene vibrations, and camera or scene straight motion. The velocity of the motion may be variable during the acquisition. This kind of blur appears namely on the pictures taken by a hand-held camera in a dark environment at long exposures (see Figure 6.19 and Figure 6.1b for two examples).

The respective PSF has the form (for the sake of simplicity, we start with the horizontal direction)

$$h(x, y) = h_1(x)\delta(y), \tag{6.38}$$

Figure 6.19   The image blurred by a camera shake blur, which is approximately a directional blur

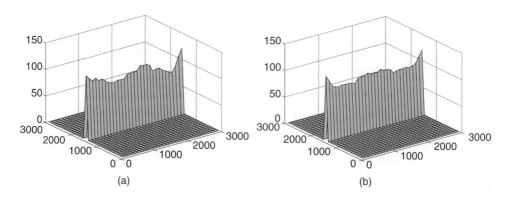

Figure 6.20   Directional projections of the original image from Figure 6.10. (a) Pf, (b) $P^s f$

where $h_1(x)$ is an arbitrary 1D image function. The space S is defined as a set of all functions of the form (6.38). It is a vector space closed under convolution. The directional projection operator P is defined as

$$(Pf)(x,y) = \delta(y) \int f(x,y)\mathrm{d}y.$$

An example of the projection is in Figure 6.20a.

Geometric moments are separated because

$$m_{p,0}^{(Pf)} = \iint x^p [\delta(y) \int f(x,y)\mathrm{d}y]\mathrm{d}x\mathrm{d}y = \iint x^p f(x,y)\mathrm{d}x\mathrm{d}y = m_{p,0}^{(f)}$$

for any p and f and

$$m_{pq}^{(Pf)} = 0$$

for any p and any $q \neq 0$. Hence, $D = \{(p,q)|q = 0\}$ and Eq. (6.32) yields the directional blur invariants

$$M(p,q) = m_{pq} - \frac{1}{m_{00}} \sum_{k=1}^{p} \binom{p}{k} M(p-k,q)m_{k,0}.$$

The same formula is valid when replacing geometric moments by central moments. Note that $M(p,0)$ vanishes for any $p > 0$.

If the directional blur is centrosymmetric (i.e. $h_1(-x) = h_1(x)$), we can incorporate this assumption into the definition of S and P. In this case, projector P^s is defined as

$$(P^s f)(x,y) = \delta(y)\frac{1}{2} \int (f(-x,y) + f(x,y))\mathrm{d}y.$$

A difference from Pf can be seen in Figure 6.20b.

The moment separation condition is again fulfilled, because we have

$$m_{p,0}^{(Pf)} = m_{p,0}^{(f)}$$

for any even p and any f and

$$m_{pq}^{(Pf)} = 0$$

otherwise. $D^s = \{(p,q)|p = 2k, q = 0\}$ and the symmetric directional blur invariants are given by the recurrence

$$M^s(p,q) = m_{pq} - \frac{1}{m_{00}} \sum_{\substack{k=2 \\ k \text{ even}}}^{p} \binom{p}{k} M^s(p-k,q)m_{k,0}.$$

When using central moments, symmetry of the PSF w.r.t. its centroid is sufficient. Comparing to the general case, the symmetry assumption reduces the number of trivial invariants to a half, since $M^s(p,0)$ vanishes only for even $p > 0$.

Traditional linear motion blur which usually appears in textbooks is modeled by eq. (6.38) with $h_1(x)$ being a unit rectangular pulse. It is a particular case of the symmetric directional blur. However, a set of rectangular functions is not closed under convolution, so we cannot use their parametric form to derive specific blur invariants.

If the direction of the blur is known but not horizontal, the above theory can be simply adapted. The space S is a set of the directional PSFs of the given angle α, and the projector $P^\alpha f$ is defined by means of a line integral along a line which is perpendicular to the blur direction.

The main problem with the directional blur invariants is similar to that of the dihedral blur. To design the invariants, we need to know the blur direction, and when comparing multiple images, this direction must be the same. We may try to estimate the blur direction from the response on a bright point in the scene, if any (see Figure 6.21 for an example). Alternatively, we may estimate the blur direction from spectral zero patterns of the blurred image. In the spectrum, characteristic zero lines perpendicular to motion direction appear (see Figures 6.22 and 6.23).

Such an estimation may not be accurate and requires additional preprocessing time. This has led to attempts to find invariants w.r.t. the blur direction. Since the set of directional blurs of variable direction is not closed under convolution, this task brings us back to the 2D centrosymmetric invariants (provided the directional PSFs are even).

Figure 6.21   The image of a bright point (a close-up of Figure 6.1b) can serve for estimation of the blur direction

Figure 6.22 Amplitude spectrum of a motion-blurred image. The zero lines are perpendicular to the motion direction

Figure 6.23 Estimated blur direction from the image spectrum in Figure 6.22

6.10 Invariants to Gaussian blur

Gaussian blur appears whenever the image has been acquired through a turbulent medium and the acquisition/exposure time is far longer than the period of Brownian motion of the particles in the medium. Ground-based astronomical imaging through the atmosphere, taking pictures through a fog, underwater imaging, and fluorescence microscopy are typical examples of such situation (in some cases, the blur may be coupled with a contrast decrease). Gaussian blur is also introduced into the images as the sensor blur which is due to a finite size of the sampling pulse; this effect is, however, mostly of low significance. Moreover, Gaussian kernel is often used as an approximation of some other blurs which are too complicated to work with them

Figure 6.24 Gaussian blur caused by the mist coupled with a contrast decrease (Vltava river, Prague, Czech Republic)

exactly. Gaussian blur is sometimes even introduced into the image intentionally, for instance to suppress additive noise, to "soften" the image or to perform local averaging before the image is down-scaled (see Figures 6.24 and 6.25 for some examples).

Numerous examples of the Gaussian convolution can be found outside the image processing area – particle transportation, diffusion processes, time-development of a heat distribution in a mass, and photon scattering in radiation physics are only few of them. Most of these phenomena are represented by 2D or 3D functions which can be visualized, that brings us back to image analysis.

Gaussian blur of unknown parameters is very difficult to remove via deconvolution because we cannot estimate the blur parameters from the spectrum, as we can do in the case of motion and circular out-of-focus blur. Although some Gaussian-deconvolution algorithms have been proposed recently [99, 100], they are sensitive to the variance overestimation and relatively time-consuming. Hence, designing Gaussian blur invariants is as desirable as the construction of the other blur invariants.

Symmetric Gaussian blur is a special case of a circular symmetric blur; asymmetric Gaussian blur is a particular case of 2-fold dihedral symmetry. So, why not to use the invariants $C_\infty(p, q)$, $C_2(p, q)$, or $L_2^\alpha(p, q)$? Of course, we can do that, but these invariants have a limited discrimination power. If we design blur invariants specifically to Gaussian blur using its parametric form, we may expect to obtain more invariants with better discriminability because we narrow their null-space.

The derivation of Gaussian blur invariants by means of projection operator must be slightly modified compared to the previous cases, because a Gaussian is not an image function in the sense of the definition introduced in Chapter 3 (its support is not bounded), and the set of

Figure 6.25   Gaussian blur as a result of denoising. The original image corrupted by a heavy noise was smoothed by a Gaussian kernel to suppress the high-frequency noise components

all Gaussians is not a vector space. Fortunately, these modifications are rather minor, and the theory can be adapted easily, as we expose in the rest of this section.

We start our explanation with a 1D case. The separability of a multidimensional Gaussian allows an easy extension of the 1D theory.

6.10.1   1D Gaussian blur invariants

In this section, we consider $\mathcal{I}$ to be a space of absolutely integrable functions of exponential decay[14].

Assuming that the blurring PSF has a traditional shape[15]

$$G_\sigma(x) = \frac{1}{\sqrt{2\pi}\sigma} e^{-\frac{x^2}{2\sigma^2}}$$

where $\sigma > 0$ is a standard deviation, we define the set S_G as the set of all unnormalized Gaussians

$$S_G = \{aG_\sigma | \sigma \geq 0, a \neq 0\}.$$

---

[14] Fourier transform of these functions exists, and all moments are finite even if the function has an infinite support. Integrable functions of bounded support are special cases of the fast decaying functions regardless of their particular shape.

[15] Let us extend this definition by setting $G_0(x) = \delta(x)$.

S_G is closed under convolution and point-wise multiplication, contains a unit element $\delta(x)$ but is not a vector space[16], as it is not closed under addition.

Let as define the operator $P_G : \mathcal{I} \to S_G$ as

$$P_G f(x) = \mu_0^{(f)} G_s(x), \qquad (6.39)$$

where

$$s^2 = \mu_2^{(f)}/\mu_0^{(f)}.$$

Hence, P_G assigns each f to a centralized Gaussian multiplied by $\mu_0^{(f)}$ such that

$$\mu_2^{(P_G f)} = \mu_2^{(f)}.$$

In other words, $P_G f$ is the "nearest" unnormalized Gaussian to f in terms of the first three moment values. In this sense, P_G can be considered a projector onto S_G, but it is not a projector in the sense of our definition since it is not a linear operator. Still, most of the properties required from projection operators are preserved. P_G is idempotent and multiplicative as $P_G(af) = aP_G f$. Any function f can be expressed as $f = P_G f + f_n$, where f_n can be considered a "non-Gaussian" part of f. The equality $P_G(f_1) = P_G(f_2)$ defines an equivalence relation on the image space. The classes of equivalence are formed by the functions of the same zeroth and second central moments.

The key property of projection operators used previously was that $P(f * h) = Pf * h$, which followed from the linearity of P. Although P_G is not linear, the same property holds well but must be proved directly from the properties of a Gaussian function. By definition of P_G we have

$$P_G(f * G_\sigma)(x) = \mu_0 G_{\sqrt{(s^2+\sigma^2)}}(x) = \frac{\mu_0}{\sqrt{2\pi(s^2+\sigma^2)}} e^{-\frac{x^2}{2(s^2+\sigma^2)}} \qquad (6.40)$$

because

$$\mu_2^{(f*G_\sigma)} = \mu_2^{(f)} + \sigma^2 \mu_0^{(f)}.$$

On the other hand, $(P_G f * G_\sigma)(x)$ is a convolution of two Gaussians, the result of which must again be a Gaussian with the variance being the sum of the input variances, which yields the right-hand side of eq. (6.40). We have just proved an analogue of Theorem 6.1. Functional I_G defined as

$$I_G(f) = \frac{\mathcal{F}(f)}{\mathcal{F}(P_G f)}$$

is an invariant to Gaussian blur, that is, $I_G(f) = I_G(f * G_\sigma)$ for any $f \in \mathcal{I}$ and for any blur parameter σ.

The meaning of $I_G(f)$ and many its properties are the same or analogous to the previous cases. We may again interpret $I_G(f)$ as a Fourier transform of the primordial image, that is, as a deconvolution of f by the kernel $P_G(f)$, which is the largest possible Gaussian kernel (larger kernels cannot exist because deblurring always monotonically decreases μ_2, reaching the limit at $\mu_2^{(\mathcal{F}^{-1}(I_G(f)))} = 0$). However, we encounter also certain differences implied by special properties of Gaussian functions. For instance, the completeness theorem can be formulated in a stronger way than in the general case.

---

[16] S_G is a commutative ring.

If f_1, f_2 are two image functions, then $I_G(f_1) = I_G(f_2)$ holds if and only if there exist $a > 0$ and $\sigma \geq 0$ such that $f_1 = f_2 * aG_\sigma$ or $f_2 = f_1 * aG_\sigma$. To prove this, let us first realize that if $I_G(f_1) = I_G(f_2)$ then obviously

$$\mathcal{F}(f_1)(u)\mathcal{F}(P_G f_2)(u) = \mathcal{F}(f_2)(u)\mathcal{F}(P_G f_1)(u),$$

which in the image domain means

$$f_1 * P_G f_2 = f_2 * P_G f_1.$$

Both $P_G f_i = a_i G_{\sigma_i}$ are unnormalized Gaussians. Assuming that $\sigma_1 \geq \sigma_2$, we define $\sigma^2 = \sigma_1^2 - \sigma_2^2$ and $a = a_1/a_2$. Since the convolution of any two Gaussians is again a Gaussian the variance of which is the sum of two input variances, we have

$$aG_\sigma * a_2 G_{\sigma_2} = a_1 G_{\sigma_1}$$

(this is the step we cannot perform in the general case). From this we immediately obtain

$$f_1 = f_2 * aG_\sigma$$

which completes the proof.

Another difference from the general case is the expansion of $I_G(f)$ in terms of moments. The geometric moments of $P_G f$ are not strictly separated. We have, for any odd p,

$$m_p^{(P_G f)} = \mu_p^{(P_G f)} = 0.$$

For $p = 0$ and $p = 2$ the moments are preserved

$$m_p^{(P_G f)} = \mu_p^{(P_G f)} = \mu_p^{(f)}$$

but in general for the other even p

$$\mu_p^{(P_G f)} \neq \mu_p^{(f)}.$$

The expansion is still possible since all moments of a Gaussian are functions of a single parameter σ. It holds

$$\mu_p^{(G_\sigma)} = \sigma^p (p-1)!! \tag{6.41}$$

for any even p. The symbol $k!!$ means a double factorial, $k!! = 1 \cdot 3 \cdot 5 \cdots k$ for odd k, and by definition $(-1)!! = 0!! = 1$.

To perform the expansion of $I_G(f)$, we first express the Taylor expansion of $\mathcal{F}(P_G f)$. It yields

$$\mathcal{F}(P_G f)(u) = \mu_0 \sum_{k=0}^{\infty} (2k-1)!! \frac{(-2\pi i)^{2k}}{(2k)!} \left(\frac{\mu_2}{\mu_0}\right)^k u^{2k} \tag{6.42}$$

(we used (6.41) to express the moments of order higher than two).

$I_G(f)$ itself is supposed to have a form

$$I_G(f)(u) = \sum_{k=0}^{\infty} \frac{(-2\pi i)^k}{k!} A_k u^k$$

where A_k are the moments of the primordial image. Substituting the power series into the definition of $I_G(f)$ and considering that

$$(2k-1)!! = \frac{(2k)!}{2^k \cdot k!}$$

we have

$$\sum_{k=0}^{\infty} \frac{(-2\pi i)^k}{k!} \mu_k u^k = \mu_0 \sum_{k=0}^{\infty} \frac{(-2\pi^2)^k}{k!} \left(\frac{\mu_2}{\mu_0}\right)^k u^{2k} \cdot \sum_{k=0}^{\infty} \frac{(-2\pi i)^k}{k!} A_k u^k.$$

Comparing the terms with the same power of u we obtain, after some algebraic manipulation, the recursive expression for each A_p

$$A_p = \frac{\mu_p}{\mu_0} - \sum_{\substack{k=2 \\ k \text{ even}}}^{p} (k-1)!! \cdot \binom{p}{k} \left(\frac{\mu_2}{\mu_0}\right)^{k/2} A_{p-k}. \tag{6.43}$$

If we restrict ourselves to a brightness-preserving blurring, we obtain from (6.43) the simplified final form of Gaussian blur invariants

$$B(p) \equiv \mu_0 A_p = \mu_p - \sum_{\substack{k=2 \\ k \text{ even}}}^{p} (k-1)!! \cdot \binom{p}{k} \left(\frac{\mu_2}{\mu_0}\right)^{k/2} B(p-k), \tag{6.44}$$

which can be equivalently expressed in a non-recursive form

$$B(p) = \sum_{\substack{k=0 \\ k \text{ even}}}^{p} (k-1)!! \cdot \binom{p}{k} \left(-\frac{\mu_2}{\mu_0}\right)^{k/2} \mu_{p-k}. \tag{6.45}$$

Investigation of the values of $B(p)$ shows that the only vanishing invariants are $B(1)$ and $B(2)$. $B(1)$ vanishes because we use central moments, while $B(2)$ vanishes because of elimination of the blurring parameter σ.

Unlike other blur invariants, where the 1D case mostly serves only as a starting point of the investigation without any practical importance, the 1D Gaussian blur invariants already found a practical application [75, 97]. Gaussian blur often acts also on the *histograms* of the images, which is totally independent of the image blurring. If the image (blurred or sharp) is corrupted by an additive noise, which is not correlated with the image, then the histogram of the noisy image is a convolution of the histograms of the original and that of the noise[17]. If the noise is Gaussian, the histogram is smoothed by a Gaussian kernel. To characterize the histogram in a robust way, for instance for the purpose of image retrieval, 1D Gaussian blur invariants may be a good choice.

---

[17] This is a consequence of a well-known theorem from probability theory, which states that the probability density function of a sum of two independent random variables is a convolution of their pdf's.

6.10.2 *Multidimensional Gaussian blur invariants*

The above theory can be extended to Gaussian blur in arbitrary dimensions, provided that the symmetry axes of the d-D Gaussian are parallel with the coordinates. The general form of a (centralized) d-D Gaussian function is

$$G_\Sigma(\mathbf{x}) = \frac{1}{\sqrt{(2\pi)^d |\Sigma|}} \exp\left(-\frac{1}{2} \mathbf{x}^T \Sigma^{-1} \mathbf{x}\right), \tag{6.46}$$

where Σ is the covariance matrix which determines the shape of the Gaussian.

Assuming that Σ is diagonal, $\Sigma = \mathrm{diag}(\sigma_1^2, \sigma_2^2, \dots, \sigma_d^2)$, the moments of the Gaussian are given as

$$\mu_\mathbf{p}^{(G_\Sigma)} = \boldsymbol{\sigma}^\mathbf{p}(\mathbf{p} - \mathbf{1})!! \tag{6.47}$$

where $\boldsymbol{\sigma} \equiv (\sigma_1, \sigma_2, \dots, \sigma_d)$, for any $\mathbf{p}$ all components of which are even, and zero otherwise. The multi-index notation used here should be read as

$$\mathbf{p}^\mathbf{k} \equiv \prod_{i=1}^d p_i^{k_i},$$

$$\mathbf{p}!! \equiv \prod_{i=1}^d p_i!!,$$

and

$$\mathbf{1} \equiv (1, 1, \dots, 1).$$

The projection operator is defined as

$$P_G f(\mathbf{x}) = \mu_\mathbf{0} G_S(\mathbf{x}),$$

where

$$S = \mathrm{diag}(\boldsymbol{\mu}_2/\mu_\mathbf{0})$$

and

$$\boldsymbol{\mu}_2 \equiv (\mu_{20\cdots0}, \mu_{02\cdots0}, \dots, \mu_{00\cdots2}).$$

$I_G(f)$ is again a blur invariant, and after expanding it into a power series we obtain the d-D versions of the moment invariants (6.44) and (6.45)

$$B(\mathbf{p}) = \mu_\mathbf{p} - \sum_{\substack{\mathbf{k}=0 \\ 0<|\mathbf{k}|}}^\mathbf{p} (\mathbf{k} - \mathbf{1})!! \cdot \binom{\mathbf{p}}{\mathbf{k}} (\boldsymbol{\mu}_2/\mu_\mathbf{0})^\mathbf{k} B(\mathbf{p} - \mathbf{k})$$

$$= \sum_{\mathbf{k}=0}^\mathbf{p} (\mathbf{k} - \mathbf{1})!! \cdot \binom{\mathbf{p}}{\mathbf{k}} (-1)^{|\mathbf{k}|} (\boldsymbol{\mu}_2/\mu_\mathbf{0})^\mathbf{k} \mu_{\mathbf{p}-2\mathbf{k}}, \tag{6.48}$$

where the summation goes only over those multi-indices $\mathbf{k}$ all elements of which are even. An example of such projection operator is in Figure 6.26.

Note that in general $\sigma_i \neq \sigma_k$, which allows us to handle the blurring PSFs which are "elongated" in some directions. This may be useful, for instance, if the sensor resolution changes

Figure 6.26   2D Gaussian projection $P_G f$ of the original image from Figure 6.10

in different directions. If Σ is a multiple of the unit matrix, which means the Gaussian is radially symmetric, we obtain $d - 1$ additional independent blur invariants of the second order $\mu_{20\cdots0} - \mu_{02\cdots0}, \mu_{20\cdots0} - \mu_{002\cdots0}, \cdots$.

Relaxing the assumption of Σ being diagonal is in principle possible only if the directions of its eigenvectors are known and fixed. The formula for the invariants would, however, look much more complicated. If the eigenvectors of Σ are not known, the projection operator cannot be defined, and the derivation of the invariants fails.

6.10.3   2D Gaussian blur invariants from complex moments

In the particular case of $d = 2$ and $\sigma_1 = \sigma_2 = \sigma$ (i.e., circularly symmetric 2D Gaussian blur), it is useful to consider also an expansion into complex moments, similarly as we did in the case of N-fold symmetric blur. The invariants we obtain in that way are of course dependent on the above $B(p, q)$'s but as we will show later, unlike the $B(p, q)$'s the new ones are convenient for designing of combined blur-rotation invariants.

The complex moments of a circularly symmetric Gaussian are

$$c_{pq}^{(G_\sigma)} = \begin{cases} (2\sigma^2)^p p! & p = q \\ 0 & p \neq q \end{cases}.$$

We define the projection operator as

$$P_G f(x, y) = c_{00}^{(f)} G_s(x, y) \equiv \frac{c_{00}^{(f)}}{2\pi s^2} e^{-\frac{x^2+y^2}{2s^2}}, \tag{6.49}$$

where

$$s^2 = c_{11}^{(f)} / 2 c_{00}^{(f)}.$$

$P_G f$ has the same c_{00} and c_{11} as f (and, of course, $c_{10} = 0$ when working in the centralized coordinates). The other moments of $P_G f$ and f are generally different from one another, but all non-trivial moments of $P_G f$ can be expressed by means of s as

$$c_{pp}^{(P_G f)} = (2s^2)^p p!.$$

Let us again use the substitution $U = u + v$ and $V = i(u - v)$. The blur invariant in Fourier domain is defined analogously to the previous cases as

$$I_G(f)(U, V) = \frac{\mathcal{F}(f)(U, V)}{\mathcal{F}(P_G f)(U, V)}.$$

Taylor expansion of the denominator is

$$\mathcal{F}(P_G f)(U, V) = c_{00} \sum_{k=0}^{\infty} \frac{(-4\pi^2)^k}{k!} \left(\frac{c_{11}}{c_{00}} \right)^k u^k v^k.$$

Using the Taylor expansion of all three factors by means of their complex moments and comparing the coefficients of the same powers, we obtain the blur invariants in the moment domain in recurrent as well as direct forms (assuming that $p \geq q$)

$$
\begin{aligned}
K_G(p, q) &= c_{pq} - \sum_{k=1}^{q} k! \binom{p}{k} \binom{q}{k} \left(\frac{c_{11}}{c_{00}} \right)^k K_G(p - k, q - k) \\
&= \sum_{k=0}^{q} k! \binom{p}{k} \binom{q}{k} \left(-\frac{c_{11}}{c_{00}} \right)^k c_{p-k, q-k}.
\end{aligned}
\tag{6.50}
$$

For the proof of the equality between the recursive and non-recursive expression in eq. (6.50) see Appendix 6.F. Note that $K_G(q, p) = K_G(p, q)^*$, $K_G(1, 0) = 0$ when working in the centralized coordinates, and $K_G(1, 1) = 0$ due to the parameter elimination.

6.11 Invariants to other blurs

The question about the blur types to which the invariants may exist had appeared already at the beginning of the blur invariant research. Thanks to the theory of projection operators, the question can be reformulated as follows. What are subspaces S of $\mathcal{I}$ such that they are closed under convolution while at the same time it is possible to explicitly construct the respective projectors? The blur types discussed above all fulfil these requirements, but are there any other such subspaces?

Without any thorough analysis, we can find many sets closed under convolution, such as the space $\mathcal{I}$ itself, the space of all compactly supported continuous functions, the space of all compactly supported infinitely differentiable functions[18], the space of all functions of exponential decay (if $\mathcal{I} = L_1(\mathbb{R}^d)$), and the space of all polynomials (if $\mathcal{I} = L_1(\Omega)$), to name a few. However, the problem with these spaces is not only that the respective projectors are difficult to construct but in fact that the spaces are too "large" (most of them are even dense in $\mathcal{I}$), which implies that none or very few non-trivial invariants exist.

---

[18] Such functions are called *bump functions*.

In this sense, there are only few spaces S which are meaningful to consider. These may be characterized either by certain symmetric properties of their elements or by their parametric form. Regarding the former case, the 2D space has been fully explored (note that there are only rotation and dihedral symmetry groups in 2D) but 3D space and multidimensional spaces still contain numerous unexplored cases. As we saw in Chapter 4, there exist seventeen symmetry groups of various properties in 3D. Although the definition of P with respect to all of them is in principle clear, the derivation of the blur invariants in terms of 3D moments or spherical harmonics is not easy and has not been published yet for most of these symmetry groups.

As for the latter case, there are very few parametric families of functions which are closed under convolution (Gaussian family is one of them). Even the class of generalized normal distributions [101] where one might intuitively expect the closure property, is not closed under convolution for exponents other than two (which is the case of Gaussian distribution).

The functions with the closure convolution property have been studied in probability theory extensively because they describe such pdf's which are closed w.r.t. to a finite sum of independent random variables. Such functions are called *alpha-stable* functions or alpha-stable distributions[19].

Unfortunately, only very few alpha-stable functions have a closed-form formula, and the others are difficult to work with. There have been reported only three closed forms expressible in terms of elementary functions (Gaussian, Cauchy, and Levy distributions) and few others which can be expressed by means of special functions. The generalized Gaussian is not an alpha-stable distribution. Among all alpha-stable functions, only the Gaussian has finite moments of all orders. This leads to the conclusion that moment invariants w.r.t. convolution with an alpha-stable kernel do not exist for other than Gaussian kernels.

There is, however, still a possibility of finding blur invariants in the Fourier domain by means of projection operators. As proved by Nolan [102], the *characteristic function* F (which is actually an inverse Fourier transform) of an arbitrary centralized and non-skewed alpha-stable function must have the form

$$F(t) = \exp(-c|t|^a)$$

where $0 < a \leq 2$ and $c > 0$.[20] This is a very strong theorem which parameterizes the set of all alpha-stable distributions. Note that for any $a \neq 2$ the respective distribution f has infinite central moments of order greater than one, so it is a *heavy-tailed* distribution.

From the above parametrization, the closure property is evident (on the other hand, it is also evident that this set is not a vector space). If h_1 and h_2 are two alpha-stable functions such that $H_1(t)$ and $H_2(t)$ have *the same* exponent a, then $h = h_1 * h_2$ is also alpha-stable with the characteristic function $H(t)$ of exponent a and with $c = c_1 + c_2$. Provided that a is fixed and known, we can define the projection operator P_a in the Fourier domain as a fit[21] of $F(u)$ by a function of the form $\exp(-c|u|^a)$. Then $F/P_a F$ is a blur invariant. Although P_a is not a linear

---

[19] Another interesting property of the alpha-stable distributions is that they are "limits" or "attractors" of the convergence of averages of heavy-tailed random variables, (see the Generalized Central Limit Theorem), while Gaussians are "attractors" of the light-tailed independent variables.

[20] This form includes Gaussian distribution for $a = 2$ and Cauchy distribution for $a = 1$.

[21] There is a question how to find the "best" fit. This question is equivalent to the problem of estimating c from sample data, which is not trivial because the least square fit leads to a non-linear equation. Detailed discussion on the estimation of c is beyond the scope of this book; we refer to any monograph on robust data fitting by heavy-tailed distributions.

operator, the proof of the invariance is simple

$$\frac{G}{P_a(G)} = \frac{F \cdot H}{P_a(F \cdot H)} = \frac{F \cdot H}{P_a(F) \cdot P_a(H)}$$

$$= \frac{F \cdot H}{P_a(F) \cdot H} = \frac{F}{P_a(F)}.$$

Alpha-stable convolution kernels may appear as blurring kernels in astronomical imaging and in some other areas where a Gaussian kernel has been commonly assumed as the rough approximation [103, 104], as filters for feature extraction [105], and as smoothing operators [106]. They may also appear as kernels which blur the histogram if the image is corrupted by additive noise with an alpha-stable distribution.

6.12   Combined invariants to blur and spatial transformations

In object recognition applications, the query image is rarely captured in a spatially rectified position. Certain geometric transformation between the query and database images is often present in addition to image blur. To handle these situations, having the *combined invariants* that are invariant simultaneously to convolution and to certain transformations of spatial coordinates is of great demand.

When designing the combined invariants, we face a principal limitation. The set S of the admissible PSFs must be closed also w.r.t. the considered geometric transformations. In other words, the geometric transformation must not violate the PSF properties which were employed for the construction of the respective blur invariants. If this is not the case, the combined invariants cannot exist. Some examples, where the combined invariants cannot exist in principle, are the following: N-FRS blur ($N > 2$) and affine transform because the affine transform does not preserve the N-fold symmetry; dihedral blur and a general rotation (the rotation changes the axis orientation); Gaussian radially symmetric blur and affine transform; Gaussian asymmetric "elongated" blur and rotation, directional blur and rotation, etc. On the other hand, combined invariants may exist for any blur coupled with translation and/or uniform scaling, and in some other cases where the closure property is preserved.

While achieving the translational and scale invariance is easy just by using central coordinates and scale-normalized moments, the construction of combined blur-rotation and blur-affine invariants (if they exist) is not as straightforward.

6.12.1   Invariants to blur and rotation

In 2D, the N-FRS blur invariants based on complex moments can be combined with rotation. The same is true for blur invariants w.r.t. radial Gaussian blur. For the purpose of this section, let us denote all $K_N(p, q)$ (including $N = \infty$) and $K_G(p, q)$ simply as $K(p, q)$. The construction of the combined invariants is very similar to the construction of pure rotation invariants from the complex moments (see Chapter 3).

Let us investigate the behavior of the invariants $K(p, q)$ under an image rotation. Let us denote the invariants of the rotated image as $K'(p, q)$. We can observe that the invariants behave under rotation in the same way as the complex moments themselves, i.e.

$$K'(p,q) = e^{-i(p-q)\theta} \cdot K(p,q), \tag{6.51}$$

where θ is the rotation angle. A formal proof of this statement can be easily accomplished by induction on the order.

Now we can repeat the deliberation from Chapter 3 when we constructed the rotation invariants. Any product of the form

$$\prod_{j=1}^{n} K(p_j, q_j)^{k_j}, \tag{6.52}$$

where $n \geq 1$ and k_j, p_j and $q_j; j = 1, \ldots, n$, are non-negative integers such that

$$\sum_{j=1}^{n} k_j(p_j - q_j) = 0$$

is a combined invariant to convolution and rotation.

To create the basis of the combined invariants up to a certain order, we apply the same approach as in Chapter 3. We choose a low-order non-trivial invariant as a basic element, let us choose $K(1,2)$. Then any non-trivial invariant $K(p,q), p > q$, generates one element of the basis as

$$K(p,q)K(1,2)^{p-q}.$$

The proof of independence and completeness (modulo rotation and convolution) of this set follows from the propositions proved earlier.

As we have already pointed out, the dihedral blur invariants cannot be combined with a general rotation. If we, however, restrict the admissible rotation angles θ to integer multiples of π/N, this constrained rotation along with the dihedral convolution forms a monoid, and we obtain the combined invariants in the same way as in the previous case. For arbitrary N and α we have

$$L_N^{\alpha}(p,q)' = e^{-i(p-q)\theta} \cdot L_N^{\alpha}(p,q) \tag{6.53}$$

and the rotation invariance can be achieved by a proper multiplication of two distinct invariants of the same parameters N and α.

Extension of the convolution-rotation invariants into 3D is rather complicated. It basically uses the methods introduced in Chapter 4 and combines them with 3D centrosymmetric blur invariants proposed earlier in this chapter. We refer to [20] for the detailed derivation.

6.12.2   *Invariants to blur and affine transformation*

After deriving the combined blur-rotation invariants in the previous section, the next logical step is an extension to affine transform (see Figure 6.27 for an example of a combined blur-affine transformation). This is meaningful to try for centrosymmetric PSFs only because the other blur types we have studied are not closed under an affine transform.

The first attempt to find the *combined blur-affine invariants* (CBAI's) was published by Zhang et al. [17]. They did not derive the combined invariants explicitly. They transformed the image into a normalized position (where the normalization constraints had been formulated

Figure 6.27 The original image and its blurred and affinely deformed version (Liberec, Czech Republic). The values of the combined blur-affine invariants are the same for both images

such that they were invariant to blur), and then they calculated the pure blur invariants (6.24) of the normalized image.

Here we present a direct approach consisting of substituting the blur invariants (6.24) for central moments in the definition formulae of the affine moment invariants that were derived in Chapter 5. The following theorem not only guarantees the existence of the CBAIs in two dimensions but also provides an explicit algorithm for how to construct them.

Theorem 6.3 Let $I(\mu_{00}, \dots, \mu_{PQ})$ be an affine moment invariant. Then $I(C(0,0), \dots, C(P,Q))$, where the $C(p,q)$'s are defined by Eq. (6.24), is a combined blur-affine invariant.

For the proof of Theorem 6.3 see Appendix 6.G. Certain care must be taken when applying this theorem because all even-order blur invariants are identically zero. Substituting them into the AMIs may lead to trivial combined invariants. In other words, not all AMIs generate non-trivial combined invariants. More details on the CBAIs and experimental studies of their recognition power can be found in [18]. The use of the CBAIs for aircraft silhouette recognition was reported in [46] and for the classification of winged insects [48]. A slightly different approach to the affine-blur invariant matching was presented in [23], where the combined invariants are constructed in Fourier domain. Most recently, Boldyš [21] proved that the "substitution rule" introduced in Theorem 6.3 holds in an arbitrary number of dimensions.

6.13   Computational issues

As we have seen, blur invariants are mostly recursive functions. Since Matlab allows recursive procedures, they are easy to implement. The computational complexity is determined by the complexity of moment calculation. As soon as the moments have been calculated, the complexity of the recurrence and other overheads are negligible if the order of the invariant is substantially less than the number of pixels (as it usually is in practice).

To make the calculation even more efficient, it is recommended to implement the recurrences as matrix operations with special matrices rather than as loops. In that way, we not only avoid the loops but also calculate all invariants up to a certain order at the same time (see [14] for the construction of special matrices). As in the case of other moment invariants, the numerical stability of blur invariants calculation can be improved by taking orthogonal moments instead of geometric or complex moments.

In case of blur invariants we face another numerical problem which is specific to this type of formulas. Among special matrices by which we multiply the vector of moments there appear matrices which contain the Pascal triangle. It is well known that the condition number of such Pascal-like triangular matrices is very poor, which negatively influences not only the solution of the corresponding linear system (which is not the case here) but also a forward calculation (i.e., the multiplication of a matrix by a vector). This effect was studied thoroughly in [107]. The authors concluded that the influence of the Pascal triangle on the numerical calculations of blur invariants is more serious than the influence of the use of non-orthogonal basis. They proposed a simple technique of modification of the power basis when a constant is added to the even powers, which suppresses the instability at least for moments up to the order around 20. This should be safely sufficient for almost all practical applications.

6.14 Experiments with blur invariants

In this section, we demonstrate the numerical properties of various blur invariants by experiments. We chose several typical scenarios in which the blur invariants can be applied. Many other real experiments and case studies can be found in the references listed in Section 6.1.3.

6.14.1 A simple test of blur invariance property

The aim of this experiment was to verify the blur invariance property under ideal conditions and to investigate the impact of boundary effect. We successively blurred a photograph of a statue (see Figure 6.28) by a simulated circularly symmetric blur of a standard deviation from

         (a)   (b)   (c)

Figure 6.28   The test image of the size 2976×3968 (the statue of Pawlu Boffa in Valletta, Malta): (a) original, (b) the image blurred by circularly symmetric blur of a standard deviation 100 pixels, (c) the same blurred image without margins

1 to 100 pixels. At first, we kept the convolution model perfectly valid, which means that the original was zero-padded and the blurred image was always larger than the input. We calculated the values of five invariants $K_\infty(p, q)$ and plotted them in Figure 6.29. We can see their values are actually almost constant: the mean relative error is only $7 \cdot 10^{-7}$, which is caused by a discrete approximation of the circular symmetry. Note that the invariance has been kept even if the blur is large (see Figure 6.28b for the heaviest blur).

Then we repeated this experiment in a more realistic scenario. We constrained the image support to stay the same under blurring, so we cut-off the "overflow", which violates the convolution model in the strip along the boundary (see Figure 6.28c for the heaviest blur with this restriction). The behavior of the invariants is depicted in Figure 6.30. The invariance is still very good, although the mean relative error now reaches 0.014.

6.14.2   Template matching in satellite images

This experiment was inspired by a remote sensing application area where the *template matching* is a frequent requirement when registering a new image into the reference coordinates.

The template matching problem is usually formulated as follows: having the templates and a digital image of a large scene, one has to find locations of the given templates in the scene. By the *template* we understand a small digital picture usually containing some significant object that was extracted previously from another image of the same scene and now is being stored in a database. In this experiment, the templates are clear while the scene image is blurred and noisy, which is a common situation in remote sensing.

Numerous image-matching techniques have been proposed in the literature (see the survey paper [108] for instance). A common denominator of all these methods is the assumption that the templates as well as the scene image had been already preprocessed and the graylevel/color degradations such as blur, additive noise, etc. and so forth had been removed.

By means of the blur invariants, we try to perform matching without any previous deblurring. We ran this experiment on real satellite images, but the blur and noise had been introduced artificially that gave us a possibility to control their amount and to quantitatively evaluate the results.

We used the invariants $C(p, q)$ w.r.t. centrosymmetric blur, composed of central geometric moments. We introduced an additional normalization to the range to ensure roughly the same range of values regardless of p and q. If we would have enough data for training, we could set up the weighting factors according to inter-class and intra-class variances of the features as proposed by Cash in [109]. However, this is not the case, because each class is represented by one template only. Thus, we used the invariants

$$\frac{C(p, q)}{\mu_{00}^{(p+q+4)/4}}$$

that yielded a satisfactory normalization of the range of values.

The matching algorithm itself was straightforward. We searched the blurred image $g(x, y)$, and for each possible position of the template we calculated the Euclidean distance in the space of the blur invariants between the template and the corresponding window in $g(x, y)$. The matching position of the template is determined by the minimum distance. We used neither a fast hierarchical implementation nor optimization algorithms; we performed a "brute-force"

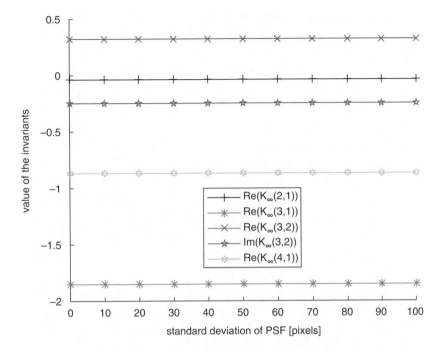

Figure 6.29   The values of the invariants, the exact convolution model

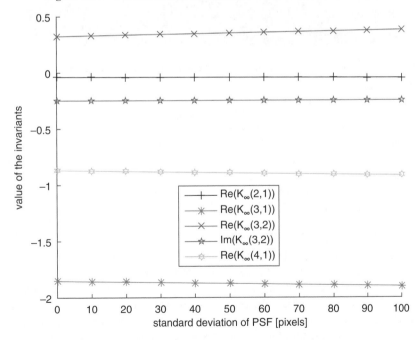

Figure 6.30   The values of the invariants violated by the boundary effect, when the realistic convolution model was applied

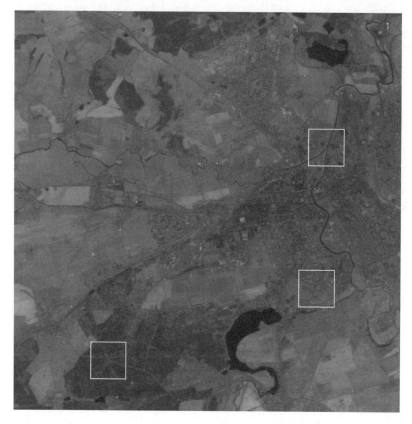

Figure 6.31   The original high-resolution satellite photograph, the City of Plzeň, Czech Republic, with three selected templates

full search. The only user-defined parameter in the algorithm is the maximum order r of the invariants used. In this experiment we used $r = 7$, that means we applied eighteen invariants from the third to the seventh order. In real experiments, if we had more data for training, we could apply more sophisticated feature-selection algorithms to ensure maximum separability between the templates.

Three templates of the size 48×48 were extracted from the 512×512 high-resolution SPOT image (City of Plzeň, Czech Republic, see Figure 6.31). The templates contain significant local landmarks – the confluence of two rivers, the apartment block, and the road crossing, see Figure 6.32.

To simulate an acquisition by another sensor with a lower spatial resolution, the image was blurred by a bell-shaped masks of the size from 3×3 to 21×21 and corrupted by Gaussian white noise of the standard deviation ranging from 0 to 40 each. In all blurred and noisy frames, the algorithm attempted to localize the templates. As one can expect, the success rate of the matching depends on the amount of blur and on the noise level. In Figure 6.33, we can see an example of a frame in which all three templates were localized correctly. Note that a visual localization would be difficult. The results are summarized in Figure 6.35. The noise standard deviation is plotted on the horizontal axis, the size of the blurring mask on the vertical axis. The

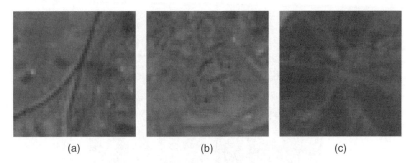

(a)   (b)   (c)

Figure 6.32   The templates: The confluence of the rivers Mže and Radbuza (a), the apartment block (b), and the road crossing (c)

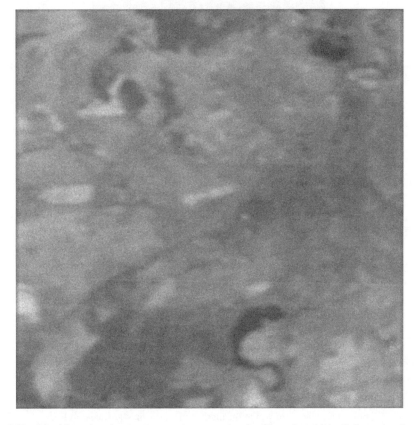

Figure 6.33   The blurred and noisy image. An example of a frame in which all three templates were localized successfully

ratio w between the size of the blurring mask and the size of the template is an important factor. Due to the blurring, the pixels lying near the boundary inside the template are affected by those pixels lying outside the template. The higher w is, the larger part of the template is involved in this *boundary effect* (see Figure 6.34 for an illustration of this phenomenon and Figure 6.36

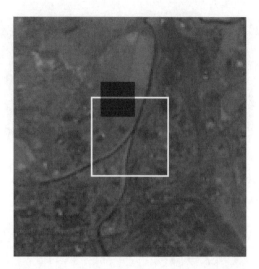

Figure 6.34 The boundary effect. Under a discrete convolution on a bounded support, the pixels near the template boundary (white square) are affected by the pixels lying outside the template. An impact of this effect depends on the size of the blurring mask (black square)

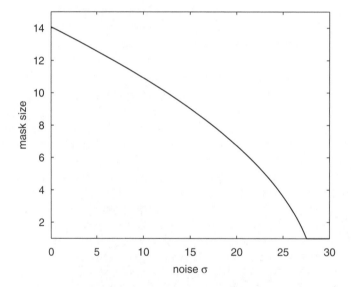

Figure 6.35 The graph summarizing the results of the experiment. The area below the curve denotes the domain in a "noise-blur space" in which the algorithm works mostly successfully

for its quantitative measurement). Since the invariants are calculated from a bounded area of the image where the blurring is not exactly a convolution, they are no longer invariant, which might lead to mismatch, especially when w is higher than 0.15. The curve in Figure 6.35 is a least-square fit of the results by a parabola, the area below the curve corresponds to the domain in which the algorithm performs mostly successfully. We can clearly see the robustness as well as the limitations of the matching by invariants.

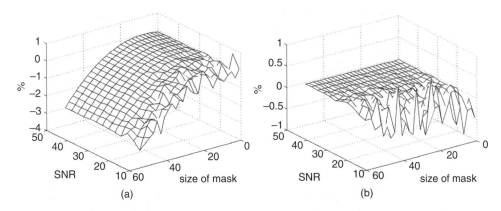

Figure 6.36 The influence of the boundary effect and noise on the numerical properties of the centrosymmetric convolution invariant $C(5,0)$. Horizontal axes: the blurring mask size and the signal-to-noise ratio, respectively. Vertical axis: the relative error in %. The invariant corrupted by a boundary effect (a) and the same invariant calculated from a zero-padded template where no boundary effect appeared (b)

6.14.3 Template matching in outdoor images

We carried out a similar template-matching experiment on outdoor images blurred by real out-of-focus blur. Since we worked with real photographs, no "perfect" clear scene was available. We took several images of the same scene (the frontal view of a house, the image size 920×1404 pixels) and deblurred them by a multichannel blind deconvolution method from [110]. This deblurred image (see Figure 6.37a) served for the template selection. The localization of the templates was done in one of the blurred frames (see Figure 6.37b). Since the out-of-focus PSF was believed to be circularly symmetric, we used $K_\infty(p,q)$ invariants up to the sixth order.

We randomly selected 200 circular templates with a radius of 100 pixels in the deblurred image. The localization error of the matching algorithm was less or equal than 5 pixels in 189 cases. In 3 cases where the matching error was greater than 100 pixels, the mismatch was caused by a periodic repetition of some structures in the image. The overall matching accuracy is probably even better than the reported one because the images may have not been perfectly registered and the error up to three pixels may actually refer to a global misregistration error.

6.14.4 Template matching in astronomical images

We performed another template matching experiment in astronomical images degraded by real atmospheric turbulence blur. We employed four images of the spot in the solar photosphere taken by a telescope with a CCD camera in a visible spectral band (the venue: Observatory Ondrejov, Czech Republic; wavelength: $\lambda = 590$ nm). Since the time interval between the two consecutive acquisitions was only few seconds, the scene can be considered still, and the images are almost perfectly registered. As the atmospheric conditions changed between the acquisitions, the amount of blur in individual images varies from one another.

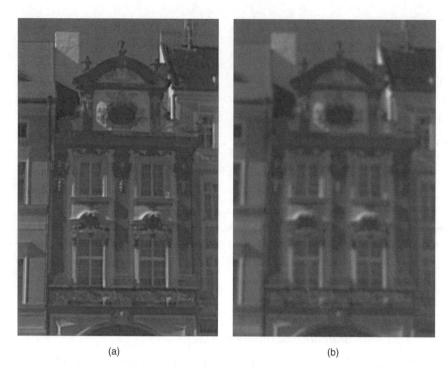

(a)   (b)

Figure 6.37   The house image: (a) the deblurred version, (b) the original blurred version

We sorted the images according to their blur level by means of the algorithm which compares the energy in low-pass and high-pass wavelet transform bands [111]. The ordered sequence can be seen and visually checked in Figure 6.38. The size of each image is 256×256 pixels. The first image is relatively sharp while the other three images, particularly the last one, are noticeably blurred. The blur kernel is believed to be approximately Gaussian (an experimental validation of this assumption can be found for instance in [112]). Mild additive noise is also present in all images, its estimated SNR is about 30 dB.

We matched 30 randomly chosen 32×32 templates extracted from the first "clear" image against each of the other three images. We used Gaussian blur invariants $K_G(p, q)$ up to the sixth order as the features, and, for a comparison, we repeated the matching also by means of a standard cross-correlation. The results are summarized in Table 6.1. While the blur invariants yield very good localization accuracy, the cross-correlation is not robust to blur, and the matching is far less accurate.

6.14.5   Face recognition on blurred and noisy photographs

Since several papers on blur invariants have mentioned face recognition as an important application area of blur-robust techniques (see [80] for instance), we include a face recognition experiment as well although the blur invariants described in this chapter have not been intended as a specialized facial descriptors and do not use specific facial features.

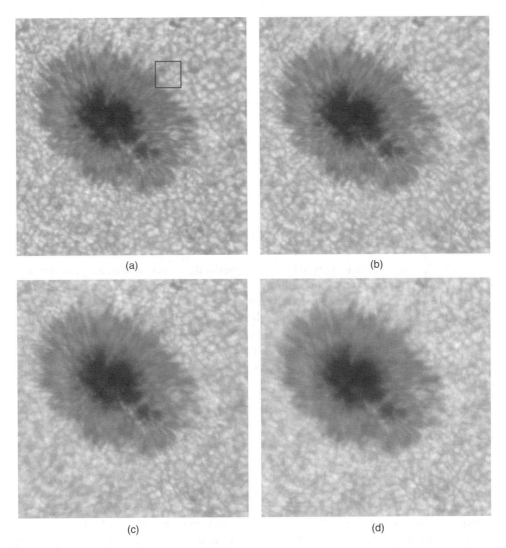

Figure 6.38 Four images of the sunspot blurred by a real atmospheric turbulence blur of various extent. The images are ordered from the less to the most blurred one. A template is depicted in the first image to illustrate its size

We used the publicly available database CASIA heterogeneous face biometrics (HFB) [113, 114]. We deliberately used a subset of twenty face images of twenty persons which are very similar to one another and even difficult to distinguish visually. In such a case a mild blur and noise might result in a number of misclassifications.

The test images are all frontal views with a neutral expression, without a head pose, and with the face being cropped (see Figure 6.39). We successively took each image of the dataset, blurred it by a Gaussian blur, with standard deviation from one to five, added a noise of SNR from 60 dB to 0 dB, and classified the image by the minimum-distance rule against the database

Table 6.1 Template matching in astronomical images

Image	Mean error	Standard deviation
	Cross-correlation	
(b)	7.30	3.82
(c)	7.29	3.81
(d)	7.28	3.80
	Gaussian blur invariants	
(b)	0.88	0.45
(c)	0.90	0.47
(d)	0.85	0.44

of the "clear" images (see Figure 6.40 for the samples of the most degraded images). For each blur and noise level we generated ten instances of the query image. Hence, we classified 7,200 images altogether and measured the success rate.

When using Gaussian blur invariants, we achieved 100% recognition rate on each blur/noise level. For a comparison, we repeated the classification also by invariant distance proposed for Gaussian-blurred images by Zhang et al. [72]. That method is apparently less robust to noise. It yields 100% recognition rate only for low noise or low blur. As both increases, the performance of the Zhang's method drops (the performance drop starts at STD = 3 and SNR = 10 dB, reaching the minimum success rate of about 50% at STD = 5 and SNR = 0 dB).

6.14.6   Traffic sign recognition

The last experiment of this chapter is aimed to demonstrate that special care is required if not only the blur but also the objects themselves are symmetric. If the fold number M of the object equals the fold number N of the PSF (or its integer multiple), then $I_N(f) = 1$ (the invariant actually treats the object as a part of the PSF), and such objects can be recognized neither by I_N nor by $K_N(p, q)$. If M is not a multiple of N, then the invariant I_N should provide enough discrimination power, but some $K_N(p, q)$ still vanish for certain p and q. The problem of symmetric objects might seem to be a marginal one, but the opposite is the case. Most of artificial objects as well as many natural objects, which are often the subjects of classification, exhibit certain types of symmetry (very often an N-FRS or N-FDS).

In this experiment, we used a collection of sixteen traffic signs from various countries (see Figure 6.41) intentionally chosen in such a way that they include seven signs of 2-FRS, 4 signs of 3-FRS, 4 signs of 4-FRS, and one sign having ∞-FRS. Each sign, the original size of which is 300×300 pixels, was available in 10 versions, starting from the non-blurred original up to a heavy blur of the radius 100 pixels (see Figure 6.42 for some examples). In all cases, a circular constant PSF, which simulated an out-of-focus blur on a circular aperture, was used. We calculated various invariants and plotted them pair wise to illustrate their discrimination power. First we plotted the feature space of the moments c_{11} and c_{22} (see Figure 6.43). These moments are not invariant to blurring and that is why the clusters of the individual traffic signs are not separable (with one exception). This graph demonstrates the need for blur invariants in general.

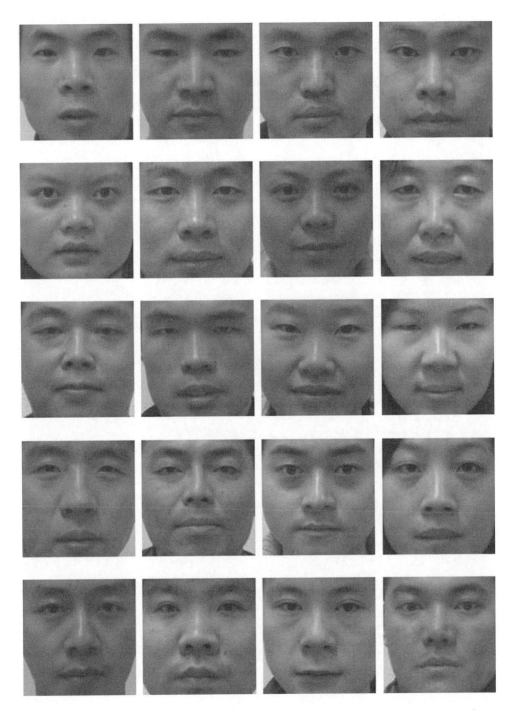

Figure 6.39   Sample "clear" images of the the CASIA HFB database. The database consists of very similar faces

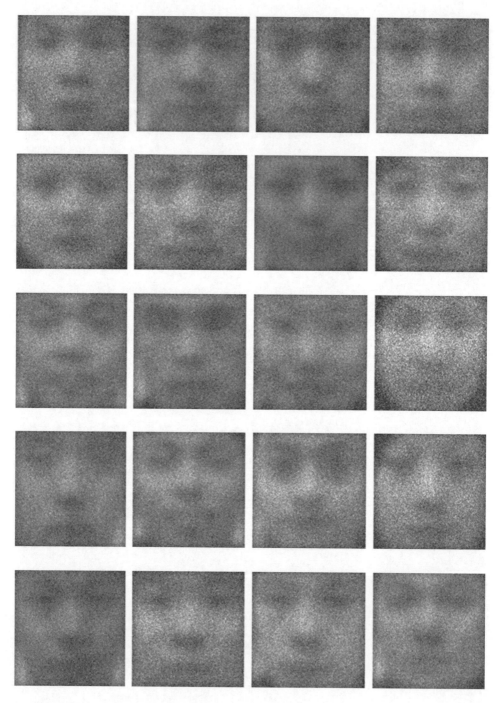

Figure 6.40   Sample test images degraded by heavy blur and noise ($\sigma = 5$ and SNR $= 0$ dB)

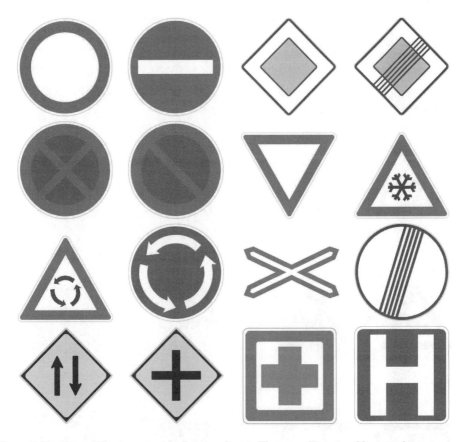

Figure 6.41 The traffic signs used in the experiment. First row: No entry, No entry into a one-way road, Main road, End of the main road. Second row: No stopping, No parking, Give way, Be careful in winter. Third row: Roundabout ahead, Roundabout, Railway crossing, End of all prohibitions. Fourth row: Two-way traffic, Intersection, First aid, Hospital

In Figure 6.44, one can see the feature space of real and imaginary parts of the invariant $K_2(3, 0)$. Since the PSF is circular, it satisfies the condition of N-FRS for any N. Thus, $K_2(3, 0)$ should be an invariant. However, the traffic signs with even-fold symmetry lie in its nullspace, so we cannot expect their separability. This is exactly what happened – the four signs having 3-FRS ("Give way", "Be careful in winter", "Roundabout ahead", and "Roundabout") create perfectly separate clusters, while all the others have $K_2(3, 0) = 0$. (Note that here as well as in most other graphs the clusters have such small variance that they appear to be a single point, but they actually contain ten samples.)

Then we investigated the "mixed-order" feature space of the invariants $K_4(2, 0)$ and $K_4(3, 0)$ (note that $K_4(2, 0) = c_{20}$ and $K_4(3, 0) = c_{30}$). The objects with a 3-FRS should have $K_4(2, 0) = 0$ while their $K_4(3, 0)$ should be different. The traffic signs with a two-fold symmetry should be distinguishable by $K_4(2, 0)$ while the signs with four and infinity-fold symmetry should all fall into zero. The experiment confirmed all these expectations. In Figure 6.45 the coarse scale was used, but the signs "Roundabout", "Railway crossing", and "End of all prohibitions" are out of the graph anyway. To show the neighborhood of zero, we plotted this graph again in

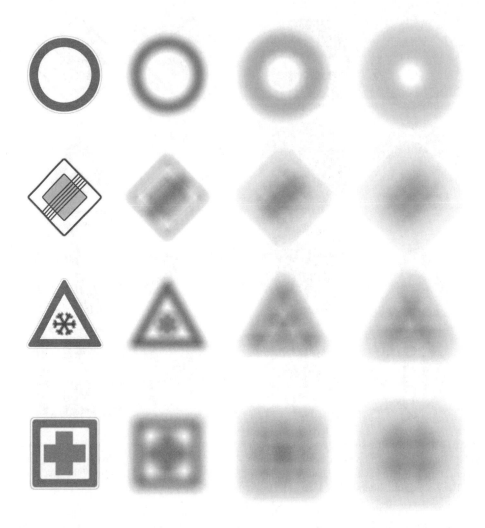

Figure 6.42   The traffic signs successively blurred by masks with radii 0, 33, 66, and 100 pixels

a finer scale; see Figure 6.46. You can see five non-separable clusters overlapping each other (the intra-cluster variances of order 10^{-15} are caused by sampling and round-off errors).

Finally, we show the situation in the feature space of the real parts of the invariants $K_\infty(3, 0) = c_{30}$ and $K_\infty(4, 0) = c_{40}$; see Figures 6.47–6.49. Theoretically, these invariants should separate all the traffic signs, which is actually true here. To see the separability clearly on the graphs, we used three scales – the coarse in Figure 6.47, the medium in Figure 6.48, and the fine scale in Figure 6.49.

This experiment clearly illustrates that the choice of proper parameters N, p, q is essential, particularly if there are symmetric objects in the database. If, for instance, only the invariants to centrosymmetric blur were applied, fourteen traffic signs out of sixteen would fall into one mixed cluster.

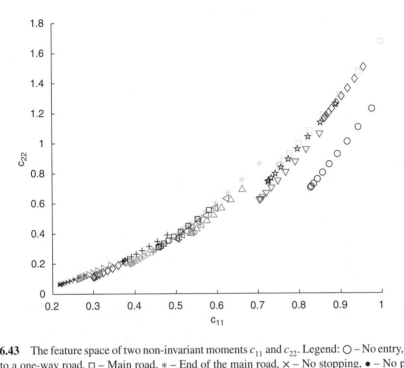

Figure 6.43   The feature space of two non-invariant moments c_{11} and c_{22}. Legend: ○ – No entry, ◇ – No entry into a one-way road, □ – Main road, ∗ – End of the main road, × – No stopping, ● – No parking, ▽ – Give way, ◁ – Be careful in winter, △ – Roundabout ahead, ▷ – Roundabout, ◇ – Railway crossing, ☆ – End of all prohibitions, □ – Two-way traffic, + – Intersection, + – First aid, ✿ – Hospital

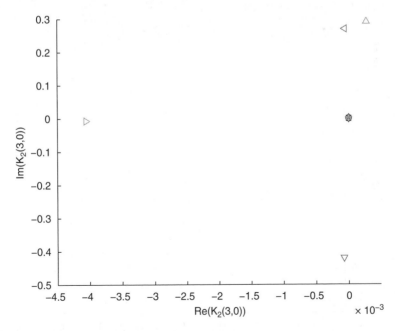

Figure 6.44   The feature space of real and imaginary parts of the invariant $K_2(3,0)$. Legend: ○ – No entry, ◇ – No entry into a one-way road, □ – Main road, ∗ – End of the main road, × – No stopping, ● – No parking, ▽ – Give way, ◁ – Be careful in winter, △ – Roundabout ahead, ▷ – Roundabout, ◇ – Railway crossing, ☆ – End of all prohibitions, □ – Two-way traffic, + – Intersection, + – First aid, ✿ – Hospital

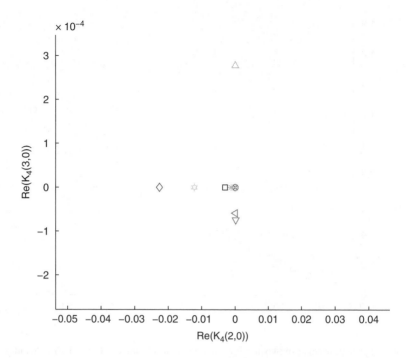

Figure 6.45 The feature space of real parts of the invariant $\mathcal{R}eK_4(2,0)$ and $\mathcal{R}eK_4(3,0)$. Legend: $\bigcirc$ – No entry, $\diamond$ – No entry into a one way road, $\square$ – Main road, $*$ – End of the main road, $\times$ – No stopping, $\bullet$ – No parking, $\triangledown$ – Give way, $\triangleleft$ – Be careful in winter, $\triangle$ – Roundabout ahead, $\square$ – Two-way traffic, $+$ – Intersection, $+$ – First aid, $\maltese$ – Hospital

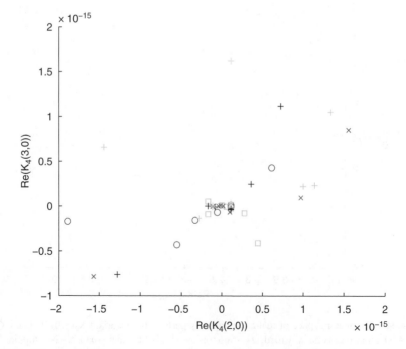

Figure 6.46 The detail of the feature space of real parts of the invariant $\mathcal{R}eK_4(2,0)$ and $\mathcal{R}eK_4(3,0)$ around zero. Legend: $\bigcirc$ – No entry, $\square$ – Main road, $\times$ – No stopping, $+$ – Intersection, $+$ – First aid

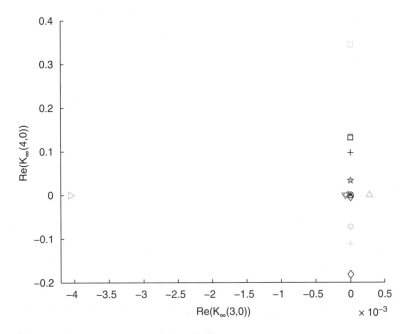

Figure 6.47 The feature space of real parts of the invariant $\mathcal{R}eK_\infty(3,0)$ and $\mathcal{R}eK_\infty(4,0)$ – the minimum zoom. Legend: ○ – No entry, ◇ – No entry into a one way road, □ – Main road, ∗ – End of the main road, × – No stopping, • – No parking, ▽ – Give way, ◁ – Be careful in winter, △ – Roundabout ahead, ▷ – Roundabout, ◇ – Railway crossing, ☆ – End of all prohibitions, ⬜ – Two-way traffic, + – Intersection, + – First aid, ✿ – Hospital

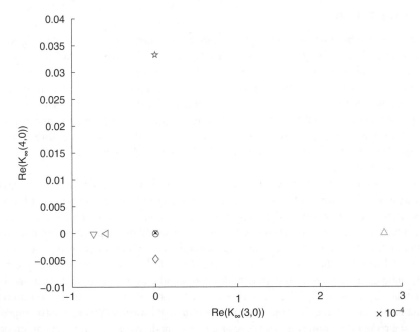

Figure 6.48 The feature space of real parts of the invariant $\mathcal{R}eK_\infty(3,0)$ and $\mathcal{R}eK_\infty(4,0)$ – the medium zoom. Legend: ○ – No entry, ◇ – No entry into a one way road, × – No stopping, • – No parking, ▽ – Give way, ◁ – Be careful in winter, △ – Roundabout ahead, ☆ – End of all prohibitions

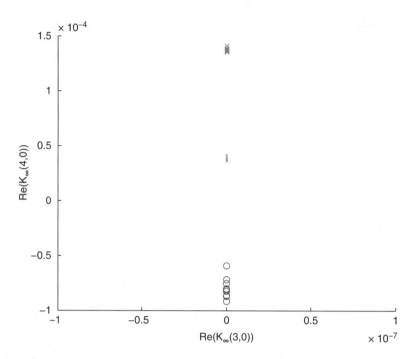

Figure 6.49 The feature space of real parts of the invariant $\mathcal{R}eK_\infty(3,0)$ and $\mathcal{R}eK_\infty(4,0)$ – the maximum zoom. Legend: ○ – No entry, × – No stopping, • – No parking

6.15 Conclusion

In this chapter, we have presented a class of moment invariants w.r.t. image blurring, which is mathematically described by convolution with the system PSF. We introduced a new unified theory of blur invariants in Fourier domain by means of projection operators. We showed that the key results follow from the properties of a projector onto a closed subspace w.r.t. convolution and that they can be achieved in a general form without specifying a particular PSF type. We demonstrated how to expand these invariants into moments, and we derived specific formulas for centrosymmetric, N-fold, dihedral, Gaussian, and directional blur invariants. The chapter is crowned by describing combined invariants to convolution and rotation/affine transform.

Although various versions of the blur invariants have been successfully used in practical tasks, one should be aware of two intrinsic limitations, in addition to the principal limitations originating from the globality of the moments, which is common to all moment invariants.

First, the blur invariants are sensitive to model errors. If the degradation is not an exact convolution, the invariance property may be violated. A boundary effect is a typical example of the model error. If the field of view is limited (as it always is), the boundary effect inevitably appears. If the size of the blurring filter is small comparing to the size of the image, the boundary effect is negligible, otherwise it may become substantial. It can be removed by isolating and zero-padding the object before the blurring; however, this is mostly impossible. Second, the recognitive power is always restricted to "modulo convolution". As an immediate consequence, we are not able to distinguish objects having the same type of symmetry as the assumed filter.

Due to the nature of the blur, using blur invariants is meaningful only if we work with original graylevel/color data. Unlike the geometric invariants introduced earlier, the application of blur invariants on binary images is useless.

Appendix 6.A

Proof of Equation (6.22). Although Eq. (6.22) was formulated in 1D, it has analogues in arbitrary dimensions. For convenience, we prove it here in 2D version because it is more relevant to imaging. For the proof in any other dimension, it is sufficient just to set a proper number of arguments.

Expanding Fourier transform of $f(x, y)$ into a power series

$$F(u, v) = \sum_{k=0}^{\infty} \sum_{j=0}^{\infty} \frac{(-2\pi i)^{k+j}}{k! \cdot j!} m_{kj}^{(f)} \cdot u^k v^j$$

implies that

$$\mathcal{Re}(F(u, v)) = \sum_{\substack{k=0 \ j=0 \\ (k+j) \text{ even}}}^{\infty} \sum_{j=0}^{\infty} \frac{(-2\pi i)^{k+j}}{k! \cdot j!} m_{kj} \cdot u^k v^j$$

$$= \sum_{\substack{k=0 \ j=0 \\ (k+j) \text{ even}}}^{\infty} \sum_{j=0}^{\infty} \frac{(-1)^{(k+j)/2}(-2\pi)^{k+j}}{k! \cdot j!} m_{kj} \cdot u^k v^j$$

and

$$i \cdot \mathcal{Im}(F(u, v)) = \sum_{\substack{k=0 \ j=0 \\ (k+j) \text{ odd}}}^{\infty} \sum_{j=0}^{\infty} \frac{(-2\pi i)^{k+j}}{k! \cdot j!} m_{kj} \cdot u^k v^j$$

$$= i \cdot \sum_{\substack{k=0 \ j=0 \\ (k+j) \text{ odd}}}^{\infty} \sum_{j=0}^{\infty} \frac{(-1)^{(k+j-1)/2}(-2\pi)^{k+j}}{k! \cdot j!} m_{kj} \cdot u^k v^j.$$

Thus, $\tan(\text{ph } F(u, v))$ is a ratio of two absolutely convergent power series, and therefore it can be also expressed as a power series

$$\tan(\text{ph } F(u, v)) = \frac{\mathcal{Im}(F(u, v))}{\mathcal{Re}(F(u, v))} = \sum_{k=0}^{\infty} \sum_{j=0}^{\infty} a_{kj} u^k v^j,$$

which must fulfill the condition

$$\sum_{\substack{k=0 \ j=0 \\ (k+j) \text{ odd}}}^{\infty} \sum_{j=0}^{\infty} \frac{(-1)^{(k+j-1)/2}(-2\pi)^{k+j}}{k! \cdot j!} m_{kj} \cdot u^k v^j$$

$$= \sum_{\substack{k=0 \ j=0 \\ (k+j) \text{ even}}}^{\infty} \sum_{j=0}^{\infty} \frac{(-1)^{(k+j)/2}(-2\pi)^{k+j}}{k! \cdot j!} m_{kj} \cdot u^k v^j \cdot \sum_{k=0}^{\infty} \sum_{j=0}^{\infty} a_{kj} u^k v^j.$$

$$(6.54)$$

Let's prove by induction that a_{kj} has the form (6.22) if $k+j$ is odd.

- $k+j=1$

  It follows from (6.54) that

  $$a_{10} = \frac{-2\pi m_{10}}{m_{00}}.$$

  On the other hand,

  $$\frac{(-2\pi)(-1)^0}{1! \cdot 0! \cdot \mu_{00}} S(1,0) = \frac{-2\pi m_{10}}{m_{00}} = a_{10}.$$

- Let's suppose the assertion has been proven for all k and j, $k+j \le r$, where r is an odd integer and let $p+q = r+2$. It follows from (6.54) that

  $$\frac{(-1)^{(p+q-1)/2}(-2\pi)^{p+q}}{p! \cdot q!} m_{pq} = m_{00}a_{pq} + \sum_{n=0}^{p}\sum_{\substack{m=0 \\ 0<n+m}}^{q} \frac{(-1)^{(n+m)/2}(-2\pi)^{n+m}}{n! \cdot m!} m_{nm}a_{p-n,q-m}.$$

  Introducing (6.22) into the right-hand side we get

  $$\frac{(-1)^{(p+q-1)/2}(-2\pi)^{p+q}}{p! \cdot q!} m_{pq} = m_{00}a_{pq} + \sum_{n=0}^{p}\sum_{\substack{m=0 \\ 0<n+m<p+q}}^{q} \frac{(-1)^{(p+q-1)/2}(-2\pi)^{p+q}}{n! \cdot m! \cdot (p-n)! \cdot (q-m)! \cdot m_{00}}$$
  $$\cdot S(p-n,q-m) \cdot m_{nm}$$

  and, consequently,

  $$a_{pq} = \frac{(-1)^{(p+q-1)/2}(-2\pi)^{p+q}}{p! \cdot q! \cdot m_{00}} \cdot (m_{pq} - \frac{1}{m_{00}}\sum_{n=0}^{p}\sum_{\substack{m=0 \\ 0<n+m<p+q}}^{q} \binom{p}{n}\binom{q}{m} S(p-n,q-m) \cdot m_{nm}).$$

  Finally, it follows from (6.24) that

  $$a_{pq} = \frac{(-1)^{(p+q-1)/2}(-2\pi)^{p+q}}{p! \cdot q! \cdot m_{00}} S(p,q). \qquad \square$$

Appendix 6.B

Proof of the moment transformation under convolution. Let us prove that the moments are transformed under convolution as stated in Eq. (6.5) and (6.23). This relation holds for geometric and complex moments, basic and centralized ones, in arbitrary dimension $d \ge 1$. Here we prove a 2D version with central geometric moments

$$\mu_{pq}^{(g)} = \sum_{k=0}^{p}\sum_{j=0}^{q} \binom{p}{k}\binom{q}{j} \mu_{kj}^{(f)} \mu_{p-k,q-j}^{(h)}$$

because it is the most relevant with respect to the blur invariants. The proof of any other modification of Eq. (6.5) is the same.

First of all, let us prove that the centroid of $g(x, y)$ is a sum of the centroids of $f(x, y)$ and $h(x, y)$:

$$(x_c^{(g)}, y_c^{(g)}) = (x_c^{(f)} + x_c^{(h)}, y_c^{(f)} + y_c^{(h)}).$$

Using the definition of a centroid we obtain

$$x_c^{(g)} = \frac{m_{10}^{(g)}}{m_{00}^{(g)}} = \frac{m_{10}^{(f)} m_{00}^{(h)} + m_{00}^{(f)} m_{10}^{(h)}}{m_{00}^{(f)} m_{00}^{(h)}} = \frac{m_{10}^{(f)}}{m_{00}^{(f)}} + \frac{m_{10}^{(h)}}{m_{00}^{(h)}} = x_c^{(f)} + x_c^{(h)}.$$

The proof for $y_c^{(g)}$ is similar.

Now we can prove the main proposition.

$$\mu_{pq}^{(g)} = \int_{-\infty}^{\infty}\int_{-\infty}^{\infty} (x - x_c^{(g)})^p (y - y_c^{(g)})^q g(x, y) dx dy$$

$$= \int_{-\infty}^{\infty}\int_{-\infty}^{\infty} (x - x_c^{(f)} - x_c^{(h)})^p (y - y_c^{(f)} - y_c^{(h)})^q (f * h)(x, y) dx dy$$

$$= \int_{-\infty}^{\infty}\int_{-\infty}^{\infty} (x - x_c^{(f)} - x_c^{(h)})^p (y - y_c^{(f)} - y_c^{(h)})^q$$

$$\left(\int_{-\infty}^{\infty}\int_{-\infty}^{\infty} h(a, b) f(x - a, y - b) da db\right) dx dy$$

$$= \int_{-\infty}^{\infty}\int_{-\infty}^{\infty} h(a, b) \left(\int_{-\infty}^{\infty}\int_{-\infty}^{\infty} (x - x_c^{(f)} - x_c^{(h)})^p (y - y_c^{(f)} - y_c^{(h)})^q\right.$$

$$\left. f(x - a, y - b) dx dy \right) da db$$

$$= \int_{-\infty}^{\infty}\int_{-\infty}^{\infty} h(a, b) \left(\int_{-\infty}^{\infty}\int_{-\infty}^{\infty} (x - x_c^{(f)} + a - x_c^{(h)})^p (y - y_c^{(f)} + b - y_c^{(h)})^q\right.$$

$$\left. f(x, y) dx dy \right) da db$$

$$= \int_{-\infty}^{\infty}\int_{-\infty}^{\infty} h(a, b) \left(\sum_{k=0}^{p}\sum_{j=0}^{q} \binom{p}{k}\binom{q}{j} (a - x_c^{(h)})^{p-k} (b - y_c^{(h)})^{q-j} \mu_{kj}^{(f)}\right) da db$$

$$= \sum_{k=0}^{p}\sum_{j=0}^{q} \binom{p}{k}\binom{q}{j} \mu_{kj}^{(f)} \mu_{p-k,q-j}^{(h)}.$$

$\square$

Appendix 6.C

A direct proof of Eq. (6.24). The invariance property of functionals $C(p, q)$ defined in Eq. (6.24) follows from the theory of projection operators. It can, however, be also proved directly without any previous theory. This proof was first published in [6] and is relatively laborious. We present it here namely for historical reasons. It can be modified for invariants $K_N(p, q)$, $K_G(p, q)$, $B(p, q)$, $L_N^\alpha(p, q)$, $M(p, q)$, and $M^s(p, q)$.

The invariance is trivial for any even r. Let us prove the statement for odd r by induction on r.

- $r = 1$

$$C(0, 1)^{(g)} = C(0, 1)^{(f)} = 0,$$

$$C(1, 0)^{(g)} = C(1, 0)^{(f)} = 0$$

regardless of f and h.

- $r = 3$

There are four invariants of the 3rd order:

$$C(1, 2) = \mu_{12},$$

$$C(2, 1) = \mu_{21},$$

$$C(0, 3) = \mu_{03},$$

$$C(3, 0) = \mu_{30}.$$

Let us prove the proposition for $C(1, 2)$; the proofs for the other three invariants are similar.

$$C(1, 2)^{(g)} = \mu_{12}^{(g)} = \sum_{k=0}^{1} \sum_{j=0}^{2} \binom{1}{k} \binom{2}{j} \mu_{kj}^{(h)} \mu_{1-k,2-j}^{(f)} \mu_{12}^{(f)} = C(1, 2)^{(f)}.$$

- Let us assume the proposition is valid for all invariants of order $1, 3, \ldots, r - 2$. Then using a 2D centralized version of eq. (6.24) we obtain

$$C(p, q)^{(g)} = \mu_{pq}^{(g)} - \frac{1}{\mu_{00}^{(g)}} \sum_{\substack{n=0 \\ 0<n+m<p+q}}^{p} \sum_{m=0}^{q} \binom{p}{n} \binom{q}{m} \cdot C(p - n, q - m)^{(g)} \cdot \mu_{nm}^{(g)}$$

$$= \sum_{k=0}^{p} \sum_{j=0}^{q} \binom{p}{k} \binom{q}{j} \mu_{kj}^{(h)} \mu_{p-k,q-j}^{(f)} - \frac{1}{\mu_{00}^{(f)}} \sum_{\substack{n=0 \\ 0<n+m<p+q}}^{p} \sum_{m=0}^{q} \binom{p}{n} \binom{q}{m}$$

$$\cdot C(p - n, q - m)^{(f)} \sum_{k=0}^{n} \sum_{j=0}^{m} \binom{n}{k} \binom{m}{j} \mu_{kj}^{(h)} \mu_{n-k,m-j}^{(f)}.$$

Using the identity

$$\binom{a}{b} \binom{b}{c} = \binom{a}{c} \binom{a - c}{b - c},$$

changing the order of the summation and shifting the indices we obtain

$$C(p,q)^{(g)} = C(p,q)^{(f)} + \sum_{\substack{k=0 \\ 0<k+j}}^{p} \sum_{j=0}^{q} \binom{p}{k}\binom{q}{j} \mu_{kj}^{(h)} \mu_{p-k,q-j}^{(f)}$$

$$- \frac{1}{\mu_{00}^{(f)}} \sum_{\substack{n=0 \\ 0<n+m<p+q}}^{p} \sum_{m=0}^{q} \sum_{\substack{k=0 \\ 0<k+j}}^{n} \sum_{j=0}^{m} \binom{p}{n}\binom{q}{m}\binom{n}{k}\binom{m}{j}$$

$$\cdot C(p-n,q-m)^{(f)} \mu_{kj}^{(h)} \mu_{n-k,m-j}^{(f)}$$

$$= C(p,q)^{(f)} + \sum_{\substack{k=0 \\ 0<k+j}}^{p} \sum_{j=0}^{q} \binom{p}{k}\binom{q}{j} \mu_{kj}^{(h)} \mu_{p-k,q-j}^{(f)}$$

$$- \frac{1}{\mu_{00}^{(f)}} \sum_{\substack{n=0 \\ 0<n+m<p+q}}^{p} \sum_{m=0}^{q} \sum_{\substack{k=0 \\ 0<k+j}}^{n} \sum_{j=0}^{m} \binom{p}{k}\binom{q}{j}\binom{p-k}{n-k}\binom{q-j}{m-j}$$

$$\cdot C(p-n,q-m)^{(f)} \mu_{kj}^{(h)} \mu_{n-k,m-j}^{(f)}$$

$$= C(p,q)^{(f)} + \sum_{\substack{k=0 \\ 0<k+j}}^{p} \sum_{j=0}^{q} \binom{p}{k}\binom{q}{j} \mu_{kj}^{(h)} \mu_{p-k,q-j}^{(f)}$$

$$- \frac{1}{\mu_{00}^{(f)}} \sum_{\substack{k=0 \\ 0<k+j}}^{p} \sum_{j=0}^{q} \sum_{\substack{n=k \\ n+m<p+q}}^{p} \sum_{m=j}^{q} \binom{p}{k}\binom{q}{j}\binom{p-k}{n-k}\binom{q-j}{m-j}$$

$$\cdot C(p-n,q-m)^{(f)} \mu_{kj}^{(h)} \mu_{n-k,m-j}^{(f)}$$

$$= C(p,q)^{(f)} + \sum_{\substack{k=0 \\ 0<k+j}}^{p} \sum_{j=0}^{q} \binom{p}{k}\binom{q}{j} \mu_{kj}^{(h)}$$

$$\cdot (\mu_{p-k,q-j}^{(f)} - \frac{1}{\mu_{00}^{(f)}} \sum_{\substack{n=k \\ n+m<p+q}}^{p} \sum_{m=j}^{q} \binom{p-k}{n-k}\binom{q-j}{m-j} \cdot C(p-n,q-m)^{(f)} \mu_{n-k,m-j}^{(f)})$$

$$= C(p,q)^{(f)} + \sum_{\substack{k=0 \\ 0<k+j}}^{p} \sum_{j=0}^{q} \binom{p}{k}\binom{q}{j} \mu_{kj}^{(h)}$$

$$\cdot (\mu_{p-k,q-j}^{(f)} - \frac{1}{\mu_{00}^{(f)}} \sum_{\substack{n=0 \\ n+m<p+q-k-j}}^{p-k} \sum_{m=0}^{q-j} \binom{p-k}{n}\binom{q-j}{m}$$

$$\cdot C(p-n-k,q-m-j)^{(f)} \mu_{nm}^{(f)}).$$

Using a shorter notation we can rewrite the last equation in the form

$$C(p,q)^{(g)} = C(p,q)^{(f)} + \sum_{\substack{k=0 \\ 0<k+j}}^{p} \sum_{j=0}^{q} \binom{p}{k} \binom{q}{j} \mu_{kj}^{(h)} \cdot D_{kj}, \qquad (6.55)$$

where

$$D_{kj} = \mu_{p-k,q-j}^{(f)} - \frac{1}{\mu_{00}^{(f)}} \sum_{\substack{n=0 \\ n+m<p+q-k-j}}^{p-k} \sum_{m=0}^{q-j} \binom{p-k}{n} \binom{q-j}{m}$$
$$\cdot C(p-n-k,q-m-j)^{(f)} \mu_{nm}^{(f)}.$$

If $k+j$ is odd then, due to the centrosymmetry, $\mu_{kj}^{(h)} = 0$. If $k+j$ is even then it follows from Eq. (6.24) that

$$C(p-k,q-j) = \mu_{p-k,q-j} - \frac{1}{\mu_{00}} \sum_{\substack{n=0 \\ 0<n+m<p+q-k-j}}^{p-k} \sum_{m=0}^{q-j} \binom{p-k}{n} \binom{q-j}{m}$$
$$\cdot C(p-k-n,q-j-m) \cdot \mu_{nm}.$$

Consequently,

$$D_{kj} = C(p-k,q-j)^{(f)} - \frac{1}{\mu_{00}^{(f)}} C(p-k,q-j)^{(f)} \mu_{00}^{(f)} = 0.$$

Thus, Eq. (6.55) implies that $C(p,q)^{(g)} = C(p,q)^{(f)}$ for every p and q. $\qquad \Box$

Appendix 6.D

Explicit forms of $C(p,q)$. We present a complete set of the centrosymmetric blur invariants of the third, fifth and seventh orders.

- Third order:

$$C(3,0) = \mu_{30},$$
$$C(2,1) = \mu_{21},$$
$$C(1,2) = \mu_{12},$$
$$C(0,3) = \mu_{03}.$$

- Fifth order:

$$C(5,0) = \mu_{50} - \frac{10\mu_{30}\mu_{20}}{\mu_{00}},$$

$$C(4,1) = \mu_{41} - \frac{2}{\mu_{00}}(3\mu_{21}\mu_{20} + 2\mu_{30}\mu_{11}),$$

$$C(3,2) = \mu_{32} - \frac{1}{\mu_{00}}(3\mu_{12}\mu_{20} + \mu_{30}\mu_{02} + 6\mu_{21}\mu_{11}),$$

$$C(2,3) = \mu_{23} - \frac{1}{\mu_{00}}(3\mu_{21}\mu_{02} + \mu_{03}\mu_{20} + 6\mu_{12}\mu_{11}),$$

$$C(1,4) = \mu_{14} - \frac{2}{\mu_{00}}(3\mu_{12}\mu_{02} + 2\mu_{03}\mu_{11}),$$

$$C(0,5) = \mu_{05} - \frac{10\mu_{03}\mu_{02}}{\mu_{00}}.$$

- Seventh order:

$$C(7,0) = \mu_{70} - \frac{7}{\mu_{00}}(3\mu_{50}\mu_{20} + 5\mu_{30}\mu_{40}) + \frac{210\mu_{30}\mu_{20}^2}{\mu_{00}^2},$$

$$C(6,1) = \mu_{61} - \frac{1}{\mu_{00}}(6\mu_{50}\mu_{11} + 15\mu_{41}\mu_{20} + 15\mu_{40}\mu_{21}$$
$$+ 20\mu_{31}\mu_{30}) + \frac{30}{\mu_{00}^2}(3\mu_{21}\mu_{20}^2 + 4\mu_{30}\mu_{20}\mu_{11}),$$

$$C(5,2) = \mu_{52} - \frac{1}{\mu_{00}}(\mu_{50}\mu_{02} + 10\mu_{30}\mu_{22} + 10\mu_{32}\mu_{20}$$
$$+ 20\mu_{31}\mu_{21} + 10\mu_{41}\mu_{11} + 5\mu_{40}\mu_{12}) + \frac{10}{\mu_{00}^2}(3\mu_{12}\mu_{20}^2 + 2\mu_{30}\mu_{20}\mu_{02}$$
$$+ 4\mu_{30}\mu_{11}^2 + 12\mu_{21}\mu_{20}\mu_{11}),$$

$$C(4,3) = \mu_{43} - \frac{1}{\mu_{00}}(\mu_{40}\mu_{03} + 18\mu_{21}\mu_{22} + 12\mu_{31}\mu_{12} + 4\mu_{30}\mu_{13}$$
$$+ 3\mu_{41}\mu_{02} + 12\mu_{32}\mu_{11} + 6\mu_{23}\mu_{20}) + \frac{6}{\mu_{00}^2}(\mu_{03}\mu_{20}^2 + 4\mu_{30}\mu_{11}\mu_{02}$$
$$+ 12\mu_{21}\mu_{11}^2 + 12\mu_{12}\mu_{20}\mu_{11} + 6\mu_{21}\mu_{02}\mu_{20}),$$

$$C(3,4) = \mu_{34} - \frac{1}{\mu_{00}}(\mu_{04}\mu_{30} + 18\mu_{12}\mu_{22} + 12\mu_{13}\mu_{21} + 4\mu_{03}\mu_{31}$$
$$+ 3\mu_{14}\mu_{20} + 12\mu_{23}\mu_{11} + 6\mu_{32}\mu_{02}) + \frac{6}{\mu_{00}^2}(\mu_{30}\mu_{02}^2 + 4\mu_{03}\mu_{11}\mu_{20}$$
$$+ 12\mu_{12}\mu_{11}^2 + 12\mu_{21}\mu_{02}\mu_{11} + 6\mu_{12}\mu_{20}\mu_{02}),$$

$$C(2,5) = \mu_{25} - \frac{1}{\mu_{00}}(\mu_{05}\mu_{20} + 10\mu_{03}\mu_{22} + 10\mu_{23}\mu_{02}$$
$$+ 20\mu_{13}\mu_{12} + 10\mu_{14}\mu_{11} + 5\mu_{04}\mu_{21}) + \frac{10}{\mu_{00}^2}(3\mu_{21}\mu_{02}^2 + 2\mu_{03}\mu_{02}\mu_{20}$$

$$+ 4\mu_{03}\mu_{11}^2 + 12\mu_{12}\mu_{02}\mu_{11}),$$

$$C(1,6) = \mu_{16} - \frac{1}{\mu_{00}}(6\mu_{05}\mu_{11} + 15\mu_{14}\mu_{02} + 15\mu_{04}\mu_{12}$$

$$+ 20\mu_{13}\mu_{03}) + \frac{30}{\mu_{00}^2}(3\mu_{12}\mu_{02}^2 + 4\mu_{03}\mu_{02}\mu_{11}),$$

$$C(0,7) = \mu_{07} - \frac{7}{\mu_{00}}(3\mu_{05}\mu_{02} + 5\mu_{03}\mu_{04}) + \frac{210\mu_{03}\mu_{02}^2}{\mu_{00}^2}. \qquad \square$$

Appendix 6.E

Explicit forms of $K_\infty(p, q)$. We present a complete set of the circular symmetric blur invariants up to the sixth order.

$$K_\infty(1,0) = c_{10}, \quad K_\infty(2,0) = c_{20}, \quad K_\infty(3,0) = c_{30},$$

$$K_\infty(4,0) = c_{40}, \quad K_\infty(5,0) = c_{50}, \quad K_\infty(6,0) = c_{60},$$

$$K_\infty(2,1) = c_{21} - 2\frac{c_{10}c_{11}}{c_{00}},$$

$$K_\infty(3,1) = c_{31} - 3\frac{c_{20}c_{11}}{c_{00}},$$

$$K_\infty(4,1) = c_{41} - 4\frac{c_{30}c_{11}}{c_{00}},$$

$$K_\infty(5,1) = c_{51} - 5\frac{c_{40}c_{11}}{c_{00}},$$

$$K_\infty(3,2) = c_{32} - 6\frac{K_\infty(2,1)c_{11}}{c_{00}} - 3\frac{c_{10}c_{22}}{c_{00}},$$

$$K_\infty(4,2) = c_{42} - 8\frac{K_\infty(3,1)c_{11}}{c_{00}} - 6\frac{K_\infty(2,1)c_{22}}{c_{00}}. \qquad \square$$

Appendix 6.F

Proof of a non-recursive form of Gaussian blur invariants. We prove the equivalence of the recursive and non-recursive definitions of $K_G(p, q)$ in Eq. (6.50). The proof for the invariants $B(\mathbf{p})$ is analogous.

Assuming $p \geq q$, the proof is via induction on q. If $q = 0$, then $K_G(p, q) = c_{pq}$. Now let $q > 0$.

$$K_G(p,q) = c_{pq} - \sum_{k=1}^{q} \binom{p}{k}\binom{q}{k}\left(\frac{c_{11}}{c_{00}}\right)^k k! K_G(p-k, q-k)$$

$$= c_{pq} - \sum_{k=1}^{q} \binom{p}{k}\binom{q}{k}\left(\frac{c_{11}}{c_{00}}\right)^k k! \sum_{j=0}^{q-k} \binom{p-k}{j}\binom{q-k}{j}\left(-\frac{c_{11}}{c_{00}}\right)^j j! \cdot c_{p-k-j, q-k-j}$$

$$= c_{pq} - \sum_{k=1}^{q} \sum_{j=0}^{q-k} \frac{p!q!}{k!j!(p-k-j)!(q-k-j)!} \left(\frac{c_{11}}{c_{00}}\right)^{k+j} (-1)^j \cdot c_{p-k-j,q-k-j}$$

$$= c_{pq} - \sum_{k=1}^{q} \sum_{j=k}^{q} \frac{p!q!}{k!(j-k)!(p-j)!(q-j)!} \left(\frac{c_{11}}{c_{00}}\right)^{j} (-1)^{j-k} c_{p-j,q-j}$$

$$= c_{pq} - \sum_{j=1}^{q} \sum_{k=1}^{j} \frac{p!q!}{k!(j-k)!(p-j)!(q-j)!} \left(\frac{c_{11}}{c_{00}}\right)^{j} (-1)^{j-k} c_{p-j,q-j}$$

$$= c_{pq} - \sum_{j=1}^{q} \binom{p}{j}\binom{q}{j} \left(\frac{c_{11}}{c_{00}}\right)^{j} (-1)^j j! c_{p-j,q-j} \sum_{k=1}^{j} \binom{j}{k}(-1)^k$$

$$= c_{pq} - \sum_{j=1}^{q} \binom{p}{j}\binom{q}{j} \left(-\frac{c_{11}}{c_{00}}\right)^{j} j! c_{p-j,q-j}((-1+1)^j - 1)$$

$$= \sum_{j=0}^{q} \binom{p}{j}\binom{q}{j} \left(-\frac{c_{11}}{c_{00}}\right)^{j} j! c_{p-j,q-j} \qquad\qquad \square$$

Appendix 6.G

Proof of Theorem 6.3. Since $I(C(0,0), \cdots, C(P,Q))$ is a function of blur invariants $C(p,q)$ only, it is also a blur invariant. To prove its invariance to affine transform, it is sufficient to prove that

$$\frac{\partial I(C(0,0),\dots,C(P,Q))}{\partial a} \equiv \sum_{p=0}^{P} \sum_{q=0}^{Q} \frac{\partial I(C(0,0),\dots,C(P,Q))}{\partial C(p,q)} \cdot \frac{\partial C(p,q)}{\partial a} = 0 \qquad (6.56)$$

for each parameter $a \in \{a_0, a_1, a_2, b_0, b_1, b_2\}$ of the affine transform.

The affine invariance of $I(\mu_{00}, \dots, \mu_{PQ})$ implies that

$$\frac{\partial I(\mu_{00},\dots,\mu_{PQ})}{\partial a} \equiv \sum_{p=0}^{P} \sum_{q=0}^{Q} \frac{\partial I(\mu_{00},\dots,\mu_{PQ})}{\partial \mu_{pq}} \cdot \frac{\partial \mu_{pq}}{\partial a} = 0. \qquad (6.57)$$

Thus, it is sufficient to prove that the partial derivatives of $C(p,q)$ are identical to the derivatives of moments μ_{pq}, when substituting $C(p,q)$ for μ_{pq}, i.e.

$$\frac{\partial C(p,q)}{\partial a} = \frac{\partial \mu_{pq}}{\partial a}\bigg|_{\mu_{pq}=C(p,q)} \qquad (6.58)$$

for each parameter a.

To prove this, we decompose the affine transform into six one-parametric transformations (similarly as in Chapter 5 when we solved the Cayley-Aronhold equation):
Horizontal translation:

$$u = x + \alpha,$$
$$v = y. \qquad\qquad\qquad (6.59)$$

Vertical translation:

$$u = x,$$
$$v = y + \beta. \tag{6.60}$$

Uniform scaling:

$$u = sx,$$
$$v = sy. \tag{6.61}$$

Stretching:

$$u = rx,$$
$$v = \frac{y}{r}. \tag{6.62}$$

Horizontal skewing:

$$u = x + ty,$$
$$v = y. \tag{6.63}$$

Vertical skewing:

$$u = x,$$
$$v = y + zx. \tag{6.64}$$

We prove that the constraint (6.58) holds for each of these simple transformations.

Eq. (6.58) holds trivially for translations (6.59) and (6.60). Without loss of generality, we assume in the rest of the proof that $(x_c, y_c) = (0, 0)$. For uniform scaling (6.61), we have

$$\frac{\partial \mu_{pq}}{\partial s} = \frac{\partial}{\partial s} \int_{-\infty}^{\infty} \int_{-\infty}^{\infty} u^p v^q f'(u, v) du dv$$

$$= \frac{\partial}{\partial s} \int_{-\infty}^{\infty} \int_{-\infty}^{\infty} s^{p+q+2} x^p y^q f(x, y) dx dy$$

$$= \int_{-\infty}^{\infty} \int_{-\infty}^{\infty} (p + q + 2) s^{p+q+1} x^p y^q f(x, y) dx dy$$

$$= \frac{p+q+2}{s} \int_{-\infty}^{\infty} \int_{-\infty}^{\infty} u^p v^q f'(u, v) du dv$$

$$= \frac{p+q+2}{s} \mu_{pq}$$

and

$$\frac{\partial C(p, q)}{\partial s} = \frac{p+q+2}{s} \mu_{pq} - \sum_{\substack{n=0 \\ 0<n+m<p+q}}^{p} \sum_{m=0}^{q} \binom{p}{n} \binom{q}{m}$$

$$\cdot \left[\frac{\mu_{nm}}{\mu_{00}} \cdot \frac{\partial C(p-n, q-m)}{\partial s} + C(p-n, q-m) \cdot \frac{\partial}{\partial s} \frac{\mu_{nm}}{\mu_{00}} \right].$$

At this moment we use the induction principle. Clearly, (6.58) holds for $p + q = 3$ because in that case $C(p, q) = \mu_{pq}$ (if $p + q < 3$ then (6.58) holds trivially). Let us assume the validity of (6.58) for all orders less than $p + q$. Using this assumption and performing the derivatives we obtain

$$
\frac{\partial C(p, q)}{\partial s} = \frac{p + q + 2}{s} \mu_{pq} - \sum_{\substack{n=0 \ m=0 \\ 0<n+m<p+q}}^{p} \sum^{q} \binom{p}{n}\binom{q}{m}
$$

$$
\cdot \left[\frac{(p + q - n - m + 2)\mu_{nm}}{s\mu_{00}} \cdot C(p - n, q - m) + \frac{(n + m + 2)\mu_{nm}}{s\mu_{00}} \cdot \right.
$$

$$
\left. \cdot C(p - n, q - m) - \frac{2\mu_{nm}}{s\mu_{00}} \cdot C(p - n, q - m) \right]
$$

$$
= \frac{p + q + 2}{s} \mu_{pq} - \sum_{\substack{n=0 \ m=0 \\ 0<n+m<p+q}}^{p} \sum^{q} \binom{p}{n}\binom{q}{m} \frac{(p + q + 2)\mu_{nm}}{s\mu_{00}} \cdot C(p - n, q - m)
$$

$$
= \frac{p + q + 2}{s} \cdot C(p, q),
$$

which proves the validity of (6.58) in case of scaling.

Similarly, for stretching (6.62) we have

$$
\frac{\partial \mu_{pq}}{\partial r} = \frac{\partial}{\partial r} \int_{-\infty}^{\infty} \int_{-\infty}^{\infty} r^{p-q} x^p y^q f(x, y) dx dy
$$

$$
= \int_{-\infty}^{\infty} \int_{-\infty}^{\infty} (p - q) r^{p-q-1} x^p y^q f(x, y) dx dy = \frac{p - q}{r} \mu_{pq}
$$

and

$$
\frac{\partial C(p, q)}{\partial r} = \frac{p - q}{r} \mu_{pq} - \frac{1}{\mu_{00}} \sum_{\substack{n=0 \ m=0 \\ 0<n+m<p+q}}^{p} \sum^{q} \binom{p}{n}\binom{q}{m}
$$

$$
\left[\mu_{nm} \frac{\partial C(p - n, q - m)}{\partial r} + \frac{n - m}{r} C(p - n, q - m)\mu_{nm} \right].
$$

Similarly to the previous case, we obtain by induction

$$
\frac{\partial C(p, q)}{\partial r} = \frac{p - q}{r} C(p, q).
$$

Finally, for horizontal skewing (6.63) we have

$$\frac{\partial \mu_{pq}}{\partial t} = \frac{\partial}{\partial t} \int\limits_{-\infty}^{\infty} \int\limits_{-\infty}^{\infty} (x + ty)^p y^q f(x, y) dx dy$$

$$= \int\limits_{-\infty}^{\infty} \int\limits_{-\infty}^{\infty} p(x + ty)^{p-1} y^{q+1} f(x, y) dx dy = p\mu_{p-1,q+1}$$

and

$$\frac{\partial C(p, q)}{\partial t} = p\mu_{p-1,q+1} - \frac{1}{\mu_{00}} \sum_{\substack{n=0 \\ 0<n+m<p+q}}^{p} \sum_{m=0}^{q} \binom{p}{n} \binom{q}{m}$$

$$\cdot \left[\mu_{nm} \frac{\partial C(p-n, q-m)}{\partial t} + C(p-n, q-m) \frac{\partial \mu_{nm}}{\partial t} \right].$$

Employing the induction principle we have

$$\frac{\partial C(p, q)}{\partial t} = p\mu_{p-1,q+1} - \frac{1}{\mu_{00}} \sum_{\substack{n=0 \\ 0<n+m<p+q}}^{p} \sum_{m=0}^{q} \binom{p}{n} \binom{q}{m}$$

$$\cdot [(p-n)\mu_{nm} C(p-n-1, q-m+1) + nC(p-n, q-m)\mu_{n-1,m+1}].$$

Shifting the indices and using the identities

$$(p-n)\binom{p}{n} = p\binom{p-1}{n} = (n+1)\binom{p}{n+1},$$

$$\binom{q}{m} + \binom{q}{m-1} = \binom{q+1}{m}$$

we obtain the final relation

$$\frac{\partial C(p, q)}{\partial t} = p\mu_{p-1,q+1} - \frac{1}{\mu_{00}} \left[\sum_{\substack{n=0 \\ 0<n+m<p+q}}^{p-1} \sum_{m=0}^{q} p\binom{p-1}{n} \binom{q}{m} \cdot C(p-n-1, q-m+1)\mu_{nm} \right.$$

$$\left. + \sum_{\substack{n=0 \\ 0<n+m<p+q}}^{p-1} \sum_{m=1}^{q+1} (n+1)\binom{p}{n+1} \binom{q}{m-1} \cdot C(p-n-1, q-m+1)\mu_{nm} \right]$$

$$= p\mu_{p-1,q+1} - \frac{1}{\mu_{00}} \sum_{\substack{n=0 \\ 0<n+m<p+q}}^{p-1} \sum_{m=0}^{q+1} p\binom{p-1}{n} \binom{q+1}{m} C(p-n-1, q-m+1)\mu_{nm}$$

$$= pC(p-1, q+1).$$

Because of symmetry, the same is true for vertical skewing (6.64). Thus, the constraint (6.58) holds for each affine parameter and, consequently, $I(C(0,0), \ldots, C(P,Q))$ is a combined invariant. □

References

[1] D. Kundur and D. Hatzinakos, "Blind image deconvolution," *IEEE Signal Processing Magazine*, vol. 13, no. 3, pp. 43–64, 1996.

[2] P. Campisi and K. Egiazarian, *Blind Image Deconvolution: Theory and Applications*. CRC Press, 2007.

[3] R. S. Blum and Z. Liu, *Multi-Sensor Image Fusion and Its Applications*. CRC, 2006.

[4] J. Flusser, T. Suk, and S. Saic, "Image features invariant with respect to blur," *Pattern Recognition*, vol. 28, no. 11, pp. 1723–1732, 1995.

[5] J. Flusser, T. Suk, and S. Saic, "Recognition of blurred images by the method of moments," *IEEE Transactions on Image Processing*, vol. 5, no. 3, pp. 533–538, 1996.

[6] J. Flusser and T. Suk, "Degraded image analysis: An invariant approach," *IEEE Transactions on Pattern Analysis and Machine Intelligence*, vol. 20, no. 6, pp. 590–603, 1998.

[7] H. Zhang, H. Shu, G.-N. Han, G. Coatrieux, L. Luo, and J. L. Coatrieux, "Blurred image recognition by Legendre moment invariants," *IEEE Transactions on Image Processing*, vol. 19, no. 3, pp. 596–611, 2010.

[8] C.-Y. Wee and R. Paramesran, "Derivation of blur-invariant features using orthogonal Legendre moments," *IET Computer Vision*, vol. 1, no. 2, pp. 66–77, 2007.

[9] X. Dai, H. Zhang, H. Shu, and L. Luo, "Image recognition by combined invariants of Legendre moment," in *Proceedings of the IEEE International Conference on Information and Automation ICIA'10*, pp. 1793–1798, IEEE, 2010.

[10] X. Dai, H. Zhang, H. Shu, L. Luo, and T. Liu, "Blurred image registration by combined invariant of Legendre moment and Harris-Laplace detector," in *Proceedings of the Fourth Pacific-Rim Symposium on Image and Video Technology PSIVT'10*, pp. 300–305, IEEE, 2010.

[11] X. Dai, T. Liu, H. Shu, and L. Luo, "Pseudo-Zernike moment invariants to blur degradation and their use in image recognition," in *Intelligent Science and Intelligent Data Engineering IScIDE'12* (J. Yang, F. Fang, and C. Sun, eds.), vol. 7751 of *Lecture Notes in Computer Science*, pp. 90–97, Springer, 2013.

[12] Q. Liu, H. Zhu, and Q. Li, "Image recognition by combined affine and blur Tchebichef moment invariants," in *Proceedings of 4th International Conference on Image and Signal Processing (CISP)*, pp. 1517–1521, 2011.

[13] X. Zuo, X. Dai, and L. Luo, "M-SIFT: A new descriptor based on Legendre moments and SIFT," in *Proceedings of the 3rd International Conference on Machine Vision ICMV'10*, pp. 183–186, 2010.

[14] J. Kautsky and J. Flusser, "Blur invariants constructed from arbitrary moments," *IEEE Transactions on Image Processing*, vol. 20, no. 12, pp. 3606–3611, 2011.

[15] B. Zitová and J. Flusser, "Invariants to convolution and rotation," in *Invariants for Pattern Recognition and Classification* (M. A. Rodrigues, ed.), pp. 23–46, World Scientific, 2000.

[16] Y. Zhang, C. Wen, and Y. Zhang, "Estimation of motion parameters from blurred images," *Pattern Recognition Letters*, vol. 21, no. 5, pp. 425–433, 2000.

[17] Y. Zhang, C. Wen, Y. Zhang, and Y. C. Soh, "Determination of blur and affine combined invariants by normalization," *Pattern Recognition*, vol. 35, no. 1, pp. 211–221, 2002.

[18] T. Suk and J. Flusser, "Combined blur and affine moment invariants and their use in pattern recognition," *Pattern Recognition*, vol. 36, no. 12, pp. 2895–2907, 2003.

[19] J. Flusser, J. Boldyš, and B. Zitová, "Invariants to convolution in arbitrary dimensions," *Journal of Mathematical Imaging and Vision*, vol. 13, no. 2, pp. 101–113, 2000.

[20] J. Flusser, J. Boldyš, and B. Zitová, "Moment forms invariant to rotation and blur in arbitrary number of dimensions," *IEEE Transactions on Pattern Analysis and Machine Intelligence*, vol. 25, no. 2, pp. 234–246, 2003.

[21] J. Boldyš and J. Flusser, "Extension of moment features' invariance to blur," *Journal of Mathematical Imaging and Vision*, vol. 32, no. 3, pp. 227–238, 2008.

[22] F. M. Candocia, "Moment relations and blur invariant conditions for finite-extent signals in one, two and n-dimensions," *Pattern Recognition Letters*, vol. 25, pp. 437–447, 2004.

[23] V. Ojansivu and J. Heikkilä, "A method for blur and affine invariant object recognition using phase-only bispectrum," in *The International Conference on Image Analysis and Recognition ICIAR'08*, vol. LNCS 5112, pp. 527–536, Springer, 2008.

[24] V. Ojansivu and J. Heikkilä, "Image registration using blur-invariant phase correlation," *IEEE Signal Processing Letters*, vol. 14, no. 7, pp. 449–452, 2007.

[25] S. Tang, Y. Wang, and Y.-W. Chen, "Blur invariant phase correlation in X-ray digital subtraction angiography," in *Preprints of the IEEE/CME International Conference on Complex Medical Engineering*, pp. 1715–1719, 2007.

[26] I. Makaremi and M. Ahmadi, "Blur invariants: A novel representation in the wavelet domain," *Pattern Recognition*, no. 43, pp. 3950–3957, 2010.

[27] I. Makaremi and M. Ahmadi, "Wavelet domain blur invariants for image analysis," *IEEE Transactions on Image Processing*, vol. 21, no. 3, pp. 996–1006, 2012.

[28] R. R. Galigekere and M. N. S. Swamy, "Moment patterns in the Radon space: invariance to blur," *Optical Engineering*, vol. 45, no. 7, pp. (077003–)1–6, 2006.

[29] V. Ojansivu and J. Heikkilä, "Blur insensitive texture classification using local phase quantization," in *Congress on Image and Signal Processing CISP'08* (A. Elmoataz, O. Lezoray, F. Nouboud, and D. Mammass, eds.), vol. 5099 of Lecture Notes in Computer Science, pp. 236–243, Springer, 2008.

[30] V. Ojansivu, E. Rahtu, and J. Heikkilä, "Rotation invariant local phase quantization for blur insensitive texture analysis," in *19th International Conference on Pattern Recognition, ICPR'08*, pp. 1–4, IEEE, 2008.

[31] E. Rahtu, J. Heikkilä, V. Ojansivu, and T. Ahonen, "Local phase quantization for blur-insensitive image analysis," *Image and Vision Computing*, vol. 30, no. 8, pp. 501–512, 2012. Special Section: Opinion Papers.

[32] Y. Bentoutou, N. Taleb, K. Kpalma, and J. Ronsin, "An automatic image registration for applications in remote sensing," *IEEE Transactions on Geoscience and Remote Sensing*, vol. 43, no. 9, pp. 2127–2137, 2005.

[33] Z. Liu, J. An, and L. Li, "A two-stage registration angorithm for oil spill aerial image by invariants-based similarity and improved ICP," *International Journal of Remote Sensing*, vol. 32, no. 13, pp. 3649–3664, 2011.

[34] S. X. Hu, Y.-M. Xiong, M. Z. W. Liao, and W. F. Chen, "Accurate point matching based on combined moment invariants and their new statistical metric," in *Proceedings of the International Conference on Wavelet Analysis and Pattern Recognition ICWAPR'07*, pp. 376–381, IEEE, 2007.

[35] Y. Bentoutou, N. Taleb, M. Chikr El Mezouar, M. Taleb, and J. Jetto, "An invariant approach for image registration in digital subtraction angiography," *Pattern Recognition*, vol. 35, no. 12, pp. 2853–2865, 2002.

[36] Y. Bentoutou and N. Taleb, "Automatic extraction of control points for digital subtraction angiography image enhancement," *IEEE Transactions on Nuclear Science*, vol. 52, no. 1, pp. 238–246, 2005.

[37] Y. Bentoutou and N. Taleb, "A 3-D space-time motion detection for an invariant image registration approach in digital subtraction angiography," *Computer Vision and Image Understanding*, vol. 97, pp. 30–50, 2005.

[38] J. Lu and Y. Yoshida, "Blurred image recognition based on phase invariants," *IEICE Transactions on Fundamentals of Electronics, Communications and Computer Sciences*, vol. E82A, no. 8, pp. 1450–1455, 1999.

[39] X.-J. Shen and J.-M. Pan, "Monocular visual servoing based on image moments," *IEICE Transactions on Fundamentals of Electronics, Communications and Computer Sciences*, vol. E87-A, no. 7, pp. 1798–1803, 2004.

[40] B. Mahdian and S. Saic, "Detection of copy-move forgery using a method based on blur moment invariants," *Forensic Science International*, vol. 171, nos. 2–3, pp. 180–189, 2007.

[41] B. Mahdian and S. Saic, "Detection of near-duplicated image regions," in *Computer Recognition Systems 2*, vol. 45 of Advances in Soft Computing, pp. 187–195, Springer, 2007.

[42] L. Li and G. Ma, "Recognition of degraded traffic sign symbols using PNN and combined blur and affine invariants," in *Fourth International Conference on Natural Computation ICNC'08*, vol. 3, pp. 515–520, IEEE, 2008.

[43] L. Li and G. Ma, "Optimizing the performance of probabilistic neural networks using PSO in the task of traffic sign recognition," in *Advanced Intelligent Computing Theories and Applications. With Aspects of Artificial Intelligence. Fourth International Conference on Intelligent Computing ICIC'08*, vol. LNAI 5227, pp. 90–98, Springer, 2008.

[44] Y. Zhang, C. Wen, and Y. Zhang, "Neural network based classification using blur degradation and affine deformation invariant features," in *Proceedings of the Thirteenth International Florida Artificial Intelligence Research Society Conference FLAIRS'00*, pp. 76–80, AAAI Press, 2000.

[45] P. Zhao, "Dynamic timber cell recognition using two-dimensional image measurement machine," *AIP Review of Scientific Instruments*, vol. 82, no. 8, pp. 083703–1, 2011.

[46] Y. Li, H. Chen, J. Zhang, and P. Qu, "Combining blur and affine moment invariants in object recognition," in *Proceedings of the Fifth International Symposium on Instrumentation and Control Technology ISICT'03*, vol. 5253, SPIE, 2003.

[47] R. Palaniappan, M. P. Paulraj, S. Yaacob, and M. S. B. Z. Azalan, "A simple sign language recognition system using affine moment blur invariant features," in *Proceedings of the International Postgraduate Conference on Engineerig IPCE'10*, p. 5. Universiti Malaysia (UniMAP), 2010.

[48] Y. Gao, H. Song, X. Tian, and Y. Chen, "Identification algorithm of winged insects based on hybrid moment invariants," *in First International Conference on Bioinformatics and Biomedical Engineering iCBBE'07*, vol. 2, pp. 531–534, IEEE, 2007.

[49] H. Ji, J. Zhu, and H. Zhu, "Combined blur and RST invariant digital image watermarking using complex moment invariants," in *Proceedings of the 2nd International Conference on Signals, Circuits and Systems, SCS'08*, (China), 2008.

[50] T. Ahonen, E. Rahtu, V. Ojansivu, and J. Heikkilä, "Recognition of blurred faces using local phase quantization," in *19th International Conference on Pattern Recognition, ICPR'08*, pp. 1–4, IEEE, 2008.

[51] Y. Zhang, Y. Zhang, and C. Wen, "A new focus measure method using moments," *Image and Vision Computing*, vol. 18, no. 12, pp. 959–965, 2000.

[52] P.-T. Yap and P. Raveendran, "Image focus measure based on Chebyshev moments," *IEE Proceedings of the Vision, Image and Signal Processing*, vol. 151, no. 2, pp. 128–136, 2004.

[53] J. Flusser and B. Zitová, "Invariants to convolution with circularly symmetric PSF," in *Proceedings of the 17th International Conference on Pattern Recognition ICPR'04*, pp. 11–14, IEEE, 2004.

[54] H. Zhu, M. Liu, H. Ji, and Y. Li, "Combined invariants to blur and rotation using Zernike moment descriptors," *Pattern Analysis and Applications*, vol. 3, no. 13, pp. 309–319, 2010.

[55] B. Chen, H. Shu, H. Zhang, G. Coatrieux, L. Luo, and J. L. Coatrieux, "Combined invariants to similarity transformation and to blur using orthogonal Zernike moments," *IEEE Transactions on Image Processing*, vol. 20, no. 2, pp. 345–360, 2011.

[56] H. Ji and H. Zhu, "Degraded image analysis using Zernike moment invariants," in *Proceedings of the International Conference on Acoustics, Speech and Signal Processing ICASSP'09*, pp. 1941–1944, 2009.

[57] Q. Liu, H. Zhu, and Q. Li, "Object recognition by combined invariants of orthogonal Fourier-Mellin moments," in *Proceedings of 8th International Conference on Information, Communications and Signal Processing ICICS'11*, pp. 1–5, IEEE, 2011.

[58] J. Flusser, B. Zitová, and T. Suk, "Invariant-based registration of rotated and blurred images," in *Proceedings of the International Geoscience and Remote Sensing Symposium IGARSS'99*, pp. 1262–1264, IEEE, 1999.

[59] B. Zitová and J. Flusser, "Estimation of camera planar motion from defocused images," in *Proceedings of the International Conference on Image Processing ICIP'02*, vol. II, pp. 329–332, IEEE, 2002.

[60] J. Flusser, T. Suk, and S. Saic, "Recognition of images degraded by linear motion blur without restoration," *Computing Supplement*, vol. 11, pp. 37–51, 1996.

[61] A. Stern, I. Kruchakov, E. Yoavi, and S. Kopeika, "Recognition of motion-blured images by use of the method of moments," *Applied Optics*, vol. 41, pp. 2164–2172, 2002.

[62] Z. Peng and C. Jun, "Weed recognition using image blur information," *Biosystems Engineering*, vol. 110, no. 2, pp. 198–205, 2011.

[63] C. Guang-Sheng and Z. Peng, "Dynamic wood slice recognition using image blur information," *Sensors and Actuators A: Physical*, vol. 176, no. April, pp. 27–33, 2012.

[64] J. Flusser, T. Suk, and B. Zitová, "On the recognition of wood slices by means of blur invariants," *Sensors and Actuators A: Physical*, vol. 198, pp. 113–118, 2013.

[65] J. Flusser, T. Suk, and B. Zitová, "Comments on 'Weed recognition using image blur information by Peng, Z. & Jun, C., Biosystems Engineering 110 (2), p. 198–205,'" *Biosystems Engineering*, vol. 126, pp. 104–108, 2014.

[66] S. Zhong, Y. Liu, Y. Liu, and C. Li, "Water reflection recognition based on motion blur invariant moments in curvelet space full text sign-in or purchase," *IEEE Transactions on Image Processing*, vol. 22, no. 11, pp. 4301–4313, 2013.

[67] B. Guan, S. Wang, and G. Wang, "A biologically inspired method for estimating 2D high-speed translational motion," *Pattern Recognition Letters*, vol. 26, pp. 2450–2462, 2005.

[68] S. Wang, B. Guan, G. Wang, and Q. Li, "Measurement of sinusoidal vibration from motion blurred images," *Pattern Recognition Letters*, vol. 28, pp. 1029–1040, 2007.

[69] J. Liu and T. Zhang, "Recognition of the blurred image by complex moment invariants," *Pattern Recognition Letters*, vol. 26, no. 8, pp. 1128–1138, 2005.

[70] B. Xiao, J.-F. Ma, and J.-T. Cui, "Combined blur, translation, scale and rotation invariant image recognition by Radon and pseudo-Fourier-Mellin transforms," *Pattern Recognition*, vol. 45, pp. 314–321, 2012.

[71] Z. Zhang, E. Klassen, A. Srivastava, P. Turaga, and R. Chellappa, "Blurring-invariant riemannian metrics for comparing signals and images," in *IEEE International Conference on Computer Vision, ICCV'11*, pp. 1770–1775, IEEE, 2011.

[72] Z. Zhang, E. Klassen, and A. Srivastava, "Gaussian blurring-invariant comparison of signals and images," *IEEE Transactions on Image Processing*, vol. 22, no. 8, pp. 3145–3157, 2013.

[73] J. Flusser, T. Suk, S. Farokhi, and C. Höschl IV, "Recognition of images degraded by Gaussian blur," in *Computer Analysis of Images and Patterns CAIP'15* (G. Azzopardi and N. Petkov, eds.), vol. 9256–9257 of Lecture Notes in Computer Science, pp. 88–99, part I, Springer, 2015.

[74] J. Flusser, S. Farokhi, C. Höschl IV, T. Suk, B. Zitová, and M. Pedone, "Recognition of images degraded by Gaussian blur," *IEEE Transactions on Image Processing*, vol. 25, no. 2, pp. 790–806, 2016.

[75] C. Höschl IV and J. Flusser, "Noise-resistant image retrieval," in *22nd International Conference on Pattern Recognition ICPR'14*, pp. 2972–2977, IEEE, 2014.

[76] J. Flusser, T. Suk, J. Boldyš, and B. Zitová, "Projection operators and moment invariants to image blurring," *IEEE Transactions on Pattern Analysis and Machine Intelligence*, vol. 37, no. 4, pp. 786–802, 2015.

[77] M. Pedone, J. Flusser, and J. Heikkilä, "Blur invariant translational image registration for N-fold symmetric blurs," *IEEE Transactions on Image Processing*, vol. 22, no. 9, pp. 3676–3689, 2013.

[78] J. Boldyš and J. Flusser, "Invariants to symmetrical convolution with application to dihedral kernel symmetry," in *Proceedings of the 17th International Conference on Image Analysis and Processing ICIAP'13* (A. Petrosino, ed.), vol. 8157 of Lecture Notes in Computer Science, pp. 369–378, part II, Springer, 2013.

[79] M. Pedone, J. Flusser, and J. Heikkilä, "Registration of images with N-fold dihedral blur," *IEEE Transactions on Image Processing*, vol. 24, no. 3, pp. 1036–1045, 2015.

[80] R. Gopalan, P. Turaga, and R. Chellappa, "A blur-robust descriptor with applications to face recognition," *IEEE Transactions on Pattern Analysis and Machine Intelligence*, vol. 34, no. 6, pp. 1220–1226, 2012.

[81] J. Flusser and T. Suk, "Classification of degraded signals by the method of invariants," *Signal Processing*, vol. 60, no. 2, pp. 243–249, 1997.

[82] J. Flusser and B. Zitová, "Combined invariants to linear filtering and rotation," *International Journal of Pattern Recognition and Artificial Intelligence*, vol. 13, no. 8, pp. 1123–1136, 1999.

[83] Y. Bentoutou, N. Taleb, A. Bounoua, K. Kpalma, and J. Ronsin, "Feature based registration of satellite images," in *Proceedings of the 15th International Conference on Digital Signal Processing DSP 2007*, pp. 419–422, IEEE, 2007.

[84] B. Zitová and J. Flusser, "Estimation of camera planar motion from blurred images," in *Proceedings of the International Conference on Image Processing ICIP'02*, pp. 329–332, IEEE, 2002.

[85] D. G. Lowe, "Distinctive image features from scale-invariant keypoints," *International Journal of Computer Vision*, vol. 60, no. 2, pp. 91–110, 2004.

[86] Z. Myles and N. V. Lobo, "Recovering affine motion and defocus blur simultaneously," *IEEE Transactions on Pattern Analysis and Machine Intelligence*, vol. 20, no. 6, pp. 652–658, 1998.

[87] A. Kubota, K. Kodama, and K. Aizawa, "Registration and blur estimation methods for multiple differently focused images," in *Proceedings of the International Conference on Image Processing ICIP'99*, vol. II, pp. 447–451, IEEE, 1999.

[88] Z. Zhang and R. S. Blum, "A categorization of multiscale-decomposition-based image fusion schemes with a performance study for a digital camera application," *Proceedings of the IEEE*, vol. 87, no. 8, pp. 1315–1326, 1999.

[89] K. V. Arya and P. Gupta, "Registration algorithm for motion blurred images," in *Proceedings of the Sixth International Conference Advances in Pattern Recognition ICAPR'07* (P. Pal, ed.), pp. 186–190, World Scientific, 2007.

[90] X. Fang, B. Luo, B. He, and H. Wu, "Feature based multi-resolution registration of blurred images for image mosaic," *International Journal of CAD/CAM*, vol. 9, no. 1, pp. 37–46, 2009.

[91] L. Yuan, J. Sun, L. Quan, and H.-Y. Shum, "Blurred/non-blurred image alignment using sparseness prior," in *Proceedings of the 11th International Conference on Computer Vision ICCV'07*, IEEE, 2007.

[92] P. Milanfar, *Super-Resolution Imaging*. CRC Press, 2011.

[93] P. Vandewalle, S. Süsstrunk, and M. Vetterli, "A frequency domain approach to registration of aliased images with application to super-resolution," *EURASIP Journal on Applied Signal Processing*, pp. 1–14, 2006.

[94] E. de Castro and C. Morandi, "Registration of translated and rotated images using finite Fourier transform," *IEEE Transactions on Pattern Analysis and Machine Intelligence*, vol. 9, no. 5, pp. 700–703, 1987.

[95] P. Vandewalle, L. Sbaiz, and M. Vetterli, "Registration for super-resolution: theory, algorithms, and applications in image and mobile video enhancement," in *Super-Resolution Imaging* (P. Milanfar, ed.), Digital Imaging and Computer Vision, pp. 155–186, CRC Press, 2010.

[96] D. Robinson, S. Farsiu, and P. Milanfar, "Optimal registration of aliased images using variable projection with applications to super-resolution," *Computer Journal*, vol. 52, no. 1, pp. 31–42, 2009.

[97] C. Höschl IV and J. Flusser, "Robust histogram-based image retrieval," *Pattern Recognition Letters*, vol. 69, no. 1, pp. 72–81, 2016.

[98] M. Pedone, *Algebraic methods for constructing blur-invariant operators and their applications*. PhD thesis, University of Oulu, Oulu, Finland, 2015.

[99] B. Honarvar and J. Flusser, "Image deconvolution in moment domain," in *Moments and Moment Invariants –Theory and Applications* (G. A. Papakostas, ed.), pp. 111–125, Science Gate Publishing, 2014.

[100] F. Xue and T. Blu, "A novel SURE-based criterion for parametric PSF estimation," *IEEE Transactions on Image Processing*, vol. 24, no. 2, pp. 595–607, 2015.

[101] S. Nadarajah, "A generalized normal distribution," *Journal of Applied Statistics*, vol. 32, no. 7, pp. 685–694, 2005.

[102] J. Nolan, *Stable Distributions: Models for Heavy-Tailed Data*. Boston: Birkhauser, 2007.

[103] A. S. Carasso, "The APEX method in image sharpening and the use of low exponent Lévy stable laws," *SIAM Journal on Applied Mathematics*, vol. 63, no. 2, pp. 593–618, 2003.

[104] A. S. Carasso, "APEX blind deconvolution of color Hubble space telescope imagery and other astronomical data," *Optical Engineering*, vol. 45, no. 10, pp. 1–15, 2006.

[105] S. Liao and A. C. S. Chung, "Feature based nonrigid brain MR image registration with symmetric alpha stable filters," *IEEE Transactions on Medical Imaging*, vol. 29, no. 1, pp. 106–119, 2010.

[106] A. Achim and E. E. Kuruoğlu, "Image denoising using bivariate α-stable distributions in the complex wavelet domain," *IEEE Signal Processing Letters*, vol. 12, no. 1, pp. 17–20, 2005.

[107] J. Kautsky and J. Flusser, "Numerical problems with the Pascal triangle in moment computation," *Journal of Computational and Applied Mathematics*, vol. 306, pp. 53–68, 2016.

[108] B. Zitová and J. Flusser, "Image registration methods: A survey," *Image and Vision Computing*, vol. 21, no. 11, pp. 977–1000, 2003.

[109] G. L. Cash and M. Hatamian, "Optical character recognition by the method of moments," *Computer Vision, Graphics, and Image Processing*, vol. 39, no. 3, pp. 291–310, 1987.

[110] F. Šroubek and J. Flusser, "Multichannel blind deconvolution of spatially misaligned images," *IEEE Transactions on Image Processing*, vol. 14, no. 7, pp. 874–883, 2005.

[111] J. Kautsky, J. Flusser, B. Zitová, and S. Šimberová, "A new wavelet-based measure of image focus," *Pattern Recognition Letters*, vol. 23, no. 14, pp. 1785–1794, 2002.

[112] F. Šroubek and J. Flusser, "Multichannel blind iterative image restoration," *IEEE Transactions on Image Processing*, vol. 12, no. 9, pp. 1094–1106, 2003.

[113] S. Z. Li, Z. Lei, and M. Ao, "The HFB face database for heterogeneous face biometrics research," in 6th IEEE Workshop on Object Classification Beyond and in the Visible Spectrum (OTCBVS'09, in conjunction with CVPR'09), pp. 1–8, June 2009.

[114] D. Yi, R. Liu, R. Chu, Z. Lei, and S. Z. Li, "Face matching between near infrared and visible light images," in *Advances in Biometrics, Proceedings of International Conference on Biometrics, (ICB'07)*, pp. 523–530, Springer, August 2007.

7

2D and 3D Orthogonal Moments

7.1 Introduction

It is well known from linear algebra that an orthogonal (OG) basis of a vector space has many
favorable numerical properties comparing to other bases. This applies to the bases of polynomials too, and, consequently, to moments, that are "projections" of an image onto a basis.

All existing bases of a given vector space are, of course, theoretically equivalent, because
each vector from one basis can be expressed as a linear combination of the vectors from the
other basis. When dealing with various polynomial bases up to a certain degree and with corresponding moments, any moment (with respect to any basis) can be expressed as a function
of moments of the same or fewer orders with respect to an arbitrary basis. In particular, any
moments can be expressed in terms of geometric moments. Taking this into consideration,
seemingly there is no reason for introducing other than geometric moments.

The above reasoning is true on a theoretical level only. When considering numerical properties and implementation issues, the motivation for introducing *orthogonal moments* becomes
apparent. OG moments are moments w.r.t. an orthogonal or weighted-orthogonal polynomial
basis. The polynomial basis $\{\pi_{\mathbf{p}}(\mathbf{x})\}$ is orthogonal (possibly with weight function $w(\mathbf{x})$) if

$$\int_{\Omega} w(\mathbf{x})\pi_{\mathbf{p}}(\mathbf{x})\pi_{\mathbf{q}}(\mathbf{x})\mathrm{d}\mathbf{x} = 0 \tag{7.1}$$

for any $\mathbf{p} \neq \mathbf{q}$. Moreover, if

$$\int_{\Omega} w(\mathbf{x})\pi_{\mathbf{p}}(\mathbf{x})\pi_{\mathbf{p}}(\mathbf{x})\mathrm{d}\mathbf{x} = 1 \tag{7.2}$$

for any $\mathbf{p}$, the polynomial basis is called *orthonormal* (ON). The set $\Omega \subset \mathbb{R}^d$ is called the *area
of orthogonality*, and it may be less than the definition domain of $\{\pi_{\mathbf{p}}(\mathbf{x})\}$. Working with OG
polynomials always means working within Ω because the favorable numerical properties of
OG polynomials are not preserved[1] outside Ω.

---

[1] The properties such as dynamic range and conditionality of OG polynomials outside Ω may be even worse than the
properties of the standard power basis.

2D and 3D Image Analysis by Moments, First Edition. Jan Flusser, Tomáš Suk and Barbara Zitová.
© 2017 John Wiley & Sons, Ltd. Published 2017 by John Wiley & Sons, Ltd.
Companion website: www.wiley.com/go/flusser/2d3dimageanalysis

The main reasons for introducing OG polynomials and OG moments are stable and fast numerical implementation. OG polynomials can be evaluated by recurrence relations,[2] which can be efficiently implemented by means of multiplication with special matrices. By using OG polynomials, we avoid a high dynamic range of moment values that may lead to the loss of precision due to overflow or underflow (unlike standard powers, the values of OG polynomials lie inside a narrow interval such as $\langle -1, 1 \rangle$ or similar). From this point of view, it is a serious mistake to evaluate OG polynomials by expanding them into standard powers. Even if these expansions are commonly presented in literature (and we do this in this chapter as well), it should serve for illustration and theoretical considerations only. They should never be used for evaluation of polynomials and for OG moment calculation.

Another reason for using OG moments frequently mentioned in literature is better image reconstruction. OG moments are, unlike geometric and other moments, coordinates of f in the polynomial basis in the common sense used in linear algebra. Thanks to this, the image reconstruction from OG moments can be performed easily as

$$f(\mathbf{x}) = \sum_{\mathbf{p}=0}^{\infty} M_{\mathbf{p}} \cdot \pi_{\mathbf{p}}(\mathbf{x}).$$

This reconstruction is "optimal" in the sense that it minimizes the mean-square error when using only a finite set of moments. On the other hand, image reconstruction from geometric moments is more complicated, less stable, and its convergence is slower. This argument, which fancies the OG moments, is correct, but at the same time we must emphasize that the task of image reconstruction from its moments is rather artificial, very rarely appearing in practice. The reconstruction problem appears most frequently in image compression when trying to recover the original from a lossy representation. However, moments are not a good tool for image compression, providing much worse compression rates than JPEG or wavelets, and are hardly ever used for compression purposes.

Numerous types of 1D OG polynomials have been described in traditional mathematical literature (see for instance the books [1] and [2]). In this chapter we present a survey of 2D and 3D OG moments that are of importance in image analysis. The literature on orthogonal moments is very broad, namely in the area of practical applications, and our survey has no claim on completeness.

For the purpose of this chapter, we divide multidimensional OG polynomials and OG moments into two basic groups. The polynomials *orthogonal on a square/cube* originate from 1D OG polynomials $\pi_n(x)$ whose 2D and 3D versions have been created as products of 1D polynomials in x, y, and z. The main advantage of the moments orthogonal on a square/cube is that they may preserve the orthogonality even on discrete images, and calculations with them may be precise. On the other hand, their main drawback is that construction of rotation invariants is complicated, indirect, and mostly inefficient.

The polynomials *orthogonal on a disk/sphere* do not have any direct ancestor in 1D. They are constructed as products of a radial factor (usually formed by 1D OG polynomials) and angular factor, which is in 2D a kind of circular harmonic and in 3D of spherical harmonic functions. They are special case of circular moments introduced in Chapter 3. When implementing the polynomials and moments OG on a disk/sphere, the image must be mapped into the disk/sphere

---

[2] Often called "three-term relations" because they allow calculating $\pi_n(x)$ by means of $\pi_{n-1}(x)$ and $\pi_{n-2}(x)$.

of orthogonality, which brings certain resampling problems and may lead to a loss of precision. The main advantage of these moments (and actually the only reason why they are used in image analysis) is that they change in a simple way under an image rotation, which makes creating rotation invariants relatively easy. In 3D, we also recognize polynomials orthogonal on a cylinder, but they represent a rather minor group.

7.2   2D moments orthogonal on a square

Let us assume we have a set of 1D OG polynomials[3] $\{\pi_n(x)\}$ the area of orthogonality of which is an interval Ω_1 (usually $\Omega_1 = \langle -1, 1 \rangle$ but other intervals including infinite ones are also possible) and the weight function is $w(x)$. Hence, we have

$$\int_{\Omega_1} w(x)\pi_p(x)\pi_q(x)\mathrm{d}x = h_p \delta_{pq}, \tag{7.3}$$

where $h_p = \|\pi_p\|^2$ and δ_{pq} is *Kronecker delta*.

Now let us consider 2D polynomials $\pi_{kj}(x, y) = \pi_k(x) \cdot \pi_j(y)$. These polynomials are OG on $\Omega = \Omega_1 \times \Omega_1$ with the weight function $w_2(x, y) = w(x)w(y)$. This follows immediately from the separability of the integrand

$$\iint_{\Omega} w_2(x, y)\pi_{kj}(x, y)\pi_{mn}(x, y)\mathrm{d}x\mathrm{d}y$$

$$= \int_{\Omega_1} w(x)\pi_k(x)\pi_m(x)\mathrm{d}x \int_{\Omega_1} w(y)\pi_j(y)\pi_n(y)\mathrm{d}y \tag{7.4}$$

$$= h_k h_j \delta_{km} \delta_{jn}.$$

The corresponding 2D OG moments are

$$M_{pq} = n_p n_q \iint_{\Omega} \pi_p(x)\pi_q(y)f(x, y)\mathrm{d}x\mathrm{d}y, \tag{7.5}$$

provided that image f has been mapped into Ω. Coefficients n_p, n_q are normalization factors.

It should be noted that theoretically we could also use 2D polynomials which are not separable. This is, however, impractical because it would increase computing complexity while not bringing any advantages. To the authors' knowledge, almost nobody has used this approach for image analysis purposes ([3] is one of very few exceptions).

---

[3] OG polynomials mostly originate from the Gram-Schmidt orthogonalization applied to standard powers $\{x^n\}$, but the way in which they have been obtained is not very important for their practical usage.

7.2.1 Hypergeometric functions

Orthogonal polynomials are related to a concept of hypergeometric series[4]. If the series converges, it defines a *hypergeometric function*. The family of hypergeometric functions includes such special functions as Bessel functions, incomplete gamma function, the error function, and the elliptic integrals, among others. Hypergeometric functions are also often used for definition of various orthogonal polynomials, and even if the OG polynomials are defined in another way, they can often be expressed in terms of hypergeometric functions. This is why we briefly introduce hypergeometric functions here.

The *hypergeometric series* $_pF_q$ is defined as

$$_pF_q(a_1,\ldots,a_p;b_1,\ldots,b_q;z) = \sum_{n=0}^{\infty} \frac{(a_1)_n(a_2)_n\cdots(a_p)_n}{(b_1)_n(b_2)_n\cdots(b_q)_n} \frac{z^n}{n!}, \tag{7.6}$$

where $(a)_n$ is a *Pochhammer symbol* (sometimes called a *rising factorial*)

$$(a)_n = a(a+1)\ldots(a+n-1) = (a+n-1)!/(a-1)! = \Gamma(a+n)/\Gamma(a). \tag{7.7}$$

Particularly,

$$_2F_1(a,b;c;z) = \sum_{n=0}^{\infty} \frac{(a)_n(b)_n}{(c)_n} \frac{z^n}{n!} \quad \text{and} \quad _3F_2(a,b,c;d,e;z) = \sum_{n=0}^{\infty} \frac{(a)_n(b)_n(c)_n}{(d)_n(e)_n} \frac{z^n}{n!}. \tag{7.8}$$

Some elementary functions can be expressed by means of the hypergeometric function $_2F_1$ (see [2] for the proofs)[5]

$$(1+z)^k = {}_2F_1(-k,1;1;-z),$$
$$\log(1+z) = z \, {}_2F_1(1,1;2;-z),$$
$$\log\left(\frac{1+z}{1-z}\right) = 2z \, {}_2F_1\left(\tfrac{1}{2},1;\tfrac{3}{2};z^2\right), \tag{7.9}$$
$$\arcsin(z) = z \, {}_2F_1\left(\tfrac{1}{2},\tfrac{1}{2};\tfrac{3}{2};z^2\right),$$
$$\arctan(z) = z \, {}_2F_1\left(\tfrac{1}{2},1;\tfrac{3}{2};-z^2\right).$$

Expressing orthogonal polynomials by means of hypergeometric functions may provide us with an insight into the complexity of the orthogonal polynomials, but recurrent formulae offer more efficient algorithms for their enumeration. We can say in simplified way the higher indices of the defining hypergeometric function, the more complicated orthogonal polynomials.

---

[4] "Hypergeometric series" is a very old term introduced probably by Wallis (1655) and studied by Euler, Gauss, Kummer, Riemann, and others.
[5] $_2F_1(z)$ is an analytic function for $|z| < 1$. For $|z| \geq 1$ it can be analytically continued along any path in the complex plane that avoids the branch points 0 and 1.

7.2.2 Legendre moments[6]

Legendre polynomials are some of the simplest and the most frequently used OG polynomials ever. There have been many papers describing the use of the Legendre moments in image processing, such as [4, 5] among many others. The 2D *Legendre moments* are defined as

$$\lambda_{mn} = \frac{(2m+1)(2n+1)}{4} \int_{-1}^{1}\int_{-1}^{1} P_m(x)P_n(y)f(x,y)\mathrm{d}x\mathrm{d}y \qquad m, n = 0, 1, 2, \ldots, \qquad (7.10)$$

where $P_n(x)$ is the n-th degree *Legendre polynomial*[7]

$$P_n(x) = \frac{1}{2^n n!}\frac{\mathrm{d}^n}{\mathrm{d}x^n}(x^2 - 1)^n. \qquad (7.11)$$

The area of orthogonality is $\Omega_1 = \langle -1, 1\rangle$; there is no effective weight ($w(x) = 1$); and the orthogonality relation is

$$\int_{-1}^{1} P_m(x)P_n(x)\mathrm{d}x = \frac{2}{2n+1}\delta_{mn}. \qquad (7.12)$$

The three-term recurrence relation, which can be used for efficient computation of Legendre polynomials, is

$$\begin{aligned}
P_0(x) &= 1, \\
P_1(x) &= x, \\
P_{n+1}(x) &= \frac{2n+1}{n+1}xP_n(x) - \frac{n}{n+1}P_{n-1}(x).
\end{aligned} \qquad (7.13)$$

From the recurrence we can immediately see the expressions of the Legendre polynomials of low degrees in terms of the standard powers

$$\begin{aligned}
P_2(x) &= \frac{1}{2}(3x^2 - 1), \\[6pt]
P_3(x) &= \frac{1}{2}(5x^3 - 3x), \\[6pt]
P_4(x) &= \frac{1}{8}(35x^4 - 30x^2 + 3), \\[6pt]
P_5(x) &= \frac{1}{8}(63x^5 - 70x^3 + 15x), \\[6pt]
P_6(x) &= \frac{1}{16}(231x^6 - 315x^4 + 105x^2 - 5).
\end{aligned} \qquad (7.14)$$

---

[6] A. M. Legendre 1752–1833, French mathematician.
[7] This definition is called the Rodrigues' formula.

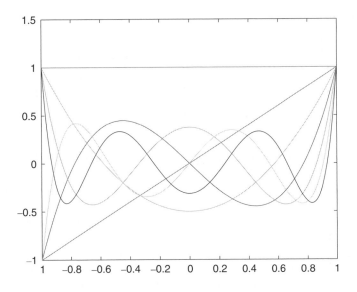

Figure 7.1 The graphs of the Legendre polynomials up to the sixth degree

The Legendre polynomials can be alternatively expressed in terms of the hypergeometric functions as

$$P_n(x) = {}_2F_1\left(-n, n+1; 1; \frac{1-x}{2}\right),$$

$$P_n(x) = \binom{2n}{n}\left(\frac{x-1}{2}\right)^n {}_2F_1\left(-n, -n; -2n; \frac{2}{1-x}\right), \tag{7.15}$$

$$P_n(x) = \binom{2n}{n}\left(\frac{x}{2}\right)^n {}_2F_1\left(-\frac{n}{2}, \frac{1-n}{2}; \frac{1}{2}-n; \frac{1}{x^2}\right).$$

The Legendre polynomials exhibit the same symmetry/antisymmetry[8] as the standard powers. We have $P_n(x) = P_n(-x)$ for any even n and $P_n(x) = -P_n(-x)$ for any odd n. This may help us when evaluating the polynomials and also indicates the vanishing moments of symmetric objects.

The graphs of the Legendre polynomials are in Figure 7.1. We can compare them with the standard powers for $n = 0, 1, \ldots, 6$ in Figure 7.2, where the phenomenon of high similarity on $\langle -1, 1 \rangle$ and of high dynamic range of x^n outside this interval can be observed. The kernel functions of the 2D Legendre moments on $\langle -1, 1 \rangle \times \langle -1, 1 \rangle$ are shown in Figure 7.3.

If we denote the coefficients of the Legendre polynomials a_{ni}

$$P_n(x) = \sum_{i=0}^{n} a_{ni} x^i, \tag{7.16}$$

---

[8] Most other OG polynomials preserve this property as well, so we will not explicitly mention it in the next sections.

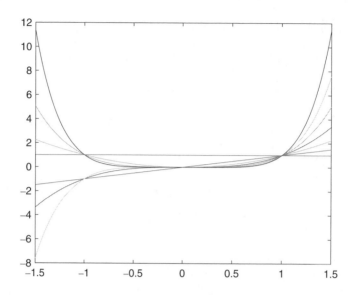

Figure 7.2   The graphs of the standard powers x^n up to the sixth degree

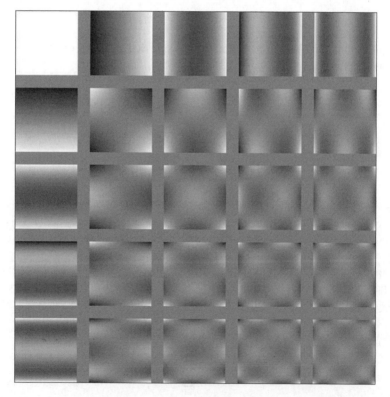

Figure 7.3   The graphs of 2D kernel functions of the Legendre moments (2D Legendre polynomials) up to the fourth degree. Black color corresponds to -1, white color to 1

then we can express the Legendre moments in terms of the geometric moments as

$$\lambda_{mn} = \frac{(2m+1)(2n+1)}{4} \sum_{p=0}^{m} \sum_{q=0}^{n} a_{mp} a_{nq} m_{pq}. \tag{7.17}$$

The coefficients a_{mi} are

$$a_{mi} = (-1)^{(m-i)/2} \frac{1}{2^m} \frac{(m+i)!}{\left(\frac{(m-i)}{2}\right)! \left(\frac{(m+i)}{2}\right)! \, i!}, \tag{7.18}$$

if $(m-i)$ is even and $a_{mi} = 0$ for $(m-i)$ being odd. The first four Legendre moments expressed by means of the geometric moments then are

$$\lambda_{00} = \frac{1}{4} m_{00},$$

$$\lambda_{10} = \frac{3}{4} m_{10},$$

$$\lambda_{20} = \frac{5}{8}(3m_{20} - m_{00}), \tag{7.19}$$

$$\lambda_{11} = \frac{9}{4} m_{11}.$$

In case of a digital image of the size $M \times N$, we calculate the *discrete Legendre moments* as

$$L_{mn} = \frac{(2m+1)(2n+1)}{4MN} \sum_{x=0}^{M-1} \sum_{y=0}^{N-1} P_m\left(-1+2\frac{x}{M-1}\right) P_n\left(-1+2\frac{y}{N-1}\right) f(x,y). \tag{7.20}$$

Legendre moments can be easily made invariant to image translation and scaling in an analogous way as geometric moments [6, 7]. They may serve as the basis for construction of invariants to image blurring [8] and to various combinations of blur and spatial transformations [9] (although other moments are more powerful for this purpose). They have been applied to image watermarking [10], to blocking artifacts removal [11], and in biometrics to hand-shape recognition [12], fingerprint recognition [13], and face recognition [14]. The use of Legendre moments in character recognition, namely for non-roman alphabets, was reported in [15–17] and for handwritten digit recognition in [18]. A modification of Legendre polynomials such that their zeros are concentrated more densely around the origin was proposed in [19].

Efficient algorithms to calculate the discrete Legendre moments can be found in [20–24].

7.2.3 Chebyshev moments[9]

Chebyshev moments were introduced into image processing by Mukundan who published a series of papers [25–28] but were also used by other authors because of their fast and stable

[9] P. L. Chebyshev, 1821–1894, Russian mathematician. There exist several Latin transcriptions of his name (П. Л. Чебышёв in Russian) e.g. Tchebichef and Tschebischeff.

implementation. We distinguish between Chebyshev polynomials of the first kind (which have been used more frequently in image analysis) and Chebyshev polynomials of the second kind.

The n-th degree *Chebyshev polynomial* of the first kind can be expressed by the recurrence relation

$$\begin{aligned} T_0(x) &= 1, \\ T_1(x) &= x, \\ T_n(x) &= 2xT_{n-1}(x) - T_{n-2}(x). \end{aligned} \tag{7.21}$$

Alternative expressions are by means of trigonometric functions

$$T_n(x) = \cos(n \arccos x) = \cosh(n \operatorname{arccosh} x) \tag{7.22}$$

in terms of standard powers (except $n = 0$)

$$\begin{aligned} T_n(x) &= \frac{(x + \sqrt{x^2 - 1})^n + (x - \sqrt{x^2 - 1})^n}{2} \\[2mm] &= \sum_{k=0}^{[n/2]} \binom{2n}{n} (x^2 - 1)^k x^{n-2k} \\[2mm] &= \frac{n}{2} \sum_{m=0}^{[n/2]} (-1)^m \frac{(n - m - 1)!}{m!(n - 2m)!} (2x)^{n-2m} \end{aligned} \tag{7.23}$$

and by hypergeometric functions

$$T_n(x) = {}_2F_1\left(-n, n; \frac{1}{2}; \frac{1 - x}{2}\right). \tag{7.24}$$

The Chebyshev polynomials can be obtained as a solution of the Chebyshev differential equation

$$(1 - x^2)\frac{d^2 y}{dx^2} - x\frac{dy}{dx} + n^2 y = 0. \tag{7.25}$$

Explicit expressions of the Chebyshev polynomials of low degrees in terms of standard powers are

$$\begin{aligned} T_2(x) &= 2x^2 - 1, \\ T_3(x) &= 4x^3 - 3x, \\ T_4(x) &= 8x^4 - 8x^2 + 1, \\ T_5(x) &= 16x^5 - 20x^3 + 5x, \\ T_6(x) &= 32x^6 - 48x^4 + 18x^2 - 1. \end{aligned} \tag{7.26}$$

The Chebyshev polynomials are orthogonal on $\Omega_1 = \langle -1, 1\rangle$ with the weight $w(x) = \frac{1}{\sqrt{1-x^2}}$. The orthogonality relation is

$$\int_{-1}^{1} \frac{1}{\sqrt{1 - x^2}} T_n(x) T_m(x) \, dx = \begin{cases} 0 & : \ n \neq m \\ \pi & : \ n = m = 0 \ , \\ \pi/2 & : \ n = m \neq 0 \end{cases} \tag{7.27}$$

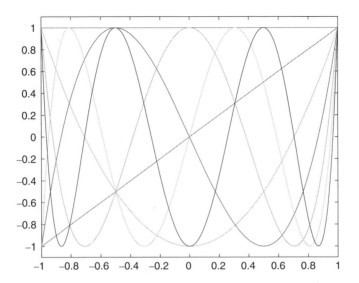

Figure 7.4   The graphs of the Chebyshev polynomials of the first kind up to the sixth degree

For the product of the Chebyshev polynomials, the following formula is sometimes useful

$$2T_n(x)T_m(x) = T_{n+m}(x) + T_{|n-m|}(x).$$

The graphs of the Chebyshev polynomials of the first kind can be seen in Figure 7.4

Chebyshev polynomials of the second kind have the same recurrence as those of the first kind but start with a different initialization

$$U_0(x) = 1,$$
$$U_1(x) = 2x, \qquad\qquad (7.28)$$
$$U_n(x) = 2xU_{n-1}(x) - U_{n-2}(x).$$

They are also orthogonal on $\langle -1, 1 \rangle$ with the inverse weight $1/w(x)$. The orthogonality relation is

$$\int_{-1}^{1} \sqrt{1-x^2}U_n(x)U_m(x)\ dx = \begin{cases} 0 & : \ n \neq m \\ \pi/2 & : \ n = m \end{cases}, \qquad (7.29)$$

The second-kind polynomials can also be expressed by hypergeometric functions as

$$U_n(x) = (n+1)\,_2F_1\left(-n, n+2; \frac{3}{2}; \frac{1-x}{2}\right). \qquad (7.30)$$

There is a close link between both kinds of Chebyshev polynomials, which may be described by several relations. One of the simplest expressions is

$$2T_n(x) = U_n(x) - xU_{n-1}(x).$$

The graphs of the Chebyshev polynomials of the second kind are depicted in Figure 7.5.

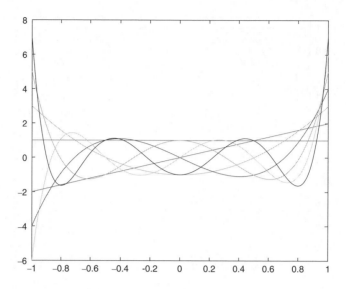

Figure 7.5 The graphs of the Chebyshev polynomials of the second kind up to the sixth degree

The 2D *Chebyshev moments* of the first kind are defined as

$$\tau_{mn} = \iint\limits_{\Omega} T_m(x)T_n(y)f(x,y)\mathrm{d}x\mathrm{d}y \qquad m,n = 0,1,2,\dots \, . \tag{7.31}$$

The respective kernel functions are visualized in Figure 7.6.

The conversion of the geometric moments to the Chebyshev moments is similar to that in the case of the Legendre moments (7.16) and (7.17). We can take the coefficients of the Chebyshev polynomials from (7.23)

$$t_{n,n-2m} = \begin{cases} 0 & m < 0 \ \ \text{or} \ \ m > n/2 \\ 1 & n = 0 \\ \dfrac{n}{2}(-1)^m \dfrac{(n-m-1)!}{m!(n-2m)!}2^{n-2m} & n \neq 0. \end{cases} \tag{7.32}$$

Then we have

$$\tau_{mn} = \sum_{p=0}^{m}\sum_{q=0}^{n} t_{mp}t_{nq}m_{pq} \qquad m,n = 0,1,2,\dots \, . \tag{7.33}$$

An important property of the Chebyshev moments, that is useful particularly in digital signal/image processing, is that they preserve orthogonality even on discrete images; see Section 7.2.6.

Chebyshev and discrete Chebyshev moments were applied to measuring of image focus [29] to recognition of deformed characters by moment matching [30], and tested for image reconstruction [31].

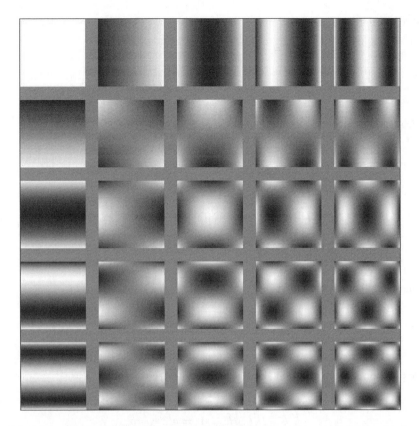

Figure 7.6 The graphs of 2D kernel functions of the Chebyshev moments of the first kind up to the fourth order. Black color corresponds to -1, white color to 1

7.2.4 Gaussian-Hermite moments[10]

Although named after Hermite, these polynomials were firstly defined by Laplace and studied in detail by Chebyshev. Hermite rediscovered them in [32].

Hermite polynomial of the n-th degree is defined as[11]

$$H_n(x) = (-1)^n e^{x^2} \frac{d^n}{dx^n} e^{-x^2}. \tag{7.34}$$

The Hermite polynomials can be expressed by means of the hypergeometric functions as

$$H_{2n}(x) = (-1)^n \frac{(2n)!}{n!} {}_1F_1\left(-n, \frac{1}{2}; x^2\right),$$

$$H_{2n+1}(x) = (-1)^n \frac{(2n+1)!}{n!} 2x \, {}_1F_1\left(-n, \frac{3}{2}; x^2\right). \tag{7.35}$$

---

[10] C. Hermite (1822–1901), French mathematician.
[11] These polynomials are sometimes called the *physicists' Hermite polynomials* in order to distinguish them from the *probabilists' Hermite polynomials*, the definition of which is slightly different – it uses $x/\sqrt{2}$ instead of x.

Their three-term recurrence relation is

$$H_0(x) = 1,$$

$$H_1(x) = 2x,$$

$$H_n(x) = 2xH_{n-1}(x) - 2(n-1)H_{n-2}(x)$$

(7.36)

which leads to the following explicit expressions for low degrees

$$H_2(x) = 4x^2 - 2,$$

$$H_3(x) = 8x^3 - 12x,$$

$$H_4(x) = 16x^4 - 48x^2 + 12,$$

$$H_5(x) = 32x^5 - 160x^3 + 120x,$$

$$H_6(x) = 64x^6 - 480x^4 + 720x^2 - 120.$$

(7.37)

The general explicit expression is

$$H_n(x) = n! \sum_{k=0}^{[n/2]} \frac{(-1)^k (2x)^{n-2k}}{k!(n-2k)!}.$$

Hermite polynomials are orthogonal on $(-\infty, \infty)$ with the weight $w(x) = e^{-x^2}$

$$\int_{-\infty}^{\infty} e^{-x^2} H_n(x) H_m(x) dx = n! 2^n \sqrt{\pi} \delta_{nm}.$$

(7.38)

The plots of the Hermite polynomials up to the sixth degree can be seen in Figure 7.7. We can observe a high dynamic range, which makes them difficult to use directly without any normalization. To overcome this, we modulate Hermite polynomials with a Gaussian function and scale them. This normalization yields *Gaussian-Hermite polynomials*

$$\hat{H}_n(x, \sigma) = e^{-\frac{x^2}{2\sigma^2}} \frac{1}{\sqrt{n! 2^n \sigma \sqrt{\pi}}} H_n(x/\sigma),$$

(7.39)

which are not only orthogonal but also orthonormal

$$\int_{-\infty}^{\infty} \hat{H}_n(x, \sigma) \hat{H}_m(x, \sigma) dx = \delta_{nm}.$$

(7.40)

For the plot of Gaussian-Hermite polynomials, see Figure 7.8. The polynomials defined exactly by (7.39) could be called weighted Hermite polynomials, but the factor $n!$ causes sometimes difficulties and must be modified, therefore the name Gaussian-Hermite polynomials is preferred.

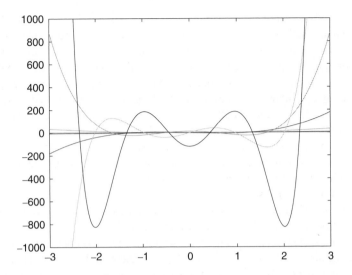

Figure 7.7   The graphs of the Hermite polynomials up to the sixth degree

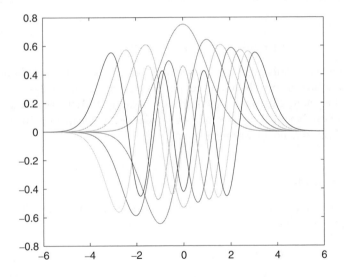

Figure 7.8   The graphs of the Gaussian-Hermite polynomials with $\sigma = 1$

2D Gaussian-Hermite (GH) moments

$$\hat{\eta}_{pq} = \int\limits_{-\infty}^{\infty} \int\limits_{-\infty}^{\infty} \hat{H}_p(x, \sigma_1)\hat{H}_q(y, \sigma_2)f(x, y)\mathrm{d}x\mathrm{d}y \tag{7.41}$$

are a very useful tool for image description. The scale parameter σ appears both in the argument of the polynomials and in the modulating Gaussian function. It should be set up such that the image is sufficiently covered by the central part the Gaussian. Large σ leads to ineffective modulation, while small σ results in too much suppression of boundary regions. The choice of σ influences the performance of the moments, but there is no exact rule how to set it up. Finding appropriate σ is a heuristic which depends on the image size and content.

The GH moments were probably introduced into image analysis area by Shen [33, 34], who also reported (with co-authors) a successful application of localized GH to detection of moving objects in a video [35]. In addition to that, the GH moments have found applications in license plate recognition [36], in image registration as landmark descriptors [37], in fingerprint classification [38], in face recognition [39, 40], and as directional feature extractors [41]. The GH moments have been proven to be very robust w.r.t. additive noise compared to other common moments, which is a significant advantage [42, 43].

In connection with their discrete implementation, the influence of the sampling error was investigated in [44] while an exact and fast algorithm for their calculation, which uses an analytic integration over a pixel area, was proposed in [45].

Probably the most significant breakthrough, which opened new application areas for the GH moments, was achieved by Yang et al. [46, 47]. They discovered that the GH moments offer a possibility of an easy and efficient design of rotation invariants. Their elegant technique was later generalized to moments of arbitrary orders [48] and even to 3D GH moments [49]. We will explain the Yang's method in detail in Section 7.2.7.

7.2.5 Other moments orthogonal on a square

There are a number of orthogonal polynomials, which can be used for definition of orthogonal moments. Many of them have been applied to various image analysis tasks but not as extensively as Legendre, Chebyshev, and Hermite polynomials. Here we give a brief overview that includes only polynomials and moments with reported applications in image analysis. We refer to [1] and [2] for a broader general survey.

Gegenbauer moments

Gegenbauer polynomial of the n-th degree with parameter λ (see also [5] and [50]) is defined by the recurrence formula

$$C_0^{(\lambda)}(x) = 1,$$

$$C_1^{(\lambda)}(x) = 2\lambda x, \tag{7.42}$$

$$(n+1)C_{n+1}^{(\lambda)}(x) = 2(n+\lambda)xC_n^{(\lambda)}(x) - (n+2\lambda-1)C_{n-1}^{(\lambda)}(x).$$

$C_n^{(\lambda)}(x)$ would be zero for $\lambda = 0$ and $n > 0$, so different initial values are used in this case to keep $C_n^{(0)}(x)$ nontrivial

$$C_0^{(0)}(x) = 1,$$

$$C_1^{(0)}(x) = 2x, \tag{7.43}$$

$$C_2^{(0)}(x) = 2x^2 - 1.$$

The explicit expression

$$C_n^{(\lambda)}(x) = \frac{1}{\Gamma(\lambda)} \sum_{m=0}^{[n/2]} (-1)^m \frac{\Gamma(\lambda + n - m)}{m! \ (n - 2m)!} (2x)^{n-2m} \qquad (7.44)$$

is valid for any $\lambda \neq 0$.

Gegenbauer polynomials are orthogonal on $\langle -1, 1 \rangle$ with the weight $w(x) = (1 - x^2)^{\lambda - 1/2}$:

$$\int_{-1}^{1} (1 - x^2)^{\lambda - 1/2} C_n^{(\lambda)}(x) C_m^{(\lambda)}(x) \ dx = \begin{cases} \dfrac{\pi 2^{1-2\lambda} \Gamma(n + 2\lambda)}{(n + \lambda) \ n! \ \Gamma^2(\lambda)} \delta_{nm} & : \ \lambda \neq 0 \\ \dfrac{2\pi}{n^2} \delta_{nm} & : \ \lambda = 0 \end{cases} . \qquad (7.45)$$

The Gegenbauer polynomials can be expressed in terms of the hypergeometric functions as

$$C_n^{(\lambda)}(x) = \frac{\Gamma(n + 2\lambda)}{n! \Gamma(2\lambda)} \ {}_2F_1 \left(-n, n + 2\lambda; \lambda + \frac{1}{2}; \frac{1 - x}{2} \right). \qquad (7.46)$$

They can be also obtained as a solution of the Gegenbauer differential equation

$$(1 - x^2)\frac{d^2 y}{dx^2} - (2\lambda + 1)x\frac{dy}{dx} + n(2\lambda + n)y = 0, \qquad (7.47)$$

where $\lambda > -1/2$.

The Gegenbauer polynomials can be understood as a generalization of the previous polynomials. We obtain the Legendre polynomials for $\lambda = 1/2$ (compare (7.42) with (7.13)) and the Chebyshev polynomials of the second kind for $\lambda = 1$. If we use special initial values (7.43) in the case of $\lambda = 0$, then we obtain a $2/n$ multiple of the Chebyshev polynomials of the first kind. So, the graphs in Figure 7.9 for $\lambda = 0.75$ are somewhere "in between" the Legendre polynomials and the Chebyshev polynomials of the second kind.

Gegenbauer moments were used for object recognition [51, 52] and for Chinese [50, 53] and Arabic character recognition [54]. An efficient algorithm for their computation was proposed in [55].

Jacobi moments

Jacobi polynomial of the n-th degree with parameters $\alpha > -1$ and $\beta > -1$ can be obtained from the recurrence

$$P_0^{(\alpha,\beta)} = 1,$$

$$P_1^{(\alpha,\beta)} = \frac{1}{2}[\alpha - \beta + (\alpha + \beta + 2)x],$$

$$2(n + 1)(n + \alpha + \beta + 1)(2n + \alpha + \beta)P_{n+1}^{(\alpha,\beta)}(x) \qquad (7.48)$$

$$= ((2n + \alpha + \beta + 1)(\alpha^2 - \beta^2) + (2n + \alpha + \beta)_3 x)P_n^{(\alpha,\beta)}(x)$$

$$-2(n + \alpha)(n + \beta)(2n + \alpha + \beta + 2)P_{n-1}^{(\alpha,\beta)}(x),$$

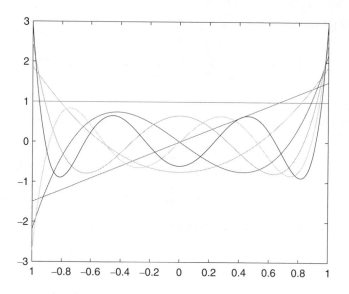

Figure 7.9   The graphs of the Gegenbauer polynomials for $\lambda = 0.75$ up to the sixth degree

where $(a)_n$ is the Pochhammer symbol (7.7). Equivalently, they can be expressed explicitly as

$$P_n^{(\alpha,\beta)}(x) = \frac{1}{2^n} \sum_{m=0}^{n} \binom{n+\alpha}{m} \binom{n+\beta}{n-m} (x-1)^{n-m}(x+1)^m, \qquad (7.49)$$

in terms of the hypergeometric functions as

$$P_n^{(\alpha,\beta)}(x) = \binom{n+\alpha}{n} {}_2F_1\left(-n, n+\alpha+\beta+1; \alpha+1; \frac{1-x}{2}\right), \qquad (7.50)$$

and as a solution of the Jacobi differential equation

$$(1-x^2)\frac{d^2y}{dx^2} + [\beta - \alpha - (\alpha+\beta+2)x]\frac{dy}{dx} + n(\alpha+\beta+n+1)y = 0. \qquad (7.51)$$

The weighted orthogonality on $\langle -1, 1\rangle$ is

$$\int_{-1}^{1} (1-x)^\alpha (1+x)^\beta P_n^{(\alpha,\beta)}(x)P_m^{(\alpha,\beta)}(x)\, dx = \frac{2^{\alpha+\beta+1}\Gamma(n+\alpha+1)\Gamma(n+\beta+1)}{(2n+\alpha+\beta+1)\quad n!\quad \Gamma(n+\alpha+\beta+1)}\,\delta_{nm}.$$
$$(7.52)$$

We can view the Jacobi polynomials as a further generalization of the Gegenbauer polynomials. We obtain the Gegenbauer polynomials from them as a special case with settings $\alpha = \beta = \lambda - 1/2$:

$$C_n^{(\lambda)}(x) = \frac{\Gamma(2\lambda+n)\Gamma(\lambda+1/2)}{\Gamma(2\lambda)\Gamma(\lambda+n+1/2)} P_n^{(\lambda-1/2,\lambda-1/2)}(x). \qquad (7.53)$$

Applications of the Jacobi moments in object recognition were reported in [56] and [57].

We can find another version of Jacobi polynomials in the literature, which are orthogonal on $\langle 0, 1 \rangle$. They are parameterized by p and q such that $p - q > -1$ and $q > 0$. The recurrence formula is

$$
\begin{aligned}
G_0^{(p,q)} &= 1, \\
G_1^{(p,q)} &= x - \frac{q}{p+1}, \\
(2n + p - 2)_4 (2n &+ p - 1) G_{n+1}^{(p,q)}(x) \\
&= [(2n + p - 2)_4(2n + p - 1)x - (2n(n+p) + q(p-1))(2n + p - 2)_3] G_n^{(p,q)}(x) \\
&\quad - n(n + p - 1)(n + q - 1)(n + p - q)(2n + p + 1) G_{n-1}^{(p,q)}(x),
\end{aligned}
\tag{7.54}
$$

and their explicit form is

$$
G_n^{(p,q)}(x) = \frac{\Gamma(q+n)}{\Gamma(p+2n)} \sum_{m=0}^{n} (-1)^m \binom{n}{m} \frac{\Gamma(p + 2n - m)}{\Gamma(q + n - m)} x^{n-m}.
\tag{7.55}
$$

The relation of orthogonality is

$$
\int_0^1 (1 - x)^{p-q} x^{q-1} G_n^{(p,q)}(x) G_m^{(p,q)}(x) \, dx = \frac{n! \Gamma(n+p)\Gamma(n+q)\Gamma(n+p-q+1)}{(2n+p)\Gamma^2(2n+p)} \delta_{nm}.
\tag{7.56}
$$

The relation between both types of Jacobi polynomials is

$$
G_n^{(p,q)}(x) = \frac{n! \Gamma(n+p)}{\Gamma(2n+p)} P_n^{(p-q,q-1)}(2x - 1)
\tag{7.57}
$$

and similarly in the opposite direction

$$
P_n^{(\alpha,\beta)}(x) = \frac{\Gamma(2n + \alpha + \beta + 1)}{n! \Gamma(n + \alpha + \beta + 1)} G_n^{(\alpha+\beta+1,\beta+1)} \left(\frac{x+1}{2} \right).
\tag{7.58}
$$

Laguerre moments

Generalized Laguerre polynomial of the n-th degree with parameter $\alpha > -1$ is defined as

$$
L_n^{(\alpha)}(x) = \frac{x^{-\alpha} e^x}{n!} \frac{d^n}{dx^n} x^{n+\alpha} e^{-x}
\tag{7.59}
$$

or explicitly as

$$
L_n^{(\alpha)}(x) = \sum_{m=0}^{n} (-1)^m \binom{n + \alpha}{n - m} \frac{1}{m!} x^m.
\tag{7.60}
$$

The Laguerre polynomials can be expressed by means of the hypergeometric functions as

$$
L_n^{(\alpha)}(x) = \frac{(\alpha + 1)_n}{n!} \, {}_1F_1(-n, \alpha + 1; x).
\tag{7.61}
$$

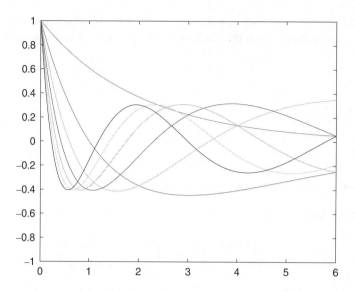

Figure 7.10   The graphs of the weighted Laguerre polynomials up to the sixth degree

The Laguerre polynomials are orthogonal on $\langle 0, \infty \rangle$ with the weight $w(x) = e^{-x} x^{\alpha}$:

$$\int_0^{\infty} e^{-x} x^{\alpha} L_n^{(\alpha)}(x) L_m^{(\alpha)}(x) \, \mathrm{d}x = \frac{\Gamma(n + \alpha + 1)}{n!} \delta_{nm}. \tag{7.62}$$

For $\alpha = 0$ we obtain standard Laguerre polynomials. The weighted Laguerre polynomials can be seen in Figure 7.10.

Since the area of orthogonality is not bounded, it is sufficient to map the image, whose moments we want to calculate, anywhere into $\langle 0, \infty \rangle$. It is not necessary to rescale the image such that it covers the whole interval $\langle 0, \infty \rangle$ as was demonstrated in [58].

7.2.6   Orthogonal moments of a discrete variable

There exist orthogonal polynomials, which fulfill the condition of *discrete orthogonality*. Given N points

$$x_1 < x_2 < \ldots < x_N \tag{7.63}$$

the discrete (weighted) orthogonality condition in 1D is

$$\sum_{\ell=1}^{N} w_{\ell} \, \pi_m(x_{\ell}) \, \pi_n(x_{\ell}) = 0 \tag{7.64}$$

for $m \neq n$. The *discrete moments* are defined as

$$M_p = \sum_{\ell=1}^{N} \pi_p(x_{\ell}) f(x_{\ell}).$$

Since the points x_ℓ may stand for sampling grid nodes, the discrete orthogonality is particularly suitable in digital signal/image analysis applications. A broad survey of polynomials of a discrete variable can be found in [59]. In the following text, we briefly review those which have found applications in image processing.

Discrete Chebyshev moments

Discrete Chebyshev polynomials [26] are defined by recurrence as

$$
\begin{aligned}
T_0(x) &= 1, \\
T_1(x) &= 2x + 1 - N, \\
(n+1)T_{n+1}(x) &= (2n+1)(2x - N + 1)T_n(x) - n(N^2 - n^2)T_{n-1}(x), \qquad n = 1, 2, \ldots
\end{aligned}
\tag{7.65}
$$

or in non-recursive form as

$$
T_n(x) = n! \sum_{k=0}^{n} \binom{N-1-k}{n-k} \binom{n+k}{n} \binom{x}{k}, \qquad n, x = 0, 1, 2, \ldots, N-1.
\tag{7.66}
$$

They can be also expressed in terms of hypergeometric functions

$$
T_n(x) = (1-N)_n \, {}_3F_2(-n, -x, 1+n; 1, 1-N; 1), \qquad n, x = 0, 1, 2, \ldots, N-1.
\tag{7.67}
$$

They satisfy the relation of orthogonality

$$
\sum_{x=0}^{N-1} T_n(x) T_m(x) = \varrho(n, N) \delta_{mn},
\tag{7.68}
$$

where the normalization factor is

$$
\varrho(n, N) = \frac{N(N^2 - 1)(N^2 - 2^2) \cdots (N^2 - n^2)}{2n + 1} = (2n)! \binom{N+n}{2n+1}, \qquad n = 0, 1, 2, \ldots, N-1.
\tag{7.69}
$$

It is advantageous to work with *orthonormal* polynomials $\hat{T}_n(x)$ whose norm equals one:

$$
\sum_{x=0}^{N-1} (\hat{T}_n(x))^2 = 1.
\tag{7.70}
$$

To find the desired normalization factor, Mukundan [26, 28] proposed normalizing $T_n(x)$ first to the magnitude by N^n and then to normalize them by $\varrho(n, N)$ to obtain the orthonormal polynomials. It resulted in the modified definition of the discrete Chebyshev polynomials

$$
\begin{aligned}
\hat{T}_0(x) &= \frac{1}{\sqrt{N}}, \\
\hat{T}_1(x) &= (2x + 1 - N)\sqrt{\frac{3}{N(N^2 - 1)}}, \\
\hat{T}_n(x) &= (\alpha_1 x + \alpha_2)\hat{T}_{n-1}(x) + \alpha_3 \hat{T}_{n-2}(x),
\end{aligned}
\tag{7.71}
$$

where

$$\alpha_1 = \frac{2}{n}\sqrt{\frac{4n^2-1}{N^2-n^2}},$$

$$\alpha_2 = \frac{1-N}{n}\sqrt{\frac{4n^2-1}{N^2-n^2}}, \qquad (7.72)$$

$$\alpha_3 = \frac{n-1}{n}\sqrt{\frac{2n+1}{2n-3}}\sqrt{\frac{N^2-(n-1)^2}{N^2-n^2}}.$$

The *discrete Chebyshev moments* suitable for digital images are then

$$\tau_{pq} = \sum_{x=0}^{M-1}\sum_{y=0}^{N-1}\hat{T}_m(x)\hat{T}_n(y)f(x,y). \qquad (7.73)$$

Efficient algorithms for calculation of discrete Chebyshev moments can be found in [28, 60–64].

Krawtchouk moments

Krawtchouk moments have become popular in image processing, because their kernel functions – Krawtchouk polynomials – can be made local by a proper setting of their parameters. In other words, if we are interested in subregion of the image only, we can choose the parameters such that the Krawtchouk moments reflect the area of interest. Other moments do not have this localization property.

Krawtchouk polynomials were proposed by Soviet Ukrainian mathematician M. P. Krawtchouk [65] as an extension of Hermite polynomials and were later generalized into a form of Meixner polynomials [66]. Krawtchouk moments were introduced into image processing area by Yap et al. [67, 68].

The *Krawtchouk polynomial* with parameter $p \in \langle 0, 1 \rangle$ is defined as

$$K_n^{(p)}(x,N) = {}_2F_1\left(-n,-x;-N;\frac{1}{p}\right), \qquad (7.74)$$

where the degree $n = 0, 1, \ldots, N$ and x is a discrete variable with the values $0, 1, \ldots, N$[12].
Krawtchouk polynomials satisfy the relation of weighted discrete orthogonality

$$\sum_{x=0}^{N}w(x;p,N)K_m^{(p)}(x,N)K_n^{(p)}(x,N) = \varrho(n;p,N)\delta_{mn}, \qquad (7.75)$$

with the weight function being

$$w(x;p,N) = \binom{N}{x}p^x(1-p)^{N-x} \qquad (7.76)$$

---

[12] Note that the number of samples is $N + 1$.

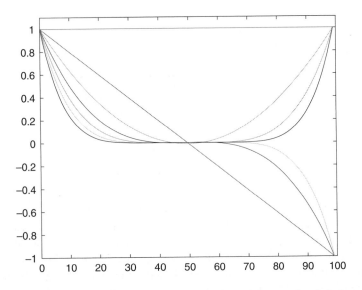

Figure 7.11   The graphs of the Krawtchouk polynomials up to the sixth degree

and with the norm

$$\varrho(n;p,N) = \left(\frac{1-p}{p}\right)^n \frac{1}{\binom{N}{n}}. \tag{7.77}$$

We can also derive the recurrence formula

$$
\begin{aligned}
K_0^{(p)}(x,N) &= 1, \\
K_1^{(p)}(x,N) &= 1 - \frac{x}{Np}, \\
K_{n+1}^{(p)}(x,N) &= \frac{Np - 2np + n - x}{(N-n)p} K_n^{(p)}(x,N) - \frac{n(1-p)}{(N-n)p} K_{n-1}^{(p)}(x,N).
\end{aligned}
\tag{7.78}
$$

The plots of the Krawtchouk polynomials for $p = 0.5$ are shown in Figure 7.11

The Krawtchouk polynomials exhibit better numerical stability if they have been normalized both to the weight and the norm as

$$\hat{K}_n^{(p)}(x,N) = K_n^{(p)}(x,N)\sqrt{\frac{w(x;p,N)}{\varrho(n;p,N)}}. \tag{7.79}$$

Then the modified recurrence formula is

$$
\begin{aligned}
\hat{K}_0^{(p)}(x,N) &= \sqrt{w(x;p,N)}, \\
\hat{K}_1^{(p)}(x,N) &= 1 - \frac{x}{Np}\sqrt{w(x;p,N)}, \\
\hat{K}_{n+1}^{(p)}(x,N) &= A\frac{Np - 2np + n - x}{(N-n)p} \hat{K}_n^{(p)}(x,N) - B\frac{n(1-p)}{(N-n)p}\hat{K}_{n-1}^{(p)}(x,N),
\end{aligned}
\tag{7.80}
$$

where

$$A = \sqrt{\frac{1-p}{p} \frac{n+1}{N-n}},$$

$$B = \sqrt{\frac{(1-p)^2}{p^2} \frac{n(n+1)}{(N-n)(N-n+1)}}.$$

The relation of orthogonality turns to

$$\sum_{x=0}^{N} \hat{K}_m^{(p)}(x, N) \hat{K}_n^{(p)}(x, N) = \delta_{mn}. \qquad (7.81)$$

The polynomials $\hat{K}_n^{(p)}(x, N)$ are called *weighted Krawtchouk polynomials*[67]. The graphs of the weighted Krawtchouk polynomials for $p = 0.5$ can be seen in Figure 7.12 and for $p = 0.2$ in Figure 7.13. The parameter p determines the "localization" of the polynomial in the interval $\langle 0, N \rangle$. If we know *a priori* where in the image our region of interest is located, we can choose such p that the central part of the polynomial is shifted to this area. Having no prior knowledge, a common choice is $p = 0.5$.

Krawtchouk moments are defined as

$$\kappa_{mn} = \sum_{x=0}^{M-1} \sum_{y=0}^{N-1} \hat{K}_m^{(p_1)}(x, M-1) \hat{K}_n^{(p_2)}(y, N-1) f(x, y), \qquad (7.82)$$

In the area of image analysis, the Krawtchouk moments have been used for image compression and reconstruction [67], extraction of local image features [69], face recognition [70], for Chinese character recognition [71] and for image watermarking [72, 73]. Slightly modified Krawtchouk moments were presented in [74] and in [75]. Computational aspects of the

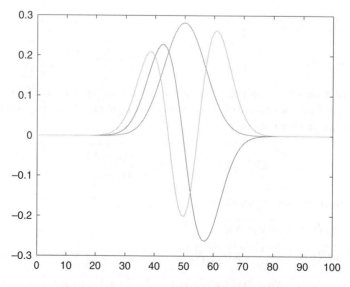

Figure 7.12   The graphs of the weighted Krawtchouk polynomials for $p = 0.5$ up to the second degree

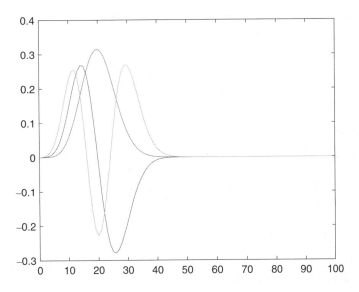

Figure 7.13   The graphs of the weighted Krawtchouk polynomials for $p = 0.2$ up to the second degree

Krawtchouk moments along with efficient algorithms were studied in [76–78] and in a more general form suitable also for Meixner moments in [79].

Discrete Laguerre moments

Discrete generalized Laguerre polynomial of the n-th degree with parameter $\alpha > -1$ is defined as

$$L_n^{(\alpha)}(x) = \frac{(\alpha + 1)_n}{n!} \, {}_1F_1(-n; \alpha + 1; x) \qquad n, x = 0, 1, 2, \dots, N - 1, \tag{7.83}$$

that is, the hypergeometric function is the same as in the continuous case. If α is an integer, then it holds

$$L_n^{(\alpha)}(x) = \frac{1}{\alpha!} \sum_{m=0}^{n} (-1)^m \binom{\alpha + 1}{n - m} \frac{n!}{m!} x^m. \tag{7.84}$$

The polynomials satisfy the orthogonality relation

$$\sum_{x=0}^{N-1} L_n^{(\alpha)}(x) L_m^{(\alpha)}(x) e^{-x} x^{\alpha} = \delta_{nm} \left(\frac{\Gamma(n + \alpha + 1)}{\Gamma(\alpha + 1)n!} \right)^2 N \qquad n = 0, 1, 2, \dots, N - 1. \tag{7.85}$$

The recurrence formula for efficient evaluation is

$$L_0^{(\alpha)}(x) = 1,$$

$$L_1^{(\alpha)}(x) = \alpha + 1 - x, \tag{7.86}$$

$$(n + 1)L_{n+1}^{(\alpha)}(x) = (2n + 1 + \alpha - x)L_n^{(\alpha)}(x) - (n + \alpha)L_{n-1}^{(\alpha)}(x), \quad n = 1, 2, \dots$$

The *discrete Laguerre moments* of image $f(x, y)$ are then [58]

$$\lambda_{mn} = \frac{\Gamma^2(\alpha + 1)m!n!}{\Gamma(m + \alpha + 1)\Gamma(n + \alpha + 1)\sqrt{MN}} \cdot \sum_{x=0}^{M-1}\sum_{y=0}^{N-1} e^{-(x+y)/2}x^{\alpha/2}y^{\alpha/2}L_m^{(\alpha)}(x)L_n^{(\alpha)}(y)f(x, y).$$

(7.87)

Hahn moments

Hahn polynomials, introduced originally by Chebyshev and rediscovered by Hahn [80] can be considered to be an extension of Chebyshev and Krawtchouk polynomials.

The n-th degree Hahn polynomial with parameters α and β of the discrete variable $x = 0, 1, \ldots N$ is defined as

$$Q_n(x, \alpha, \beta, N) = {}_3F_2(-n, n + \alpha + \beta + 1, -x; \alpha + 1, -N; 1),$$

(7.88)

where $\alpha > -1 \wedge \beta > -1 \vee \alpha < -N \wedge \beta < -N$ and $n = 0, 1, \ldots, N$. The Hahn polynomials satisfy the relation of discrete orthogonality

$$\sum_{x=0}^{N} w(x)Q_m(x)Q_n(x) = \varrho(n)\delta_{mn},$$

(7.89)

where $Q_n(x)$ stands for $Q_n(x, \alpha, \beta, N)$. The weight function is

$$w(x) = \binom{\alpha + x}{x}\binom{\beta + N - x}{N - x}$$

(7.90)

and the norm is

$$\varrho(n) = \frac{(-1)^n n!(n + \alpha + \beta + 1)_{N+1}(\beta + 1)_n}{N!(2n + \alpha + \beta + 1)(\alpha + 1)_n(-N)_n}.$$

(7.91)

The weighted Hahn polynomials are then given as

$$\hat{Q}_n(x) = Q_n(x)\sqrt{\frac{w(x)}{\varrho(n)}}.$$

(7.92)

Their recurrence formula is

$$Q_0(x) = 1,$$

$$Q_1(x) = 1 - \frac{\alpha + \beta + 2}{\alpha + 1}\frac{x}{N},$$

(7.93)

$$A_nQ_{n+1}(x) = (A_n + C_n - x)Q_n(x) - C_nQ_{n-1}(x),$$

where

$$A_n = \frac{(n + \alpha + \beta + 1)(n + \alpha + 1)(N - n)}{(2n + \alpha + \beta + 1)(2n + \alpha + \beta + 2)},$$

$$C_n = \frac{n(n + \beta)(n + \alpha + \beta + N + 1)}{(2n + \alpha + \beta)(2n + \alpha + \beta + 1)}.$$

If we set $\alpha = pt$ and $\beta = (1-p)t$ and let $t \to \infty$, we obtain the weighted Krawtchouk polynomials

$$\lim_{t \to \infty} \hat{Q}_n(x, \alpha, \beta, N) = \hat{K}_n^{(p)}(x, N). \tag{7.94}$$

For $\alpha = 0$ and $\beta = 0$ we obtain the discrete Chebyshev polynomials

$$\hat{Q}_n(x, 0, 0, N) = T_n(x). \tag{7.95}$$

In this sense, the Hahn polynomials are generalization both of the Krawtchouk and discrete Chebyshev polynomials. If the discrete Chebyshev polynomials are too "global" and the Krawtchouk polynomials are too "local" for our application, we can look for a suitable trade-off by tuning the parameters. If $\alpha + \beta = 0$, the moments are totally global. If, on the other hand, $\alpha + \beta$ are high ($\alpha + \beta > 20N$), the moments become completely local with $p = \alpha/(\alpha + \beta)$ being the position parameter. If we choose the value of $\alpha + \beta$ between 0 and $20N$, we obtain a reasonable trade-off between globality and locality.

Hahn moments were firstly used for image analysis purposes in [81], where the main topic was image compression and reconstruction. More detailed description of the properties of Hahn moments was given in [82]. Translation and scale-invariant Hahn moments were proposed in [83]. In [84] the author proposed 2D *non-separable* Hahn polynomials and moments, and used them for image compression and reconstruction. Among several applications of Hahn moments, face recognition [85], image watermarking [86], and estimation of the orientation of simple binary patterns [87] are worth mentioning. Attention has been paid also to fast algorithms [88, 89].

Dual Hahn moments

Dual Hahn polynomials were obtained by interchanging the variable x and parameter n in the definition of Hahn polynomials. This action leads to a non-uniform lattice $x(s) = s(s + \gamma + \delta + 1)$, where s is a serial number of the sample and $x(s)$ is its distance from the origin. If we change the notation such that s is used instead of the previous x, we obtain the definition

$$W_n(x(s), \gamma, \delta, N) = {}_3F_2(-n, -s, s + \gamma + \delta + 1; \gamma + 1, -N; 1). \tag{7.96}$$

Zhu and al. [90] adapted them for use in image processing such that s stands for the pixel coordinates. Here we use their approach although their definition differs slightly from some other papers, such as [59]. The lattice is now $x(s) = s(s + 1)$. The n-th degree dual Hahn polynomials with parameter c on the interval $\langle a, b - 1 \rangle$ are then defined as

$$\begin{aligned} & W_n^{(c)}(s, a, b) \\ &= \frac{(a - b + 1)_n (a + c + 1)_n}{n!} \, {}_3F_2(-n, a - s, a + s + 1; a - b + 1, a + c + 1; 1), \end{aligned} \tag{7.97}$$

where $n = 0, 1, \ldots, N - 1$ and $s = a, a + 1, \ldots, b - 1$. The parameters are assumed to fulfill the conditions

$$-\frac{1}{2} < a < b, \qquad |c| < 1 + a, \qquad b = a + N. \tag{7.98}$$

The dual Hahn polynomials satisfy the relation of discrete orthogonality

$$\sum_{s=a}^{b-1} W_m^{(c)}(s, a, b) W_n^{(c)}(s, a, b) \varrho(s)(2s + 1) = d_n^2 \delta_{mn}. \tag{7.99}$$

The weight function ϱ is

$$\varrho(s) = \frac{\Gamma(a + s + 1)\Gamma(c + s + 1)}{\Gamma(s - a + 1)\Gamma(b - s)\Gamma(b + s + 1)\Gamma(s - c + 1)} \tag{7.100}$$

and the norm is

$$d_n^2 = \frac{\Gamma(a + c + n + 1)}{n!(b - a - n - 1)!\Gamma(b - c - n)}. \tag{7.101}$$

Dual Hahn polynomials have also a weighted (normalized) version with better numerical stability

$$\hat{W}_n^{(c)}(s, a, b) = W_n^{(c)}(s, a, b)\sqrt{\frac{\varrho(s)}{d_n^2}(2s + 1)}. \tag{7.102}$$

The recurrence formula of the weighted polynomials is

$$\hat{W}_0^{(c)}(s, a, b) = \sqrt{\frac{\varrho(s)}{d_n^2}(2s + 1)},$$

$$\hat{W}_1^{(c)}(s, a, b) = -\frac{\varrho(s + 1)(s + 1 - a)(s + 1 + b)(s + 1 - c) - \varrho(s)(s - a)(s + b)(s - c)}{\varrho(s)(2s + 1)}$$

$$\cdot \sqrt{\frac{\varrho(s)}{d_n^2}(2s + 1)}, \tag{7.103}$$

$$\hat{W}_{n+1}^{(c)}(s, a, b) = A\frac{d_n}{d_{n+1}}\hat{W}_n^{(c)}(s, a, b) - B\frac{d_{n-1}}{d_{n+1}}\hat{W}_{n-1}^{(c)}(s, a, b),$$

where

$$A = \frac{1}{n + 1}[s(s + 1) - ab + ac - bc - (b - a - c - 1)(2n + 1) + 2n^2],$$

$$B = \frac{1}{n + 1}(a + c + n)(b - a - n)(b - c - n).$$

The orthogonality relation obtains the form

$$\sum_{s=a}^{b-1} \hat{W}_m^{(c)}(s, a, b)\hat{W}_n^{(c)}(s, a, b) = \delta_{mn}. \tag{7.104}$$

If the parameters $a = (\alpha + \beta)/2$, $b = a + N$, and $c = (\beta - \alpha)/2$, then the dual Hahn polynomials become Hahn polynomials $Q_n(x, \alpha, \beta, N)$. If $\alpha = \beta = 0$, then the dual Hahn polynomials reduce to the discrete Chebyshev polynomials.

After the first study on using dual Hahn moments in image processing had been published [90], their applications to image contrast enhancement [91] and to image classification [92] appeared.

Racah moments

Racah polynomials were defined by Wilson [93] and named after an Israeli physicist and mathematician G. Racah. Racah polynomials have many common properties with dual Hahn polynomials, including a non-uniform lattice $x(s) = s(s + 1)$. One may find several versions of Racah polynomials in the literature. Here we use the adaptation from [94], which is suitable for image processing.

The n-th degree Racah polynomial with parameters α and β on the interval $\langle a, b - 1 \rangle$ is defined by hypergeometric functions as

$$U_n^{(\alpha,\beta)}(s, a, b) = \frac{(a - b + 1)_n(\beta + 1)_n(a + b + \alpha + 1)_n}{n!}$$

$$\cdot {}_4F_3(-n, \alpha + \beta + n + 1, a - s, a + s + 1; \beta + 1, a - b + 1, a + b + \alpha + 1; 1),$$

$$(7.105)$$

where $n = 0, 1, \ldots, L - 1$ and $s = a, a + 1, \ldots, b - 1$. We assume that

$$-\frac{1}{2} < a < b, \qquad \alpha > -1, \qquad -1 < \beta < 2a + 1, \qquad b = a + N. \tag{7.106}$$

The Racah polynomials satisfy the relation of discrete orthogonality

$$\sum_{s=a}^{b-1} U_m^{(\alpha,\beta)}(s, a, b) U_n^{(\alpha,\beta)}(s, a, b) \varrho(s)(2s + 1) = d_n^2 \delta_{mn}. \tag{7.107}$$

The weight function ϱ is

$$\varrho(s) = \frac{\Gamma(a + s + 1)\Gamma(s - a + \beta + 1)\Gamma(b + \alpha - s)\Gamma(b + \alpha + s + 1)}{\Gamma(a - \beta + s + 1)\Gamma(s - a + 1)\Gamma(b - s)\Gamma(b + s + 1)} \tag{7.108}$$

and the norm is

$$d_n^2 = \frac{\Gamma(\alpha + n + 1)\Gamma(\beta + n + 1)\Gamma(b - a + \alpha + \beta + n + 1)\Gamma(a + b + \alpha + n + 1)}{(\alpha + \beta + 2n + 1)n!(b - a - n - 1)!\Gamma(\alpha + \beta + n + 1)\Gamma(a + b - \beta - n)}. \tag{7.109}$$

As in the case of dual Hahn polynomials, the Racah polynomials also have a weighted version which exhibits better numerical properties

$$\hat{U}_n^{(\alpha,\beta)}(s, a, b) = U_n^{(\alpha,\beta)}(s, a, b)\sqrt{\frac{\varrho(s)}{d_n^2}(2s + 1)}. \tag{7.110}$$

The recurrence formula of weighted Racah polynomials is

$$\hat{U}_0^{(\alpha,\beta)}(s, a, b) = \sqrt{\frac{\varrho(s)}{d_n^2}(2s + 1)},$$

$$\hat{U}_1^{(\alpha,\beta)}(s, a, b) = -\left(\frac{\varrho(s + 1)(s + 1 - a)(s + 1 + b)(s + 1 + a - \beta)(b + \alpha - s - 1)}{\varrho(s)(2s + 1)} \right.$$

$$\left. - \frac{\varrho(s)(s - a)(s + b)(s + a - \beta)(b + \alpha - s)}{\varrho(s)(2s + 1)} \right)\sqrt{\frac{\varrho(s)}{d_n^2}(2s + 1)}, \tag{7.111}$$

$$A\,\hat{U}_{n+1}^{(\alpha,\beta)}(s, a, b) = B\frac{d_n}{d_{n+1}}\,\hat{U}_n^{(\alpha,\beta)}(s, a, b) - C\frac{d_{n-1}}{d_{n+1}}\,\hat{U}_{n-1}^{(\alpha,\beta)}(s, a, b),$$

where

$$A = \frac{(n+1)(\alpha+\beta+n)}{(\alpha+\beta+2n+1)(\alpha+\beta+2n+2)},$$

$$B = s(s+1) - \frac{a^2+b^2+(a-\beta)^2+(b+\alpha)^2}{4} + \frac{(\alpha+\beta+2n)(\alpha+\beta+2n+2)}{8}$$

$$- \frac{(\beta^2-\alpha^2)[(b+\alpha/2)^2-(a-\beta/2)^2]}{2(\alpha+\beta+2n)(\alpha+\beta+2n+2)},$$

$$C = \frac{(\alpha+n)(\beta+n)}{(\alpha+\beta+2n)(\alpha+\beta+2n+1)}$$

$$\cdot \left[\left(a+b+\frac{\alpha-\beta}{2} \right)^2 - \left(n+\frac{\alpha+\beta}{2} \right)^2 \right] \left[\left(b-a+\frac{\alpha+\beta}{2} \right)^2 - \left(n+\frac{\alpha+\beta}{2} \right)^2 \right].$$

The weighted Racah polynomials satisfy the relation of orthogonality

$$\sum_{s=a}^{b-1} \hat{U}_m^{(\alpha,\beta)}(s,a,b)\hat{U}_n^{(\alpha,\beta)}(s,a,b) = \delta_{mn}. \tag{7.112}$$

Applications of Racah moments in face recognition [95], handwritten digit recognition [54], Chinese character recognition [96], and skeletonization of craft images [97] have been reported.

7.2.7 Rotation invariants from moments orthogonal on a square

As we already mentioned, the moments orthogonal on a square are inherently inappropriate for the design of rotation invariants, which substantially limits their use in practical object recognition tasks. Basically, the only way to create rotation invariants from these kind of OG moments is to take the formulas for geometric rotation invariants (which are known from Chapter 3) and substitute OG moments for the geometric moments. To do so, we have to express geometric moments in terms of the OG moments. This leads to complicated clumsy expressions (at least each sum is replaced by a double sum) of questionable numerical properties[13]. This is why only a few papers have followed this approach. Yap et al. [67] did it for Krawtchouk moments and Hosny [98] and Deepika et al. [99] for Legendre moments (Hosny calls this technique "indirect approach").

The groundbreaking idea by Yang et al. [46, 47] can be summarized as follows. They discovered that the 2D Hermite polynomials $H_{pq}(x,y) \equiv H_p(x)H_q(y)$ change under an in-plane rotation by angle α in the same way as do the monomials $x^p y^q$. More precisely, if we denote

$$(x^p y^q)' = \sum_{r=0}^{p+q} k(r,p,q,\theta)x^{p+q-r}y^r, \tag{7.113}$$

---

[13] There is, of course, still a possibility of applying normalization and then calculating OG moments of the normalized image but the transformation equations for OG moments are of the same difficulty.

where $k(r, p, q, \alpha)$ are certain coefficients, then the Hermite polynomials are transformed by exactly the same linear combination

$$H_{pq}(x,y)' = \sum_{r=0}^{p+q} k(r,p,q,\alpha)H_{p+q-r}(x)H_r(y). \tag{7.114}$$

This property propagates itself from polynomials to moments. The geometric moments are transformed under rotation according to the rule

$$m'_{pq} = \sum_{k=0}^{p}\sum_{j=0}^{q}(-1)^{p-k}\binom{p}{k}\binom{q}{j}(\cos\alpha)^{q+k-j}(\sin\alpha)^{p-k+j}m_{k+j,p+q-k-j}; \tag{7.115}$$

see also (3.16). The same rule applies to Hermite moments

$$\eta_{pq} = \int_{-\infty}^{\infty}\int_{-\infty}^{\infty} H_p(x)H_q(y)f(x,y)\mathrm{d}x\mathrm{d}y \tag{7.116}$$

which yields

$$\eta'_{pq} = \sum_{k=0}^{p}\sum_{j=0}^{q}(-1)^{p-k}\binom{p}{k}\binom{q}{j}(\cos\alpha)^{q+k-j}(\sin\alpha)^{p-k+j}\eta_{k+j,p+q-k-j}. \tag{7.117}$$

Hermite moments are the only OG moments with this property among the commonly used moments[14]. From this property, we can derive the assertion of the *Yang's theorem*: If there exist a rotation invariant of geometric moments $I(m_{p_1q_1}, m_{p_2q_2}, \ldots, m_{p_dq_d})$, the same function of the corresponding Hermite moments $I(\eta_{p_1q_1}, \eta_{p_2q_2}, \ldots, \eta_{p_dq_d})$ is also a rotation invariant (see [47] for the detailed proof).

As we explained in Section 7.2.4, we prefer working with Gaussian-Hermite polynomials (and moments) rather than directly with Hermite ones. Fortunately, the Gaussian weighting and scaling do not violate Yang's theorem provided that the scale parameter σ is the same for x and y. However, the weighting coefficient must be modified a little, as explained bellow.

If we define the Gaussian-Hermite moment in a "standard" way as

$$\hat{\eta}_{pq} = \int_{-\infty}^{\infty}\int_{-\infty}^{\infty} \hat{H}_p(x,\sigma)\hat{H}_q(y,\sigma)f(x,y)\mathrm{d}x\mathrm{d}y$$

$$= \frac{1}{\sigma\sqrt{\pi p!q!2^{p+q}}}\int_{-\infty}^{\infty}\int_{-\infty}^{\infty} e^{-\frac{(x^2+y^2)}{2\sigma^2}}H_p(x/\sigma)H_q(y/\sigma)f(x,y)\mathrm{d}x\mathrm{d}y, \tag{7.118}$$

then the factor $p!q!$ violates the proposition. If we replace it by $(p+q)!$ such that we have[15]

$$\hat{\eta}_{pq} = \frac{1}{\sigma\sqrt{\pi(p+q)!2^{p+q}}}\int_{-\infty}^{\infty}\int_{-\infty}^{\infty} e^{-\frac{(x^2+y^2)}{2\sigma^2}}H_p(x/\sigma)H_q(y/\sigma)f(x,y)\mathrm{d}x\mathrm{d}y, \tag{7.119}$$

---

[14] One can find other OG moments with the rotation property but they all are similar to Hermite moments.

[15] For the normalization, we can use any function of $(p+q)$ which grows similarly as $p!q!$. Other possible choices are $\left(\frac{p+q}{2}!\right)^2$ or $\sqrt{(p+q)!}\frac{p+q}{2}!$. The choice of $(p+q)!$ exactly preserves orthonormality for $\hat{\eta}_{p0}$ and $\hat{\eta}_{0q}$; the second choice for $\hat{\eta}_{p/2,p/2}$; and the third choice is a trade-off between them.

Yang's theorem holds well, and the functionals $I(\hat{\eta}_{p_1q_1}, \hat{\eta}_{p_2q_2}, \ldots, \hat{\eta}_{p_dq_d})$ are rotation invariants of the Gaussian-Hermite moments.

The practical applicability of Yang's theorem depends on our ability to find a set (preferably complete) of rotation invariants from geometric moments. As we know from Chapter 3, such invariants exist in explicit forms for low orders only. When we constructed the rotation invariants of high orders, we used complex moments instead of geometric ones (see Chapter 3), which helped a lot thanks to their simple rotation property. We use the same trick here, too. However, no "complex GH moments" exist, so we shall create them first.

Let us define

$$d_{pq} = \sum_{k=0}^{p} \sum_{j=0}^{q} \binom{p}{k} \binom{q}{j} (-1)^{q-j} i^{p+q-k-j} \hat{\eta}_{k+j,p+q-k-j}. \qquad (7.120)$$

Note that d_{pq} has been intentionally defined such that it resembles the complex moment. Hence, d_{pq} works as a bridge between Gaussian-Hermite moments and the expressions of rotation invariants expressed by complex moments. As follows from Yang's theorem, d_{pq} must change under rotation by α in the same way as c_{pq}. Hence,

$$d'_{pq} = e^{-i(p-q)\alpha} d_{pq}. \qquad (7.121)$$

Thanks to that, any product of the form

$$\prod_{i=1}^{n} d_{p_iq_i}^{k_i}, \qquad (7.122)$$

where k_i, p_i, and q_i, $(i = 1, \ldots, n)$ are non-negative integers such that

$$\sum_{i=1}^{n} k_i(p_i - q_i) = 0, \qquad (7.123)$$

cancels the phase shift and provides the invariance to rotation.

Now we can repeat all the steps we performed in Chapter 3 with the complex moments. We construct an independent and complete system of the GH invariants of arbitrary order exactly as we did for the complex moments in Theorem 3.2:

$$\Psi(p,q) \equiv d_{pq}d_{q_0p_0}^{p-q}, \text{ with } p \geq q, p_0 - q_0 = 1 \qquad (7.124)$$

where p_0 and q_0 are fixed user-defined indices such that $d_{p_0q_0} \neq 0$ on the dataset we work with. We can also identify the vanishing invariants due to the N-fold symmetry of the image – if $(p - q)/N$ is not an integer, then $d_{pq} = 0$.

For a detailed discussion on high-order GH invariants, their numerical properties, and the reconstruction power, see [48]. Here we present only an illustrative experiment which verifies their rotation invariance.

We scanned several pexeso cartoon cards (see Figure 7.19 for some examples) in various orientations on the table and calculated the GH invariants of the orders from 12 to 15. Their values are plotted in Figure 7.14 (one card, six orientations, six GH invariants denoted $W_1 = d_{66}$, $W_2 = d_{67}d_{76}$, $W_3 = \mathcal{R}e(d_{76}d_{12})$, $W_4 = \mathcal{R}e(d_{87}d_{12})$, $W_6 = \mathcal{R}e(d_{13,12}d_{12})$.

We can observe some noticeable variances there due to the illumination changes and the sampling effect but still the corresponding mean relative errors are reasonably low: 2.14%,

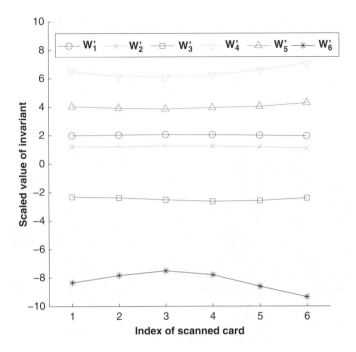

Figure 7.14   The values of the selected invariants computed from the rotated pexeso card

3.48%, 7.71%, 5.11%, 2.96%, and 7.56%, which demonstrates the invariance property under real conditions.

The recognition power of the GH rotation invariants is tested in Section 7.4, along with its comparison to that of Zernike moments.

7.3   2D moments orthogonal on a disk

One of the main motivations to employ moments as the features is easy construction of rotation invariants. As already observed in Chapter 3, an efficient way is working in polar coordinates, where rotation is transformed into a shift. In the case of 2D complex moments, this shift causes a change of the phase that is further eliminated by a multiplication of proper moments. The same idea applies for any circular moments (see Chapter 3). The need for stable numerical computation has led to introducing OG circular moments, often referred to as the moments *orthogonal on a disk*. Unlike the OG moments on a square, the moments orthogonal on a disk do not have an analogy in 1D; their generalization to 3D is possible but not straightforward.

The moments orthogonal on a disk are mostly defined in the form

$$v_{pq} = n_{pq} \int\limits_{0}^{2\pi} \int\limits_{0}^{1} R_{pq}(r)e^{-iq\theta}f(r,\theta)r\mathrm{d}r\mathrm{d}\theta \qquad p = 0, 1, 2, \ldots, q = -p, \ldots, p, \qquad (7.125)$$

where n_{pq} is some normalizing factor, r, θ are polar coordinates (3.17), $R_{pq}(r)$ is a 1D radial polynomial, and $e^{-iq\theta}$ is an angular part of the respective basis function. The area of orthogonality is usually a unit disk $\Omega = \{(x,y)|x^2 + y^2 \leq 1\}$ (after a rescaling, the basis functions can be made orthogonal on an arbitrary disk of a finite radius). To calculate the moments, image f must be appropriately scaled and shifted such that it is fully contained in Ω.

7.3.1 Zernike and Pseudo-Zernike moments[16]

Zernike polynomials were originally developed to describe the diffracted wavefront in phase contrast imaging [100] and have found numerous applications in mathematics and optics. Among others characteristics they efficiently represent common errors of optical systems such as coma, astigmatism, and spherical aberrations.

The idea of using Zernike moments (ZMs) for image description comes from Teague [101], who used Zernike moments to construct rotation invariants. He used the fact that Zernike moments keep their magnitude constant under rotation. He also showed that Zernike invariants of the second and third orders are equivalent to the Hu invariants when expressed in terms of geometric moments. He presented the invariants up to eighth order in explicit form but no general rule for deriving them was given. Later on, Wallin [102] described an algorithm for formation of rotation invariants of any order.

Zernike polynomial of the n-th degree with repetition ℓ is defined as

$$V_{n\ell}(r,\theta) = R_{n\ell}(r)\, e^{i\ell\theta} \qquad n = 0, 1, 2, \ldots, \ell = -n, -n+2, \ldots, n, \qquad (7.126)$$

that is, the difference $n - |\ell|$ is always even. The radial part is given as

$$R_{n\ell}(r) = \sum_{s=0}^{(n-|\ell|)/2} (-1)^s \frac{(n-s)!}{s!\left(\frac{n+|\ell|}{2} - s\right)!\left(\frac{n-|\ell|}{2} - s\right)!} r^{n-2s} = \sum_{k=|\ell|,|\ell|+2,\ldots}^{n} B_{n\ell k} r^k. \quad (7.127)$$

The radial function is symmetric with respect to the repetition, that is, $R_{n,-\ell}(r) = R_{n\ell}(r)$. The graphs of the radial polynomials are in Figure 7.15. Zernike polynomials $V_{n\ell}(r,\theta)$ on the unit disk are visualized in Figure 7.16.

Zernike polynomials satisfy the relation of orthogonality

$$\int_0^{2\pi} \int_0^1 V_{n\ell}^*(r,\theta) V_{mk}(r,\theta) r\, dr\, d\theta = \frac{\pi}{n+1} \delta_{mn} \delta_{k\ell}, \qquad (7.128)$$

which is due to the orthogonality of the radial polynomials

$$\int_0^1 R_{n\ell}(r) R_{m\ell}(r) r\, dr = \frac{1}{2(n+1)} \delta_{mn}. \qquad (7.129)$$

---

[16] F. Zernike (1888–1966), Dutch physicist and the Nobel Prize winner (1953).

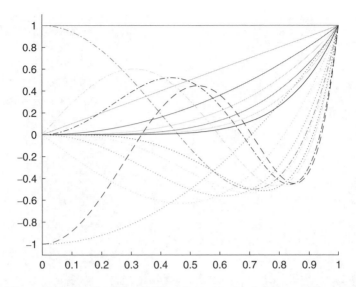

Figure 7.15   The graphs of the Zernike radial functions up to the sixth degree. The graphs of the polynomials of the same degree but different repetition are drawn by the same type of the line

Explicit forms of selected radial polynomials are listed below.

$$R_{00}(r) = 1,$$
$$R_{11}(r) = r,$$
$$R_{20}(r) = 2r^2 - 1,$$
$$R_{22}(r) = r^2,$$
$$R_{31}(r) = 3r^3 - 2r,$$
$$R_{33}(r) = r^3,$$
$$R_{40}(r) = 6r^4 - 6r^2 + 1,$$
$$R_{42}(r) = 4r^4 - 3r^2,$$
$$R_{44}(r) = r^4,$$
$$R_{51}(r) = 10r^5 - 12r^3 + 3r,$$
$$R_{53}(r) = 5r^5 - 4r^3,$$
$$R_{55}(r) = r^5,$$
$$R_{60}(r) = 20r^6 - 30r^4 + 12r^2 - 1,$$
$$R_{62}(r) = 15r^6 - 20r^4 + 6r^2,$$
$$R_{64}(r) = 6r^6 - 5r^4,$$

$$R_{66}(r) = r^6,$$
$$R_{71}(r) = 35r^7 - 60r^5 + 30r^3 - 4r,$$
$$R_{73}(r) = 21r^7 - 30r^5 + 10r^3,$$
$$R_{75}(r) = 7r^7 - 6r^5,$$
$$R_{77}(r) = r^7,$$
$$R_{80}(r) = 70r^8 - 140r^6 + 90r^4 - 20r^2 + 1,$$
$$R_{82}(r) = 56r^8 - 105r^6 + 60r^4 - 10r^2,$$
$$R_{84}(r) = 28r^8 - 42r^6 + 15r^4,$$
$$R_{86}(r) = 8r^8 - 7r^6,$$
$$R_{88}(r) = r^8,$$
$$R_{91}(r) = 126r^9 - 280r^7 + 210r^5 - 60r^3 + 5r,$$
$$R_{93}(r) = 84r^9 - 168r^7 + 105r^5 - 20r^3,$$
$$R_{95}(r) = 36r^9 - 56r^7 + 21r^5,$$
$$R_{97}(r) = 9r^9 - 8r^7,$$
$$R_{99}(r) = r^9.$$

$$(7.130)$$

The radial factor of Zernike polynomials is linked to Jacobi polynomials as

$$R_{m+2s,|m|}(r) = (-1)^{m+2s} \binom{m+s}{s} r^m G_s^{(m+1,m+1)}(r^2) \tag{7.131}$$

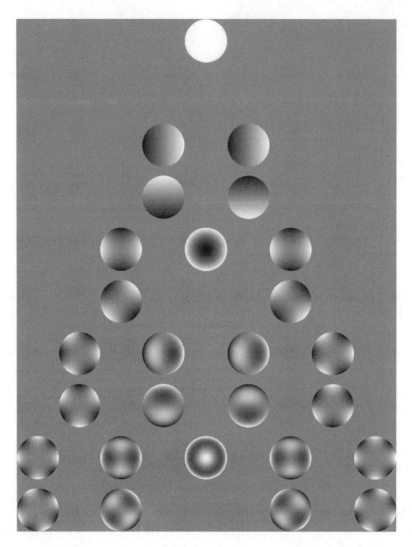

Figure 7.16 The graphs of the Zernike polynomials up to the fourth degree. Black $= -1$, white $= 1$. Real parts: 1st row: $n = 0, \ell = 0$, 3rd row: $n = 1, \ell = -1, 1$, 5th row: $n = 2, \ell = -2, 0, 2$, 7th row: $n = 3$, $\ell = -3, -1, 1, 3$, 9th row: $n = 4, \ell = -4, -2, 0, 2, 4$. Imaginary parts: 2nd, 4th, 6th, 8th and 10th row, respectively. The indices are the same as above

and to Legendre polynomials as

$$R_{2n,0}(r) = P_n(2r^2 - 1).\tag{7.132}$$

The recurrence formula for fast calculation of the radial part is

$$R_{n\ell}(r) = \frac{2rn}{n+\ell}R_{n-1,\ell-1}(r) - \frac{n-\ell}{n+\ell}R_{n-2,\ell}(r).\tag{7.133}$$

The computation by means of this formula starts with

$$R_{nn}(r) = r^n, \qquad R_{n,-n}(r) = r^n, \qquad n = 0, 1, \ldots \qquad (7.134)$$

and then it proceeds with $n = 2, 3, \ldots, \ell = -n + 2, -n + 4, \ldots, n - 2$. This formula, sometimes called the Prata's method, is the simplest but not the fastest recurrence formula. Some other methods for fast evaluation of Zernike moments are discussed in Chapter 8.

Zernike moments are defined as

$$A_{n\ell} = \frac{n+1}{\pi} \int_0^{2\pi} \int_0^1 V_{n\ell}^*(r, \theta) f(r, \theta) r \, dr d\theta. \qquad (7.135)$$

They can be expressed in terms of geometric moments as

$$A_{n\ell} = \frac{n+1}{\pi} \sum_{k=|\ell|,|\ell|+2, \ldots}^{n} \sum_{j=0}^{(k-|\ell|)/2} \sum_{m=0}^{|\ell|} \binom{(k-|\ell|)/2}{j} \binom{|\ell|}{m} w^m B_{n\ell\,k} m_{k-2j-m,2j+m}, \qquad (7.136)$$

where

$$w = \begin{cases} -i & \ell > 0 \\ i & \ell \leq 0 \end{cases} \qquad (7.137)$$

and

$$B_{n\ell\,k} = \frac{(-1)^{\frac{n-k}{2}} \left(\frac{n+k}{2}\right)!}{\left(\frac{n-k}{2}\right)! \left(\frac{k+\ell}{2}\right)! \left(\frac{k-\ell}{2}\right)!}. \qquad (7.138)$$

The behavior of Zernike moments under rotation is similar to that of the complex moments. If we rotate the coordinates by an angle α, the values of the Zernike moments change as

$$A'_{n\ell} = A_{n\ell} e^{i\ell\alpha}. \qquad (7.139)$$

Thus, rotation invariants can be constructed either by taking the magnitude $|A_{n\ell}|$, that leads to an incomplete system with limited recognition power, or by taking proper moment products as in case of the complex moments.

Alternatively, we can apply the normalization approach. The algorithm for determining a proper non-zero normalizing moment is analogous to that from Chapter 3. First, we search the Zernike moments with repetition 1, that is, $A_{31}, A_{51}, \ldots A_{n_{\max}1}$. If they are all zero (i.e., under-threshold), we suppose the object is rotationally symmetric, and then we search the moments with successively increased repetitions 2,3, etc. For non-symmetric objects, the suitable normalizing moment is A_{31}. If the normalizing moment $A_{m_r\ell_r}$ has a phase

$$\theta = \frac{1}{\ell_r} \arctan\left(\frac{Im(A_{m_r\ell_r})}{Re(A_{m_r\ell_r})}\right), \qquad (7.140)$$

then

$$Z_{m\ell} = A_{m\ell} e^{-i\ell\theta} \qquad (7.141)$$

is a rotation invariant. It is an analogy with eq. (3.36). Another method is using the ratios

$$\overline{Z}_{m\ell} = A_{m\ell}/(A_{m_r\ell_r})^{\ell/\ell_r}, \tag{7.142}$$

which is equivalent to the phase cancellation by multiplication according to Theorem 3.1.

If we want to avoid using one specific Zernike moment for normalization to the rotation, we can use the method from [103]. If the Zernike moments of two images f and g are to be compared, the authors look for an angle θ minimizing the sum of square differences

$$d_{f,g}^2(\theta) = \sum_{n=0}^{n_{max}} \sum_{\ell=-n,-n+2,\,...}^{n} \frac{\pi}{n+1}\left|A_{n\ell}^{(f)} - A_{n\ell}^{(g)}e^{i\ell\theta}\right|^2. \tag{7.143}$$

The value of $d_{f,g}^2(\theta)$ in this minimum is used as a dissimilarity measure between f and g.

Zernike moments can also be normalized to the affine transform. The method is similar to the affine normalization from Chapter 5. This approach was used for instance in [104] for pseudo-Zernike moments.

Zernike moments are not inherently invariant to scaling. However, the image must be mapped into the unit disk before the Zernike moments can be calculated. This mapping provides the scaling invariance.

If the image has the size $M \times N$, then the discrete approximation of the Zernike moments is

$$A_{n\ell} = \frac{n+1}{\pi} \sum_{x=0}^{M-1}\sum_{y=0}^{N-1} V_{n\ell}^*(r,\theta)f(x,y), \quad n = 0,1,2,\dots,\ell = -n,-n+2,\dots,n. \tag{7.144}$$

There are several methods of converting Cartesian coordinates (x,y) to polar coordinates (r,θ); their suitability depends on the specific application. One of the most common ones supposes the center of rotation in the centroid of the image

$$r = \frac{\sqrt{(x-x_c)^2 + (y-y_c)^2}}{r_{max}},$$

where

$$r_{max} = \sqrt{\frac{m_{00}}{2}}\sqrt{\frac{M}{N} + \frac{N}{M}}$$

and

$$\theta = \arctan\left(\frac{y-y_c}{x-x_c}\right).$$

Another method uses the center of the image $(M-1)/2, (N-1)/2$ instead of the centroid x_c, y_c and MN instead of m_{00}. The values of the Zernike moments should also be normalized to the sampling density. We can divide the values by the number of pixels in the unit disk (approximately πr_{max}^2), but their division by the A_{00} yields slightly more stable results.

Mapping into the unit disk and normalization to the sampling density can also be provided by the formula (7.135), where we substitute the geometric moments m_{pq} by the central moments normalized to scaling

$$v_{pq} = \frac{\mu_{pq}}{\mu_{00}^{\frac{p+q}{2}+1}}. \tag{7.145}$$

Pseudo-Zernike moments (PZM) differ from the Zernike moments by releasing from the condition $n - |\ell|$ is even. It means the definition (7.134) is valid also for pseudo-Zernike moments, but with $\ell = -n, -n + 1, \ldots, n$. The radial part can then be rewritten as

$$R_{n\ell}(r) = \sum_{s=0}^{n-|\ell|} (-1)^s \frac{(2n+1-s)!}{s!(n+|\ell|+1-s)!(n-|\ell|-s)!} r^{n-s} = \sum_{k=|\ell|}^{n} S_{n|\ell|k} r^k. \qquad (7.146)$$

The consequence is that the number of the Zernike moments up to the n-th order $(n+1)$ $(n+2)/2$ increases to $(n+1)^2$ for the pseudo-Zernike moments. The radius r appears also in odd powers in (7.145), so we must use a special method for conversion of the geometric moments to the pseudo-Zernike moments [104]

$$A_{n\ell} = \frac{n+1}{\pi} \sum_{\substack{s=0 \\ n-\ell-s \text{ even}}}^{n-|\ell|} S_{n|\ell|s} \sum_{j=0}^{(n-|\ell|-s)/2} \sum_{m=0}^{|\ell|} \binom{(n-|\ell|-s)/2}{j} \binom{|\ell|}{m} (-i)^m$$

$$\cdot \, m_{n-s-2j-m,2j+m}$$

$$+ \frac{n+1}{\pi} \sum_{\substack{s=0 \\ n-\ell-s \text{ odd}}}^{n-|\ell|} S_{n|\ell|s} \sum_{j=0}^{(n-|\ell|-s+1)/2} \sum_{m=0}^{|\ell|} \binom{(n-|\ell|-s+1)/2}{j} \binom{|\ell|}{m} (-i)^m$$

$$\cdot \, \overline{m}_{n-s+1-2j-m,2j+m}, \qquad (7.147)$$

where $\overline{m}_{pq}$ are *geometric moments with radial component*

$$\overline{m}_{pq} = \int\limits_{-\infty}^{\infty} \int\limits_{-\infty}^{\infty} x^p y^q \sqrt{x^2 + y^2} f(x, y) \, dx dy \qquad p, q = 0, 1, 2, \ldots. \qquad (7.148)$$

The radius $\sqrt{x^2 + y^2}$ can be expanded into a power series. Therefore, if we have an infinite number of moments of all orders, we can express the moments with a radial component in terms of the geometric moments. From this point of view, the independence of the pseudo-Zernike moments claimed in [4] is questionable. On the other hand, if we had only a limited number of the geometric moments, we would be able to compute the moments with radial component only approximately.

The PZMs preserve the same rotation property as the ZMs, so it is easy to generate PZM rotation invariants. Both ZM and PZM are robust w.r.t. additive noise [105]. In some tests the PZMs exhibited even slightly higher robustness than ZMs. The robustness to noise is considered the main advantage of the PZMs.

Zernike and pseudo-Zernike moments have attracted significant attention in image analysis, because of their favorable transformation under rotation [4, 101, 102, 106]. Several modifications which extend the Zernike/pseudo Zernike moment invariance property to translation [107], affine transformation [108], and image blurring [109, 110] have been published. Other modifications aimed at improving numerical properties, localization, and robustness [103, 111].

Both Zernike and pseudo-Zernike moments have found numerous applications. It is impossible to mention all of them here. We have chosen a few representative ones. In image registration, ZM and PZM may serve as landmark neighborhood descriptors and may be used for rotation-invariant landmark matching [112–114]. In face recognition, they are mostly used as global features, often accompanied by some kind of local features [39, 115–119]. In addition to face recognition, ZM and PZM have been used also in other biometric applications, such as iris recognition [120, 121], fingerprint recognition [122–124], and hand-shape recognition [125]. Other applications include image watermarking [126–128], image retrieval [129–131], and various examples of shape recognition.

Great effort has been made to develop efficient algorithms for fast ZM calculation [132–136]. It is hard job to evaluate objectively which algorithm is the best one because one has to compare not only the speed but also, at the very least, the accuracy and memory requirements. A comparison of some of the above-cited algorithms can be found in [137]. A detailed error analysis of discrete Zernike moments is provided in [138] and [5].

7.3.2   Fourier-Mellin moments

Fourier-Mellin moments (FM) got their name from the Fourier-Mellin transformation. They were first proposed for object recognition by Sheng and Duvernoy in [139]. We may find several definitions of FM; probably the simplest one is

$$M_{s\ell} = \int_0^\infty \int_0^{2\pi} r^s \, e^{-i\ell\theta} \, f(r,\theta) \, d\theta \frac{dr}{r} \, , \tag{7.149}$$

where both s and ℓ are integers and $s \geq 0$. If the image is supposed to be mapped into the unit disk, the outer integration goes from 0 to 1 only. Clearly, FM moments are a particular case of circular moments introduced in Chapter 3. They are closely linked to the complex moments as

$$M_{s\ell} = c_{(s-\ell)/2,(s+\ell)/2}.$$

Sheng and Arsenault realized that the use of orthogonal radial polynomials may increase the performance [140] which later led to introducing *Orthogonal Fourier-Mellin moments* (OFM) by Sheng and Shen [141]. They are defined as

$$\Phi_{n\ell} = \frac{n+1}{\pi} \int_0^{2\pi} \int_0^1 Q_n(r) \, e^{-i\ell\theta} \, f(r,\theta) \, r \, dr d\theta \qquad n = 0,1,2,\ldots, \ell = 0,\pm 1,\pm 2,\ldots, \tag{7.150}$$

where the radial function $Q_n(r)$ is

$$Q_n(r) = \sum_{s=0}^n (-1)^{n+s} \frac{(n+s+1)!}{(n-s)!s!(s+1)!} \, r^s \tag{7.151}$$

is in fact nothing but the Jacobi polynomial

$$Q_n(r) = (-1)^n \binom{n+1}{n} G_n^{(2,2)}(r). \tag{7.152}$$

The relation of orthogonality of the radial functions is

$$\int_0^1 Q_n(r) Q_m(r) r \, dr = \frac{1}{2(n+1)} \quad \delta_{mn}. \tag{7.153}$$

The first six radial polynomials in explicit forms are

$$\begin{aligned}
Q_0(r) &= 1, \\
Q_1(r) &= 3r - 2, \\
Q_2(r) &= 10r^2 - 12r + 3, \\
Q_3(r) &= 35r^3 - 60r^2 + 30r - 4, \\
Q_4(r) &= 126r^4 - 280r^3 + 210r^2 - 60r + 5, \\
Q_5(r) &= 462r^5 - 1260r^4 + 1260r^3 - 560r^2 + 105r - 6.
\end{aligned} \tag{7.154}$$

Their graphs are plotted in Figure 7.17. The kernel functions $Q_n(r) \, e^{-i\ell\theta}$ of the orthogonal Fourier-Mellin moments on the unit disk are shown in Figure 7.18.

The main difference between the orthogonal Fourier-Mellin moments and the Zernike moments is the absence of the constraint $\ell \leq n$ and the independence of the radial function $Q_n(r)$ on the repetition factor ℓ. The different distributions of zeros over the unit disk is also worth noting, compare Figure 7.17 with Figure 7.15. Still, the radial functions of OFM and ZM are closely related to each other:

$$r Q_n(r^2) = R_{2n+1,1}(r). \tag{7.155}$$

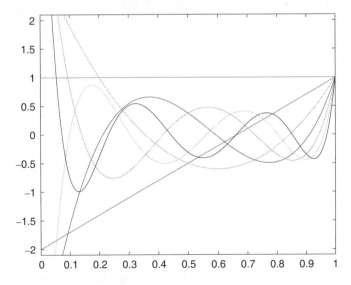

Figure 7.17   The graphs of the radial functions of the orthogonal Fourier-Mellin moments up to the sixth degree

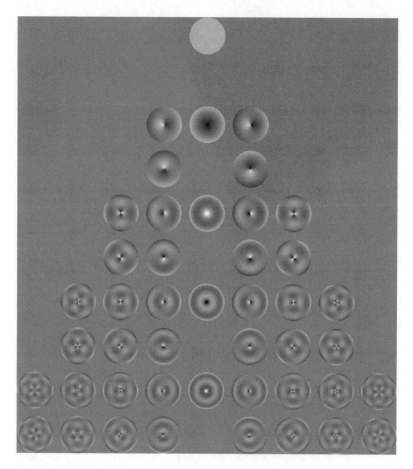

Figure 7.18 The graphs of 2D kernel functions of the orthogonal Fourier-Mellin moments up to the fourth order. Black $= -2$, white $= 2$. Real parts: 1st row: $n = 0$, $\ell = 0$, 3rd row: $n = 1$, $\ell = -1, 0, 1$, 5th row: $n = 2$, $\ell = -2, -1, 0, 1, 2$, 7th row: $n = 3$, $\ell = -3, -2, -1, 0, 1, 2, 3$, 9th row: $n = 4$, $\ell = -4, -3, -2, -1, 0, 1, 2, 3, 4$. Imaginary parts: 2nd, 4th, 6th, 8th and 10th rows. The indices are the same as above

Rotation invariance of the OFM can be achieved analogously as in the case of ZM. We have again basically three options – taking the moment magnitudes, phase cancellation by a multiplication of several moments, and image normalization.

OFM moments can be expressed as a linear combination of FM moments as

$$\Phi_{n\ell} = \frac{n+1}{\pi} \sum_{s=0}^{n} (-1)^{n+s} \frac{(n+s+1)!}{(n-s)!s!(s+1)!} M_{s\ell}, \tag{7.156}$$

but this formula is not suitable for numerical calculation. We can replace the FM moments by the complex moments, which theoretically allows the computation of OFM moments in Cartesian coordinates. However, current efficient algorithms for OFM moment computation used different schemes; see [142–146]. OFMs were proven to be very efficient in image reconstruction from a limited set of moments [147, 148]. Extension of OFM to color images was proposed in [149]. In [150], blur invariance was added to the rotation invariance of the OFM moments.

Since OFM moments are natural competitors of ZMs, several comparative tests on various data have been published. A test of the performance in character recognition showed that OFMs are better than ZM and PZM (at least on the given database) [151].

7.3.3 Other moments orthogonal on a disk

We can construct plenty of moments of the form (7.125) which differ from one another just by the radial part. In principle, any set of 1D orthogonal polynomials can be substitute for $R_{pq}(r)$. Thanks to the angular part, all such moments possess rotational invariance. On the other hand, according to Ping et al. [152] moments of a different form providing rotation invariance do not exist. In the sequel, we present a short survey of moments orthogonal on a disk which are relevant to image analysis.

Jacobi-Fourier moments

Jacobi-Fourier moments with parameters p and q are defined as

$$
\Phi_{n\ell}^{(p,q)} = \int_0^{2\pi} \int_0^1 J_n^{(p,q)}(r)\, e^{-i\ell\theta} f(r,\theta)\, r\, \mathrm{d}r\mathrm{d}\theta,
\tag{7.157}
$$

the n and ℓ are arbitrary integers. $J_n^{(p,q)}(x)$ is a normalized Jacobi polynomial

$$
J_n^{(p,q)}(r) = \sqrt{\frac{w(p,q,r)}{b_n(p,q)}}\; G_n^{(p,q)}(r),
\tag{7.158}
$$

where the normalization constant is

$$
b_n(p,q) = \frac{n!\,\Gamma(n+p)\Gamma(n+q)\Gamma(n+p-q+1)}{(2n+p)\Gamma^2(2n+p)}
\tag{7.159}
$$

and the weight function is

$$
w(p,q,r) = (1-r)^{p-q} r^{q-1}.
\tag{7.160}
$$

Jacobi-Fourier moments were introduced by Ping et al. [56] and then studied in [153–156]. Note that in [56] a slightly different normalization was used, which may lead to certain misunderstanding as exposed in [157]. Applications in gait recognition [158, 159] and in image watermarking [160] have been reported. Two efficient numerical algorithms for calculation of the Jacobi-Fourier moments can be found in [161, 162].

Chebyshev-Fourier moments

Ping et al. [152] introduced *Chebyshev-Fourier moments*

$$\Psi_{n\ell} = \int_0^{2\pi} \int_0^1 R_n(r)\, e^{-i\ell\theta} f(r,\theta)\, r\, dr d\theta, \qquad (7.161)$$

where $R_n(r)$ are normalized shifted Chebyshev polynomials of the second kind

$$R_n(r) = \sqrt{\frac{8}{\pi}}\; U_n^*(r)(r-r^2)^{1/4} r^{-1/2}$$

$$= \sqrt{\frac{8}{\pi}} \left(\frac{1-r}{r}\right)^{1/4} \sum_{k=0}^{(n+2)/2} (-1)^k \frac{(n-k)!}{k!(n-2k)!}[2(2r-1)]^{n-2k}. \qquad (7.162)$$

$U_n^*(r)$ are shifted Chebyshev polynomials of the second kind

$$U_n^*(r) = U_n(2r-1) \qquad (7.163)$$

that are orthogonal on the interval $\langle 0,1\rangle$ satisfying the relation

$$\int_0^1 U_m^*(r)U_n^*(r)(r-r^2)^{1/2} dr = \frac{8}{\pi}\delta_{mn}. \qquad (7.164)$$

After [152], the use of Chebyshev-Fourier moments for image description was studied in [163]. An algorithm for their fast implementation was proposed in [164].

Moments with a non-polynomial radial function

Another type of orthogonal moments called *radial harmonic Fourier moments* (RHFM) was proposed in [165]

$$\Lambda_{n\ell} = \int_0^{2\pi} \int_0^1 T_n(r)\, e^{-i\ell\theta} f(r,\theta)\, r\, dr d\theta, \qquad (7.165)$$

where $T_n(r)$ is

$$T_n(r) = \begin{cases} \dfrac{1}{\sqrt{r}} & \text{if } n=0 \\[2ex] \sqrt{\dfrac{2}{r}}\sin[(n+1)\pi r] & \text{if } n \text{ is odd} \\[2ex] \sqrt{\dfrac{2}{r}}\cos(n\pi r) & \text{if } n \text{ is even.} \end{cases} \qquad (7.166)$$

The radial harmonic Fourier moments satisfy the relation of orthogonality

$$\int\limits_{0}^{2\pi} \int\limits_{0}^{1} T_n(r)\, e^{-i\ell\theta} T_m(r)\, e^{-ik\theta} r\, \mathrm{d}r\mathrm{d}\theta = \delta_{mn}\delta_{k\ell}. \tag{7.167}$$

We can see that $T_n(r)$ is not a polynomial in r, so the radial harmonic Fourier moments are not moments in the strict sense. Nevertheless, they originate from similar ideas as other orthogonal moments, and the authors have used them in a similar way. The RHFMs were used for character description [166], for classification of cells [167, 168], and were adapted by means of a quaternion formalism to color images [169].

Some authors proposed other OG radial "moments", the basis functions of which are not polynomials. Very close to the RHFMs is the *Polar harmonic transform* by Yap et al. [170] where in the radial factor of the basis are harmonic functions of r^2 (the basis functions look like 2D spirals). The authors claimed a fast and stable computation. Dominguez [171] used binary stepwise radial functions similar to Walsh functions. Their values can only be 0 or 1, so the calculation of a product is extremely fast. Feng [172] used wavelets as the radial functions to obtain locality of the features in radial directions.

Xiao introduced the *Bessel–Fourier moments* that used the Bessel function of the first kind as a radial factor [173]. The motivation was that the Bessel function has more zeros than Zernike and OFM radial factors, and these zeros are more evenly distributed, which improves the description of the image as was demonstrated for image retrieval [174] and robust watermarking [175]. Quaternion Bessel-Fourier moments for description of color images were proposed in [176].

7.4 Object recognition by Zernike moments

The goal of this experiment is to demonstrate the recognition power of Zernike moments and compare it to that of Gaussian-Hermite moments, which are orthogonal on a square, and also to the invariants from complex moments, which are not orthogonal.

The experiment was performed on real data, that is, on real photographs and real rotations; no computer simulation was used. We intentionally chose circular objects to avoid problems with the definition domain of Zernike polynomials. We photographed a set of pexeso cards from the popular "Ferdy the Ant" cartoon collection;[17] see Figure 7.19, to create our database.

We put the cards on the floor and rotated the hand-held camera. Each card was captured eight times with different rotation angles. The cards were photographed on a dark background allowing easy segmentation. The first snapshot of each card was used as a representative of its class, and the other seven images were recognized by the minimum-distance classifier in the space of invariants.

We tried to recognize graylevel versions of the images first. There have been several parameters which can influence the recognition rate – the number of invariants, the precision of their calculation, and the way the invariants of different orders are normalized to the same dynamic range. Normalization to the range is in fact, the same as the setting of weights when using weighted Euclidean distance in the space of invariants. In practice this step is of great

---

[17] Fairy tales by Ondřej Sekora, a Czech writer.

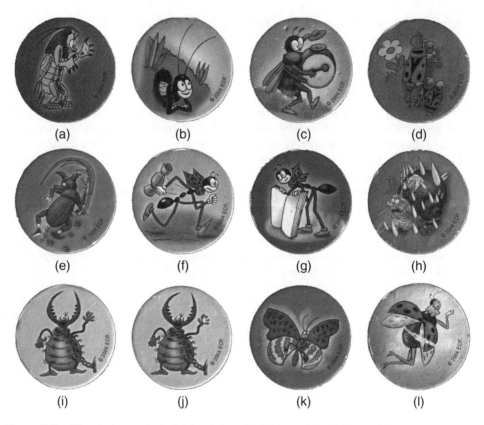

(a)   (b)   (c)   (d)

(e)   (f)   (g)   (h)

(i)   (j)   (k)   (l)

Figure 7.19 The playing cards: (a) Mole cricket, (b) Cricket, (c) Bumblebee, (d) Heteropter, (e) Poke the Bug, (f) Ferdy the Ant 1, (g) Ferdy the Ant 2, (h) Snail, (i) Ant-lion 1, (j) Ant-lion 2, (k) Butterfly, and (l) Ladybird

importance, and the features are usually normalized by their standard deviations. If we did not normalize the values, the features with high range would override the others, which is not desirable.

First, we tested Zernike moments of the normalized image and Zernike rotation invariants. As one can expect, the differences between them were not significant. When using the normalized Zernike moments (7.140) we obtained 100% correct recognition when we took the moments at least up to the third order (the rotation normalization was done by setting A_{31} real and positive). A 2D subspace of the feature space is in Figure 7.20. Note that two features are too few to provide enough separability. When using Zernike invariants (7.141) up to the fifth order with range normalization by standard deviations, we got two misclassifications. Avoiding range normalization led to two misclassifications for the normalized Zernike moments up to fifth order and four misclassifications for the Zernike rotation invariants up to the 9th order.

When using the Gaussian-Hermite invariants (7.124), we have to carefully choose the parameter σ, which influences the performance of the method. The best value depends on the data, so it should be found by trial and error. Here we tested the recognition rate of the Gaussian-Hermite invariants up to the third order and then up to the sixth order for σ from 0.05

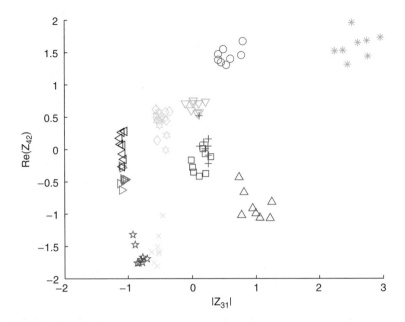

Figure 7.20 The feature space of two Zernike normalized moments $|Z_{31}|$ and $\mathcal{R}e(Z_{42})$. $\triangledown$– Ferdy the Ant 1, $\triangle$– Ferdy the Ant 2, $\square$– Ladybird, $\diamondsuit$– Poke the Bug, $\triangleright$– Ant-lion 1, $\triangleleft$– Ant-lion 2, $\times$– Mole cricket, $+$– Snail, $\stackrel{\star}{}$– Butterfly, $*$– Cricket, $\stackrel{\star}{}$– Bumblebee, $\bigcirc$– Heteropter

to 1 (note that the images have been mapped into the unit disk). The results are in Figure 7.21. We can see there are two good options – the first one is around 0.2, the second one is around 0.7. In this setting, the recognition rate is 100%. Comparing to Zernike moments, the GH invariants work well even without the normalization to the range, so from this point of view their usage is easier.

We classified the images also by the complex moment invariants for a comparison. To achieve the 100% success rate, we had to use the moments up to the fifth order at least. This illustrates slightly higher discrimination power of Zernike and GH invariants due to their orthogonality.

Finally, we re-run the above experiments with color images of the cards. The moments were computed from each color channel separately, which yielded three-times more features of each order. The recognition rates were better than before. Misclassifications occurred only if no normalization to the range was performed. For all other settings mentioned above, including the complex moment invariants, we obtained the 100% success rate. Moreover, fewer features than before were sufficient in most cases (see Figure 7.22 showing separable clusters in the space of two Zernike normalized moments).

7.5 Image reconstruction from moments

Reconstruction of the original image from a set of its moments has been discussed in the literature quite often. Although it has attracted considerable attention, its practical importance

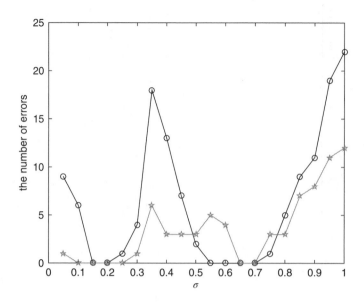

Figure 7.21 The error rate in dependency on parameter σ of the Gaussian-Hermite moments. There were used the moments up to: $\bigcirc$– 3rd order, ☆– 6th order

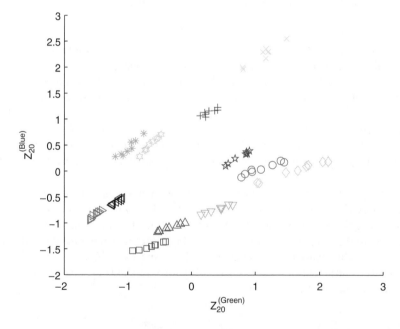

Figure 7.22 The feature space of two Zernike normalized moments of color images $Z_{20}^{(Green)}$ and $Z_{20}^{(Blue)}$. $\triangledown$– Ferdy the Ant 1, $\triangle$– Ferdy the Ant 2, $\square$– Ladybird, $\lozenge$– Poke the Bug, $\triangleright$– Ant-lion 1, $\triangleleft$– Ant-lion 2, $\times$– Mole cricket, +– Snail, ☆– Butterfly, *– Cricket, ✿– Bumblebee, $\bigcirc$– Heteropter

is not very high because moments in general are not a good tool for image compression, so the reconstruction problem does not appear very frequently in practice. On the other hand, reconstruction abilities of moments are connected with their recognition power. Studying the reconstruction problem provides an insight into those moment properties that are important for classification without actually performing any recognition experiments. This is why so many authors have discussed the reconstruction from moments in their papers.

In a continuous domain, we have a uniqueness theorem saying that each image function (which is supposed to be compactly supported and piece wise continuous by definition[18]) can be precisely reconstructed from a complete set of its geometric moments. Using the equivalence among different polynomial bases, this theorem holds for other moments as well.

The reconstruction problem can be formulated in a more general way: can any series of numbers be considered a set of moments of an image? This question is well known from statistics and probability as the *Hausdorff moment problem* and the general answer to it is "No"[19].

In the discrete domain, the uniqueness theorem says that the image can be precisely reconstructed from the same number of its moments as the number of pixels. In applied digital image processing, we do not work with a such high number of moments. The reconstruction problem converts into two questions: what is the optimal number of moments for the reconstruction and what is the error of this reconstruction? The "optimality" of the reconstruction is usually measured by the quality, memory requirements, and the speed.

In image processing, reconstruction is sometimes considered not directly from moments but from certain moment invariants. This task can be easily transformed to the previous one keeping in mind that we never can "reconstruct" those image properties the invariants are insensitive to (e.g., it is impossible to recover the orientation of the object from its rotation invariants).

In the sequel, we review three approaches to the image reconstruction from moments – direct calculation, reconstruction in a Fourier domain, and reconstruction from OG moments. By experiments we show their major advantages and limitations.

7.5.1   Reconstruction by direct calculation

Let us have a discrete image $f(x, y)$ of a size $M \times N$. Its discrete geometric moments are defined as

$$m_{pq} = \sum_{x=0}^{M-1} \sum_{y=0}^{N-1} x^p y^q f(x, y) \qquad p = 0, 1, 2, \ldots, M-1, q = 0, 1, 2, \ldots, N-1 \qquad (7.168)$$

(this is exactly equivalent to the definition (3.8), if we imagine the discrete image as an array of Dirac δ-functions). We can consider (7.167) to be a system of linear equations for unknowns $f(x, y)$, where the moments arranged to a vector of size MN create the right-hand side and kernel functions $x^p y^q$ arranged to an $MN \times MN$ matrix create the matrix of the system. For very small images, this system can be resolved by Gaussian elimination, and in this way an original image is obtained. For larger images, the system becomes ill-conditioned. Double-precision floating

---

[18] The reconstruction problem on non-compact supports is not considered in this book.

[19] A sufficient condition is a so-called complete monotonicity of the moment values, see [5] for details.

Figure 7.23 The collapse of the image reconstruction from geometric moments by the direct calculation. Top row: original images 12×12 and 13×13, bottom row: the reconstructed images from the moments

point arithmetic is no longer sufficient, higher-order moments as well as the kernel functions $x^p y^q$ lose their precision, and the whole algorithm collapses.

We can slightly improve the algorithm by translation of the coordinate origin to the center of the image, but the same problem appears a little bit later. Of course we can use variables with higher precision, but that is not a principal improvement. Moreover, the set of moments would demand much more memory.

Summarizing, the direct calculation is convenient for very small images only. To demonstrate this, we carried out the following experiment. We created a series of images of the capital letter E of the size $N \times N$ for $N = 7, 8, \ldots, 13$, computed the geometric moments m_{pq} for $p, q = 0, 1, \ldots, N - 1$ and then solved the system (7.167). The reconstruction of images up to 11×11 was precise, Figure 7.23 catches the collapse of the algorithm. The original image 12×12 is top-left, its reconstruction (bottom-left) is almost precise (the maximum error is 5 graylevels from 255). The reconstruction of the image 13×13 (top-right) totally collapsed (bottom-right), the range of values of the reconstructed image was $\langle-857, 1460\rangle$ instead of the original $\langle0,255\rangle$. The difference of these two intervals is an attribute of the depth of the reconstruction error.

7.5.2   Reconstruction in the Fourier domain

As we already know from Chapter 3, the continuous Fourier transform and the geometric moments are linked by the relation

$$F(u, v) = \sum_{p=0}^{\infty} \sum_{q=0}^{\infty} \frac{(-2\pi i)^{p+q}}{p!q!} m_{pq} u^p v^q.$$

In the discrete $M \times N$ case, we have a similar relation

$$F(u, v) = \sum_{p=0}^{M-1} \sum_{q=0}^{N-1} \frac{(-2\pi i)^{p+q}}{p!q!} \left(\frac{u}{M}\right)^p \left(\frac{v}{N}\right)^q m_{pq}. \tag{7.169}$$

We can compute the Fourier transform from the moments by (7.168) and then obtain the original image f by inverse Fourier transform. The main computational problem of this method is that we are not able to calculate enough geometric moments with reasonable precision. We face the problem well-known from the previous method, the overflow of the factors $x^p y^q$. We can see there must be a certain optimal moment order such that the higher order moments do not improve the image reconstruction.

If the number of moments is not sufficient, the most affected frequencies are the highest ones. We must erase high frequencies from the spectrum $F(u, v)$ to obtain an acceptable result. So, we must search for two parameters in this method: the optimal moment order and the optimal threshold in the frequency domain.

An analogous method can be used when reconstructing the image from the complex moments instead of the geometric ones [177]. Then the formula (7.168) reads

$$F(u, v) = \sum_{p=0}^{M+N-2} \frac{(-\pi i)^p}{p!} \sum_{k=0}^{p} \binom{p}{k} \left(\frac{u+iv}{N}\right)^{p-k} \left(\frac{u-iv}{M}\right)^k c_{k,p-k}. \tag{7.170}$$

We tested the Fourier domain reconstruction in the experiment similar to the previous one. We used an image of the capital letter E of size 32×32 pixels, computed its geometric moments, reconstructed its spectrum by (7.168), and computed the inverse Fourier transform. We employed the moments up to the order 21, 32, 43, 54, 65, 76, and 87, respectively. Higher orders were meaningless because the loss of precision was so high that the reconstruction failed completely. We found optimal frequency thresholds ω as 0.12, 0.18, 0.21, 0.25, 0.34, 0.4, and 0.43, for the respective orders, and set $F(u, v) = 0$ for such u, v that $\sqrt{(u/M)^2 + (v/N)^2} > \omega$. The reconstructions are visualized in Figure 7.24. One can see this method does not yield very good results even if a redundant number of moments is used. The reasons are that neither the implementation nor the higher-order moments themselves can be calculated with sufficient precision.

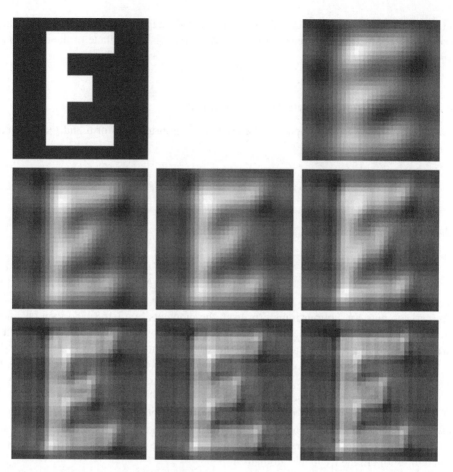

Figure 7.24   Image reconstruction from geometric moments in the Fourier domain. The original image
32 × 32 and the reconstructed images with maximum moment orders 21, 32, 43, 54, 65, 76, and 87,
respectively

7.5.3   Reconstruction from orthogonal moments

Image reconstruction from orthogonal moments is easier and more stable. Having discrete
moments v_{pq} orthogonal on a rectangle $M \times N$

$$v_{pq} = \sum_{x=0}^{M-1} \sum_{y=0}^{N-1} \pi_p(x)\pi_q(y)f(x,y), \qquad p = 0, 1, 2, \dots, M-1, q = 0, 1, 2, \dots, N-1, \quad (7.171)$$

we can reconstruct the original image directly as

$$f(x,y) = \sum_{p=0}^{M-1} \sum_{q=0}^{N-1} v_{pq}\pi_p(x)\pi_q(y), \qquad x = 0, 1, 2, \dots, M-1, y = 0, 1, 2, \dots, N-1. \quad (7.172)$$

For a precise reconstruction, the orthogonality of the polynomial basis should be a discrete one, that is,

$$\sum_{x=0}^{N-1} \pi_p(x)\pi_q(x) = \delta_{pq},$$ (7.173)

but a basis which exhibits a "continuous" orthogonality only may also yield satisfactory (although not precise) reconstruction results.

A similar formula can be written for the moments orthogonal on a disk. For instance, the reconstruction from the Zernike moments is accomplished by

$$f(x, y) = \sum_{n=0}^{\infty} \sum_{\ell=-n,-n+2,\dots}^{n} A_{n\ell} V_{n\ell}(x, y),$$ (7.174)

where $V_{n\ell}(x, y)$ was obtained from $V_{n\ell}(r, \theta)$ by the same mapping as in the computation of the Zernike moments. In reality, the first sum is truncated to a finite number of terms.

The above idea can be used when reconstructing the image from geometric moments as well. If we express the orthogonal polynomial in powers

$$\pi_p(x) = \sum_{s=0}^{p} g_{ps} x^s,$$ (7.175)

then for the corresponding OG moment holds

$$\upsilon_{pq} = \sum_{s=0}^{p} \sum_{t=0}^{q} g_{ps} g_{qt} m_{st}.$$ (7.176)

Now we can use (7.171) directly for the reconstruction. In this way we seemingly avoid any numerically instable operations that appeared in the direct solution of (7.167). Nevertheless, as the geometric moments have already lost their precision, then the orthogonal moments calculated from them are also imprecise and the reconstruction is not perfect. This illustrates what we already mentioned in connection with evaluation of OG polynomials. Expressing OG polynomials in terms of powers cannot be used for a stable computation.

Image reconstruction from moments is still a numerically delicate process, we must pay attention to the numerical stability in each step of the algorithm. Therefore it is better to use the moments orthogonal exactly on a discrete set of points rather than the discrete versions of moments with "continuous" orthogonality. The moments orthogonal on a rectangle are more suitable in a rectangular raster, and the moments orthogonal on a disk are more suitable in a polar raster (see Figure 7.25). The problems with the polar raster and the errors of the reconstruction are analyzed in detail in [5].

Let us explain, on an example of discrete Chebyshev moments, how to organize the calculations. Suppose $N = M$ for simplicity, and suppose the discrete Chebyshev polynomials (7.67) have already been made orthonormal.

We could employ the formulae (7.71) and (7.72) separately for each x for the computation of the values of the polynomials, but repeated multiplication by large x ($x = N - 1$ in maximum) would cause large magnitude variations of the $\hat{T}_n(x)$. This can violate the numerical stability for N somewhere about 100, and the reconstruction fails.

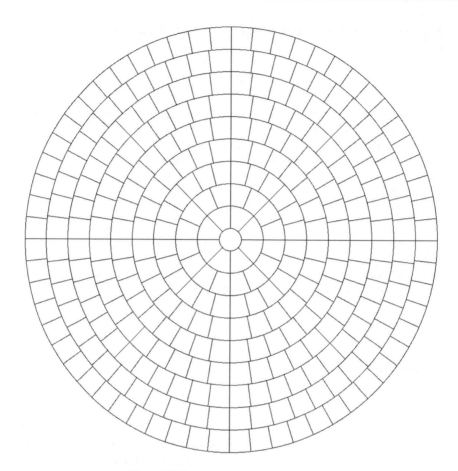

Figure 7.25 An example of the polar raster

We can avoid this problem by using formulae recurrent to x. The initial values are

$$\hat{T}_0(x) = \frac{1}{\sqrt{N}}, \qquad x = 0, 1, \dots, N-1, \tag{7.177}$$

then

$$\hat{T}_n(0) = -\sqrt{\frac{N-n}{N+n}}\sqrt{\frac{2n+1}{2n-1}}\hat{T}_{n-1}(0), \qquad n = 1, 2, \dots, N-1, \tag{7.178}$$

and from that

$$\hat{T}_n(1) = \left(1 + \frac{n(n+1)}{1-N}\right)\hat{T}_n(0), \qquad n = 1, 2, \dots, N-1. \tag{7.179}$$

Other values are computed by the recurrence

$$\hat{T}_n(x) = \gamma_1\hat{T}_n(x-1) + \gamma_2\hat{T}_n(x-2), \qquad n = 1, 2, \dots, N-1; \quad x = 2, 3, \dots, N/2, \tag{7.180}$$

where

$$\gamma_1 = \frac{-n(n+1) - (2x-1)(x-N-1) - x}{x(N-x)}, \qquad \gamma_2 = \frac{(x-1)(x-N-1)}{x(N-x)}. \tag{7.181}$$

The values for $x > N/2$ can be computed using the symmetry condition

$$\hat{T}_n(N-1-x) = (-1)^n \hat{T}_n(x). \tag{7.182}$$

Small numerical errors can be corrected by renormalization

$$\hat{T}_n(x) \leftarrow \frac{\hat{T}_n(x)}{\sqrt{\sum_{i=0}^{N-1} (\hat{T}_n(i))^2}}, \qquad x = 0, 1, \ldots, N-1. \tag{7.183}$$

This algorithm can be used for precise reconstruction of images up to about 1000×1000 pixels. If we continue increasing the size of the image, then we face a floating-point under-flow[20], as will be demonstrated in Section 7.5.5.

7.5.4 Reconstruction from noisy data

In practice, the moment values are often calculated from a noisy version of the original image. In such cases, even if we had a "perfect" reconstruction algorithm, we would never obtain the noise-free original but would end up with the noisy image the moments have been calculated from. If we minimize the mean-square error between the reconstructed and noise-free original images, we find out that an optimal order of the moments exists. Moments of higher orders contribute more to reconstruction of the noise rather than of the image. The optimal order decreases as the noise increases and may be surprisingly low for heavy noise. Since the problem is ill-posed, the reconstruction error from noisy data may be very high even if we use this optimal moment order. If we replace the sequence of original moments $\{m_{pq}\}$ by $\{m_{pq}+\varepsilon_{pq}\}$, where $\{\varepsilon_{pq}\}$ are arbitrarily small numbers, the reconstruction result $f_\varepsilon(x,y)$ may be arbitrarily far from $f(x,y)$. For detailed analysis of this phenomenon, we refer to [5].

7.5.5 Numerical experiments with a reconstruction from OG moments

The experiments presented in this section aim to compare various kinds of OG moments in terms or reconstruction quality and to show the limits of the reconstruction algorithms.

In the first experiment, a binary image of the capital letter E of the size 32×32 (Figure 7.24 top left) was reconstructed from four kinds of orthogonal moments – Legendre, continuous Chebyshev of the first kind, Gaussian-Hermite, and Zernike, respectively. The goal of this experiment was to show the impact of using moments orthogonal in a continuous domain. We used the moments up to thirty-first order[21] (in case of the Zernike moments up to forty-fourth order), so the precise reconstruction is theoretically possible. The reconstructed images can be

---

[20] Since computers have certain minimum positive floating-point numbers that can be stored, all lower values are treated as zeros.
[21] Solely in this section, we take the moment order as a maximum of $\{p,q\}$.

seen in Figure 7.26. All four reconstructions are not perfect, although the errors and artifacts are different. The errors have been caused by using continuous orthogonal moments on the discrete image. In the case of the Zernike moments, the error is larger because the sampling error on the unit disk is larger than that on the square. It is interesting to note that the normalization of the Legendre moments brings almost no improvement over the non-normalized Chebyshev moments. When evaluating the quality visually, probably the best reconstruction was achieved by the Gaussian-Hermite moments.

When we used the discrete Chebyshev moments, the reconstruction was precise, exhibiting no difference from the original. In this case, it is interesting to investigate the influence of the number of moments on the reconstruction quality. In Figure 7.27, we can see the the reconstructions from discrete Chebyshev moments up to the limited order only. Obviously, the reconstruction quality grows as the moment order increases, reaching the precise reconstruction for $p = 31$ (Figure 7.27 bottom right). This experiment illustrates the well-known fact that if we use moments up to (let us say) the third order for recognition, we cannot distinguish among the fourth, fifth, and further images in Figure 7.27 (even if they are easy to

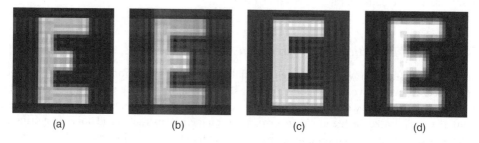

(a)   (b)   (c)   (d)

Figure 7.26   Image reconstruction from the orthogonal moments: (a) Legendre moments, (b) Chebyshev moments of the first kind, (c) Gaussian-Hermite moments, and (d) Zernike moments

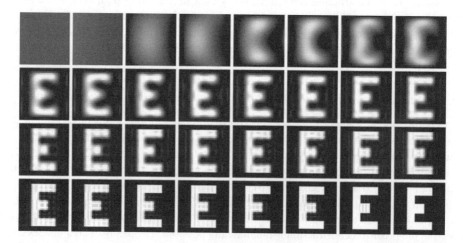

Figure 7.27   Image reconstruction from the incomplete set of discrete Chebyshev moments. The maximum moment order is $0, 1, \ldots, 31$, respectively. The last image (bottom-right) is a precise reconstruction of the original image

Figure 7.28 Image reconstruction from the orthogonal moments: (a) Legendre moments, (b) Chebyshev moments of the first kind, (c) Gaussian-Hermite moments, and (d) Zernike moments

discriminate visually) since their third-order moments are the same. On the other hand, using thirty-first-order moments for recognition would not be a good choice because of its low robustness to noise. To find an optimal order of moments (or invariants) for object recognition, we have to perform a discriminative analysis on the given training set.

In the next experiment, we reconstructed the graylevel Lena image of the size 256×256. We again used the Legendre moments, the continuous Chebyshev moments of the first kind, the Gaussian-Hermite moments, and the Zernike moments. We applied the moments up to the order 255 to give them a chance for precise reconstruction. The reconstruction results are in Figure 7.28. We can see disturbing oscillation artifacts in the case of the Legendre and Chebyshev moments and also mild artifacts in the case of the Zernike moments. The reconstruction from the Gaussian-Hermite moments looks visually the best. Actually, the average

Figure 7.29 Image reconstruction from the discrete Chebyshev moments. There are no errors; it is identical with the original Lena image

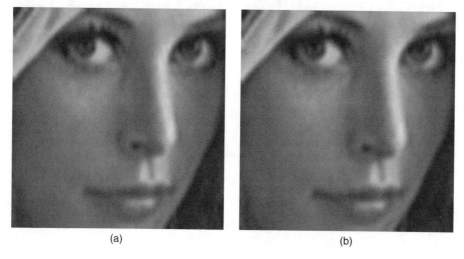

(a)   (b)

Figure 7.30 Detail of the reconstruction from the Gaussian-Hermite moments: (a) reconstruction and (b) original

error is 30, 32, 1, and 5 graylevels, respectively. In spite of a very high visual quality, a detailed examination of the Gaussian-Hermite reconstruction discovered some slight oscillation errors even there (see Figure 7.30, right cheek). When using discrete Chebyshev moments, the reconstruction is precise without any errors in graylevels (see Figure 7.29).

The last experiment was performed on a large-size graylevel image. Its aim was to explore the limits of the reconstruction abilities of OG moments. We used a standard photograph of the size 2048 × 1536 pixels (see Figure 7.31) converted to grayscale. We cropped it to

Figure 7.31 The test image for the reconstruction experiment with discrete Chebyshev moments (Astronomical Clock, Prague, Czech Republic)

1024×1024, then to 1025×1025, 1026×1026, etc. For each crop we calculated the normalized discrete Chebyshev moments with all available algorithmic improvements (7.176) to (7.182) and reconstructed the image. Up to 1076×1076, the reconstruction was precise. From 1077×1077, the errors started appearing due to the floating point underflow. For 1077×1077 the errors are negligible (6 pixels by 1 gray level), but they increase slowly as the image size increases.

For 1100×1100, about 8% of pixels were reconstructed with errors. The mean square error of the reconstruction is only 0.1 gray level, what is 0.004% of the range from black to white, (see Figure 7.32)[22]. The magnitude and the distribution of errors can be seen in Figures 7.32 and 7.33. The reconstruction of the image 1536×1536 is worse in terms of the mean square error but still visually fully acceptable; see Figure 7.34. The mean square error is twenty-five gray levels, that is, 9.8% of the range from black to white. We can see on the error map (see Figure 7.34b and a close-up in Figure 7.35) that there are not only random oscillations, but also systematic errors.

7.6 3D orthogonal moments

Similarly as in 2D, in 3D as well OG polynomials and moments can be categorized according to their area of orthogonality Ω. We recognize polynomials orthogonal on a cube, on a sphere,

---

[22] These values may be slightly different for other images, but in general this behavior is determined by the fact that both in C and Matlab, the floating-point variables use a 48-bit mantissa that underflows as the size of the image reaches a certain limit.

(a)   (b)

Figure 7.32 The reconstruction experiment: (a) the reconstructed cropped image 1100×1100 and (b) the error map. The range of errors from -8 to 9 is visualized in the range black – white. The relative mean square error is 0.004%

Figure 7.33 The close-up of the 1100×1100 error map with typical oscillations

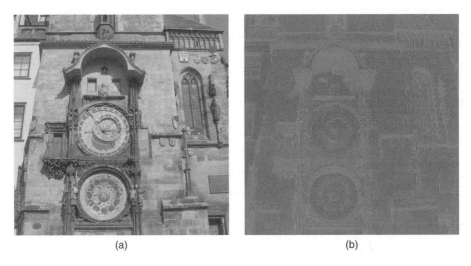

(a)   (b)

Figure 7.34   The reconstruction experiment: (a) the reconstructed cropped image 1536×1536 and (b) the error map. The range of errors from -185 to 198 is visualized in the range black – white. The relative mean square error is 9.8%

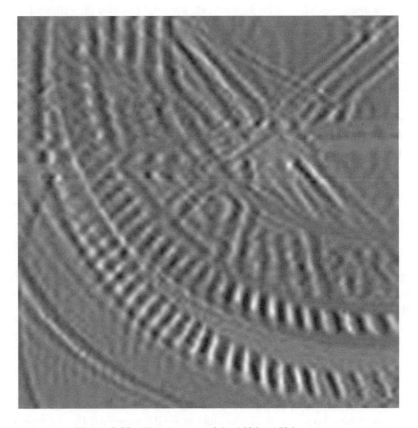

Figure 7.35   The close-up of the 1536×1536 error map

and on a cylinder. The polynomials OG on a cube are very easy to construct and calculate, while the polynomials OG on a sphere are generally more appropriate for construction of 3D rotation invariants.

7.6.1 3D moments orthogonal on a cube

As we already mentioned at the beginning of this chapter, 3D polynomials OG on a cube are mostly constructed as the products

$$\pi_{pqr}(x, y, z) = \pi_p(x)\pi_q(y)\pi_r(z),$$

where $\{\pi_n(x)\}$ is an arbitrary set of 1D polynomials orthogonal on interval Ω_1. Then the area of orthogonality of $\{\pi_{pqr}(x, y, z)\}$ is $\Omega = \Omega_1 \times \Omega_1 \times \Omega_1$ and the corresponding moments are defined as

$$v_{pqr} = n_p n_q n_r \iiint\limits_{\Omega} \pi_p(x)\pi_q(y)\pi_r(z)f(x, y, z)dxdydz \qquad p, q, r = 0, 1, 2, \dots . \qquad (7.184)$$

Although this can be performed for any OG polynomials mentioned in the previous sections, Legendre and Gaussian-Hermite moments are the most popular choices (and the only ones used in image analysis). The Legendre moments are used because of their simplicity [178] and because they can be calculated efficiently. Algorithms for fast calculation of 3D Legendre moments mostly rely on 1D recurrent formulas but were also adapted to particular classes of object type or shape [20, 21, 179, 180].

The main advantage of the Gaussian-Hermite moments is the same as in 2D – an easy construction of rotation invariants. Yang's theorem holds in 3D, too (see [49] for the proof). Hence, if there exists a 3D rotation invariant of geometric moments $I(m_{p_1q_1r_1}, m_{p_2q_2r_2}, \dots, m_{p_dq_dr_d})$, the same function of the corresponding Hermite moments $I(\eta_{p_1q_1r_1}, \eta_{p_2q_2r_2}, \dots, \eta_{p_dq_dr_d})$, is also a rotation invariant. This theorem holds even if we replace Hermite moments η_{pqr} with GH moments

$$\hat{\eta}_{pqr} = \frac{1}{\sqrt{\sqrt{(p+q+r)!}((p+q+r)/2)!2^{p+q+r}\sigma^3(\sqrt{\pi})^3}}$$
$$\times \int\limits_{-\infty}^{\infty} \int\limits_{-\infty}^{\infty} e^{-\frac{(x^2+y^2+z^2)}{2\sigma^2}} H_p(x/\sigma)H_q(y/\sigma)H_r(z/\sigma)f(x, y, z)dxdydz. \qquad (7.185)$$

In Chapter 4 we presented a number of 3D rotation invariants of geometric moments derived by the tensor method. We can now substitute the weighted GH moments into any of them to obtain 3D GH rotation invariants. For instance, we have

$$\Phi_1 = \hat{\eta}_{200} + \hat{\eta}_{020} + \hat{\eta}_{002}$$

and

$$\Phi_2 = \hat{\eta}_{200}^2 + \hat{\eta}_{020}^2 + \hat{\eta}_{002}^2 + 2\hat{\eta}_{110}^2 + 2\hat{\eta}_{101}^2 + 2\hat{\eta}_{011}^2.$$

The 3D rotation invariants of geometric moments have been explicitly derived up to the sixteenth order, which limits our ability to create the GH invariants. This is, however, not a

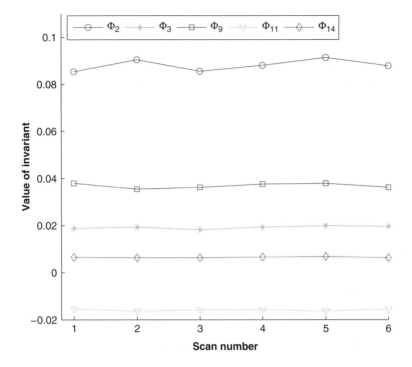

Figure 7.36   The values of five selected invariants of the teddy bear

serious limitation for practical purposes because the order 16 is usually completely sufficient for object recognition. The main advantage of the GH invariants is their better numerical precision compared to geometric invariants of the same order.

To verify Yang's 3D theorem experimentally, we performed the following simple test with a real 3D object. We used the teddy bear scans from Chapter 4 obtained by Kinect. We calculated the GH invariants of six scans differing from one another by a physical rotation in the space and plotted their values in Figure 7.36. The values of the invariants have low mean relative errors between 3% and 4%, which demonstrates the desirable invariance in a real environment.

7.6.2   3D moments orthogonal on a sphere

Orthogonal polynomials on a unit sphere $S = \{(x, y, z)|x^2 + y^2 + z^2 \leq 1\}$ are natural generalizations of 2D polynomials orthogonal on a disk. The respective moments are in general defined as

$$v_{n\ell}^{m} = n_{n\ell m} \int_{0}^{2\pi} \int_{0}^{\pi} \int_{0}^{1} R_{n\ell}(\varrho)(Y_{\ell}^{m}(\theta, \varphi))^{*} f(\varrho, \theta, \varphi) \varrho^2 \sin\theta \, d\varrho d\theta d\varphi$$

$$n = 0, 1, 2, \ldots, \ell = \mathrm{rem}(n, 2), \mathrm{rem}(n, 2) + 2, \ldots, n - 2, n, m = -\ell, \ldots, \ell, \quad (7.186)$$

where $R_{n\ell}$ is an arbitrary radial function of moments orthogonal on a disk (in other words, it can be an arbitrary 1D orthogonal polynomial); $\mathrm{rem}(n, 2)$ is the reminder after division of n by 2.

Although the definition (7.185) offers many choices, only the 3D Zernike moments have been described in the image analysis literature. When using spherical harmonics (4.21), the *3D Zernike polynomials* have the form

$$Z_{n\ell}^m(\varrho, \theta, \varphi) = R_{n\ell}(\varrho)Y_\ell^m(\theta, \varphi),\qquad(7.187)$$

where Y_ℓ^m is the spherical harmonics of degree ℓ and order m. Their relation of orthogonality on a unit sphere is

$$\frac{3}{4\pi}\int_0^{2\pi}\int_0^\pi\int_0^1 Z_{n\ell}^m(\varrho, \theta, \varphi)(Z_{n'\ell'}^{m'}(\varrho, \theta, \varphi))^*\varrho^2\sin\theta\,\mathrm{d}\varrho\mathrm{d}\theta\mathrm{d}\varphi = \delta_{nn'}\delta_{\ell\ell'}\delta_{mm'}.\qquad(7.188)$$

The *3D Zernike moments* are then defined as

$$A_{n\ell}^m = \int_0^{2\pi}\int_0^\pi\int_0^1 (Z_{n\ell}^m(\varrho, \theta, \varphi))^*f(\varrho, \theta, \varphi)\varrho^2\sin\theta\,\mathrm{d}\varrho\mathrm{d}\theta\mathrm{d}\varphi$$

$$n = 0, 1, 2, \ldots, \ell = -n, -n+2, \ldots, n, m = -\ell, \ldots, \ell.\qquad(7.189)$$

They were proposed by Canterakis [181] and used by for shape retrieval [182, 183] and particularly for description of the structure of protein molecules and measurement of their similarity [184–188]. A modification to 3D pseudo-Zernike moments appeared in [189]. Two algorithms for fast evaluation were published in [190, 191].

Both the 3D Zernike moments and 3D complex moments are based on the spherical harmonics, therefore we can use the same approach to the construction of the rotation invariants. This leads to the same formulas as in the case of complex moments. For instance, an invariant of complex moments is

$$\phi_5 = \frac{1}{\sqrt{3}}(2c_{31}^1 c_{31}^{-1} - (c_{31}^0)^2).$$

Repeating its derivation with Zernike moments, we obtain the analogous function

$$\psi_5 = \frac{1}{\sqrt{3}}(2A_{31}^1 A_{31}^{-1} - (A_{31}^0)^2)$$

which is a 3D rotation invariant, too.

Canterakis [181] proposed a normalization approach, very similar to the one we used in Chapter 5. His constraints were formulated in terms of Zernike moments. In this way, he reached invariance to 3D affine transformation.

7.6.3   3D moments orthogonal on a cylinder

Polynomials and moments orthogonal on a cylinder are of a limited importance in image analysis. To make them rotation invariant, we would have to know the axis of rotation and have to make the z-axis identical with it. Then we could construct the rotation invariants in the same way as in 2D. Since the rotation axis is usually unknown and/or may change, this approach is not commonly applicable.

The *cylindrical moments* are defined as

$$v_{pqr} = n_{pqr} \int\limits_{0}^{2\pi} \int\limits_{0}^{1} \int\limits_{-1}^{1} \pi_r(z) R_{pq}(\varrho) e^{-iq\varphi} f(\varrho, \varphi, z) \varrho \, dz d\varrho d\varphi$$

$$p, r = 0, 1, 2, \ldots, q = -p, \ldots, p, \tag{7.190}$$

where $R_{pq}(\varrho)e^{-iq\varphi}$ is a kernel function of some moments orthogonal on a disk and $\pi_r(z)$ are 1D orthogonal polynomials.

7.6.4   Object recognition of 3D objects by orthogonal moments

Recognition of real objects from ancient Greece

In this experiment, we compare the recognitive abilities of 3D Zernike and Gaussian-Hermite rotation invariants.

We used a part of the collection of archeological findings from ancient Greece; see Figure 7.37. We converted them from triangulated to the volumetric representation, because GH moments, unlike geometric moments, cannot be computed directly from the triangulated surface (at least such an algorithm has not been published yet). The resolution was 100 voxels per the maximum diameter of the object; see Figure 7.38.

(a)   (b)   (c)   (d)   (e)

Figure 7.37   Archeological findings used for the experiment

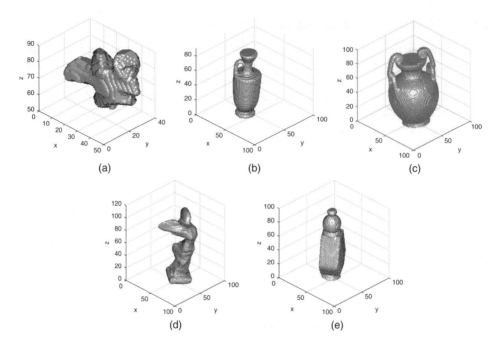

Figure 7.38   Volumetric forms of the archeological findings

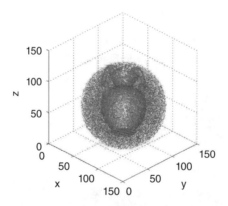

Figure 7.39   The rotated and noisy version of the object from Figure 7.38c. The noise of SNR = 12 dB was added to the circumscribed sphere

Then we randomly rotated and translated each object nine times and added Gaussian white noise (see Figure 7.39 for a sample degraded object). These degraded versions were classified to the clear objects in initial positions by a simple minimum distance classifier in the space of invariants up to the fifth order. We increased the noise level from 54 dB to −18 dB by the steps of 6 dB.

On each level, we repeated the experiment twice with different realizations of the noise and the average success rate was calculated.

The Gaussian-Hermite moments were computed by the recurrence formula (7.36) and normalized by (7.184). The Zernike moments were computed by the Kintner method (see Chapter 8). Finally, we normalized the invariants magnitudes w.r.t. the degree

$$\hat{I} = \text{sign}(I)|I|^{\frac{1}{d}}, \tag{7.191}$$

where d is the degree of the invariant, that is, the number of moments in a single term.

The success rates of both types of invariants are in Figure 7.40. Both the Gaussian-Hermite and the Zernike invariants perform perfectly if the SNR is greater than 10 dB. For SNR less than zero, both perform equally poorly. On the intermediate interval, the Zernike invariants exhibited higher robustness to noise.

Recognition of objects from the Princeton shape benchmark

In this experiment, we demonstrate that the better numerical stability of the Gaussian-Hermite moments actually increases the recognition power of the respective invariants in some cases. We again used the objects from the Princeton Shape Benchmark [192]. To demonstrate the phenomenon, we selected several pairs of objects that are visually very similar to each other (see Figure 7.41 for an example). For each object we generated ninety "shaky" instances by adding a zero-mean Gaussian noise to the coordinates of the triangle vertices of the object surface (see Figure 7.42 for a sample degradation). The amount of the noise was measured by

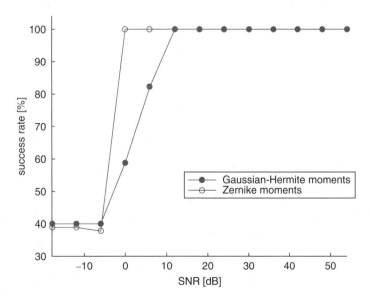

Figure 7.40 The success rates of the recognition of the archeological artifacts by Gaussian-Hermite moments and Zernike moments

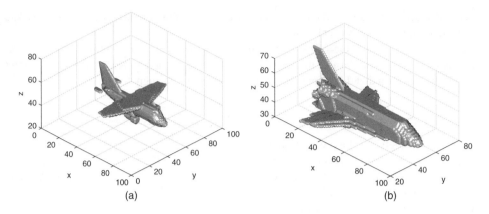

Figure 7.41 Two different airplanes from the PSB. (a) Object No. 1245, (b) Object No. 1249

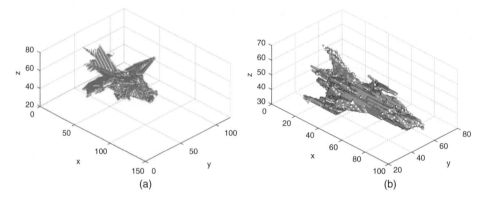

Figure 7.42 The noisy (10 dB) versions of the airplanes. (a) Object No. 1245, (b) Object No. 1249

the signal-to-noise ratio (SNR), defined in this case as

$$SNR = 10\log(\sigma_s^2/\sigma_n^2),\qquad(7.192)$$

where σ_s is the standard deviation of the object coordinates and σ_n is the standard deviation of the noise. We gradually changed the SNR from 55 dB to 10 dB, so we had nine degraded instances on each noise level. Now we let the algorithm recognize the shaky objects by means of five geometric invariants (according to the notation introduced in [193], we used $I_{1181}, I_{1182},$ $I_{1183}, I_{1184},$ and I_{1185}) and by the corresponding Gaussian-Hermite invariants. We intentionally used these higher-order invariants (from thirteenth to sixteenth order) because low-order invariants cannot distinguish similar objects.

The success curves are shown in Figure 7.43. If the noise is mild, both types of invariants can discriminate the objects perfectly. On medium levels of noise, the performance of the Gaussian-Hermite invariants dominates. The explanation is in numerical stability – geometric moments of orders 13–16 lose precision due to the necessity of working with very high values while Gaussian-Hermite moments can be computed in a stable way with much less precision loss. For a heavy noise 10 dB, the shaky objects are so difficult to distinguish that the

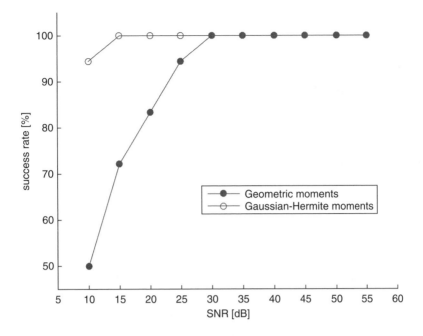

Figure 7.43   The success rate of the noisy airplane recognition

performance of both methods decreases. However, the performance of the Gaussian-Hermite invariants is still about 90% while that of the geometric invariants drops to 50%, which is in the case of two classes equal to a random guess. This experiment clearly illustrates the superiority of the Gaussian-Hermite invariants when recognizing similar objects. If the objects (classes) are significantly different, they can be discriminated by low-order invariants, and the described effect does not show up.

7.6.5   Object reconstruction from 3D moments

This experiment shows the reconstruction of a 3D binary object from Gaussian-Hermite and geometric moments. We chose a helicopter (object No. 1321; see Figure 7.44a) from the Princeton Shape Benchmark for this experiment because its propeller is a good indicator of the quality of the reconstruction.

We computed the normalized Gaussian-Hermite moments up to the order 100 and reconstructed the helicopter by means of the formula

$$f(x, y, z) = \sum_{p=0}^{R} \sum_{q=0}^{R} \sum_{r=0}^{R} \eta_{pqr} \hat{H}_p(x; \sigma) \hat{H}_q(y; \sigma) \hat{H}_r(z; \sigma), \qquad (7.193)$$

where $x, y, z = 0, 1, 2, \ldots, N - 1$. We chose the maximum index R successively as $R = 8, 46$, and 84. There were 1534, 340, and 0 erroneously reconstructed voxels (out of 1,000,000 voxels of the object) The results are in Figure 7.44b–d. The reconstruction for $R = 84$ is precise including all details.

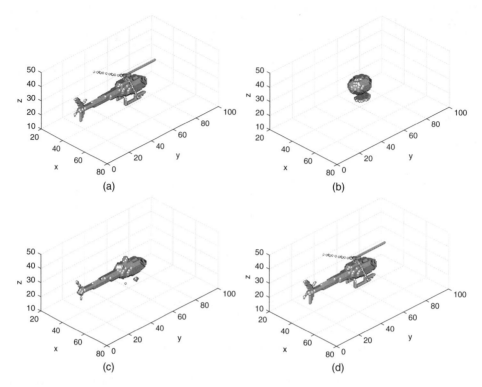

Figure 7.44   The helicopter: (a) the original converted to volumetric data, (b–d) the reconstruction from the weighted Hermite moments up to the order: (b) 8, (c) 46 and (d) 84

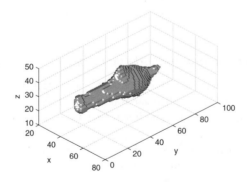

Figure 7.45   The reconstruction of the helicopter from the seventy-first-order geometric moments

For a comparison, we reconstructed the object also from geometric moments via Fourier transform. The best result was obtained for the moments up to the seventy-first order filtered at 0.12 of maximum frequency; see Figure 7.45. It contains 3662 erroneous voxels. Regardless of the chosen R, we never get a precise reconstruction from the geometric moments.

Clearly, the GH moments are superior to the geometric moments in object reconstruction, thanks to their better numerical properties.

7.7 Conclusion

In this chapter we have presented an overview of 2D and 3D orthogonal moments with relevance to image analysis. Orthogonal moments are a useful tool for overcoming certain drawbacks of the geometric and complex moments. The primary motivation for using them is their stable numerical implementation and improved reconstruction abilities.

From the theoretical point of view, all moment systems up to the given order are equivalent. This equivalence implies that one cannot expect any significant differences between geometric and OG moments when used for recognition. However, certain distinctions may appear in the experiments due to better numerical properties of the OG moments or if we use incomplete sets of moments of the given order. In that case, the sets of OG moments and geometric moments are not equivalent, and even if the number of elements is the same, the OG moments capture more information about the image than the geometric moments, as those are highly correlated due to the standard powers being nearly dependent.

We explained that OG moments are basically of two types, which differ from one another by the area of orthogonality. The moments OG on a square/cube are very good for image description if no rotation or affine transformation is present, and are powerful in image reconstruction. They are easy to calculate on a rectangular raster. On the other hand, creation of rotation invariants is possible only in an indirect way. The only exceptions are Gaussian-Hermite moments, a fact that makes them prominent moments for image analysis and object recognition, both in 2D and 3D.

The moments OG on a disk/sphere are easy to be made rotation invariant by phase cancellation. On the other hand, their computation in polar raster is not precise and may also require more time.

Selected efficient algorithms for OG moment computation will be presented in Chapter 8 and some practical applications of OG moments can be found in Chapter 9.

References

[1] T. S. Chihara, *An Introduction to Orthogonal Polynomials*. New York: Gordon and Breach, 1978.

[2] M. Abramowitz and I. A. Stegun, *Handbook of Mathematical Functions with Formulas, Graphs and Mathematical Tables*. Washington, D.C.: National Bureau of Standards, 1964.

[3] H. Wu and S. Yan, "Bivariate Hahn moments for image reconstruction," *International Journal of Applied Mathematics and Computer Science*, vol. 24, no. 2, pp. 417–428, 2014.

[4] C.-H. Teh and R. T. Chin, "On image analysis by the method of moments," *IEEE Transactions on Pattern Analysis and Machine Intelligence*, vol. 10, no. 4, pp. 496–513, 1988.

[5] M. Pawlak, *Image Analysis by Moments: Reconstruction and Computational Aspects*. Wrocław, Poland: Oficyna Wydawnicza Politechniki Wrocławskiej, 2006.

[6] C.-W. Chong, P. Raveendran, and R. Mukundan, "Translation and scale invariants of Legendre moments," *Pattern Recognition*, vol. 37, no. 1, pp. 119–129, 2004.

[7] K. M. Hosny, "Refined translation and scale Legendre moment invariants," *Pattern Recognition Letters*, vol. 31, no. 7, pp. 533–538, 2010.

[8] H. Zhang, H. Shu, G.-N. Han, G. Coatrieux, L. Luo, and J. L. Coatrieux, "Blurred image recognition by Legendre moment invariants," *IEEE Transactions on Image Processing*, vol. 19, no. 3, pp. 596–611, 2010.

 [9] X. Dai, H. Zhang, T. Liu, H. Shu, and L. Luo, "Legendre moment invariants to blur and affine transformation and their use in image recognition," *Pattern Analysis and Applications*, vol. 17, no. 2, pp. 311–326, 2012.

[10] H. Zhang, H. Shu, G. Coatrieux, J. Zhu, J. Wu, Y. Zhang, H. Zhu, and L. Luo, "Affine Legendre moment invariants for image watermarking robust to geometric distortions," *IEEE Transactions on Image Processing*, vol. 20, no. 8, pp. 2189–2199, 2011.

[11] Z. Bahaoui, K. Zenkouar, H. Fadili, H. Qjidaa, and A. Zarghili, "Blocking artifact removal using partial over-lapping based on exact Legendre moments computation," *Journal of Real-Time Image Processing*, pp. 1–19, 2014. doi= 0.1007/s11554-014-0465-3.

[12] A. Boucetta and K. E. Melkemi, "Hand shape recognition using Hu and Legendre moments," in *Proceedings of the 6th International Conference on Security of Information and Networks SIN'13*, pp. 272–276, ACM, 2013.

[13] R. Luo and T. Lin, "Finger crease pattern recognition using Legendre moments and principal component analysis," *Chinese Optics Letters*, vol. 5, no. 3, pp. 160–163, 2007.

[14] S. Dasari and I. V. Murali Krishna, "Combined classifier for face recognition using Legendre moments," *Computer Engineering and Applications*, vol. 1, no. 2, pp. 107–118, 2012.

[15] S. X. Liao and M. Pawlak, "Chinese character recognition via orthogonal moments," in *Information Theory and Applications II* (J.-Y. Chouinard, P. Fortier, and T. A. Gulliver, eds.), vol. 1133 of *Lecture Notes in Computer Science*, pp. 296–308, Berlin, Heidelberg, Germany, Springer, 1996.

[16] K. Kale, S. Chavan, M. Kazi, and Y. Rode, "Handwritten Devanagari compound character recognition using Legendre moment: An artificial neural network approach," in *International Symposium on Computational and Business Intelligence ISCBI'13*, pp. 274–278, IEEE, 2013.

[17] A. Wahi, S. Sundaramurthy, and P. Poovizhi, "Handwritten Tamil character recognition using Zernike moments and Legendre polynomial," in *Artificial Intelligence and Evolutionary Algorithms in Engineering Systems ICAEES'14* (L. P. Suresh, S. S. Dash, and B. K. Panigrahi, eds.), vol. 325-2 of *Advances in Intelligent Systems and Computing*, pp. 595–603, Springer, 2015.

[18] R. R. Bailey and M. Srinath, "Orthogonal moment features for use with parametric and non-parametric classifiers," *IEEE Transactions on Pattern Analysis and Machine Intelligence*, vol. 18, pp. 389–398, 1996.

[19] B. Fu, J. Zhou, Y. Li, G. Zhang, and C. Wang, "Image analysis by modified Legendre moments," *Pattern Recognition*, vol. 40, no. 2, pp. 691–704, 2007.

[20] H. Shu, L. Luo, X. Bao, W. Yu, and G. Han, "An efficient method for computation of Legendre moments," *Graphical Models*, vol. 62, no. 4, pp. 237–262, 2000.

[21] K. M. Hosny, "Exact Legendre moment computation for gray level images," *Pattern Recognition*, vol. 40, no. 12, pp. 3597–3605, 2007.

[22] R. Mukundan and K. R. Ramakrishnan, "Fast computation of Legendre and Zernike moments," *Pattern Recognition*, vol. 28, no. 9, pp. 1433–1442, 1995.

[23] P.-T. Yap and R. Paramesran, "An efficient method for the computation of Legendre moments," *IEEE Transactions on Pattern Analysis and Machine Intelligence*, vol. 27, no. 12, pp. 1996–2002, 2005.

[24] G. Wang and S. Wang, "Parallel recursive computation of the inverse Legendre moment transforms for signal and image reconstruction," *IEEE Signal Processing Letters*, vol. 11, no. 12, pp. 929–932, 2004.

[25] R. Mukundan and K. R. Ramakrishnan, *Moment Functions in Image Analysis*. Singapore: World Scientific, 1998.

[26] R. Mukundan, S. H. Ong, and P. A. Lee, "Image analysis by Tchebichef moments," *IEEE Transactions on Image Processing*, vol. 10, no. 9, pp. 1357–1364, 2001.

[27] R. Mukundan, "Improving image reconstruction accuracy using discrete orthonormal moments," in *Proceedings of the International Conference on Imaging Systems, Science and Technology CISST'03*, pp. 287–293, CSREA Press, 2003.

[28] R. Mukundan, "Some computational aspects of discrete orthonormal moments," *IEEE Transactions on Image Processing*, vol. 13, no. 8, pp. 1055–1059, 2004.

[29] P.-T. Yap and P. Raveendran, "Image focus measure based on Chebyshev moments," *IEE Proceedings of the Vision, Image and Signal Processing*, vol. 151, no. 2, pp. 128–136, 2004.

[30] J. Flusser, J. Kautsky, and F. Šroubek, "Implicit moment invariants," *International Journal of Computer Vision*, vol. 86, no. 1, pp. 72–86, 2010.

[31] P.-T. Yap, P. Raveendran, and S.-H. Ong, "Chebyshev moments as a new set of moments for image reconstruction," in *International Joint Conference on Neural Networks IJCNN '01*, vol. 4, pp. 2856–2860, 2001.

[32] C. Hermite, *Sur un nouveau développement en série des fonctions*, vol. 58. Gauthier-Villars, 1864. (in French).

[33] J. Shen, "Orthogonal Gaussian-Hermite moments for image characterization," in *Intelligent Robots and Computer Vision XVI: Algorithms, Techniques, Active Vision, and Materials Handling* (D. P. Casasent, ed.), vol. 3208, pp. 224–233, SPIE, 1997.

[34] J. Shen, W. Shen, and D. Shen, "On geometric and orthogonal moments," *International Journal of Pattern Recognition and Artificial Intelligence*, vol. 14, no. 7, pp. 875–894, 2000.

[35] Y. Wu and J. Shen, "Properties of orthogonal Gaussian-Hermite moments and their applications," *EURASIP Journal on Applied Signal Processing*, no. 4, pp. 588–599, 2005.

[36] X. Ma, R. Pan, and L. Wang, "License plate character recognition based on Gaussian-Hermite moments," in *Second International Workshop on Education Technology and Computer Science ETCS'10*, vol. 3, pp. 11–14, IEEE, 2010.

[37] B. Yang, T. Suk, M. Dai, and J. Flusser, "2D and 3D image analysis by Gaussian-Hermite moments," in *Moments and Moment Invariants – Theory and Applications* (G. A. Papakostas, ed.), pp. 143–173, Xanthi, Greece, Science Gate Publishing, 2014.

[38] L. Wang and M. Dai, "Application of a new type of singular points in fingerprint classification," *Pattern Recognition Letters*, vol. 28, no. 13, pp. 1640–1650, 2007.

[39] S. Farokhi, U. U. Sheikh, J. Flusser, and B. Yang, "Near infrared face recognition using Zernike moments and Hermite kernels," *Information Sciences*, vol. 316, pp. 234–245, 2015.

[40] N. Belghini, A. Zarghili, and J. Kharroubi, "3D face recognition using Gaussian Hermite moments," *International Journal of Computer Applications, Special Issue on Software Engineering, Databases and Expert Systems*, vol. SEDEX, no. 1, pp. 1–4, 2012.

[41] B. Yang, J. Flusser, and T. Suk, "Steerability of Hermite kernel," *International Journal of Pattern Recognition and Artificial Intelligence*, vol. 27, no. 4, pp. 1–25, 2013.

[42] L. Wang, Y. Wu, and M. Dai, "Some aspects of Gaussian-Hermite moments in image analysis," in *Third International Conference on Natural Computation ICNC'07* (L. P. Suresh, S. S. Dash, and B. K. Panigrahi, eds.), vol. 2 of *Advances in Intelligent Systems and Computing*, pp. 450–454, IEEE, 2007.

[43] J. Shen, W. Shen, and D. Shen, "On geometric and orthogonal moments," in *Multispectral Image Processing and Pattern Recognition* (J. Shen, P. S. P. Wang, and T. Zhang, eds.), vol. 44 of *Machine Perception Artificial Intelligence*, pp. 17–36, World Scientific Publishing, 2001.

[44] C. X. Zhang and P. Xi, "Analysis of Gaussian-Hermite moment invariants on image geometric transformation," in *Computer and Information Technology* (P. Yarlagadda, S.-B. Choi, and Y.-H. Kim, eds.), vol. 519–520 of *Applied Mechanics and Materials*, pp. 557–561, Trans Tech Publications, 2014.

[45] K. M. Hosny, "Fast computation of accurate Gaussian-Hermite moments for image processing applications," *Digital Signal Processing*, vol. 22, no. 3, pp. 476–485, 2012.

[46] B. Yang and M. Dai, "Image analysis by Gaussian–Hermite moments," *Signal Processing*, vol. 91, no. 10, pp. 2290–2303, 2011.

[47] B. Yang, G. Li, H. Zhang, and M. Dai, "Rotation and translation invariants of Gaussian–Hermite moments," *Pattern Recognition Letters*, vol. 32, no. 2, pp. 1283–1298, 2011.

[48] B. Yang, J. Flusser, and T. Suk, "Design of high-order rotation invariants from Gaussian-Hermite moments," *Signal Processing*, vol. 113, no. 1, pp. 61–67, 2015.

[49] B. Yang, J. Flusser, and T. Suk, "3D rotation invariants of Gaussian–Hermite moments," *Pattern Recognition Letters*, vol. 54, no. 1, pp. 18–26, 2015.

[50] S. X. Liao, A. Chiang, Q. Lu, and M. Pawlak, "Chinese character recognition via Gegenbauer moments," in *Proceedings of the 16th International Conference on Pattern Recognition ICPR'02*, vol. 3, pp. 485–488, IEEE, 2002.

[51] K. M. Hosny, "New set of Gegenbauer moment invariants for pattern recognition applications," *Arabian Journal for Science and Engineering*, vol. 39, no. 10, pp. 7097–7107, 2014.

[52] S. Liao and J. Chen, "Object recognition with lower order Gegenbauer moments," *Lecture Notes on Software Engineering*, vol. 1, no. 4, pp. 387–391, 2012.

[53] A. Chiang, S. Liao, Q. Lu, and M. Pawlak, "Gegenbauer moment-based applications for Chinese character recognition," in *Proceedings of the Canadian Conference on Electrical Computer Engineering CCECE'02* (W. Kinsner, A. Sebak, and K. Ferens, eds.), vol. 2, pp. 908–911, IEEE, 2002.

[54] R. Salouan, S. Safi, and B. Bouikhalene, "Handwritten Arabic characters recognition using methods based on Racah, Gegenbauer, Hahn, Tchebychev and orthogonal Fourier-Mellin moments," *International Journal of Advanced Science and Technology*, vol. 78, pp. 13–28, 2015.

[55] K. M. Hosny, "Image representation using accurate orthogonal Gegenbauer moments," *Pattern Recognition Letters*, vol. 32, no. 6, pp. 795–804, 2011.

[56] Z. Ping, H. Ren, J. Zou, Y. Sheng, and W. Bo, "Generic orthogonal moments: Jacobi-Fourier moments for invariant image description," *Pattern Recognition*, vol. 40, no. 4, pp. 1245–1254, 2007.

[57] P.-T. Yap and R. Paramesran, "Jacobi moments as image features," in *Proceedings of the Region 10 Conference on TENCON'04*, vol. 1, pp. 594–597, IEEE, 2004.

[58] H. Qjidaa, "Image reconstruction by Laguerre moments," in *Proceedings of the Second International Symposium on Communications, Control and Signal Processing ISCCSP'06*, IEEE, 2006.

[59] R. Koekoek and R. F. Swarttouw, "The askey-scheme of hypergeometric orthogonal polynomials and its q-analogue," Report 98-17, Technische Universiteit Delft, Faculty of Technical Mathematics and Informatics, 1996.

[60] G. Wang and S. Wang, "Recursive computation of Tchebichef moment and its inverse transform," *Pattern Recognition*, vol. 39, no. 1, pp. 47–56, 2006.

[61] L. Kotoulas and I. Andreadis, "Fast computation of Chebyshev moments," *IEEE Transactions on Circuits and Systems for Video Technology*, vol. 16, no. 7, pp. 884–888, 2006.

[62] H. Shu, H. Zhang, B. Chen, P. Haigron, and L. Luo, "Fast computation of Tchebichef moments for binary and grayscale images," *IEEE Transactions on Image Processing*, vol. 19, no. 12, pp. 3171–3180, 2010.

[63] K.-H. Chang, R. Paramesran, B. Honarvar, and C.-L. Lim, "Efficient hardware accelerators for the computation of Tchebichef moments," *IEEE Transactions on Circuits and Systems for Video Technology*, vol. 22, no. 3, pp. 414–425, 2012.

[64] B. Honarvar, R. Paramesran, and C.-L. Lim, "The fast recursive computation of Tchebichef moment and its inverse transform based on Z-transform," *Digital Signal Processing*, vol. 23, no. 5, pp. 1738–1746, 2013.

[65] M. Krawtchouk, "Sur une généralisation des polynômes d'Hermite," *Comptes Rendus Hebdomadaires des Séances de l'Académie des Sciences, Paris*, vol. 189, no. 1, pp. 620–622, 1929. (in French).

[66] J. Meixner, "Orthogonale Polynomsysteme mit einer besonderen Gestalt der erzeugenden Funktion," *Journal of London Mathematical Society*, vol. s1–9, no. 1, pp. 6–13, 1934. (in German).

[67] P.-T. Yap, R. Paramesran, and S.-H. Ong, "Image analysis by Krawtchouk moments," *IEEE Transactions on Image Processing*, vol. 12, no. 11, pp. 1367–1377, 2003.

[68] P.-T. Yap, P. Raveendran, and S.-H. Ong, "Krawtchouk moments as a new set of discrete orthogonal moments for image reconstruction," in *Proceedings of the 2002 International Joint Conference on Neural Networks IJCNN '02*, vol. 1, p. 5–, 2002.

[69] B. Hu and S. Liao, "Local feature extraction property of Krawtchouk moment," *Lecture Notes on Software Engineering*, vol. 1, no. 4, pp. 356–359, 2013.

[70] J. Sheeba Rani and D. Devaraj, "Face recognition using Krawtchouk moment," *Sādhanā*, vol. 37, no. 4, pp. 441–460, 2012.

[71] B. Hu and S. Liao, "Chinese character recognition by Krawtchouk moment features," in *10th International Conference on Image Analysis and Recognition, ICIAR'13* (M. Kamel and A. Campilho, eds.), vol. 7950 of *Lecture Notes in Computer Science*, pp. 711–716, Springer, 2013.

[72] A. Venkataramana and P. Ananth Raj, "Image watermarking using Krawtchouk moments," in *International Conference on Computing: Theory and Applications ICCTA'07*, pp. 676–680, 2007.

[73] L. Zhang, W.-W. Xiao, and Z. Ji, "Local affine transform invariant image watermarking by Krawtchouk moment invariants," *IET Information Security*, vol. 1, no. 3, pp. 97–105, 2007.

[74] L. Zhu, J. Liao, X. Tong, L. Luo, B. Fu, and G. Zhang, "Image analysis by modified Krawtchouk moments," in *Advances in Neural Networks ISNN'09* (W. Yu, H. He, and N. Zhang, eds.), vol. 5553 of *Lecture Notes in Computer Science*, pp. 310–317, Springer, 2009.

[75] A. Venkataramana and P. Ananth Raj, "Combined scale and translation invariants of Krawtchouk moments," in *National Conference on Computer Vision, Pattern Recognition, Image Processing and Graphics NCVPRIPG'08*, pp. 1–5, IEEE, 2008.

[76] G. Zhang, Z. Luo, B. Fu, B. Li, J. Liao, X. Fan, and Z. Xi, "A symmetry and bi-recursive algorithm of accurately computing Krawtchouk moments," *Pattern Recognition Letters*, vol. 31, no. 7, pp. 548–554, 2010.

[77] A. Venkataramana and P. Ananth Raj, "Fast computation of inverse Krawtchouk moment transform using Clenshaw's recurrence formula," in *International Conference on Image Processing ICIP'07*, vol. 4, pp. 37–40, IEEE, 2007.

[78] B. Honarvar and J. Flusser, "Fast computation of Krawtchouk moments," *Information Sciences*, vol. 288, pp. 73–86, 2014.

[79] M. Sayyouri, A. Hmimid, and H. Qjidaa, "A fast computation of novel set of Meixner invariant moments for image analysis," *Circuits Systems and Signal Processing*, vol. 34, no. 3, pp. 875–900, 2014.

[80] W. Hahn, "Über Orthogonalpolynome, die q-Differenzengleichungen genügen," *Mathematische Nachrichten*, vol. 2, no. 1–2, pp. 4–34, 1949.

[81] J. Zhou, H. Shu, H. Zhu, C. Toumoulin, and L. Luo, "Image analysis by discrete orthogonal Hahn moments," in *The International Conference on Image Analysis and Recognition ICIAR'05* (M. Kamel and A. Campilho, eds.), vol. 3656 of *Lecture Notes in Computer Science*, pp. 524–531, Springer, 2005.

[82] P.-T. Yap, R. Paramesran, and S.-H. Ong, "Image analysis using Hahn moments," *IEEE Transactions on Pattern Analysis and Machine Intelligence*, vol. 29, no. 11, pp. 2057–2062, 2007.

[83] H.-A. Goh, C.-W. Chong, R. Besar, F. S. Abas, and K. S. Sim, "Translation and scale invariants of Hahn momnets," *International Journal of Image and Graphics*, vol. 9, no. 2, pp. 271–285, 2009.

[84] H. Wu and S. Yan, "Bivariate Hahn moments for image reconstruction," *International Journal of Applied Mathematics and Computer Science*, vol. 24, no. 2, pp. 417–428, 2014.

[85] F. Akhmedova and S. Liao, "Face recognition using discrete orthogonal Hahn moments," *International Conference on Image Analysis and Processing ICIAP'15*, vol. 2, no. 5, p. 1269–, 2015.

[86] S. Ahmad, Q. Zhang, Z.-M. Lu, and M. W. Anwar, "Feature-based watermarking using discrete orthogonal Hahn moment invariants," in *Proceedings of the 7th International Conference on Frontiers of Information Technology FIT'09*, pp. 38:1–38:6, ACM, 2009.

[87] C. Camacho-Bello and J. J. Báez-Rojas, "Angle estimation using Hahn moments for image analysis," in *Progress in Pattern Recognition, Image Analysis, Computer Vision, and Applications. Proceedings of the 19th Iberoamerican Congress CIARP'14* (E. Bayro-Corrochano and E. Hancock, eds.), vol. 8827 of *Lecture Notes in Computer Science*, pp. 127–134, Springer, 2014.

[88] A. Venkataramana and P. Ananth Raj, "Computation of Hahn moments for large size images," *Journal of Computer Science*, vol. 6, no. 9, pp. 1037–1041, 2010.

[89] M. Sayyouri, A. Hmimid, and H. Qjidaa, "Improving the performance of image classification by Hahn moment invariants," *Journal of the Optical Society of America A*, vol. 30, no. 11, pp. 2381–2394, 2013.

[90] H. Zhu, H. Shu, J. Zhou, L. Luo, and J.-L. Coatrieux, "Image analysis by discrete orthogonal dual Hahn moments," *Pattern Recognition Letters*, vol. 28, no. 13, pp. 1688–1704, 2007.

[91] P. Ananth Raj, "Image contrast enhancement using discrete dual Hahn moments," in *International Conference on Machine Vision Applications MVA'13*, pp. 206–209, IAPR, 2013.

[92] E. G. Karakasis, G. Papakostas, D. E. Koulouriotis, and V. Tourassis, "Generalized dual Hahn moment invariants," *Pattern Recognition*, vol. 46, no. 7, pp. 1998–2014, 2013.

[93] J. A. Wilson, *Hypergeometric Series Recurrence Relations and Some New Orthogonal Functions*. PhD thesis, University of Wisconsin, Madison, WI, USA, 1978.

[94] H. Zhu, H. Shu, J. Liang, L. Luo, and J.-L. Coatrieux, "Image analysis by discrete orthogonal Racah moments," *Signal Processing*, vol. 87, no. 4, pp. 687–708, 2007.

[95] J. P. Ananth and V. Subbiah Bharathi, "Face image retrieval system using discrete orthogonal moments," in *4th International Conference on Bioinformatics and Biomedical Technology ICBBT'12*, vol. 29, pp. 218–223, IACSIT, 2012.

[96] Y. Wu and S. Liao, "Chinese characters recognition via Racah moments," in *International Conference on Audio, Language and Image Processing ICALIP'14*, pp. 691–694, IEEE, 2014.

[97] K. Fardousse, Annass. El affar, H. Qjidaa, and Abdeljabar. Cherkaoui, "Robust skeletonization of hand written craft motives of "zellij" using Racah moments," *International Journal of Computer Science and Network Security*, vol. 9, no. 8, pp. 216–221, 2009.

[98] K. M. Hosny, "New set of rotationally Legendre moment invariants," *International Journal of Electrical, Computer, and Systems Engineering*, vol. 4, no. 3, pp. 176–180, 2010.

[99] C. L. Deepika, A. Kandaswamy, C. Vimal, and B. Satish, "Palmprint authentication using modified Legendre moments," *Procedia Computer Science*, vol. 2, pp. 164–172, 2010. Proceedings of the International Conference and Exhibition on Biometrics Technology ICEBT'10.

[100] F. Zernike, "Beugungstheorie des Schneidenverfarhens und seiner verbesserten Form, der Phasenkontrastmethode," *Physica*, vol. 1, no. 7, pp. 689–704, 1934 (in German).

[101] M. R. Teague, "Image analysis via the general theory of moments," *Journal of the Optical Society of America*, vol. 70, no. 8, pp. 920–930, 1980.

[102] Å. Wallin and O. Kübler, "Complete sets of complex Zernike moment invariants and the role of the pseudoin-variants," *IEEE Transactions on Pattern Analysis and Machine Intelligence*, vol. 17, no. 11, pp. 1106–1110, 1995.

[103] J. Revaud, G. Lavoué, and A. Baskurt, "Improving Zernike moments comparison for optimal similarity and rotation angle retrieval," *IEEE Transactions on Pattern Analysis and Machine Intelligence*, vol. 31, no. 4, pp. 627–636, 2009.

[104] J. Haddadnia, M. Ahmadi, and K. Faez, "An efficient feature extraction method with pseudo-Zernike moment in RBF neural network-based human face recognition system," *EURASIP Journal on Applied Signal Processing*, vol. 9, pp. 890–901, 2003.

[105] S. O. Belkasim, M. Shridhar, and M. Ahmadi, "Pattern recognition with moment invariants: A comparative study and new results," *Pattern Recognition*, vol. 24, no. 12, pp. 1117–1138, 1991.

[106] A. Khotanzad and Y. H. Hong, "Invariant image recognition by Zernike moments," *IEEE Transactions on Pattern Analysis and Machine Intelligence*, vol. 12, no. 5, pp. 489–497, 1990.

[107] C.-W. Chong, P. Raveendran, and R. Mukundan, "Translation invariants of Zernike moments," *Pattern Recognition*, vol. 36, no. 3, pp. 1765–1773, 2003.

[108] G. Amayeh, S. Kasaei, G. Bebis, A. Tavakkoli, and K. Veropoulos, "Improvement of Zernike moment descriptors on affine transformed shapes," in *9th International Symposium on Signal Processing and Its Applications ISSPA'07*, pp. 1–4, IEEE, 2007.

[109] X. Dai, T. Liu, H. Shu, and L. Luo, "Pseudo-Zernike moment invariants to blur degradation and their use in image recognition," in *Intelligent Science and Intelligent Data Engineering IScIDE'12* (J. Yang, F. Fang, and C. Sun, eds.), vol. 7751 of *Lecture Notes in Computer Science*, pp. 90–97, Springer, 2013.

[110] B. Chen, H. Shu, H. Zhang, G. Coatrieux, L. Luo, and J. L. Coatrieux, "Combined invariants to similarity transformation and to blur using orthogonal Zernike moments," *IEEE Transactions on Image Processing*, vol. 20, no. 2, pp. 345–360, 2011.

[111] B. Chen, H. Shu, H. Zhang, G. Chen, C. Toumoulin, J. Dillenseger, and L. Luo, "Quaternion Zernike moments and their invariants for color image analysis and object recognition," *Signal Processing*, vol. 92, no. 2, pp. 308–318, 2012.

[112] J. Sarvaiya, S. Patnaik, and H. Goklani, "Image registration using NSCT and invariant moment," *International Journal of Image Processing*, vol. 4, no. 2, pp. 119–130, 2010.

[113] Z.-L. Yang and B.-L. Guo, "Image registration using feature points extraction and pseudo-Zernike moments," in *International Conference on Intelligent Information Hiding and Multimedia Signal Processing IIH-MSP'08*, pp. 752–755, IEEE, 2008.

[114] S. Gillan and P. Agathoklis, "Image registration using feature points, Zernike moments and an M-estimator," in *53rd International Midwest Symposium on Circuits and Systems MWSCAS'10*, pp. 434–437, IEEE, 2010.

[115] C. Singh, N. Mittal, and E. Walia, "Face recognition using Zernike and complex Zernike moment features," *Pattern Recognition and Image Analysis*, vol. 21, no. 1, pp. 71–81, 2011.

[116] E. Sariyanidi, V. Dagli, S. C. Tek, B. Tunc, and M. Gokmen, "Local Zernike moments: A new representation for face recognition," in *19th IEEE International Conference on Image Processing ICIP'12*, pp. 585–588, IEEE, 2012.

[117] S. Farokhi, S. M. Shamsuddin, J. Flusser, U. U. Sheikh, M. Khansari, and K. Jafari-Khouzani, "Rotation and noise invariant near-infrared face recognition by means of Zernike moments and spectral regression discriminant analysis," *Journal of Electronic Imaging*, vol. 22, no. 1, pp. 1–11, 2013.

[118] S. Farokhi, S. M. Shamsuddin, U. U. Sheikh, J. Flusser, M. Khansari, and K. Jafari-Khouzani, "Near infrared face recognition by combining Zernike moments and undecimated discrete wavelet transform," *Digital Signal Processing*, vol. 31, pp. 13–27, 2014.

[119] C. Singh, E. Walia, and N. Mittal, "Discriminative Zernike and pseudo Zernike moments for face recognition," *Pattern Recognition and Image Analysis*, vol. 2, no. 2, pp. 12–35, 2012.

[120] C.-W. Tan and A. Kumar, "Accurate iris recognition at a distance using stabilized iris encoding and Zernike moments phase features," *IEEE Transactions on Image Processing*, vol. 23, no. 9, pp. 3962–3974, 2014.

[121] C. Lu and Z. Lu, "Zernike moment invariants based iris recognition," in *Advances in Biometric Person Authentication, 5th Chinese Conference on Biometric Recognition, SINOBIOMETRICS'04* (S. Z. Li, J. Lai, T. Tan, G. Feng, and Y. Wang, eds.), vol. 3338 of *Lecture Notes in Computer Science*, pp. 554–561, Springer, 2005.

[122] C. L. Deepika, A. Kandaswamy, C. Vimal, and B. Sathish, "Invariant feature extraction from fingerprint biometric using pseudo Zernike moments," in *Proceedings of the International Joint Journal Conference on Engineering and Technology IJJCET'10*, pp. 104–108, gopalax Publications & TCET, 2010.

[123] I. El-Feghi, A. Tahar, and M. Ahmadi, "Efficient features extraction for fingerprint classification with multi layer perceptron neural network," in *10th International Symposium on Signals, Circuits and Systems ISSCS'11*, pp. 1–4, IEEE, 2011.

[124] H.-U. Jang, D.-K. Hyun, D.-J. Jung, and H.-K. Lee, "Fingerprint – PKI authentication using Zernike moments," in *International Conference on Image Processing ICIP'14*, pp. 5022–5026, IEEE, 2014.

[125] G. Amayeh, G. Bebis, A. Erol, , and M. Nicolescu, "Peg-free hand shape verification using high order Zernike moments," in *Conference on Computer Vision and Pattern Recognition Workshop CVPRW'06*, pp. 1–8, IEEE, 2006.

[126] H. S. Kim and H.-K. Lee, "Invariant image watermark using Zernike moments," *IEEE Transactions on Circuits and Systems for Video Technology*, vol. 13, no. 8, pp. 766–775, 2003.

[127] N. Singhal, Y.-Y. Lee, C.-S. Kim, and S.-U. Lee, "Robust image watermarking using local Zernike moments," *Journal of Visual Communication and Image Representation*, vol. 20, no. 6, pp. 408–419, 2009.

[128] X.-C. Yuan, C.-M. Pun, and C.-L. P. Chen, "Geometric invariant watermarking by local Zernike moments of binary image patches," *Signal Processing*, vol. 93, no. 7, pp. 2087–2095, 2013.

[129] W.-Y. Kim and Y.-S. Kim, "A region-based shape descriptor using Zernike moments," *Signal Processing: Image Communication*, vol. 16, no. 1–2, pp. 95–102, 2000.

[130] S. Li, M.-C. Lee, and C.-M. Pun, "Complex Zernike moments features for shape-based image retrieval," *IEEE Transactions on Systems, Man, and Cybernetics – Part A: Systems and Humans*, vol. 39, no. 1, pp. 227–237, 2009.

[131] S. A. Bakar, M. S. Hitam, and W. N. J. Hj Wan Yussof, "Investigating the properties of Zernike moments for robust content based image retrieval," in *International Conference on Computer Applications Technology ICCAT'13*, pp. 1–6, IEEE, 2013.

[132] S.-K. Hwang and W.-Y. Kim, "A novel approach to the fast computation of Zernike moments," *Pattern Recognition*, vol. 39, no. 11, pp. 2065–2076, 2006.

[133] J. Gu, H. Z. Shu, C. Toumoulin, and L. M. Luo, "A novel algorithm for fast computation of Zernike moments," *Pattern Recognition*, vol. 35, no. 12, pp. 2905–2911, 2002.

[134] C.-W. Chong, P. Raveendran, and R. Mukundan, "An efficient algorithm for fast computation of pseudo-Zernike moments," *International Journal of Pattern Recognition and Artificial Intelligence*, vol. 17, no. 6, pp. 1011–1023, 2003.

[135] S. O. Belkasim, M. Ahmadi, and M. Shridhar, "Efficient algorithm or fast computation of Zernike moments," *Journal of the Franklin Institute*, vol. 333(B), no. 4, pp. 577–581, 1996.

[136] K. M. Hosny, "Fast computation of accurate Zernike moments," *Journal of Real-Time Image Processing*, vol. 3, no. 1, pp. 97–107, 2010.

[137] C.-W. Chong, R. Paramesran, and R. Mukundan, "A comparative analysis of algorithms for fast computation of Zernike moments," *Pattern Recognition*, vol. 36, no. 3, pp. 731–742, 2003.

[138] S. X. Liao and M. Pawlak, "On the accuracy of Zernike moments for image analysis," *IEEE Transactions on Pattern Analysis and Machine Intelligence*, vol. 20, no. 12, pp. 1358–1364, 1998.

[139] Y. Sheng and J. Duvernoy, "Circular-Fourier-radial-Mellin transform descriptors for pattern recognition," *Journal of the Optical Society of America A*, vol. 3, no. 6, pp. 885–888, 1986.

[140] Y. Sheng and H. H. Arsenault, "Experiments on pattern recognition using invariant Fourier-Mellin descriptors," *Journal of the Optical Society of America A, Optics and image science*, vol. 3, no. 6, pp. 771–776, 1986.

[141] Y. Sheng and L. Shen, "Orthogonal Fourier-Mellin moments for invariant pattern recognition," *Journal of the Optical Society of America A*, vol. 11, no. 6, pp. 1748–1757, 1994.

[142] G. A. Papakostas, Y. S. Boutalis, D. A. Karras, and B. G. Mertzios, "Fast numerically stable computation of orthogonal Fourier-Mellin moments," *IET Computer Vision*, vol. 1, no. 1, pp. 11–16, 2007.

[143] B. Fu, J.-Z. Zhou, Y.-H. Li, B. Peng, L.-Y. Liu, and J.-Q. Wen, "Novel recursive and symmetric algorithm of computing two kinds of orthogonal radial moments," *Image Science Journal*, vol. 56, no. 6, pp. 333–341, 2008.

[144] E. Walia, C. Singh, and A. Goyal, "On the fast computation of orthogonal Fourier-Mellin moments with improved numerical stability," *Journal of Real-Time Image Processing*, vol. 7, no. 4, pp. 247–256, 2012.

[145] C. Singh and R. Upneja, "Accurate computation of orthogonal Fourier-Mellin moments," *Journal of Mathematical Imaging and Vision*, vol. 44, no. 3, pp. 411–431, 2012.

[146] K. M. Hosny, M. A. Shouman, and H. M. A. Salam, "Fast computation of orthogonal Fourier-Mellin moments in polar coordinates," *Journal of Real-Time Image Processing*, vol. 6, no. 2, pp. 73–80, 2011.

[147] X. Wang and S. Liao, "Image analysis by orthogonal Fourier-Mellin moments," *Journal of Theoretical and Applied Computer Science*, vol. 7, no. 1, pp. 70–78, 2013.

[148] X. Wang and S. Liao, "Image reconstruction from orthogonal Fourier-Mellin moments," in *Image Analysis and Recognition, 10th International Conference, ICIAR'13* (M. Kamel and A. Campilho, eds.), vol. 7950 of *Lecture Notes in Computer Science*, pp. 687–694, Springer, 2013.

[149] L.-Q. Guo and M. Zhu, "Quaternion Fourier-Mellin moments for color images," *Pattern Recognition*, vol. 44, no. 2, pp. 187–195, 2011.

[150] Q. Liu, H. Zhu, and Q. Li, "Object recognition by combined invariants of orthogonal Fourier-Mellin moments," in *Proceedings of 8th International Conference on Information, Communications and Signal Processing ICICS'11*, pp. 1–5, IEEE, 2011.

[151] C. Kan and M. D. Srinath, "Invariant character recognition with Zernike and orthogonal Fourier-Mellin moments," *Pattern Recognition*, vol. 35, no. 1, pp. 143–154, 2002.

[152] Z. Ping, R. Wu, and Y. Sheng, "Image description with Chebyshev-Fourier moments," *Journal of the Optical Society of America A*, vol. 19, no. 9, pp. 1748–1754, 2002.

[153] B. Xiao, J. Feng Ma, and J.-T. Cui, "Invariant image recognition using radial Jacobi moment invariants," in *Sixth International Conference on Image and Graphics ICIG'11*, pp. 280–285, IEEE, 2011.

[154] A. Padilla-Vivanco, C. Toxqui-Quitl, and C. Santiago-Tepantlán, "Gray level image reconstruction using Jacobi-Fourier moments," in *Optical and Digital Image Processing* (P. Schelkens, T. Ebrahimi, G. Cristóbal, and F. Truchetet, eds.), vol. 7000, (Bellingham, Washington, USA), pp. 1–10, SPIE, 2008.

[155] C. Toxqui-Quitl, A. Padilla-Vivanco, and C. Santiago-Tepantlán, "Minimum image resolution for shape recognition using the generic Jacobi Fourier moments," in *22nd Congress of the International Commission for Optics: Light for the Development of the World* (R. Rodríguez-Vera and R. Díaz-Uribe, eds.), vol. 8011, (Bellingham, Washington, USA), pp. 1–10, SPIE, 2011.

[156] G. Amu, S. Hasi, X. Yang, and Z. Ping, "Image analysis by pseudo-Jacobi ($p = 4$, $q = 3$)–Fourier moments," *Applied Optics*, vol. 43, no. 10, pp. 2093–2101, 2004.

[157] T. V. Hoang and S. Tabbone, "Errata and comments on 'Generic orthogonal moments: Jacobi-Fourier moments for invariant image description,'" *Pattern Recognition*, vol. 46, no. 11, pp. 3148–3155, 2013.

[158] A. Padilla-Vivanco, "Gait recognition by Jacobi-Fourier moments," in *Frontiers in Optics 2011/Laser Science XXVII*, p. JTuA19, Optical Society of America, 2011.

[159] C. Camacho-Bello, C. Toxqui-Quitl, and A. Padilla-Vivanco, "Generic orthogonal moments and applications," in *Moments and Moment Invariants – Theory and Applications* (G. A. Papakostas, ed.), pp. 175–204, Science Gate Publishing, 2014.

[160] J.-l. Dong, G.-r. Yin, and Z.-l. Ping, "Geometrically robust image watermarking based on Jacobi-Fourier moments," *Optoelectronics Letters*, vol. 5, no. 5, pp. 387–390, 2009.

[161] R. Upneja and C. Singh, "Fast computation of Jacobi-Fourier moments for invariant image recognition," *Pattern Recognition*, vol. 48, no. 5, pp. 1836–1843, 2015.

[162] C. Camacho-Bello, C. Toxqui-Quitl, A. Padilla-Vivanco, and J. J. Báez-Rojas, "High-precision and fast computation of Jacobi-Fourier moments for image description," *Journal of the Optical Society of America A*, vol. 31, no. 1, pp. 124–134, 2014.

[163] B. Li, G. Zhang, and B. Fu, "Image analysis using radial Fourier-Chebyshev moments," in *International Conference on Multimedia Technology ICMT'11*, pp. 3097–3100, IEEE, 2011.

[164] Y. Jiang, Z. Ping, and L. Gao, "A fast algorithm for computing Chebyshev-Fourier moments," in *International Seminar on Future Information Technology and Management Engineering FITME'10*, pp. 425–428, IEEE, 2010.

[165] H. Ren, Z. Ping, W. Bo, W. Wu, and Y. Sheng, "Multidistortion-invariant image recognition with radial harmonic Fourier moments," *Journal of the Optical Society of America A*, vol. 20, no. 4, pp. 631–637, 2003.

[166] H. Ren, A. Liu, J. Zou, D. Bai, and Z. Ping, "Character reconstruction with radial-harmonic-Fourier-moments," in *Fourth International Conference on Fuzzy Systems and Knowledge Discovery FSKD'07*, vol. 3, pp. 307–310, IEEE, 2007.

[167] H.-P. Ren, Z.-L. Ping, W.-R.-G. Bo, Y.-L. Sheng, S.-Z. Chen, and W.-K. Wu, "Cell image recognition with radial harmonic Fourier moments," *Chinese Physics*, vol. 12, no. 6, pp. 610–614, 2003.

[168] H. Ren, A. Liu, Z. Ping, and D. Bai, "Study on a novel tumor cell recognition system based on orthogonal image moments," in *7th Asian-Pacific Conference on Medical and Biological Engineering APCMBE'08* (Y. Peng and X. Weng, eds.), vol. 19 of *IFMBE Proceedings*, pp. 290–292, Springer, 2013.

[169] W. Xiang-yang, L. Wei-yi, Y. Hong-ying, N. Pan-pan, and L. Yong-wei, "Invariant quaternion radial harmonic Fourier moments for color image retrieval," *Optics & Laser Technology*, vol. 66, pp. 78–88, 2015.

[170] P.-T. Yap, X. Jiang, and A. C. Kot, "Two-dimensional polar harmonic transforms for invariant image representation," *IEEE Transactions on Pattern Analysis and Machine Intelligence*, vol. 32, no. 7, pp. 1259–1270, 2010.

[171] S. Dominguez, "Image analysis by moment invariants using a set of step-like basis functions," *Pattern Recognition Letters*, vol. 34, no. 16, pp. 2065–2070, 2013.

[172] Z. Feng, L. Shang-Qian, W. Da-Bao, and G. Wei, "Aircraft recognition in infrared image using wavelet moment invariants," *Image and Vision Computing*, vol. 27, no. 4, pp. 313–318, 2009.

[173] B. Xiao, J.-F. Ma, and X. Wang, "Image analysis by Bessel-Fourier moments," *Pattern Recognition*, vol. 43, no. 8, pp. 2620–2629, 2010.

[174] Z.-P. Ma, B.-S. Kang, and B. Xiao, "A study of Bessel Fourier moments invariant for image retrieval and classification," in *International Conference on Image Analysis and Signal Processing IASP'11*, pp. 316–320, IEEE, Oct 2011.

[175] G. Gao and G. Jiang, "Bessel-Fourier moment-based robust image zero-watermarking," *Multimedia Tools and Applications*, vol. 74, no. 3, pp. 841–858, 2015.

[176] Z. Shao, H. Shu, J. Wu, B. Chen, and J. L. Coatrieux, "Quaternion Bessel Fourier moments and their invariant descriptors for object reconstruction and recognition," *Pattern Recognition*, vol. 47, no. 2, pp. 603–611, 2014.

[177] F. Ghorbel, S. Derrode, R. Mezhoud, T. Bannour, and S. Dhahbi, "Image reconstruction from a complete set of similarity invariants extracted from complex moments," *Pattern Recognition Letters*, vol. 27, no. 12, pp. 1361–1369, 2006.

[178] H. Zhang, H. Shu, L. Luo, and J. L. Dillenseger, "A Legendre orthogonal moment based 3D edge operator," *Science in China Series G: Physics, Mechanics and Astronomy*, vol. 48, no. 1, pp. 1–13, 2005.

[179] H. Shu, L. Luo, W. Yu, and J. Zhou, "Fast computation of Legendre moments of polyhedra," *Pattern Recognition*, vol. 34, no. 5, pp. 1119–1126, 2001.

[180] K. M. Hosny, "Fast and low-complexity method for exact computation of 3D Legendre moments," *Pattern Recognition Letters*, vol. 32, no. 9, pp. 1305–1314, 2011.

[181] N. Canterakis, "3D Zernike moments and Zernike affine invariants for 3D image analysis and recognition," in *Proceedings of the 11th Scandinavian Conference on Image Analysis SCIA'99* (B. K. Ersbøll and P. Johansen, eds.), DSAGM, 1999.

[182] M. Novotni and R. Klein, "3D Zernike descriptors for content based shape retrieval," in *Proceedings of the Eighth symposium on Solid modeling and applications SM'03*, pp. 216–225, ACM, 2003.

[183] M. Novotni and R. Klein, "Shape retrieval using 3D Zernike descriptors," *Computer-Aided Design*, vol. 36, no. 11, pp. 1047–1062, 2004.

[184] S. Grandison, C. Roberts, and R. J. Morris, "The application of 3D Zernike moments for the description of "model-free" molecular structure, functional motion, and structural reliability," *Journal of Computational Biology*, vol. 16, no. 3, pp. 487–500, 2009.

[185] V. Venkatraman, P. R. Chakravarthy, and D. Kihara, "Application of 3D Zernike descriptors to shape-based ligand similarity searching," *Journal of Cheminformatics*, vol. 1, no. 19, pp. 1–20, 2009.

[186] V. Venkatraman, L. Sael, and D. Kihara, "Potential for protein surface shape analysis using spherical harmonics and 3D Zernike descriptors," *Cell Biochemistry and Biophysics*, vol. 54, no. 1–3, pp. 23–32, 2009.

[187] V. Venkatraman, Y. D. Yang, L. Sael, and D. Kihara, "Protein-protein docking using region-based 3D Zernike descriptors," *BMC Bioinformatics*, vol. 10, no. 407, pp. 1–21, 2009.

[188] A. Sit, J. C. Mitchell, G. N. Phillips, and S. J. Wright, "An extension of 3D Zernike moments for shape description and retrieval of maps defined in rectangular solids," *Molecular Based Mathematical Biology*, vol. 1, no. 1, pp. 75–89, 2013.

[189] M. S. Al-Rawi, "3D (pseudo) Zernike moments: Fast computation via symmetry properties of spherical harmonics and recursive radial polynomials," in *19th International Conference on Image Processing ICIP'12*, pp. 2353–2356, IEEE, 2012.

[190] K. M. Hosny and M. A. Hafez, "An algorithm for fast computation of 3D Zernike moments for volumetric images," *Mathematical Problems in Engineering*, vol. 2012, pp. 1–17, 2012.

[191] J. M. Pozo, M.-C. Villa-Uriol, and A. F. Frangi, "Efficient 3D geometric and Zernike moments computation from unstructured surface meshes," *IEEE Transactions on Pattern Analysis and Machine Intelligence*, vol. 33, no. 3, pp. 471–484, 2011.

[192] P. Shilane, P. Min, M. Kazhdan, and T. Funkhouser, "The Princeton Shape Benchmark," 2004. http://shape.cs.princeton.edu/benchmark/.

[193] DIP, "3D rotation moment invariants," 2011. http://zoi.utia.cas.cz/3DRotationInvariants.

8

Algorithms for Moment Computation

8.1 Introduction

This chapter is devoted to the computational aspects of moments of digital images. In the previous chapters, we have been working mostly with moments and moment invariants in a continuous domain. In digital image processing, all quantities have to be converted from the continuous to the discrete domain, and efficient algorithms for dealing with discrete quantities must be developed. The discrete algorithms sometimes follow their continuous ancestors in a straightforward manner, but sometimes the development of a computationally efficient algorithm requires new inventions. The same holds true for moments and moment invariants.

When speaking about algorithms for moment computation, we understand in general algorithms for binary as well as graylevel images; for geometric, complex, and various orthogonal moments; algorithms which yield precise values and also approximation algorithms; algorithms which speed up the computation and make it more robust to numerical errors; and algorithms which act in 2D and 3D. The categorization of the algorithms is very difficult not only because of different ideas, kind of data, purposes, and required properties as indicated above, but also due to a high number of papers that have been published on this topic[1]. We can, however, observe several trends.

- Many algorithms are devoted to the computation of moments of binary images. The reasons are quite clear. First, any binary object is fully determined by its boundary, which is supposed to contain many fewer pixels than the whole object, so there is a significant potential for reducing computational complexity. Second, most of the practical applications, especially in shape recognition, work with binary images. These algorithms mostly calculate geometric moments because when working with (reasonably small) binary images, the danger of numerical errors is not serious, and the use of OG moments is usually not necessary.

---

[1] We can find more than 1000 references on Google Scholar.

2D and 3D Image Analysis by Moments, First Edition. Jan Flusser, Tomáš Suk and Barbara Zitová.
© 2017 John Wiley & Sons, Ltd. Published 2017 by John Wiley & Sons, Ltd.
Companion website: www.wiley.com/go/flusser/2d3dimageanalysis

- The algorithms for graylevel images are mainly aimed at improving numerical stability of the calculation rather than increasing speed. Many algorithms of this category are focused on efficient calculation of various OG moments.
- It is not necessary to develop particular algorithms for the calculation of moment invariants. Since all invariants are functions of a small number of moments (relatively to the image dimension), the computing complexity of the invariants is determined solely by the complexity of the moment computation. As soon as we have calculated the moments, we can calculate any invariant in $\mathcal{O}(1)$ time.

8.2 Digital image and its moments

In this section, we show how the definition of image moments can be converted from the continuous domain to the discrete one.

8.2.1 Digital image

Although the notion of a *digital* or *discrete image* is intuitive, we start with its formal definition.

Definition 8.1 By a *digital image* we understand any $N \times M$ matrix $\mathbf{f} \equiv (f_{ij})$, the elements of which are from the finite set of integers $\{0, 1, 2, \ldots, L - 1\}$, where $L > 1$.

In practice usually $L = 256$, and each pixel value is stored in one byte. If $L = 2$, the image is called *binary*.

The conversion from a "continuous" image $f(x, y)$ into a digital image $\mathbf{f}$ is accomplished by *sampling* (spatial discretization) followed by *quantization* (graylevel discretization). The model of these processes in real imaging systems is very complicated and comprises many device-dependent parameters. However, an approximative model of digital image formation is well known from standard textbooks [1, 2].

Ideal sampling of image $f(x, y)$ is modelled as a multiplication with the *sampling function* $s(x, y)$

$$s(x, y) = \sum_{i=-\infty}^{\infty} \sum_{j=-\infty}^{\infty} \delta(x - i\Delta x, y - j\Delta y),$$

where Δx and Δy are the sampling steps (see Figure 8.1). The result is an image $d(x, y)$

$$d(x, y) = f(x, y) \cdot s(x, y),$$

which is still real-valued and equals $f(x, y)$ on the sampling grid and is zero otherwise.

It is well known that if f is band-limited, and the sampling steps satisfy the Nyquist inequalities

$$\Delta x \leq \frac{1}{2U},$$

$$\Delta y \leq \frac{1}{2V},$$

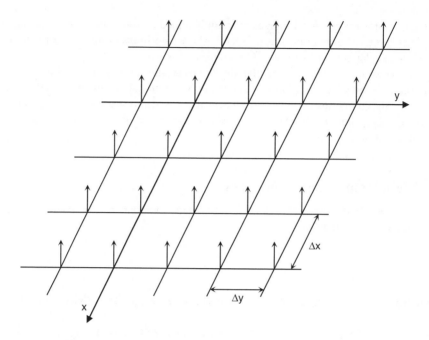

Figure 8.1   Sampling function $s(x, y)$ with the steps Δx and Δy

where U and V are the frequency bounds[2] of f, the exact reconstruction of f from d is guaranteed. This reconstruction can be accomplished in spatial domain by an interpolation between the samples by means of the $\text{sinc}(x) \cdot \text{sinc}(y)$ interpolation kernel[3].

After image f has been sampled, it is quantized pixel-wise by quantizer[4] Q such that $f_{ij} = Q(d(i\Delta x, j\Delta y))$.

8.2.2   Discrete moments

Now we can start thinking of the moments of digital images. A philosophic question arises immediately: for which image shall we calculate the moments? There are several different answers, all of them being meaningful and justifiable.

One possible answer, which originates from the above sampling model, is to say that the moment of discrete image $\mathbf{f}$ equals the "continuous" moments of d (we neglect the quantization), which is

$$M_{pq}^{(d)} = \int_{-\infty}^{\infty} \int_{-\infty}^{\infty} \pi_{pq}(x, y) d(i\Delta x, j\Delta y) \, \mathrm{d}x \, \mathrm{d}y = \sum_{i=1}^{N} \sum_{j=1}^{M} \pi_{pq}(i\Delta x, j\Delta y) f_{ij}. \qquad (8.1)$$

---

[2] This means that $F(u, v) = 0$ for all (u, v) such that $|u| \geq U$ or $|v| \geq V$.
[3] Function $\text{sinc}(x) = \sin(x)/x$ is called the *ideal interpolant*.
[4] Quantizer Q is a function from $\mathbb{R}$ to $\{0, 1, 2, \dots, L - 1\}$.

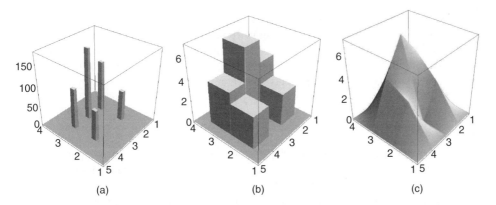

Figure 8.2 The concept of a digital image (a) as a sum of Dirac δ-functions, (b) as a nearest neighbor interpolation of the samples, and (c) as a bilinear interpolation of the samples

In case of geometric moments, assuming that $\Delta x = \Delta y = 1$, this leads to the expression

$$m_{pq}^{(d)} = \sum_{i=1}^{N} \sum_{j=1}^{M} i^p j^q f_{ij}, \tag{8.2}$$

where i,j are the coordinates of the pixel centers. This formula can be understood as the *exact* formula for geometric moments of the uniformly sampled image.

We can, however, also think differently. We may want to calculate (or at least estimate) the moments of the *original* image f from its digital version. To do so, we interpolate the discrete values by a proper interpolation function to recover an image defined in a continuous domain. Since the ideal interpolant $\text{sinc}(x) \cdot \text{sinc}(y)$ is not compactly supported, we commonly replace it by some simpler kernel. Nearest-neighbor interpolation leads to a piece-wise constant image $\hat{f}$ (see Figure 8.2 (b)); bilinear interpolation produces a more smooth surface (Figure 8.2 (c)).

Let us denote the moments of the piece-wise constant image $\hat{f}$ as $\hat{M}_{pq}$. We have

$$\hat{M}_{pq} = \int_{-\infty}^{\infty} \int_{-\infty}^{\infty} \pi_{pq}(x, y)\hat{f}(x, y)\, dx\, dy. \tag{8.3}$$

If we approximate the basis polynomials on each pixel by a constant, we obtain an estimate $\bar{M}_{pq}$ of $\hat{M}_{pq}$ (assuming that the pixels are of a unit area)

$$\bar{M}_{pq} = \sum_{i=1}^{N} \sum_{j=1}^{M} \pi_{pq}(i, j)f_{ij}. \tag{8.4}$$

For geometric moments we have

$$\bar{m}_{pq} = \sum_{i=1}^{N} \sum_{j=1}^{M} i^p j^q f_{ij}. \tag{8.5}$$

We can see that $m_{pq}^{(d)} = \bar{m}_{pq}$, but the interpretation of both moments is different. While $m_{pq}^{(d)}$ is an exact "traditional" continuous moment of d, $\bar{m}_{pq}$ is a double zero-order approximation of the moment of original image f.

We can, however, obtain a precise formula for $\hat{M}_{pq}$. There is no need to approximate the polynomials locally by a constant. As originally proposed by Lin and Wang [3] and later generalized by Flusser [4], we can use a pixel-wise integration of the polynomials. Since the polynomials are integrable through the Newton-Leibnitz formula by means of their primitive functions (which always exist and are again polynomials), we have

$$\hat{M}_{pq} = \sum_{i=1}^{N} \sum_{j=1}^{M} f_{ij} \iint_{S_{ij}} \pi_{pq}(x, y) \, \mathrm{d}x \, \mathrm{d}y \tag{8.6}$$

where S_{ij} denotes the area of the pixel (i, j). This formula can be viewed as a zero-order approximation of a moment of original image f where only the image is approximated, or, alternatively, as an exact continuous moment of image $\hat{f}$.

For geometric moments, we can actually perform the integration of the monomials over the pixels which leads to

$$\hat{m}_{pq} = \sum_{i=1}^{N} \sum_{j=1}^{M} f_{ij} \iint_{S_{ij}} x^p y^q \, \mathrm{d}x \, \mathrm{d}y$$

$$= \sum_{i=1}^{N} \sum_{j=1}^{M} f_{ij} \cdot \frac{(i+0.5)^{p+1} - (i-0.5)^{p+1}}{p+1} \cdot \frac{(j+0.5)^{q+1} - (j-0.5)^{q+1}}{q+1}. \tag{8.7}$$

Now we have three kinds of discrete moments – the moment $M_{pq}^{(d)}$, the moment $\hat{M}_{pq}$ of $\hat{f}$, and the moment $\bar{M}_{pq}$ being a zero-order approximation of $\hat{M}_{pq}$ and a double zero-order approximation of $M_{pq}^{(f)}$. Liao and Pawlak [5] analyzed the differences between them in detail and proved that they are negligible for common images. However, for certain specific artificial images these differences may be arbitrary high.

We can continue this process by taking higher-order image approximations which should yield even smaller approximation errors, but this is meaningless for practical purposes. It should be noted that most of the application papers skip the above analysis and use only (8.5) for moment calculation without any reasoning.

8.3 Moments of binary images

Since the direct evaluation of both (8.5) and (8.7) is time-consuming (it requires $\mathcal{O}(MN)$ operations), much effort has been spent in recent years to develop more efficient algorithms. As we already mentioned, most authors have focused on binary images because of their importance in practical pattern recognition applications.

8.3.1 Moments of a rectangle

Let us first show how the evaluation of (8.5) and (8.7) can be organized if the binary object is an $N \times M$ rectangle, the lower left corner pixel of which has the central coordinates $(1, 1)$.

Eq. (8.5) reads as

$$\bar{m}_{pq} = \sum_{i=1}^{N} \sum_{j=1}^{M} i^p j^q = \sum_{i=1}^{N} i^p \sum_{j=1}^{M} j^q. \tag{8.8}$$

To evaluate the sums, we employ the formulas for the sums of the finite power series

$$\sum_{i=1}^{N} i = \frac{N(N+1)}{2} \qquad \sum_{i=1}^{N} i^2 = \frac{N(N+1)(2N+1)}{6}$$

$$\sum_{i=1}^{N} i^3 = \frac{N^2(N+1)^2}{4} \qquad \sum_{i=1}^{N} i^4 = \frac{N(N+1)(2N+1)(3N^2+3N+1)}{30}. \tag{8.9}$$

The formulas for higher powers are given recursively as

$$\sum_{t=1}^{m} \binom{m+1}{t} S_N^t = (N+1)((N+1)^m - 1), \tag{8.10}$$

where

$$S_N^t = \sum_{i=1}^{N} i^t. \tag{8.11}$$

One can see that the computational complexity does not depend on N and M, but still depends on p and q – it is $\mathcal{O}(\max(p,q))$.

A substantial improvement can be reached when using (8.7) and when integrating the monomials not over separate pixels but on the whole rectangle. In that way we obtain the formula

$$\hat{m}_{pq} = \int_{0.5}^{N+0.5} x^p \, dx \int_{0.5}^{M+0.5} y^q \, dy$$

$$= \frac{(N+0.5)^{p+1} - 0.5^{p+1}}{p+1} \cdot \frac{(M+0.5)^{q+1} - 0.5^{q+1}}{q+1}, \tag{8.12}$$

the complexity of which is $\mathcal{O}(1)$ independently of M, N, p, and q. We totally avoided using the summation formulas. This is why using the moments $\hat{m}_{pq}$ is preferred when working with binary images.

The above result can be generalized to any polynomial basis defined in Cartesian coordinates. We may, however, face certain difficulties when using orthogonal polynomials. Sometimes the basis polynomials may not be defined explicitly, and finding their primitive functions may be complicated. Even if we find them, they may not belong to the given OG basis, and we cannot use the recurrent relations for calculating of the moments.

8.3.2 Moments of a general-shaped binary object

The methods for fast computation of the moments of binary images can be divided into two groups referred to as the *boundary-based* methods and the *decomposition methods*.

The boundary-based methods employ the property that the boundary of a binary object contains complete information about the object; all other pixels are redundant. Provided that the boundary consists of far fewer pixels than the whole object (which uses to be true for "common" shapes but does not hold in general[5]), an algorithm that calculates the

---

[5] There are objects, such as regular and random chessboards, all pixels of which are boundary pixels.

moments from the boundary pixels only should be more efficient than the direct calculation by definition.

The idea behind the decomposition methods is different and quite simple. Having a binary object Ω, we decompose it into $K \geq 1$ rectangular blocks $B_1, B_2, \ldots, B_K$ such that $B_i \cap B_j = \emptyset$ for any $i \neq j$ and $\Omega = \bigcup_{k=1}^{K} B_k$ (i.e., the set of blocks is a partition of Ω). Then

$$M_{pq}^{(\Omega)} = \sum_{k=1}^{K} M_{pq}^{(B_k)}.$$

We have already seen that the moment of each block can be calculated in $\mathcal{O}(1)$ time. Then the overall complexity of $M_{pq}^{(\Omega)}$ is $\mathcal{O}(K)$. If $K \ll MN$ the speed-up compared to the definition may be significant. This sum-rule cannot be directly used for central moments because the centroids of individual blocks are distinct, but we can either shift the object to the central position or to express the central moments in terms of the standard ones.

The power of any decomposition method depends on our ability to decompose the object into a small number of blocks in a reasonable time. Simple decomposition algorithms produce a relatively high number of blocks but perform fast, while more sophisticated decomposition methods end up with a small number of blocks but require more time. Some authors ignore the complexity of the decomposition itself and do not include it in the overall complexity estimation, claiming that "the fewer blocks the better method". This is a serious mistake. Even if the decomposition is performed only once and can be used for the calculation of all required moments, the time needed for the decomposition of the object is often so long that it substantially influences the efficiency of the whole method (especially if only a few moments are required to calculate). It should also be noted that some images cannot be efficiently decomposed by any algorithm and always $K \sim MN$. A chessboard is an extreme example.

Among the methods that do not fall into the two above-mentioned categories, parallel algorithms play an important role. Since moment computation is easy to parallelize, several authors have proposed parallel implementations on special processors that may be very efficient. Being dependent on the availability of a particular hardware, they are not reviewed here in detail. Two frequently cited representatives of this class are described in references [6] and [7]. Chen [6] proposed a recursive algorithm for a SIMD processor array; Chung [7] presented a constant-time algorithm on reconfigurable meshes.

In the next two sections, we review the boundary-based method and the decomposition methods in detail. We mostly show only the formulas for geometric moments, but the generalization to other moments can be done easily. In some cases, where the generalization is not straightforward or if the method is of a particular interest, we show the version for OG moments explicitly.

8.4 Boundary-based methods for binary images

8.4.1 The methods based on Green's theorem

A majority of the boundary-based methods use the *Green's theorem*, which evaluates the double integral over the object by means of a line integration along the object boundary.

Calculating geometric moments

The Green's theorem has several versions. An appropriate one for calculating geometric moments (in the continuous domain) is

$$\iint\limits_{\Omega} \frac{\partial}{\partial x} g(x, y) \, dx \, dy = \oint\limits_{\partial \Omega} g(x, y) \, dy, \tag{8.13}$$

where $g(x, y)$ is an arbitrary function having continuous partial derivatives and $\partial \Omega$ is a piecewise smooth, simple closed-curve boundary of object Ω. By setting

$$g(x, y) = \frac{x^{p+1} y^q}{p + 1}$$

we obtain a relation for the geometric moments

$$m_{pq}^{(\Omega)} = \frac{1}{p + 1} \oint\limits_{\partial \Omega} x^{p+1} y^q \, dy.$$

In the discrete domain, the individual methods differ from each other by the procedure used to discretize the boundary and calculate the line integral. Probably the first method of this kind was published by Li and Shen [8]. However, their results depend on the choice of discrete approximation of the boundary and differ from the theoretical values. Jiang and Bunke [9] had approximated the object first by a polygon, and then they applied Green's theorem. Therefore, they calculated only single integrals along line segments. Unfortunately, due to the two-stage approximation, their method produces inaccurate results.

Philips [10] proposed using the discrete version of Green's theorem [11] instead of the discretized continuous one. The Philips' method yields exact results as it does not use any approximation. Philips started from the continuous-domain Green's theorem in a more general version than (8.13)

$$\oint\limits_{\partial \Omega} h(x, y) \, dx + g(x, y) \, dy = \iint\limits_{\Omega} \left(\frac{\partial g}{\partial x} - \frac{\partial h}{\partial y} \right) dx \, dy, \tag{8.14}$$

where $g(x, y)$ and $h(x, y)$ are arbitrary functions having continuous partial derivatives. By means of a discrete analogy of this theorem, Philips proved that the discrete moment (8.5) can be calculated from the boundary pixels as

$$\bar{m}_{pq}^{(\Omega)} = \sum_{(x,y) \in \partial \Omega+} y^q \sum_{i=1}^{x} i^p - \sum_{(x,y) \in \partial \Omega-} y^q \sum_{i=1}^{x} i^p, \tag{8.15}$$

where $\partial \Omega-$ and $\partial \Omega+$ are so-called *left-hand-side* and *right-hand-side boundaries*, respectively. They are defined as

$$\partial \Omega- = \{(x, y) | (x, y) \notin \Omega, (x + 1, y) \in \Omega\},$$

$$\partial \Omega+ = \{(x, y) | (x, y) \in \Omega, (x + 1, y) \notin \Omega\}.$$

The Philips' method can also be used to calculate the moment approximation (8.7)

$$
\hat{m}_{pq}^{(\Omega)} = \frac{1}{(p+1)(q+1)} \left(\sum_{(x,y)\in\partial\Omega+} (x+0.5)^{p+1}((y+0.5)^{q+1} - (y-0.5)^{q+1}) \right.
$$

$$
\left. - \sum_{(x,y)\in\partial\Omega-} (x+0.5)^{p+1}((y+0.5)^{q+1} - (y-0.5)^{q+1}) \right). \tag{8.16}
$$

It is worth noting that for convex shapes the Philips' approach leads to the same formulas as the simplest decomposition method – the delta method (see the next section).

Yang and Albregtsen [12], Sossa-Azuela et al. [13], and Flusser [14] further improved the speed of the Philips' method. A significant speed-up was achieved by precalculating certain variables which do not depend on the object and are used repeatedly, such as

$$
\sum_{n=1}^{j} n^{i-1}
$$

for all possible values of i and j and by storing them in properly organized matrices (see reference [14] for details).

Calculating Legendre moments

A modification of the above method for Legendre moments was published by Mukundan [15]. He used Green's theorem in the form (8.13) and set

$$
\frac{\partial}{\partial x} g(x, y) = P_m(x)P_n(y), \tag{8.17}
$$

where $P_m(x)$ is the Legendre polynomial. Since its primitive function, expressed by means of the recurrence, is

$$
\int P_m(x)dx = \frac{1}{m+1}(xP_m(x) - P_{m-1}(x)), \tag{8.18}
$$

we can substitute into (8.13) and obtain the expression for Legendre moment λ_{mn}

$$
\lambda_{mn} = \frac{(2m+1)(2n+1)}{4(m+1)} \oint_{\partial\Omega} P_n(y)(xP_m(x) - P_{m-1}(x)) \, dy. \tag{8.19}
$$

In the discrete domain, the curve integral is approximated by a sum. We can either track the boundary sequentially pixel by pixel or pass through the image row by row and search for the boundary points. In the second version, the algorithm turns out to be close to the Philips' method.

8.4.2   The methods based on boundary approximations

Another group of boundary-based methods exploits approximation of the boundary by a polygon[6]. Any polygon is fully determined by its vertices (corners) only. Then its moments can be calculated by means of the corner points only [16, 17]; this is particularly efficient for

---

[6] Although other piece-wise smooth curves can also be theoretically used, only polygons are of practical importance.

simple shapes with few corners. However, the results obtained by methods of this kind are only approximate. Their accuracy depends on the accuracy of the boundary approximation. Recall that the method by Jiang and Bunke [9], already mentioned in connection with Green's theorem, also approximates the boundary by a polygon, and thus it belongs at least partially to this category.

Let us assume that the approximating polygon has n vertices $A_0, A_1, A_2, ..., A_{n-1}, A_n$, where $A_0 = A_n$, with coordinates $A_i = (x_i, y_i)$. We can create a set of triangles, each triangle with two vertices equal to two adjacent vertices of the polygon A_i, A_{i+1} and with the third vertex in the origin of the coordinates $O = (0,0)$. The moment of the polygon is then

$$\hat{m}_{pq} = \sum_{i=0}^{n-1} \hat{m}_{pq}^{(i)}, \qquad (8.20)$$

where $\hat{m}_{pq}^{(i)}$ is the moment of the triangle $A_i A_{i+1} O$ (note that for some triangles it may be negative). It can be calculated as

$$\hat{m}_{pq}^{(i)} = \alpha_i \sum_{k=0}^{p} \sum_{\ell=0}^{q} C_{k\ell}(p,q) x_i^k x_{i+1}^{p-k} y_i^{\ell} y_{i+1}^{q-\ell}, \qquad (8.21)$$

where $\alpha_i = (x_i y_{i+1} - x_{i+1} y_i)/2$ is the area of the triangle $A_i A_{i+1} O$ and the coefficient $C_{k\ell}(p,q)$ is

$$C_{k\ell}(p,q) = \frac{2\, p!q!(k+\ell)!(p+q-k-\ell)!}{(p-k)!(q-\ell)!k!\ell!(p+q+2)!}. \qquad (8.22)$$

The complete proof of (8.21) can be found in reference [17], but its meaning can be understood intuitively at least for low p and q. The formula (8.21) can be evaluated efficiently by means of a proper recurrence.

8.4.3   Boundary-based methods for 3D objects

As already explained in Chapter 4, binary 3D objects are mostly represented by their triangulated surface. Compared to volumetric representation, which is highly redundant in the case of binary objects, this surface representation is much less memory demanding. In all public databases, including the PSB, binary objects are stored in this format.

3D binary objects in volumetric representation also exist. They originate from 3D graylevel volumetric images as a result of segmentation. We can, of course, convert one representation to the other one, but such a conversion takes some time and introduces errors along the surface. This is why we have to study computation of moments in both representations. However, boundary-based methods are inherently more appropriate for objects given by their triangulated surface.

Methods for objects in a volumetric representation

When calculating moments of binary 3D objects, the 2D boundary-based methods mentioned above can be adopted. Green's theorem holds in arbitrary dimensions. In 3D, it is called the *divergence theorem* or the *Gauss-Ostrogradsky theorem*, but its meaning is basically the same as in 2D. It states that the outward flux of a vector field through a closed surface is equal to the volume integral of the divergence of the region inside the surface. Consequently, the methods based on Green's theorem have 3D and even d-D modifications which are easy to derive (see reference [18] for a d-D version of the Philips' method and reference [19] for another example).

We can find a few methods in the literature that were developed specifically for 3D moment computation without having a counterpart in 2D. For instance, Li and Ma [20] introduced special linear transformations allowing 3D binary moments to be calculated by means of 2D moments of an artificially created graylevel image.

Methods for objects in a surface representation

If the object is represented by its triangulated surface, it is fully determined by the vertices of the triangulation. The algorithms working with triangular surface representation use only these vertices. Since the number of the vertices is usually by one order less than the number of the surface voxels and by two orders less than the number of the volume voxels, these algorithms are very efficient.

The triangulated surface is stored in a form of two lists: the list of vertices $\mathbf{V} = (v_{ij})$ and the list of triangles $\mathbf{F} = (f_{ij})$. Matrix $\mathbf{V}$ is of the size of the $P \times 3$, where P is the number of vertices. Matrix $\mathbf{F}$ is of the size of the $N \times 3$, where N is the number of triangles. Element f_{ij} is the index of the jth vertex of the ith triangle in the first list. Hence, the coordinates of the vertices of the ith triangle are $(v_{f_{i,1},1}, v_{f_{i,1},2}, v_{f_{i,1},3})$, $(v_{f_{i,2},1}, v_{f_{i,2},2}, v_{f_{i,2},3})$, and $(v_{f_{i,3},1}, v_{f_{i,3},2}, v_{f_{i,3},3})$. The triangles are oriented, that is, a cyclic shift of the vertices does not change the triangle, but swapping of two vertices results in a change of the sign of the triangle area.

Probably the most efficient algorithm for calculating geometric moments from triangular surface representation was proposed by Sheynin and Tuzikov [21]. We already used this algorithm in the experiments in Chapter 4. Here we explain it in more detail. Some other methods of this kind have been published (see [22], for example), but their performance is lower, and/or they are limited to special shapes only (for instance, Mamistvalov [23] derived analytical $\mathcal{O}(1)$ expressions for a few specific shapes).

The core formula proposed by Sheynin and Tuzikov [21] is

$$m_{pqr} = \frac{p!q!r!}{(p+q+r+d)!} \sum_{(k_{ij}) \in \mathcal{K}} \frac{\prod\limits_{j=1}^{3}\left(\left(\sum\limits_{i=1}^{3} k_{ij}\right)!\right)}{\prod\limits_{i,j=1}^{3} (k_{ij}!)} \sum_{\ell=1}^{N} A_\ell \prod_{i,j=1}^{3} \left(a_{ij}^{(\ell)}\right)^{k_{ij}}, \qquad (8.23)$$

where $\mathcal{K}$ is a set of 3×3 matrices k_{ij} of non-negative integers such that $\sum_{j=1}^{3} k_{1j} = p$, $\sum_{j=1}^{3} k_{2j} = q$ and $\sum_{j=1}^{3} k_{3j} = r$; $\left(a_{ij}^{(\ell)}\right)$ is a matrix of the vertex coordinates of the ℓ-th triangle; i is the index of the coordinate, j is the index of the vertex, that is, $a_{ij}^{(\ell)} = v_{f_{\ell,j},i}$; $d = 3$ and $A_\ell = \det\left(a_{ij}^{(\ell)}\right)$.

An efficient implementation of the formula (8.23) in Matlab requires not implementing its inner part (the sum over all triangles) as a loop, but as a single-instruction matrix operation.

As explained already in Chapter 4, the formula (8.23) can easily be adapted to calculation of surface geometric moments.

The computational complexity of all moments up to the order s directly from the definition in a volumetric representation is $\mathcal{O}(N_1 N_2 N_3\, s^3)$, it means it is linear both to the number of voxels and to the number of moments. The computational complexity of the same task by means of (8.23) is $\mathcal{O}(N \cdot s!)$; that is, it is linear to the number of triangles, and it grows exponentially with the number of required moments. Hence, the computation via (8.23) is much faster if we do not require the calculation up to a very high order. Otherwise, the direct computation by definition may be faster.

Unfortunately, it is not known how to adapt the formula (8.23) to calculation of other than geometric moments. If we want to calculate 3D complex moments (their kernel functions contain spherical harmonics; see Chapter 4), we can do so via geometric moments.

First, we express the spherical harmonics in Cartesian coordinates as a polynomial

$$(x^2 + y^2 + z^2)^{s/2} Y_\ell^m(x, y, z) = \sum_{\substack{k_x, k_y, k_z = 0 \\ k_x + k_y + k_z = s}}^{s} a_{k_x k_y k_z} x^{k_x} y^{k_y} z^{k_z}. \tag{8.24}$$

Then we can compute the complex moments as

$$c_{s\ell}^m = \sum_{\substack{k_x, k_y, k_z = 0 \\ k_x + k_y + k_z = s}}^{s} a_{k_x k_y k_z} m_{k_x k_y k_z}, \tag{8.25}$$

where the geometric moments can be evaluated by means of (8.23).

This approach could be seemingly extended to any OG moments because in principle all of them can be expressed in terms of the geometric moments. Although this is formally true, such an algorithm would be practically useless because of the loss of precision when using geometric moments. We would lose the main advantage of orthogonal moments. We can proceed in an alternative way. We convert the triangular representation into the volumetric one, where the OG moments can be calculated either by means of the Gauss-Ostrogradsky theorem and the Philips' method or, even more efficiently, via decomposition into rectangular blocks (see the next section).

Below we present an algorithm for such a conversion. It can be advantageous not only for OG moment calculation but also in case we need to compute the geometric or complex moments up to a very high order, where the formula (8.23) is slow.

Conversion of the triangulated surface representation to the volumetric representation

The basic principle of the conversion is straightforward. We choose (possibly randomly) the major axis (one of x, y, and z), choose a size of the voxels we want to obtain, and construct a 2D array of rays parallel with the major axis. For each ray, we find its intersections with the surface. The ray fills the voxels by zeros at the beginning. Each intersection swaps the value from zero to one and vice versa. After accomplishing this with all rays, we obtain the binary object in the volumetric representation of the chosen resolution.

Even though the algorithm looks easy, it contains some hidden parameters which must be tuned. The resolution must be fine enough to capture all important parts of the object but if the resolution has been chosen too high, the moment calculation becomes slow.

When the ray intersects the surface precisely in an edge or even in a vertex, we have to test carefully if it has passed through or if it just has touched the surface. The easiest heuristic solution is to slightly translate the raster.

A serious practical problem appears when converting the object, whose triangulated surface representation has not been created correctly. This is a common problem in public databases including the PSB. The surfaces of many objects contain unwanted holes, which originate from errors when modelling or scanning the object. The conversion algorithm then cannot distinguish what is inside and what is outside the object, and this creates strange artifacts; see Figure 8.3b for an example. To prevent these artifacts from appearing, the triangulated surface should undergo preprocessing that detects and closes the holes before the conversion into voxels has been carried out.

The detection of the holes is not difficult. If the object is closed, then each triangle edge must be included $2k$-times, where k is an arbitrary positive integer (usually $k = 1$). The k edges must be oriented in one direction and the other k ones in the opposite direction. If this test fails, the edge lies on the boundary of a hole. The hole should be filled up before the conversion by artificially constructed triangles. An algorithm for filling up the holes can be found in Appendix 8.A.

The object in Figure 8.3 consists of 5490 points and 3339 triangles; it includes 76 holes with 780 open edges altogether. It was converted to the volumetric data with resolution 200 voxels in the longest direction. First, the conversion was applied without any preprocessing (see Figure 8.3b). Then we filled up the holes first and repeated the conversion. The differences may be observed in Figure 8.3c, where all artifacts have disappeared.

8.5 Decomposition methods for binary images

The methods belonging to this group decompose the object into rectangular blocks. As we have already seen, the computational complexity of the moment of a single block is $\mathcal{O}(1)$ (neither the size nor the shape of the block matters), so the decomposition methods should minimize the number of blocks K. In practice, however, we rather look for a reasonable trade-off between K and the decomposition time.

Numerous decomposition methods have been published in the literature. Their number is much higher than the number of boundary-based methods. Decomposition methods may be very efficient, especially if many moments of the object are to be calculated, because the decomposition is performed only once.

Decomposition methods are suitable for objects in pixel or volumetric representation and can be applied both to 2D and 3D objects.

It should be noted that some decomposition methods were developed particularly for moment calculation while others originate from computer geometry, binary image compression, manufacturing, and other areas, but any method can be used for any purpose. Many image compression methods and formats (RLE, TIFF, BMP and others) share the same decomposition idea.

A comparison of the efficiency of various decomposition methods in image compression can be found in [24]. In the same paper, the decomposition of the convolution kernel was proposed to speed up the numerical convolution. Convolution with a constant rectangular mask can be performed in $\mathcal{O}(1)$ time per pixel if the matrix of partial sums is precalculated. If the mask is

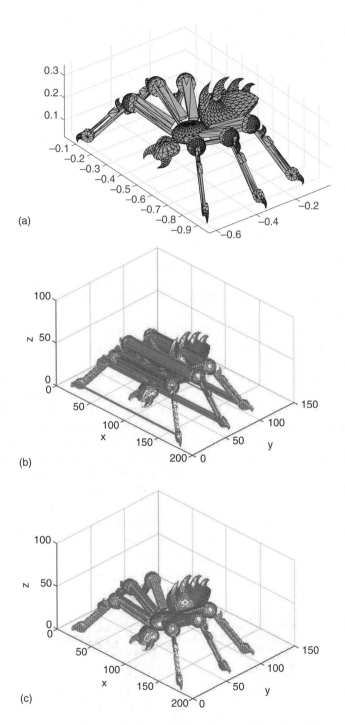

Figure 8.3 The spider (PSB No. 19): (a) the original triangulated version, (b) a conversion to the volumetric representation without any preprocessing, and (c) the same with previous detecting and filling-up the holes

constant but not rectangular, it can be decomposed into rectangles, and the convolution in one pixel can be calculated in $\mathcal{O}(K)$ time as a sum of partial $\mathcal{O}(1)$ convolutions. This speed-up may be very useful wherever multiple convolutions with the same kernel are to be calculated, such as in iterative deconvolution methods [25] and in the *coded aperture* imaging [26].

It is interesting that the decomposition problem appeared many years ago in VLSI design. The circuit masks are rectilinear polygons (often very complex ones), which should be decomposed into rectangles in such a way that the pattern generator can effectively generate the mask. The time needed for a mask generation is proportional to the number of rectangles, so it is highly desirable to minimize it [27, 28].

Below we present a survey of those decomposition methods which have been applied to moment calculation. We start with the 2D case.

8.5.1   The "delta" method

Decomposition of an object into rows or columns is the most straightforward and the oldest method. The blocks are continuous row segments for which only the coordinate of the beginning and the length are stored (the same principle has been known in image compression as the *run-length encoding*).

In moment calculation, Zakaria et al. [29] used the same representation for fast computation of image moments of convex shapes and called it *delta-method* (DM) since the lengths of the row segments were labeled by the Greek letter δ. The original Zakaria's method worked for convex shapes only and dealt with geometric moments up to the third order. The moments were approximated by (8.5). Dai et al. [30] extended Zakaria's method by employing (8.7) and Li [31] generalized it for non-convex shapes. These decomposition algorithms are very fast, but the number of blocks is usually (much) higher than the minimal decomposition.

A simple but powerful improvement of the delta method was proposed by Spiliotis and Mertzios [32] and further improved by Flusser [4]. This *generalized delta-method* (GDM) employs a rectangular-wise object representation instead of the row-wise one. The adjacent rows are compared, and if there are some segments with the same beginning and end, they are unified into a rectangle (see Figure 8.4). For each rectangle, the coordinates of its upper-left corner, the length and the width are stored. The GDM is only slightly slower than DM while producing a (sometimes significantly) smaller number of blocks, see Figure 8.5.

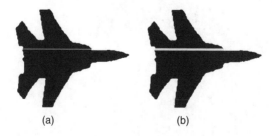

(a)   (b)

Figure 8.4   The delta method. In the basic version the object is decomposed into rows (a). The generalized version unifies the adjacent rows of the same length into a rectangle (b)

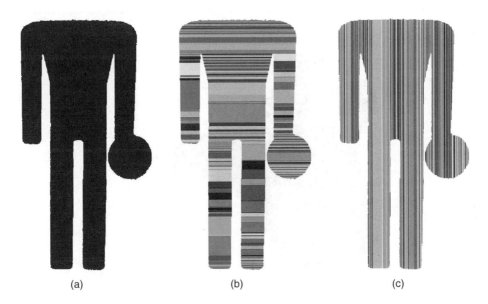

Figure 8.5 Generalized delta method: (a) the headless figure (black=1) (b) the generalized delta method applied row-wise (395 blocks) (c) the generalized delta method applied column-wise (330 blocks). The basic delta method generated 1507 blocks

8.5.2 *Quadtree decomposition*

The *quadtree decomposition* (QTD) is a popular hierarchical decomposition scheme used in several image processing areas including representation and compression [33], spatial transformations [34], and moment calculation [35].

In its basic version, the QTD works with square images of a size of a power of two. If this is not the case, the image is zero-padded to the nearest such size. The image is iteratively divided into four quadrants. Homogeneity of each quadrant is checked, and if the whole quadrant lies either in the object or in the background, it is not further divided. If it contains both object and background pixels, it is divided into quadrants, and the process is repeated until all blocks are homogeneous (see Figure 8.6). The decomposition can be efficiently encoded into three-symbol string, where 2 means division, 1 means a part of the object and 0 stands for the background.

The algorithm always yields square blocks, which may be advantageous for some purposes, but usually it leads to a higher number of blocks than necessary. It would be possible to implement backtracking and to unify the adjacent blocks of the same size into a rectangle, but this would increase complexity. Since speed is the main advantage of this method, backtracking is mostly not employed here. A drawback of this decomposition algorithm is that the division scheme is not adapted to the content of the image, but it is defined by absolute spatial coordinates. Hence, the decomposition is not translation-invariant and may lead to absurd results when, for instance, a large single square is uselessly decomposed up to individual pixels. We may use a bintree or other trees producing non-square blocks but it does not overcome this principal weakness.

Two examples of a quadtree decomposition can be seen in Figure 8.7.

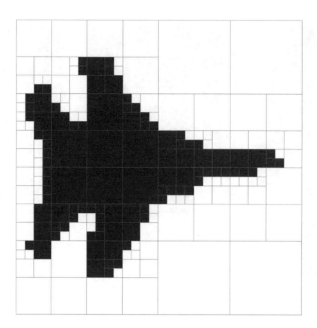

Figure 8.6   Quadtree decomposition of the image

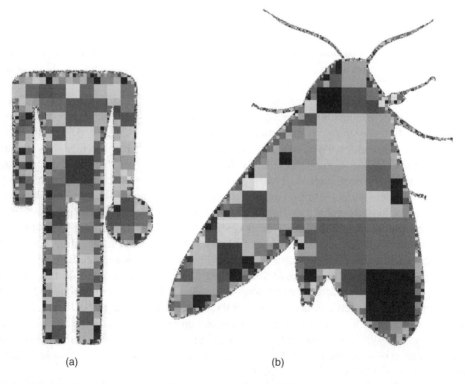

(a)   (b)

Figure 8.7   Quadtree decomposition. (a) the headless figure, 3843 square blocks and (b) the moth, 5328 square blocks

8.5.3 Morphological decomposition

In order to better adapt the decomposition to the image content, Sossa-Azuela et al. [36] published an algorithm based on a *morphological erosion*. The erosion is an operation, where a small structural element (here 3×3 square is used) moves over the image, and when the whole element lies inside the object, then its central pixel is assigned to the object, otherwise it is assigned to the background. So, each erosion shrinks the object by a one-pixel boundary layer.

The decomposition works in an iterative manner. It finds the largest square inscribed in the object, removes it and looks for the largest square inscribed in the rest of the object. This outer loop is repeated until the object has been completely decomposed.

The inner loop serves for finding the center and the size of the largest inscribed square. We repeat the erosion until the whole object disappears and count the number of erosions s. Then a $(2s - 1) \times (2s - 1)$ square can be inscribed into the object, and it forms one block of the decomposition. The pixels of the object before the last erosion are potential centers of the inscribed square. Theoretically, we can choose one of them randomly, but the "corner" pixels provide better odds to a more compact remainder of the object. If the potential square centers create a line, then the corresponding inscribed squares can be unified into a rectangle, which reduces the number of output blocks.

An intermediate object decomposition after two outer loops can be seen in Figure 8.8.

Compared to other algorithms, the morphological decomposition is not very efficient. It is slow, and the final number of blocks is usually higher than that produced by the generalized delta method. This is because a sequence of locally optimal steps that this "greedy" algorithm applies to an image (placing always the largest inscribed rectangle) does not yield an optimal solution. As soon as a block has been created, it cannot be removed anymore, because the method does not include any backtracking.

The speed of the morphological decomposition can be increased substantially when the centers of the inscribed squares/rectangles are found by means of the *distance transformation (DT)* instead of the erosion. This improvement was proposed by Suk in [37].

In morphological decomposition, we must repeat the erosions s-times for finding $(2s - 1) \times (2s - 1)$ inscribed square, while the distance transformation with a suitable metric can be calculated only once. DT of a binary image is an image, where each pixel shows the distance to the nearest background pixel [38].

Figure 8.8   Partial object decomposition after two outer loops of the morphological method

The DT strongly depends on the metric used for the distance measurement. We use a simplified version of Seaidoun's algorithm [39] for the chessboard metric

$$d(\mathbf{x}, \mathbf{y}) = \max\{|x_1 - y_1|, |x_2 - y_2|\}. \tag{8.26}$$

To calculate the distance transformation, we successively search the image from the left, right, top, and bottom, count distances from the last boundary pixel, and calculate the minimum from the four directions. If the maximum of the DT equals s, then the pixels with this maximum value are potential centers of the largest inscribed squares of the size $(2s - 1) \times (2s - 1)$.

The calculation can be accelerated when using an improved version of DT inspired by [40]. If only a small part of the original image has been changed, then upgrading the DT in its close neighborhood is sufficient. In our case, if we remove a rectangle from the image, then the rectangle is zeroed, and the DT is recomputed in a small frame around it.

Both morphological and DT decompositions end up with the same set of blocks. The DT decomposition is faster thanks to its simple upgrading, but still it performs slower comparing to the GDM. Two examples of the decompositions are in Figure 8.9.

8.5.4   Graph-based decomposition

A large group of decomposition algorithms appeared in the 1980s in computational geometry, where they were studied in a broader context [41]. Until 2011, they had received almost no

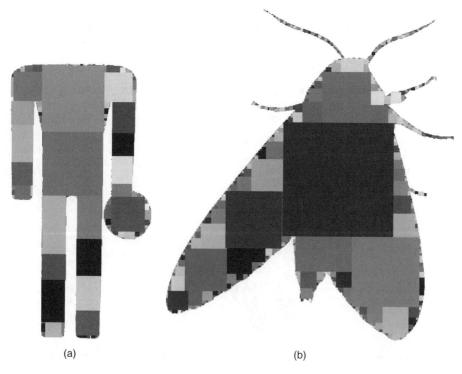

(a)   (b)

Figure 8.9   Distance transformation decomposition: (a) the headless figure, 421 blocks and (b) the moth, 1682 blocks

attention from the image analysis community. Their formulation was usually much more general than ours. They tried to decompose general polygons into specific polygonal components (convex polygons, star-shape polygons, triangles, generally oriented rectangles, etc.). A common feature of these methods was that they transformed the decomposition problem to a graph partitioning problem and employed known tools from graph theory. Several authors [42–44] independently proposed basically the same algorithm (later discussed and improved in [45, 46]) for a decomposition of a digital polygon into rectilinear rectangles, which exactly matches the decomposition required for moment calculation. Their algorithm was proved to be optimal since it actually minimizes the number of blocks. It works without limitations for any finite object even if it contains holes. The only difference between various versions is a rather technical one in implementation of one step, which may influence the complexity but not the total number of the blocks. The version by Ferrari et al. [44] was adapted to image analysis purposes by Suk et al. [24]. We refer to this method as *graph-based decomposition* (GBD).

The method performs hierarchically on two levels. On the first level, we detect all "concave" vertices (i.e., those having the inner angle 270°) of the input object and identify pairs of "cogrid" concave vertices (i.e., those having the same horizontal or vertical coordinates). Then we divide the object into subpolygons by constructing chords which connect certain cogrid concave vertices. It is proved in [44] and other papers that the optimal choice of the chord set is such that the chords are pair-wise disjoint, and their number is the maximum possible.

The problem of optimal selection of the chords is equivalent to the problem of finding the maximal set of independent nodes in a graph, where each node corresponds to a chord and two nodes are connected with an edge if the two chords have a common point (either a vertex or an intersection). Generally, this problem is NP-complete, but our graph is a bipartite one, since any two horizontal (vertical) chords cannot intersect one another. In a bipartite graph, this task can be efficiently resolved. We find a maximal matching, which is a classical problem in graph theory, whose algorithmic solution in a polynomial time has been published in various versions. Some of them are optimized with respect to the number of edges, the others with respect to the number of the nodes (see [45, 47–49] for some examples of particular algorithms), but all of them are polynomial in both. It is impossible to choose one that would be time-optimal for any object, because the number of vertices and/or edges of the graph depends on the shape of the object. We implemented the algorithm by Edmonds and Karp [49], which is based on the *maximum network flow* and is linear in the number of the graph nodes (i.e., in the number of the chords) and quadratic in the number of edges (which means in the number of the chord intersections). Its complexity is in the worst case $\mathcal{O}(b^5)$, where b is the number of the concave vertices, which is at the same time the total complexity of the decomposition.

As soon as the maximal matching has been constructed, the maximal set of independent nodes can be found much faster than the maximal matching itself – roughly speaking, the maximal independent set contains one node of each matching pair plus all isolated nodes plus some other nodes, which are not included in the matching but still independent. As a result, we obtain a set of nodes that is unique in terms of the number but ambiguous in terms of the particular nodes involved. However, this ambiguity does not matter – although each set led to a different object partition, the number of the rectangles is always the same. Hence, at the end of the first level, the object is decomposed into subpolygons, which do not contain any cogrid concave vertices (see Figure 8.10).

The second level is very simple. Each subpolygon coming from the first level is either a rectangle or a concave polygon. In the latter case, it is further divided. From each of its concave

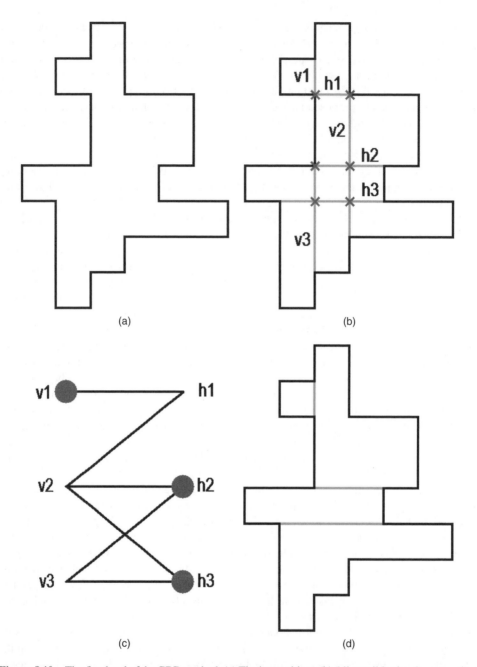

(a)

(b)

(c)

(d)

Figure 8.10   The first level of the GBD method. (a) The input object. (b) All possible chords connecting the cogrid concave vertices. The crosses indicate the chord intersections. (c) The corresponding bipartite graph with a maximum independent set of three vertices. Other choices are also possible, such as $\{h_1, h_2, h_3\}$ or $\{v_1, v_2, v_3\}$. (d) The first-level object decomposition

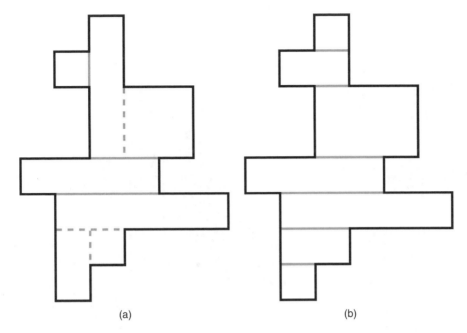

Figure 8.11   The second level of the GBD method. (a) The first-level decomposition (solid line) and a possible second-level decomposition (dashed line). From each concave vertex a single chord of arbitrary direction is constructed. (b) If on the first level the chords h_1, h_2, and h_3 were chosen, then both GBD and GDM would yield the same decomposition

vertices, a single chord is constructed such that this chord terminates either on the boundary of the subpolygon or on the chord that has been constructed earlier. This is a sequential process in which each concave vertex is visited only once. The order of the concave vertices may be chosen arbitrarily. Similarly, we may choose randomly between two possible chords offered in each concave vertex. This choice does not influence the final number of blocks. After that, the subpolygon is divided into rectangles, because the rectangle is the only polygon having no concave vertices (see Figure 8.11a).

Let us explain that this algorithm actually minimizes the number of blocks. When using C chords, the number of blocks is always $K = C + 1$. If the object has P concave vertices, at most P chords are sufficient to divide the object into rectangles. This is clear because each concave vertex must lie on a chord (in order to get rectangles) and the chords that would not pass through any concave vertex are useless. Hence, each concave vertex must lie on just one chord. A chord can pass through one or two concave vertices, not through more of them. This implies that maximizing the number of chords passing through two concave vertices (which is what the algorithm does in its first stage) is the same as minimizing the total number of the chords needed, which is equivalent to minimizing the number of the resulting blocks[7].

The strength of this algorithm is in the fact that it guarantees minimizing the number of decomposing rectangles regardless of the particular choices on both levels. Another attractive property is the polynomial computational complexity.

The ambiguity of the decomposition is not a serious drawback. If we want to guarantee a unique decomposition, we may introduce additional constraints, such as minimum length of

---

[7] For a more formal proof we refer to [44].

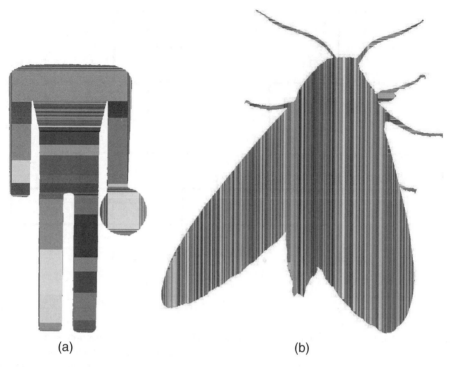

(a)   (b)

Figure 8.12 Graph-based decomposition: (a) the headless figure, 302 blocks and (b) the moth, 1092 blocks. In the case of the moth, the result is very similar to the GDM

the inner boundary or close-to-square block shapes, but this would lead to increasing the time complexity, and it is not important for moment calculation (although it might be essential for some other applications).

If the object does not have any cogrid concave vertices or if the maximum independent set of nodes (or, more precisely, at least one of such sets) contains only the nodes corresponding to the horizontal (or vertical) chords, then also all chords in the second level may be constructed in the same direction, and, consequently, GBD decomposition leads exactly to the same partition as GDM applied in that direction (see Figure 8.11b and Figure 8.12b). This helps us not only to understand when the GDM is strong, but also to interpret the idea behind the GBD algorithm—it may be viewed as a "local GDM" properly switching between the directions. Examples of the GBD decomposition are in Figure 8.12.

A comparison of the efficiency of the four decomposition algorithms on two test images is in Table 8.1. A more representative comparison of these methods in terms of the moment calculation complexity can be found in the next section.

8.5.5 Computing binary OG moments by means of decomposition methods

To calculate moments orthogonal on a rectangle we can, in principle, take any decomposition scheme used previously and simply replace the monomials $x^p y^q$ by respective OG polynomials.

Table 8.1 Comparison of the numbers of blocks and of the decomposition time

Method	GDM	QT	DT	GBD
	The number of blocks			
Headless figure	395	3843	421	302
Moth	1192	5328	1682	1092
	The decomposition time [ms]			
Headless figure	18	148	354	52
Moth	18	201	918	182

Particularly attractive are Legendre moments because primitive functions of Legendre polynomials are again Legendre polynomials, as we already saw in Eq. (8.18). This means that a Legendre moment of a rectangle can be exactly computed in $\mathcal{O}(1)$ time by integration, where we need simply to evaluate $P_{p+1}(x)$ and $P_{p-1}(x)$ at the corner pixels. This idea was, for instance, used by Zhou et al. [50], who employed the delta method decomposition.

For Zernike moments (and for other moments orthogonal on a disk) the decomposition methods cannot be used in the Cartesian coordinates but can be employed in polar coordinates, where "rectangular blocks" are actually angular-circular segments. Since polar resampling requires additional time and causes sampling errors on the object boundary, this approach is rare.

Anyway, if we decide to convert the algorithm to the polar coordinates, the "polar delta method" works as follows. The outer loop goes over the radii, the inner loop is over the angles in one ring with pixels in the same distance from the center; see Figure 7.25. The Zernike moment of a ring segment of radius r bounded by angles θ_1, θ_2 is

$$v_{pq}(r, \theta_1, \theta_2) = \int_{\theta_1 - \Delta\theta/2}^{\theta_2 + \Delta\theta/2} \int_{r - \Delta r/2}^{r + \Delta r/2} R_{pq}(r) e^{-iq\theta} f(r, \theta) r \, dr \, d\theta, \tag{8.27}$$

where Δr is the radial size and $\Delta\theta$ the angular size of one pixel.

If we compute the segment moment as the sum of delta functions, we can express it as the partial sum of geometric series

$$v_{pq}(r, \theta_1, \theta_2) = R_{pq}(r) \frac{e^{-iq\theta_2} - e^{-iq\theta_1}}{e^{-iq\Delta\theta} - 1}. \tag{8.28}$$

and for zero repetition q

$$v_{p0}(r, \theta_1, \theta_2) = R_{p0}(r)(\theta_2 - \theta_1 + \Delta\theta)/\Delta\theta.$$

The angular part can be easily integrated. If we are able to integrate also the radial function, we can compute the moment of the constants in the individual pixels. For instance, for the complex moments (for $p \neq q$) we have

$$
\begin{aligned}
\upsilon_{pq}(r, \theta_1, \theta_2) &= \int\limits_{\theta_1 - \Delta\theta/2}^{\theta_2 + \Delta\theta/2} \int\limits_{r - \Delta r/2}^{r + \Delta r/2} \rho^{p+q+1} e^{i(p-q)\theta} \mathrm{d}\rho \mathrm{d}\theta \\
&= \frac{1}{p+q+1}((r + \Delta r/2)^{p+q+2} - (r - \Delta r/2)^{p+q+2}) \\
&\quad \cdot \frac{1}{i(p-q)}(e^{i(p-q)(\theta_2 + \Delta\theta/2)} - e^{i(p-q)(\theta_1 - \Delta\theta/2)}).
\end{aligned}
\tag{8.29}
$$

The total object moment υ_{pq} is then the sum of all segment moments $\upsilon_{pq}(r, \theta_1, \theta_2)$. Xin et al. [51] used integration of the explicit formula for computation of Zernike moments, Hosny et al. [52] did the same for Fourier-Mellin moments. Both used the integration formulas on individual polar pixels for computation on graylevel images, not for larger arc segments in binary images.

8.5.6   Experimental comparison of decomposition methods

We carried out an illustrative comparison of the four decomposition methods described in the previous sections. We also included a direct computation from the definition (i.e., without any decomposition) as the fifth method. The particular choice of the moments is not essential because it influences neither the decomposition algorithms themselves nor their mutual comparison. We used discrete Chebyshev moments to prevent overflow for high orders. We performed the comparison for two test images that we already had used before – the headless figure and the moth.

In the case of the figure, we computed the moments with indices from 0 to 620, because the image size was 621×621. The total time of the direct calculation was 5:01 minutes, the time of QTD, DT, GDM, and GBD was 19 s, 2.78 s, 2.3 s and 1.96 s, respectively. The decompositions and the number of blocks can be seen in Figures 8.7a, 8.9a, 8.5b (we used the row-wise GDM) and 8.12a.

The graphs in Figure 8.13 provide a transparent summary of the experiment. The horizontal axis shows the number of the moments calculated, while the vertical axis shows the total time of both parts – decomposition and moment calculation[8]. The initial time at the beginning is the time of the decomposition (which was already reported in Table 8.1). Direct computation does not require any initialization. The GDM is the fastest method followed by the GBD (note the slight difference between them), the QDT, and, finally, by the DT with a big delay. Theoretically, the graphs should be straight lines, which is not exactly the case here because internal processes in the computer slightly influenced the time measurement. The slope of the line is proportional to the number of blocks. Since for this object GBD, GDM, and DT yield similar number of blocks, the corresponding lines are almost parallel. Starting from the moment order 240 (which is equivalent to approx 58,000 moments), the graph-based method becomes the fastest thanks to the minimum number of blocks. This is, of course, an extremely huge number of moments. For almost all situations, the GDM would be the optimal choice

---

[8] The overheads, such as loading the image and pre-computing of Chebyshev polynomials, are common for all of the methods, so they are not displayed on the graph.

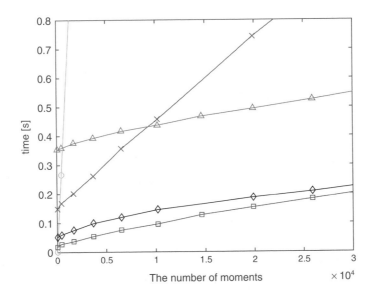

Figure 8.13 The time complexity of the moment computation of the headless figure image. Legend: ○ - definition, □ - generalized delta method, × - quadtree, ◇ - graph-based decomposition, △ - distance transformation

for this object. If we need only a few moments (up to the fifth order), the direct computation is the best choice. The QDT and DT will never pay.

We repeated the experiment with the moth image of the size 829×829 and moments up to the order 828. The results were qualitatively similar; only the particular numbers were different because of a different shape and size of the object (check the graphs in Figure 8.14). The decompositions can be seen in Figures 8.7b, 8.9b, and 8.12b. The total computation time by definition was 19:33 minutes; the other times are QTD—66.9 s, DT—21.6 s, GDM—15.2 s, and GBD—13.4 s. The generalized delta method is the best solution for a wide range of moment orders from 5 to 460.

The above results can be generalized for most "reasonably simple" shapes. However, it is easy to find a counterexample. The results for a chessboard image are dramatically different, see Figure 8.15. The chessboard-like images (including random pixel mosaics) are the worst possible case, because they cannot be decomposed efficiently. The decomposition is just a waste of time, and a direct calculation exhibits the best performance. We meet this kind of images when working with individual bit or intensity slices of graylevel images (see Section 8.6.2) and also when heavy impulse noise is presented.

8.5.7 3D decomposition methods

Decomposition of binary 3D objects may be on the one hand even more useful than the decomposition in 2D because there is a chance that the number of blocks will be much lower than the number of voxels. On the other hand, the decomposition algorithms are more time-consuming. Decomposition into rectangular blocks can only be applied to the objects in

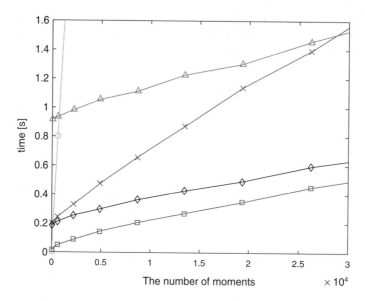

Figure 8.14 The time complexity of the moment computation of the moth image. Legend: O - definition, □ - generalized delta method, × - quadtree, ◇ - graph-based decomposition, △ - distance transformation

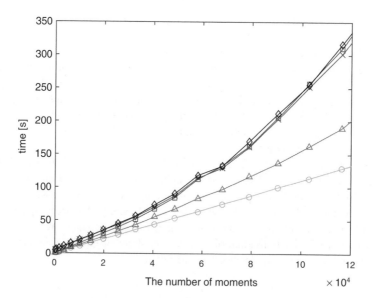

Figure 8.15 The time complexity of the moment computation of the chessboard image. Legend: O - definition, □ - generalized delta method, × - quadtree, ◇ - graph-based decomposition, △ - distance transformation

volumetric representation. If the object is given by a triangulated surface, we could convert it to the voxels and then decompose it, but for such objects the boundary-based methods are usually more efficient because they are applied directly without any conversion.

Some 2D decomposition algorithms can be easily and intuitively extended into 3D. The basic delta method is a typical example—it simply decomposes the object into the lines of adjacent voxels. The 3D generalized delta method (3GDM) unifies adjacent "voxel lines" of the same length into blocks. In 2D the GDM always produces rectangles, but this is not the case in 3D, where we generally obtain a prism the base of which may be an arbitrary polygon. This polygon is decomposed by 2D GDM to obtain rectangular blocks. Hence, the 3GDM has six possible combinations (in 2D, we have only two options) of the directions—three directions for voxel lines construction and two for the base decomposition. The 3GDM tests all combinations and chooses the one with the minimum number of blocks. As in 2D, the 3GDM is very fast and provides a sufficiently small (although not minimal) number of blocks, which makes them a good compromise between the speed and decomposition quality (see Figure 8.16 for two sample objects and Figure 8.17 for the results of the 3GDM decomposition and for the comparison with two other methods). The 3GDM algorithm also works well for complex objects which are highly non-convex, contain holes, and are "sparse" in the space (see Figure 8.18).

The quadtree decomposition turns to the *octree* in 3D. It works exactly in the same manner and keeps all pros and cons of its 2D ancestor. The same is true for morphologic decomposition, where the only change is replacing the square structure element with a cube, and also for the faster version using the distance transformation.

The generalization of the optimal graph-based decomposition to 3D is much more difficult. Let us try to do so in a "naive" way first. In 3D, the role of concave vertices is played by the *concave edges*, the concave vertices do not have any importance in 3D. Instead the chords, we analogically construct the *dividers*. The chord can connect only two points while the divider can connect an unlimited number of concave edges. This is a principal difference from the 2D version. Hence, the 3D algorithm has p levels, $p \geq 1$, where p is called the *significance* of the divider. It equals the number of the concave edges the divider connects. Intuitively, for the

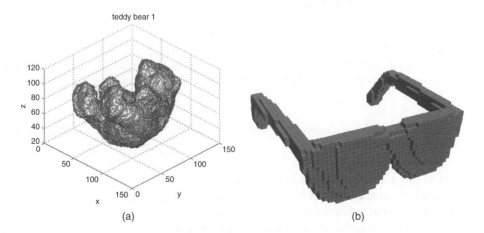

Figure 8.16   Examples of 3D binary objects in a volumetric representation: (a) Teddy bear – 85068 voxels, (b) Glasses – 3753 voxels

(a)   (b)

(c)   (d)

(e)   (f)

Figure 8.17 Various methods of 3D decomposition: (a) glasses – octree – 6571 blocks, (b) airplane – octree – 4813 blocks, (c) glasses – 3GDM – 776 blocks, (d) airplane – 3GDM – 586 blocks, (e) glasses – suboptimal algorithm – 732 blocks, (f) airplane – suboptimal algorithm – 573 blocks

same reasons as in 2D, we should prefer the dividers of high significance. So, the algorithm could work as follows.

1. Find all concave edges of the object.
2. Find the maximum number of coplanar edges p. If $p = 1$, go to step 5.
3. Find the maximum set of disjoint dividers of significance p.
4. Add these dividers to the object surface and go to step 2.
5. From each concave edge, construct a divider in an arbitrary of two directions.
6. Stop.

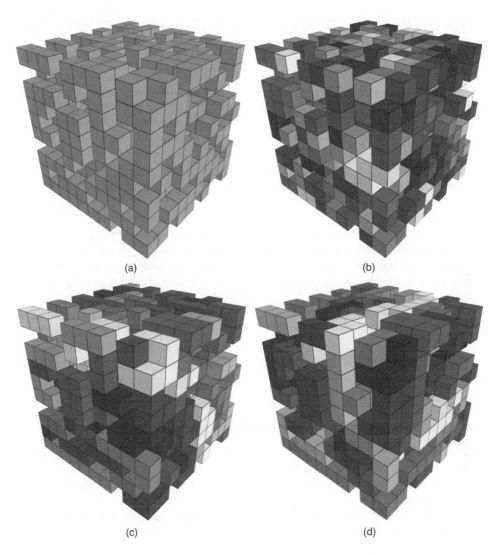

Figure 8.18 The random $10 \times 10 \times 10$ cube: (a) original (518 voxels), (b) octree decomposition into 504 blocks, (c) 3GDM decomposition into 227 blocks, (d) suboptimal decomposition into 208 blocks

The most difficult part of this algorithm is Step 3. We can again reformulate it as a graph problem. All dividers of significance p are depicted as graph nodes. Two nodes are connected with an edge if the respective dividers intersect each other or share the same concave edge. Now we look for the maximum independent set of nodes in this graph.

Although the graph is a tri-partite one, which is much simpler than a general graph, finding the maximum independent set is NP-hard w.r.t. the number of dividers of the same significance. Theoretically, this number may be proportional to the number of all boundary voxels, but in practice it is much lower, so the algorithm is feasible for simple objects even if it is exponential.

However, for objects with hundreds of dividers of the same significance we have to approx-imate this exponential algorithm by a suboptimal polynomial heuristics (see [53] for various approximations). The performance of the suboptimal algorithm is illustrated in Figures 8.17 and 8.18.

Unfortunately, the above algorithm does *not* produce the minimum number of blocks. The topology of 3D space is so different from 2D that a straightforward extension of the algorithm which is optimal (and polynomial) in 2D yields only a sub-optimal but NP-hard 3D algorithm. In 3D, a divider connects certain concave edges (let us denote their number as e_1) but also may *intersect* some other e_2 concave edges which are perpendicular to the divider. Hence, using this divider we not only decrease the total number of the concave edges by e_1 but also increase it by e_2 (this cannot happen in 2D where a chord never creates a new concave vertex). This can be overcome by changing the definition of the significance as $p = e_1 - e_2$ and applying the above algorithm as it is [53].

This modification improves the decomposition but still does not yield the minimum number of blocks. Placing a divider which is optimal on the given level (i.e., has the highest significance) may prevent the placing of certain divider(s) of lower significance that would lead to fewer blocks. For each divider, the optimal algorithm should search the complete "sub-tree" up to $p = 1$ before the divider is placed. This is equivalent to a "brute-force" approach where all possible combinations of the dividers, regardless of their significance, are considered. This intuitive conjecture that the optimal decomposition in 3D cannot be polynomial was formally confirmed by Dielissen and Kaldewaij [54], who proved that the optimal 3D decomposition is equivalent to a variant of the *Boolean three satisfiability* problem called 3SAT3 and hence is NP-complete.

8.6   Geometric moments of graylevel images

When calculating moments of a graylevel image, we cannot hope for a significant speed-up over the direct calculation without a loss of accuracy. Clearly, any exact algorithm must read each pixel value at least once, which is itself of the same complexity $\mathcal{O}(MN)$ as the direct evaluation of the definition.

It should be noted that the geometric moments can be calculated from the definition rela-tively fast in Matlab. We can rewrite Eq. (8.5)

$$\bar{m}_{pq} = \sum_{i=1}^{N} \sum_{j=1}^{M} i^p j^q f_{ij} \tag{8.30}$$

in the form

$$\bar{m}_{pq} = (\mathbf{i}^p)^T \cdot \mathbf{f} \cdot \mathbf{j}^q \tag{8.31}$$

where $\mathbf{i}^p = (1, 2^p, 3^p, \dots, N^p)^T$ and $\mathbf{j}^q = (1, 2^q, 3^q, \dots, M^q)^T$. This can be implemented in Matlab as a single-line instruction with two vector-matrix multiplications only. The vectors $\mathbf{i}^p, \mathbf{j}^q$ can be precalculated in advance.

If we want to calculate all moments up to a certain order, which is more common than the calculation of a single moment, the computation can be arranged even more efficiently. The complete moment matrix $\mathbf{m}$ of the size $(P + 1) \times (Q + 1)$ (i.e., of the moments with the indices from zero to P and Q, respectively) can be obtained again as a single-line instruction with two matrix multiplications

$$\mathbf{m} = (\mathbf{I}^P)^T \cdot \mathbf{f} \cdot \mathbf{J}^Q, \tag{8.32}$$

where $\mathbf{I}^P = (\mathbf{1}, \mathbf{i}, \mathbf{i}^2, \dots, \mathbf{i}^P)$ and $\mathbf{J}^Q = (\mathbf{1}, \mathbf{j}, \mathbf{j}^2, \dots, \mathbf{j}^Q)$.

We can use the same implementation "trick" when computing central moments and/or moments $\hat{m}_{pq}$ given by (8.7), only the precalculated vectors/matrices must be slightly modified[9]. Note that in all cases the precalculated quantities are completely independent of the image, so they can be used repeatedly for all images we work with.

When calculating 3D moments, we may proceed in a similar way and employ a 3D analogue of (8.31). We precalculate 3D arrays $\mathbf{I}^P$, $\mathbf{J}^q$ and $\mathbf{K}^r$ of the size of the original $N_1 \times N_2 \times N_3$. The array $\mathbf{I}^P$ is formed by vector $\mathbf{i}^p$ along the first dimension, which is $N_2 \times N_3$ times copied unaltered to create a 3D array. The arrays $\mathbf{J}^q$ and $\mathbf{K}^r$ are created accordingly. Then the 3D geometric moment $\bar{m}_{pqr}$ is computed as the sum of the element-wise multiplications of these three arrays and $\mathbf{f}$. This algorithm contains only three loops from zero to the moment order. No loops over the coordinates are required. Eq. (8.32) could also be modified to 3D, but it would lead to a general tensor product that has not been implemented in Matlab as a single instruction.

From the above, we can see that it is really hard to overcome the speed of the direct moment calculation in case of graylevel images. Yet certain fast algorithms do still exist. However, they are meaningful for specific images only and/or do not yield exact results. In the following, we mention two basic groups—*graylevel slicing*, which includes intensity and bit slicing, and *graylevel approximation*. All methods belonging to these groups were originally proposed for geometric moments but could be in principle used for moments orthogonal on a rectangle as well. They can also readily extended into 3D, but practical applicability of these methods in 3D is very limited.

8.6.1 Intensity slicing

The intensity slicing method came from Papakostas et al. [55]. The idea behind this method is a decomposition of the image into so-called *slices*. A slice is a set of all pixels having the same intensity value. The original image can be considered to be a sum of the slices multiplied by the respective intensity:

$$f(x, y) = \sum_{k=1}^{L-1} k f_k(x, y), \tag{8.33}$$

where L is the number of graylevels (usually $L = 256$) and the slice $f_k(x, y)$ contains pixels of the graylevel k only, i.e.

$$f_k(x, y) = \begin{cases} 1 & \text{if } f(x, y) = k \\ 0 & \text{if } f(x, y) \neq k. \end{cases} \tag{8.34}$$

In this way we transformed the problem of calculating graylevel moments to the previous task of calculating binary moments. Each slice $f_k(x, y)$ is nothing but a binary image, and we have

---

[9] This is possible thanks to the separability of the basis functions.

for the moments of $f(x, y)$

$$m_{pq}^{(f)} = \sum_{k=1}^{L-1} k \, m_{pq}^{(f_k)}.$$

Note that the contribution of the slice $f_0(x, y)$ is zero, so we can ignore it. The moments of the individual slices can be evaluated by any method designed for binary images. In reference [54], the GDM decomposition was used, but we can use an arbitrary method.

The computational complexity of the intensity slicing depends greatly on the given image. The method is appropriate for images with only few graylevels, where the slices are not "fragmented" and consist of large compact areas. In such cases we can save a significant amount of time. If these conditions are not met, the moment computation by intensity slicing may be even slower than a direct calculation from the definition.

For an example of an image with ten graylevels and its decomposition into slices see Figures 8.19 and 8.20.

8.6.2   Bit slicing

Spiliotis and Boutalis [56] proposed another slicing technique, where the *bit slices* are used instead of the intensity slices. The graylevel image is decomposed as

$$f(x, y) = \sum_{k=0}^{\log_2 L - 1} 2^k \, f_k(x, y),$$

where $f_k(x, y)$ is the binary image formed by the kth bit plane of the original image $f(x, y)$. Then we have, for its moments,

$$m_{pq}^{(f)} = \sum_{k=0}^{\log_2 L - 1} 2^k \, m_{pq}^{(f_k)}.$$

Figure 8.19   An image containing ten graylevels only

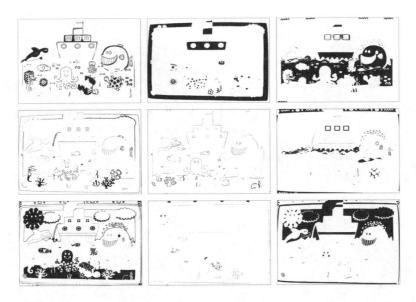

Figure 8.20   Nine intensity slices of the image from Figure 8.19 (the zeroth slice has been omitted)

The advantage of this method compared to the intensity slicing is a low number of binary slices ($\log_2 L$ comparing to $L - 1$). On the other hand, the bitplanes are usually very fragmented (particularly those of the low bits) and hence less appropriate for decomposition. An example of the bit slicing technique is in Figure 8.21, where we can see the bit planes of the original image from Figure 8.22a.

We decomposed the individual bit planes from Figure 8.21 by the GDM. Then we computed discrete Chebyshev moments[10] up to the order 100. The results are summarized in Table 8.2. The second column shows the number of blocks of each bit slice, the time in the third column means a cumulative time of the decomposition and moment calculation of the respective slice. The direct calculation from the definition had to go over 4,915,200 pixels and took 416 seconds, so the bit slicing is faster by two seconds only.

A significant speed-up may be achieved if we do not insist on exact values of the moments. Then we can omit some of the low bit planes because they contribute to the moment value insignificantly but substantially increase computational time. The last column of the table shows the relative (w.r.t. the original) mean square error of the moments if all lower bits than the current bit have been ignored. For instance, if we throw away two lowest bit planes, the moment computation via bit slicing takes only 213 seconds, but the precision loss of the moments is insignificant—only 0.4% In terms of the visual quality of the image, we cannot see any difference from the original (see Figure 8.22b). Even if we keep only the two highest bits, the relative moment error is 30%, (which still may be sufficient for classification in some cases), but the computation time is twenty times shorter.

---

[10] We did not use the geometric moments to prevent the overflow. The type of the moments does not influence the results of this experiment significantly.

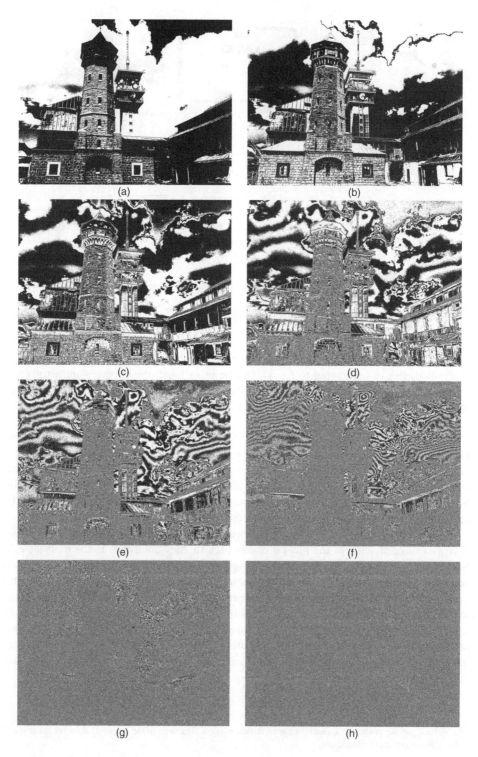

Figure 8.21 Bit slicing of the Klínovec image: from (a) to (h) the bit planes from 7 to 0

(a)   (b)

Figure 8.22   (a) The original graylevel photograph of the size 2560×1920 (lookout and telecommunication tower at Klínovec, Ore Mountains, Czech Republic), (b) the image after the bit planes 0 and 1 have been removed. To human eyes, the images look the same

Table 8.2   The numbers of blocks in the individual bit planes, computation time, and the relative mean square error of the moments due to omitting less significant bit planes

Bit plane	No. of blocks	Time [s]	RMSE [%]
0	1 115 252	104	0
1	1 039 142	97	0.2
2	833 911	79	0.4
3	578 658	56	2.6
4	365 708	36	16.1
5	208 370	22	19.6
6	106 414	12	30.7
7	61 584	8	168.0
Total	4 309 039	414	

8.6.3   Approximation methods

Approximation methods are algorithms for computation of moments of graylevel images that reach low complexity at the expense of accuracy. They decompose the image into rectangular blocks such that the graylevels in each block can be approximated by a "simply integrable" function with a user-defined tolerance. In this sense, the approximation methods can be considered a generalization of the decomposition methods for binary images.

Like the binary decomposition methods, the computational complexity of a moment is given by the number of blocks since the image moment is a sum of all block moments. The approximating function should be symbolically integrable—we can use a constant, linear/bilinear function, and also higher-degree polynomials. In such cases, the block moments can be obtained in $\mathcal{O}(1)$ time by means of the Newton-Leibnitz formula.

There is a trade-off between the number of blocks and the degree of the approximating polynomial, on the one hand, and the accuracy of the moments, on the other hand. For common images, higher-degree approximating polynomials generate fewer blocks than lower-degree ones at the same accuracy level. On the other hand, for a given degree, a request for higher accuracy leads to a higher number of blocks and, consequently, to a slower algorithm. Since both accuracy and the degree of approximating polynomials are user-defined parameters that significantly influence the complexity, one has to choose them carefully with respect to the image and to the particular application.

Individual approximation methods differ from each other mainly in the decomposition scheme (in principle, any decomposition method for binary images can be adapted to this purpose) and by the approximating functions.

Chung and Chen [57] proposed a method whereby the image is decomposed by a *bintree* to rectangular blocks. The bintree is similar to the quadtree, but the block is divided into two rectangles only in one step, and the algorithm alternates between horizontal and vertical directions. The image function $f(x, y)$ in the block is approximated by the bilinear function

$$b(x, y) = a_0 + a_1 x + a_2 y + a_3 xy$$

such that $f(x, y) = b(x, y)$ at the block corners (this is always possible for any block). If $|f(x, y) - b(x, y)| \le e$ (where e is a user-defined tolerance parameter) for all pixels of the block, this block is not further divided. If $|f(x, y) - b(x, y)| > e$, the block is divided into two sub-blocks, and the procedure repeats until the required approximation accuracy is reached. In reference [57], the block moments are then computed by the formulas for the sums of finite power series (8.9) which is inefficient. A better approach is to use integration over the blocks by means of the Newton-Leibnitz formula.

Chung's method [57] can be improved in several ways. The bintree decomposition is not optimal for this purpose; it should be replaced by a method that would respect "natural" homogeneous blocks in the image. There is no theoretical reason to apply only bilinear approximation. In principle, any polynomial (and even any "reasonable" function) can be applied to obtain larger blocks.

Let $p_n(x, y)$ be a bivariate polynomial of degree n

$$p_n(x, y) = \sum_{\substack{k=0 \\ \ell+k \le n}}^{n} \sum_{\ell=0}^{n} a_{k\ell} x^k y^\ell$$

and let $C = (n + 1)(n + 2)/2$ be the number of its coefficients. To find the coefficients of the approximating $p_n(x, y)$ on the block, we need C points (x_i, y_i) in which the interpolation constraint

$$p_n(x_i, y_i) = f(x_i, y_i)$$

is fulfilled. These points can be selected regularly or randomly within the block. Since polynomials are well known to be numerically unstable interpolants, it is better to release the interpolating conditions and to calculate the coefficients of $p_n(x, y)$ by a least-square fit over more than C points. The degree of the polynomials can be set up by the user in advance, or it can be automatically selected during the decomposition (note that different blocks may be approximated by polynomials of different degrees). The moment of an $N \times M$ block is then estimated as

$$\hat{m}_{pq}^{(B)} = \int_{0.5}^{N+0.5} \int_{0.5}^{M+0.5} x^p y^q p_n(x, y) \, \mathrm{d}x \, \mathrm{d}y$$

which can be evaluated in $\mathcal{O}(1)$ time.

8.7   Orthogonal moments of graylevel images

Algorithms for efficient calculation of OG moments are mainly aimed at stable and fast evaluation of the respective OG polynomials. As already explained in Chapter 7, for numerical reasons OG polynomials should not be calculated via expanding them into standard powers (although it is theoretically possible, it would lead to precision loss). We obtain much more stable evaluation when we use the recurrent tree-term relations.

It was discovered that in some cases we can, by properly re-ordering the recurrent calculations, reach a certain speed-up of the algorithm or increase its numerical precision while preserving an acceptable computing complexity. The main idea of several algorithms has been exactly that one.

Another group of fast algorithms employs specific properties of particular OG polynomials, namely their symmetry of various kinds, to speed up the calculation. A similar idea is to precalculate some parts of the kernel functions and use them repeatedly for many moments and/or for several images.

In this section, we present typical representatives of both groups. Note that all these algorithms can be further combined with general principles presented earlier for geometric moments, such as with slicing and approximation methods. Extension of these methods into 3D is also possible. It is trivial in case of the moments orthogonal on a cube. The moments orthogonal on a sphere require more adaptation, but generalizing the main idea is straightforward.

8.7.1   Recurrent relations for moments orthogonal on a square

Let us illustrate this approach by an algorithm for calculating the discrete Chebyshev moments introduced by Mukundan in reference [58], where the original relations recurrent to the degree of the orthogonal polynomials are substituted by relations recurrent to the variable of the polynomials.

The usual form of the recurrence relation is

$$T_n(x) = (\alpha_1 x + \alpha_2)T_{n-1}(x) - \alpha_3 T_{n-2}(x) \tag{8.35}$$

for $n = 2, 3, \ldots, N - 1$ and $x = 0, 1, \ldots, N - 1$. The multiplications by big values of x near $N - 1$ may lead to numerical instability that can be overcome by the recurrence relation in x

$$T_n(x) = \gamma_1 T_n(x - 1) + \gamma_2 T_n(x - 2). \tag{8.36}$$

For more details about these formulas see Chapter 7, equations (7.71) to (7.72) and equations (7.177) to (7.182). It is worth noting that similar formulas recurrent in x can also be derived for other discrete orthogonal moments, such as for dual Hahn moments [59] and Racah moments [59].

This recurrence re-ordering does not change the speed – the computational complexity stays $\mathcal{O}(N^2 p^2)$ where p is the maximum moment order – while it increases the stability of the calculation and consequently the accuracy of the moments.

8.7.2   Recurrent relations for moments orthogonal on a disk

Thanks to their importance in object recognition, moments orthogonal on a disk and particularly Zernike and pseudo-Zernike moments have attracted significant attention from designers of efficient algorithms. Most of these algorithms look for a better utilization of the recurrence relations for the radial component [60, 61].

The ZMs (together with the PZMs) differ from other moments orthogonal on a disk by dependency of their radial function $R_{n\ell}(r)$ not only on the degree n, but also on the repetition factor ℓ.

If we need to compute all ZMs up to the order p, then the direct method using the definition (7.127) needs $\mathcal{O}(p^3)$ multiplications. If we arrange the recurrences properly, we can decrease the complexity substantially.

The recurrent relation (7.133) initialized by (7.134), referred to as the Prata method [63], needs $\mathcal{O}(p^2)$ multiplications only for the computation of all ZMs up to the order p. The computational flow of the Prata method is shown in Figure 8.23.

The Kintner method [61, 63] uses another recurrence formula

$$R_{n\ell}(r) = \frac{(K_2 r^2 + K_3)R_{n-2,\ell}(r) + K_4 R_{n-4,\ell}(r)}{K_1}, \tag{8.37}$$

where

$$\begin{aligned}
K_1 &= (n+\ell)(n-\ell)(n-2)/2, \\
K_2 &= 2n(n-1)(n-2), \\
K_3 &= -\ell^2(n-1) - n(n-1)(n-2), \\
K_4 &= -n(n+\ell-2)(n-\ell-2)/2.
\end{aligned} \tag{8.38}$$

This formula cannot be used for $|\ell| = n$ and $|\ell| = n - 2$. Originally, the direct method (7.127) was proposed for these cases, but the computation can also start with the initial conditions

$$\begin{aligned}
R_{nn}(r) &= r^n, \\
R_{n+2,n}(r) &= (n+2)R_{n+2,n+2}(r) - (n+1)R_{nn}(r),
\end{aligned} \tag{8.39}$$

that overcome this problem. The computational flow of the Kintner method is depicted in Figure 8.24.

The Chong repetition-recursive method [62] uses yet another recurrence formula

$$R_{n,\ell-4}(r) = H_1 R_{n\ell}(r) + \left(H_2 + \frac{H_3}{r^2}\right) R_{n,\ell-2}(r), \tag{8.40}$$

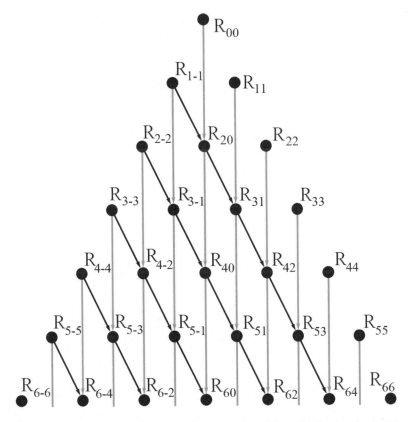

Figure 8.23 The computational flow of the Prata method. The initialization is $R_{nn}(r) = r^n$ and $R_{n,-n}(r) = r^n$

where

$$H_3 = \frac{-4(\ell - 2)(\ell - 3)}{(n + \ell - 2)(n - \ell + 4)},$$

$$H_2 = \frac{H_3(n + \ell)(n - \ell + 2)}{4(\ell - 1)} + \ell - 2, \tag{8.41}$$

$$H_1 = \frac{\ell(\ell - 1)}{2} - \ell H_2 + \frac{H_3(n + \ell + 2)(n - \ell)}{8}.$$

The computation starts with

$$R_{nn}(r) = r^n,$$
$$R_{n,n-2}(r) = nR_{nn}(r) - (n - 1)R_{n-2,n-2}(r). \tag{8.42}$$

The computational flow of the Chong method is shown in Figure 8.25.

Both the Kintner method and the Chong method require $\mathcal{O}(p^2)$ multiplications for the computation of all ZMs up to the order p, but their actual complexity is lower than that of the Prata method. The Chong method is appropriate if we need only the ZMs of a particular

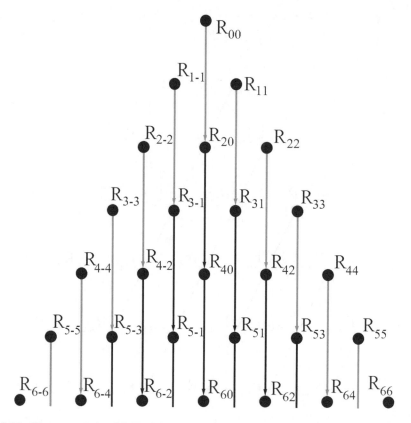

Figure 8.24 The computational flow of the Kintner method. The first two values in each sequence, i.e., $R_{n,-n}$, $R_{n+2,-n}$ and R_{nn}, $R_{n+2,n}$ must be computed directly from equation (8.39)

order, because the recurrence goes over the repetition. If we need all moments up to a certain order, then the Kintner method is more efficient.

8.7.3   Other methods

An example of a method that uses precalculations of certain quantities to speed up the calculation of Zernike moments can be found in [64].

The formula for ZMs (7.135) can be rewritten as

$$A_{n\ell} = \frac{n+1}{\pi} \int_0^{2\pi} \int_0^1 \sum_{k=|\ell|,|\ell|+2,\dots}^{n} B_{n\ell k} r^k e^{-i\ell\varphi} f(r,\varphi) r \, dr \, d\varphi$$

$$= \frac{n+1}{\pi} \sum_{k=|\ell|,|\ell|+2,\dots}^{n} B_{n\ell k} \int_0^{2\pi} \int_0^1 r^k e^{-i\ell\varphi} f(r,\varphi) r \, dr \, d\varphi \qquad (8.43)$$

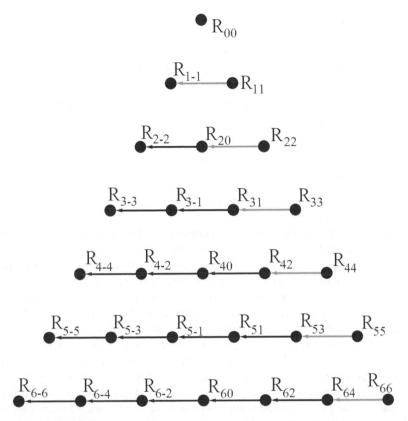

Figure 8.25 The computational flow of the Chong method. The first two values in each sequence, i.e., $R_{n,n-2}$, R_{nn} must be computed directly from equation (8.42)

$$= \frac{n+1}{\pi} \sum_{k=|\ell|,|\ell|+2,\dots}^{n} B_{n\ell k} \chi_{k\ell},$$

where $\chi_{k\ell}$ are Fourier-Mellin moments that do not depend on n. Thus, they can be precalculated, stored, and used repeatedly for various n.

Hwang and Kim [66] proposed another improvement of the computation of the ZMs that is based on their symmetry. Zernike polynomials are symmetrical not only across the x and y axes, but also with respect to the axes $x = y$ and $x = -y$. Thanks to this, it is sufficient to evaluate them in one octant only. If we denote $h_1 = f(x, y)$ and the values in the symmetric points as $h_2 = f(y, x)$, $h_3 = f(-y, x)$, $h_4 = f(-x, y)$, $h_5 = f(-x, -y)$, $h_6 = f(-y, -x)$, $h_7 = f(y, -x)$ and $h_8 = f(x, -y)$, we have for the ZM

$$A_{n\ell} = \frac{n+1}{\pi} \iint_{\substack{x^2+y^2\le 1 \\ 0\le x\le 1 \\ 0\le y\le x}} R_{n\ell}(r)(g_\ell^r(x, y) - ig_\ell^i(x, y)) \, dx \, dy, \qquad (8.44)$$

where $g_\ell^r(x, y)$ and $g_\ell^i(x, y)$ depend on the modulus m of ℓ after dividing by four:

$$
\begin{aligned}
g_{4k}^r(x, y) &= (h_1 + h_2 + h_3 + h_4 + h_5 + h_6 + h_7 + h_8)\cos(\ell\varphi), \\
g_{4k}^i(x, y) &= (h_1 - h_2 + h_3 - h_4 + h_5 - h_6 + h_7 - h_8)\sin(\ell\varphi), \\
g_{4k+1}^r(x, y) &= (h_1 - h_4 - h_5 + h_8)\cos(\ell\varphi) + (h_2 - h_3 - h_6 + h_7)\sin(\ell\varphi), \\
g_{4k+1}^i(x, y) &= (h_1 + h_4 - h_5 - h_8)\sin(\ell\varphi) + (h_2 + h_3 - h_6 - h_7)\cos(\ell\varphi), \\
g_{4k+2}^r(x, y) &= (h_1 - h_2 - h_3 + h_4 + h_5 - h_6 - h_7 + h_8)\cos(\ell\varphi), \\
g_{4k+2}^i(x, y) &= (h_1 + h_2 - h_3 - h_4 + h_5 + h_6 - h_7 - h_8)\sin(\ell\varphi), \\
g_{4k+3}^r(x, y) &= (h_1 - h_4 - h_5 + h_8)\cos(\ell\varphi) + (-h_2 + h_3 + h_6 - h_7)\sin(\ell\varphi), \\
g_{4k+3}^i(x, y) &= (h_1 + h_4 - h_5 - h_8)\sin(\ell\varphi) + (-h_2 - h_3 + h_6 + h_7)\cos(\ell\varphi),
\end{aligned}
\tag{8.45}
$$

where $\ell = 4k + m$. Experimental results reported in [66] show that this algorithm actually increases the speed almost eight times.

In the literature, we can find numerous algorithms developed for particular OG moments. They basically share the same ideas as those described above – re-arranging the recurrences, employing symmetries of the kernels, and precalculations of some values. Sometimes these principles are combined with various approximations and image decomposition. The algorithms differ from each other only slightly; very often the same algorithm is used in an unaltered form for various moments just by replacing the kernel functions.

In addition to the references that we have already cited in this chapter, the following papers are worth mentioning. Efficient algorithms for the discrete Legendre moments can be found in [66–68] and for calculation of discrete Chebyshev moments in [69–73]. Hosny published a series of papers on "exact" calculation of various moments, where he employed the idea of an analytic integration over a pixel area [74–77]. Algorithms for fast calculation of Krawtchouk moments were proposed in [78–80] and for more general Meixner moments in [81]. Algorithms for computation of Hahn moments were proposed in [82, 83].

Concerning the moments orthogonal on a disk, efficient algorithms for Zernike moments were proposed in [84–86]. Attention has also been paid to algorithms for other moments, such as orthogonal Fourier-Mellin [52, 87–90], Jacobi-Fourier [91, 92], and Chebyshev-Fourier moments [93].

Few algorithms have been designed for fast calculation of 3D Legendre moments [94, 95] and 3D Zernike moments [96, 97].

8.8   Conclusion

In this chapter, we have reviewed existing efficient techniques for calculating image moments. We mentioned all the main categories of the method—algorithms for fast computation of geometric moments of binary and graylevel images and algorithms for efficient evaluation of orthogonal moments. The reader may now ask the question whether or not the sophisticated "efficient" algorithms are actually better than a direct calculation from the definition formulas. The answer is not so simple and should be found for each category separately.

As for the decomposition methods of binary images, one should bear in mind that the decomposition itself takes considerable time, depending neither on the number of the moments to be calculated nor on their orders. These methods may pay off only when a large number of moments of the image are calculated or if the blockwise representation is also used for another purpose (for instance, for compression). The power of the decomposition methods depends also on the image itself. Our comparative experiments showed that for common objects and "normal" moment orders the generalized delta method usually provides the best compromise between time complexity and decomposition efficiency. If the image does not contain large "natural" blocks, any decomposition method can perform even worse than the direct calculation (a chessboard is an illustrative example of an object where decomposition cannot help at all).

The boundary-based methods also require a certain constant-time overhead needed for detecting and representing the boundary, but this time is usually much lower than the time of decomposition. The efficiency depends substantially on the boundary shape—for simple boundaries the speed-up could be really significant. Again, a chessboard is an example where these methods perform poorly. Some of these methods allow certain image-independent remedial variables to be pre-computed that can be further used for any image in the application.

The methods of intensity slicing and bit slicing have limited use for graylevel images with only a few intensity levels and with large areas inside them. The approximation methods for graylevel images have even less practical importance. The overhead time for decomposing and piece-wise approximating the image is high, and the moment values are not calculated accurately, which mostly prevents these methods from being used in practice.

In the case of orthogonal moments, our primary goal is to find numerically stable algorithms. The use of proper recurrent relations (along with possible improvements due to the kernel symmetry) is highly recommended. If necessary, these algorithms can be used together with the previous speed-up schemes.

Most of the methods can be easily extended from 2D into 3D. One of the few exceptions is the optimal decomposition algorithm, which is polynomial in 2D bud turns to NP-complete in 3D. Due to this fact, only sub-optimal 3D decompositions have been used.

Appendix 8.A

In this Appendix, we present three possible algorithms for filling up the unwanted holes in the triangulated body surface. One of them should always be used as a post-processing after the triangular surface representation has been created or as a pre-processing before the conversion from triangulated to voxel representation.

The detection of the unwanted holes is easy. For each edge of the triangulated surface we check its parity. If the number of occurrences of an edge in the list of triangles is odd, the edge lies on the boundary of an unwanted hole[11]. We track the boundary of the hole until the starting point has been reached. The boundary forms a closed non-planar polygon of n vertices $A_1, \ldots, A_n$, which must be filled by artificially created triangles. If $n = 3$ the solution is unique, for any $n > 3$, there are various possibilities how to do it.

---

[11] Each hole in the surface is unwanted, there are no intentional holes. If there is an actual hole through the object, the surface should still be closed.

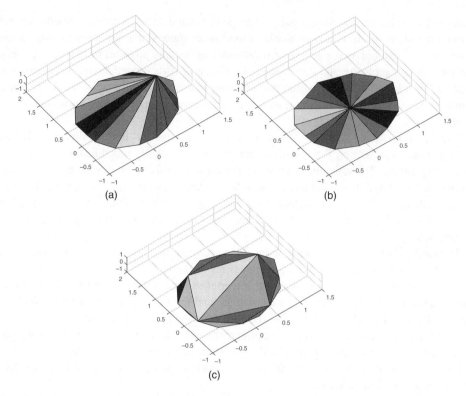

Figure 8.26   The hole filled-up with artificial triangles. (a) The asymmetric method, (b) using the artificial centroid, (c) the iterative "cutting off" method. The rest of the object is not displayed

Easy asymmetric solution

This method chooses one vertex of the polygon (let us say A_1) as a "leading" vertex and creates $n - 2$ artificial triangles A_1, A_k, A_{k+1}, where $k = 2, 3, \ldots, n - 1$ (see Figure 8.26a). This method yields long thin triangles, which may cause problems if the triangulation is used for numerical computation.

Solution with an artificial vertex

This method calculates the centroid of the hole C and creates $n - 1$ triangles C, A_k, A_{k+1}, where $k = 1, 2, \ldots, n - 1$, and the triangle C, A_n, A_1 (see Figure 8.26b). The advantage of this method is more regular triangulation, achieved however at the expense of the higher number of triangles.

Iterative solution

This algorithm fills the hole iteratively. In the first loop, it creates triangles A_k, A_{k+1}, A_{k+2}, where $k = 1, 3, 5, \ldots, n - 2$ (the last triangle is either A_{n-2}, A_{n-1}, A_n or A_{n-1}, A_n, A_1 depending on n). This is like a "cutting off" the hole along the boundary. After this loop, the hole is either completely filled up or its boundary is a smaller polygon with $[(n + 1)/2]$ vertices. We repeat this loop until the whole hole is completely filled.

The advantage of this method is that it creates only $n - 2$ triangles and does not insert any artificial vertex. If the hole is regular, then the area of the triangles gradually increases with the number of loops. For $n \leq 5$, this method is equivalent with the asymmetric one. The result of the algorithm can be seen in Figure 8.26c. We used this algorithm in all experiments with 3D bodies.

More formally, the algorithm is described by the following pseudocode. The input is 1D array *oelist* with *ph* vertex indices on the hole boundary. The output is the $pp \times 3$ array *ffh* with the triangle indices.

```
step ← 1
pp ← 0
while ph ≥ 2 * step + 1 do
    for i from 1 to ph − 2 * step with step 2 * step do
        pp ← pp + 1
        f fh(pp, :) ← [oelist(i), oelist(i + step), oelist(i + 2 * step)]
    end for
    if i ≥ ph − 3 * step then
        pp ← pp + 1
        f fh(pp, :) ← [oelist(i + 2 * step), oelist(i + 3 * step), oelist(1)]
    end if
    step ← step * 2
end while
```

References

[1] W. K. Pratt, *Digital Image Processing*. Wiley Interscience, 4th ed., 2007.

[2] M. Šonka, V. Hlaváč, and R. Boyle, *Image Processing, Analysis and Machine Vision*. Thomson, 3rd ed., 2007.

[3] W. G. Lin and S. Wang, "A note on the calculation of moments," *Pattern Recognition Letters*, vol. 15, no. 11, pp. 1065–1070, 1994.

[4] J. Flusser, "Refined moment calculation using image block representation," *IEEE Transactions on Image Processing*, vol. 9, no. 11, pp. 1977–1978, 2000.

[5] S. X. Liao and M. Pawlak, "On image analysis by moments," *IEEE Transactions on Pattern Analysis and Machine Intelligence*, vol. 18, pp. 254–266, 1996.

[6] K. Chen, "Efficient parallel algorithms for the computation of two-dimensional image moments," *Pattern Recognition*, vol. 23, no. 1–2, pp. 109–119, 1990.

[7] K. L. Chung, "Computing horizontal/vertical convex shape's moments on reconfigurable meshes," *Pattern Recognition*, vol. 29, no. 10, pp. 1713–1717, 1996.

[8] B.-C. Li and J. Shen, "Fast computation of moment invariants," *Pattern Recognition*, vol. 24, no. 8, pp. 807–813, 1991.

[9] X. Y. Jiang and H. Bunke, "Simple and fast computation of moments," *Pattern Recognition*, vol. 24, no. 8, pp. 801–806, 1991.

[10] W. Philips, "A new fast algorithm for moment computation," *Pattern Recognition*, vol. 26, no. 11, pp. 1619–1621, 1993.

[11] G. Y. Tang, "A discrete version of Green's theorem," *IEEE Transactions on Pattern Analysis and Machine Intelligence*, vol. 4, no. 3, pp. 242–249, 1982.

[12] L. Yang and F. Albregtsen, "Fast and exact computation of Cartesian geometric moments using discrete Green's theorem," *Pattern Recognition*, vol. 29, no. 11, pp. 1061–1073, 1996.

[13] J. H. Sossa-Azuela, I. Mazaira, and J. I. Zannatha, "An extension to Philip's algorithm for moment calculation," *Computación y Sistemas*, vol. 3, no. 1, pp. 5–16, 1999.

[14] J. Flusser, "Fast calculation of geometric moments of binary images," in *Proceedings of the 22nd Workshop on Pattern Recognition and Medical Computer Vision OAGM'98*, pp. 265–274, ÖCG, 1998.

[15] R. Mukundan and K. R. Ramakrishnan, "Fast computation of Legendre and Zernike moments," *Pattern Recognition*, vol. 28, no. 9, pp. 1433–1442, 1995.

[16] J. G. Leu, "Computing a shape's moments from its boundary," *Pattern Recognition*, vol. 24, no. 10, pp. 949–957, 1991.

[17] M. H. Singer, "A general approach to moment calculation for polygons and line segments," *Pattern Recognition*, vol. 26, no. 7, pp. 1019–1028, 1993.

[18] J. Flusser, "Effective boundary-based calculation of object moments," in *Proceedings of the International Conference on Image and Vision Computing '98 New Zealand IVCNZ '98*, pp. 369–374, The University of Auckland, 1998.

[19] L. Yang, F. Albregtsen, and T. Taxt, "Fast computation of three-dimensional geometric moments using a discrete divergence theorem and a generalization to higher dimensions," *Graphical Models and Image Processing*, vol. 59, no. 2, pp. 97–108, 1997.

[20] B. C. Li and S. D. Ma, "Efficient computation of 3D moments," in *Proceedings of the 12th International Conference on Pattern Recognition ICPR'94*, vol. I, pp. 22–26, IEEE, 1994.

[21] S. A. Sheynin and A. V. Tuzikov, "Explicit formulae for polyhedra moments," *Pattern Recognition Letters*, vol. 22, no. 10, pp. 1103–1109, 2001.

[22] B. C. Li, "The moment calculation of polyhedra," *Pattern Recognition*, vol. 26, no. 8, pp. 1229–1233, 1993.

[23] A. G. Mamistvalov, "*n*-dimensional moment invariants and conceptual mathematical theory of recognition *n*-dimensional solids," *IEEE Transactions on Pattern Analysis and Machine Intelligence*, vol. 20, no. 8, pp. 819–831, 1998.

[24] T. Suk, C. Höschl IV,, and J. Flusser, "Decomposition of binary images – a survey and comparison," *Pattern Recognition*, vol. 45, no. 12, pp. 4279–4291, 2012.

[25] P. Campisi and K. Egiazarian, *Blind Image Deconvolution: Theory and Applications*. CRC Press, 2007.

[26] A. Levin, R. Fergus, F. Durand, and W. T. Freeman, "Image and depth from a conventional camera with a coded aperture," in *Special Interest Group on Computer Graphics and Interactive Techniques Conference SIGGRAPH'07*, (New York, NY, USA), ACM, 2007.

[27] W. Liou, J. Tan, and R. Lee, "Minimum partitioning simple rectilinear polygons in O(*n* log log *n*)-time," in *Proceeding of the fifth Annual ACM symposium on Computational Geometry SoCG'89*, (New York, NY, USA), pp. 344–353, ACM, 1989.

[28] K. D. Gourley and D. M. Green, "A polygon-to-rectangle conversion algorithm," *IEEE Computer Graphics and Applications*, vol. 3, no. 1, pp. 31–36, 1983.

[29] M. F. Zakaria, L. J. Vroomen, P. Zsombor-Murray, and J. M. van Kessel, "Fast algorithm for the computation of moment invariants," *Pattern Recognition*, vol. 20, no. 6, pp. 639–643, 1987.

[30] M. Dai, P. Baylou, and M. Najim, "An efficient algorithm for computation of shape moments from run-length codes or chain codes," *Pattern Recognition*, vol. 25, no. 10, pp. 1119–1128, 1992.

[31] B. C. Li, "A new computation of geometric moments," *Pattern Recognition*, vol. 26, no. 1, pp. 109–113, 1993.

[32] I. M. Spiliotis and B. G. Mertzios, "Real-time computation of two-dimensional moments on binary images using image block representation," *IEEE Transactions on Image Processing*, vol. 7, no. 11, pp. 1609–1615, 1998.

[33] E. Kawaguchi and T. Endo, "On a method of binary-picture representation and its application to data compression," *IEEE Transactions on Pattern Analysis and Machine Intelligence*, vol. 2, no. 1, pp. 27–35, 1980.

[34] J. Flusser, "An adaptive method for image registration," *Pattern Recognition*, vol. 25, no. 1, pp. 45–54, 1992.

[35] C.-H. Wu, S.-J. Horng, and P.-Z. Lee, "A new computation of shape moments via quadtree decomposition," *Pattern Recognition*, vol. 34, no. 7, pp. 1319–1330, 2001.

[36] J. H. Sossa-Azuela, C. Yáñez-Márquez, and J. L. Díaz de León Santiago, "Computing geometric moments using morphological erosions," *Pattern Recognition*, vol. 34, no. 2, pp. 271–276, 2001.

[37] T. Suk and J. Flusser, "Refined morphological methods of moment computation," in *20th International Conference on Pattern Recognition ICPR'10*, pp. 966–970, IEEE, 2010.

[38] G. Borgefors, "Distance transformations in digital images," *Computer Vision, Graphics, and Image Processing*, vol. 34, no. 3, pp. 344–371, 1986.

[39] M. Seaidoun, *A fast exact euclidean distance transform with application to computer vision and digital image processing*. PhD thesis, Northeastern University, Boston, Massachusetts, USA, 1993. Advisor John Gauch.

[40] T. E. Schouten and E. L. van den Broek, "Incremental distance transforms (IDT)," in *20th International Conference on Pattern Recognition ICPR'10*, pp. 237–240, IEEE, 2010.

[41] J. M. Keil, "Polygon decomposition," in *Handbook of Computational Geometry*, pp. 491–518, Elsevier, 2000.

[42] W. Lipski Jr., E. Lodi, F. Luccio, C. Mugnai, and L. Pagli, "On two-dimensional data organization II," in *Fundamenta Informaticae*, vol. II of *Annales Societatis Mathematicae Polonae, Series IV*, pp. 245–260, 1979.

[43] T. Ohtsuki, "Minimum dissection of rectilinear regions," in *Proceedings of the IEEE International Conference on Circuits and Systems ISCAS'82*, pp. 1210–1213, IEEE, 1982.

[44] L. Ferrari, P. V. Sankar, and J. Sklansky, "Minimal rectangular partitions of digitized blobs," *Computer Vision, Graphics, and Image Processing*, vol. 28, no. 1, pp. 58–71, 1984.

[45] H. Imai and T. Asano, "Efficient algorithms for geometric graph search problems," *SIAM Journal on Computing*, vol. 15, no. 2, pp. 478–494, 1986.

[46] D. Eppstein, "Graph-theoretic solutions to computational geometry problems," in *35th International Workshop on Graph-Theoretic Concepts in Computer Science WG'09*, vol. LNCS 5911, pp. 1–16, Springer, 2009.

[47] A. V. Goldberg and S. Rao, "Beyond the flow decomposition barrier," *Journal of the Association for Computing Machinery*, vol. 45, no. 5, pp. 783–797, 1998.

[48] A. V. Goldberg and R. E. Tarjan, "A new approach to the maximum-flow problem," *Journal of the Association for Computing Machinery*, vol. 35, no. 4, pp. 921–940, 1988.

[49] J. Edmonds and R. M. Karp, "Theoretical improvements in algorithmic efficiency for network flow problems," *Journal of the Association for Computing Machinery*, vol. 19, no. 2, pp. 248–264, 1972.

[50] J. D. Zhou, H. Z. Shu, L. M. Luo, and W. X. Yu, "Two new algorithms for efficient computation of Legendre moments," *Pattern Recognition*, vol. 35, no. 5, pp. 1143–1152, 2002.

[51] Y. Xin, M. Pawlak, and S. Liao, "Image reconstruction with polar Zernike moments," in *Proceedings of the International Conference on Advances in Pattern Recognition ICAPR'05* (S. Singh, M. Singh, C. Apté, and P. Perner, eds.), vol. 3687 of *Lecture Notes in Computer Science*, pp. 394–403, Springer, 2005.

[52] K. M. Hosny, M. A. Shouman, and H. M. A. Salam, "Fast computation of orthogonal Fourier-Mellin moments in polar coordinates," *Journal of Real-Time Image Processing*, vol. 6, no. 2, pp. 73–80, 2011.

[53] C. Höschl IV and J. Flusser, "Close-to-optimal algorithm for decomposition of 3D shapes," *Computational Geometry*, 2016 (to appear).

[54] V. J. Dielissen and A. Kaldewaij, "Rectangular partition is polynomial in two dimensions but NP-complete in three," *Information Processing Letters*, vol. 38, no. 1, pp. 1–6, 1991.

[55] G. A. Papakostas, E. G. Karakasis, and D. E. Koulouriotis, "Efficient and accurate computation of geometric moments on gray–scale images," *Pattern Recognition*, vol. 41, no. 6, pp. 1895–1904, 2008.

[56] I. M. Spiliotis and Y. S. Boutalis, "Parameterized real-time moment computation on gray images using block techniques," *Journal of Real-Time Image Processing*, vol. 6, no. 2, pp. 81–91, 2011.

[57] K.-L. Chung and P.-C. Chen, "An efficient algorithm for computing moments on a block representation of a grey-scale image," *Pattern Recognition*, vol. 38, no. 12, pp. 2578–2586, 2005.

[58] R. Mukundan, "Some computational aspects of discrete orthonormal moments," *IEEE Transactions on Image Processing*, vol. 13, no. 8, pp. 1055–1059, 2004.

[59] H. Zhu, H. Shu, J. Zhou, L. Luo, and J.-L. Coatrieux, "Image analysis by discrete orthogonal dual Hahn moments," *Pattern Recognition Letters*, vol. 28, no. 13, pp. 1688–1704, 2007.

[60] H. Zhu, H. Shu, J. Liang, L. Luo, and J.-L. Coatrieux, "Image analysis by discrete orthogonal Racah moments," *Signal Processing*, vol. 87, no. 4, pp. 687–708, 2007.

[61] M. Al-Rawi, "Fast Zernike moments," *Journal of Real–Time Image Processing*, vol. 3, no. 1–2, pp. 89–96, 2008.

[62] C.-W. Chong, R. Paramesran, and R. Mukundan, "A comparative analysis of algorithms for fast computation of Zernike moments," *Pattern Recognition*, vol. 36, no. 3, pp. 731–742, 2003.

[63] A. Prata and W. V. T. Rusch, "Algorithm for computation of Zernike polynomials expansion coefficients," *Applied Optics*, vol. 28, no. 4, pp. 749–754, 1989.

[64] E. C. Kintner, "On the mathematical properties of Zernike polynomials," *Journal of Modern Optics*, vol. 23, no. 8, pp. 679–680, 1976.

[65] G. R. Amayeh, A. Erol, G. Bebis, and M. Nicolescu, "Accurate and efficient computation of high order Zernike moments," in *Proceedings of the First International Symposium on Advances in Visual Computing ISVC'05*, vol. LNCS 3804, pp. 462–469, Springer, 2005.

[66] S.-K. Hwang and W.-Y. Kim, "A novel approach to the fast computation of Zernike moments," *Pattern Recognition*, vol. 39, no. 11, pp. 2065–2076, 2006.

[67] H. Shu, L. Luo, X. Bao, W. Yu, and G. Han, "An efficient method for computation of Legendre moments," *Graphical Models*, vol. 62, no. 4, pp. 237–262, 2000.

[68] P.-T. Yap and R. Paramesran, "An efficient method for the computation of Legendre moments," *IEEE Transactions on Pattern Analysis and Machine Intelligence*, vol. 27, no. 12, pp. 1996–2002, 2005.

[69] G. Wang and S. Wang, "Parallel recursive computation of the inverse Legendre moment transforms for signal and image reconstruction," *IEEE Signal Processing Letters*, vol. 11, no. 12, pp. 929–932, 2004.

[70] G. Wang and S. Wang, "Recursive computation of Tchebichef moment and its inverse transform," *Pattern Recognition*, vol. 39, no. 1, pp. 47–56, 2006.

[71] L. Kotoulas and I. Andreadis, "Fast computation of Chebyshev moments," *IEEE Transactions on Circuits and Systems for Video Technology*, vol. 16, no. 7, pp. 884–888, 2006.

[72] H. Shu, H. Zhang, B. Chen, P. Haigron, and L. Luo, "Fast computation of Tchebichef moments for binary and grayscale images," *IEEE Transactions on Image Processing*, vol. 19, no. 12, pp. 3171–3180, 2010.

[73] K.-H. Chang, R. Paramesran, B. Honarvar, and C.-L. Lim, "Efficient hardware accelerators for the computation of Tchebichef moments," *IEEE Transactions on Circuits and Systems for Video Technology*, vol. 22, no. 3, pp. 414–425, 2012.

[74] B. Honarvar, R. Paramesran, and C.-L. Lim, "The fast recursive computation of Tchebichef moment and its inverse transform based on Z-transform," *Digital Signal Processing*, vol. 23, no. 5, pp. 1738–1746, 2013.

[75] K. M. Hosny, "Exact Legendre moment computation for gray level images," *Pattern Recognition*, vol. 40, no. 12, pp. 3597–3605, 2007.

[76] K. M. Hosny, "Fast computation of accurate Zernike moments," *Journal of Real-Time Image Processing*, vol. 3, no. 1, pp. 97–107, 2010.

[77] K. M. Hosny, "Fast computation of accurate Gaussian-Hermite moments for image processing applications," *Digital Signal Processing*, vol. 22, no. 3, pp. 476–485, 2012.

[78] K. M. Hosny, "Image representation using accurate orthogonal Gegenbauer moments," *Pattern Recognition Letters*, vol. 32, no. 6, pp. 795–804, 2011.

[79] G. Zhang, Z. Luo, B. Fu, B. Li, J. Liao, X. Fan, and Z. Xi, "A symmetry and bi-recursive algorithm of accurately computing Krawtchouk moments," *Pattern Recognition Letters*, vol. 31, no. 7, pp. 548–554, 2010.

[80] A. Venkataramana and P. Ananth Raj, "Fast computation of inverse Krawtchouk moment transform using Clenshaw's recurrence formula," in *International Conference on Image Processing ICIP'07*, vol. 4, pp. 37–40, IEEE, 2007.

[81] B. Honarvar and J. Flusser, "Fast computation of Krawtchouk moments," *Information Sciences*, vol. 288, pp. 73–86, 2014.

[82] M. Sayyouri, A. Hmimid, and H. Qjidaa, "A fast computation of novel set of Meixner invariant moments for image analysis," *Circuits Systems and Signal Processing*, vol. 34, no. 3, pp. 875–900, 2014.

[83] A. Venkataramana and P. Ananth Raj, "Computation of Hahn moments for large size images," *Journal of Computer Science*, vol. 6, no. 9, pp. 1037–1041, 2010.

[84] M. Sayyouri, A. Hmimid, and H. Qjidaa, "Improving the performance of image classification by Hahn moment invariants," *Journal of the Optical Society of America A*, vol. 30, no. 11, pp. 2381–2394, 2013.

[85] J. Gu, H. Z. Shu, C. Toumoulin, and L. M. Luo, "A novel algorithm for fast computation of Zernike moments," *Pattern Recognition*, vol. 35, no. 12, pp. 2905–2911, 2002.

[86] C.-W. Chong, P. Raveendran, and R. Mukundan, "An efficient algorithm for fast computation of pseudo-Zernike moments," *International Journal of Pattern Recognition and Artificial Intelligence*, vol. 17, no. 06, pp. 1011–1023, 2003.

[87] S. O. Belkasim, M. Ahmadi, and M. Shridhar, "Efficient algorithm or fast computation of Zernike moments," *Journal of the Franklin Institute*, vol. 333(B), no. 4, pp. 577–581, 1996.

[88] G. A. Papakostas, Y. S. Boutalis, D. A. Karras, and B. G. Mertzios, "Fast numerically stable computation of orthogonal Fourier-Mellin moments," *IET Computer Vision*, vol. 1, no. 1, pp. 11–16, 2007.

[89] B. Fu, J.-Z. Zhou, Y.-H. Li, B. Peng, L.-Y. Liu, and J.-Q. Wen, "Novel recursive and symmetric algorithm of computing two kinds of orthogonal radial moments," *Image Science Journal*, vol. 56, no. 6, pp. 333–341, 2008.

[90] E. Walia, C. Singh, and A. Goyal, "On the fast computation of orthogonal Fourier-Mellin moments with improved numerical stability," *Journal of Real-Time Image Processing*, vol. 7, no. 4, pp. 247–256, 2012.

[91] C. Singh and R. Upneja, "Accurate computation of orthogonal Fourier-Mellin moments," *Journal of Mathematical Imaging and Vision*, vol. 44, no. 3, pp. 411–431, 2012.

[92] R. Upneja and C. Singh, "Fast computation of Jacobi-Fourier moments for invariant image recognition," *Pattern Recognition*, vol. 48, no. 5, pp. 1836–1843, 2015.

[93] C. Camacho-Bello, C. Toxqui-Quitl, A. Padilla-Vivanco, and J. J. Báez-Rojas, "High-precision and fast compu-tation of Jacobi-Fourier moments for image description," *Journal of the Optical Society of America A*, vol. 31, no. 1, pp. 124–134, 2014.

[94] Y. Jiang, Z. Ping, and L. Gao, "A fast algorithm for computing Chebyshev-Fourier moments," in *International Seminar on Future Information Technology and Management Engineering FITME'10*, pp. 425–428, IEEE, 2010.

[95] H. Shu, L. Luo, W. Yu, and J. Zhou, "Fast computation of Legendre moments of polyhedra," *Pattern Recognition*, vol. 34, no. 5, pp. 1119–1126, 2001.

[96] K. M. Hosny, "Fast and low-complexity method for exact computation of 3D Legendre moments," *Pattern Recognition Letters*, vol. 32, no. 9, pp. 1305–1314, 2011.

[97] K. M. Hosny and M. A. Hafez, "An algorithm for fast computation of 3D Zernike moments for volumetric images," *Mathematical Problems in Engineering*, vol. 2012, pp. 1–17, 2012.

[98] J. M. Pozo, M.-C. Villa-Uriol, and A. F. Frangi, "Efficient 3D geometric and Zernike moments computation from unstructured surface meshes," *IEEE Transactions on Pattern Analysis and Machine Intelligence*, vol. 33, no. 3, pp. 471–484, 2011.

9

Applications

9.1 Introduction

The main categories of the image processing problems suited for application of moments are introduced in this chapter. The aim is to delineate practical applicability of moments, and to provide an extensive collection of references covering various approaches. There is a selection of illustrative experiments described in detail to illustrate the core ideas behind them. For each topic, several published papers are listed, together with a short overview of the proposed methods. They represent distinctive solutions for given problems. At the same time, advice for successful application of moments is discussed. For any utilization of moments, it is important to be aware of their limitations; their high sensitivity to segmentation error, especially in case of higher-order moments, their global characteristics and tied negative implications for the applicability on potentially occluded objects and on objects with a non-uniform background. These facts should be always considered whenever moments are chosen as the method of solving a problem.

The described classes of algorithms include image understanding focused on different classes of analyzed objects, image registration, robot navigation and visual servoing, measures of image focus and image quality in general, image retrieval, image watermarking, medical imaging, image forensic applications, and finally the "miscellaneous" category. Among the latter, problems like optical flow estimation and edge detection are described. The key issues of individual classes of algorithms are discussed.

The aim of this chapter is not to give a comprehensive explanation of the methodology in its full complexity, nor a full overview of all existing approaches. The references listed were included to illustrate the variability of the application areas as well as to show many interesting ideas. We did not implement and test most of the methods proposed by other authors, so we cannot guarantee their full functionality. All sections cover solutions based only on moments and their modified versions, even though approaches based on different ideas could be more feasible in particular cases.

9.2 Image understanding

An invariant description of an object and its application for the object representation in order to understand what was actually captured in the image is probably the most common utilization

2D and 3D Image Analysis by Moments, First Edition. Jan Flusser, Tomáš Suk and Barbara Zitová.
© 2017 John Wiley & Sons, Ltd. Published 2017 by John Wiley & Sons, Ltd.
Companion website: www.wiley.com/go/flusser/2d3dimageanalysis

of the moments and the corresponding invariants. Such recognition of objects and patterns independent of their position, size, orientation, and other variations in their geometry and colors has been the aim of much recent research. Finding efficient invariant object descriptors is the key to solving this problem. Various groups of features have been used for this purpose, such as simple visual features (edges, contours, textures, etc.), Fourier and Hadamard coefficients, differential invariants, and moment invariants, among others, as it was introduced in Section 2.2, where the theory and possible pitfalls of efficient object representation were discussed in detail. The moment invariants and studies concerning their behavior for various experimental setting show that they can be profitably used for object description and within reasonable scope fulfill given conditions. The fundamental importance has the quality of object segmentation, due to the fact that especially higher-order moments are sensitive to segmentation errors.

As stated previously, it is important to choose the proper class of invariants with respect to expected image degradations and, eventually, take into consideration a method for the moment computation acceleration depending on the type and amount of data to be analyzed (see Chapter 8). The matching part of the method can then be realized using various approaches for the correspondence estimation, starting from standard minimum distance classification, clustering methods, SVM, up to neural networks, hidden Markov models (HMM), or the Bayesian classifier. The moment description is used stand-alone or can be combined together with other descriptive features (for example geometric, Fourier, or differential invariants).

There is one broad application area of the second-order moments for the object description and classification which is often executed before chosen object description starts. The moments up to the second order are used to bring selected parts of the scene (segmented objects, detected affine invariant regions) into the normalized position. Then other types of feature descriptors can be applied, which are not necessarily moment based. Usually it is the question of the numerical properties of involved descriptors if the object normalization or the choice of higher order invariance is the best solution. Examples of methods for region detection which are subjects to such object normalization can be found in [1, 2], or [3].

No dedicated experiment of object recognition is described here, because many illustrative examples can be found in previous chapters. There are applications of several classes of the moment invariants, tested under simple as well as complex degradations. Moreover, even experiments in this chapter described in other sections can be often interpreted as object classification tasks. An illustrative list of methods when moment invariants were used for the image understanding task sorted by application areas follows.

9.2.1   Recognition of animals

The fish recognition system was designed by Zion et al. in [4]. They proposed to evaluate Hu moment invariants and their combinations to be able to distinguish between three fish species. The images were acquired while the fish were swimming in an aquarium with their sides to the camera. The moments were computed both on the whole body of the fish as well as on its parts (the head, the tail). This in-vivo identification of fish species regardless of size and two-dimensional orientation produced satisfying results. The fish identification was the aim of [5], too; however here White et al. implemented only the moment-based estimation of the fish orientation. Gope et al. [6] utilized affine moment invariants for the classification of marine mammals such as sea lions, gray whales, and dolphins. They were involved as well in [7], where the identification of humpback and gray whales based on images of their flukes is addressed. The classification is here based on the white fluke patches and their description by means of the affine moment invariants. The analysis of fish trajectories in underwater videos is demonstrated in [8]. They want to detect new/rare fish behaviors and thus to detect possible environmental changes which can be derived from the abnormal behavior of fish. Fish trajectories are distinguished by moment descriptors. They utilize moments, central moments, translation and scale moments, and affine moment invariants, fifty-five features in total.

Regarding other classes of animals, group-housed pigs and their thermal comfort state (cold/comfortable/warm) is evaluated by means of the first two Hu moment invariants in [9]. The proposed system works real time, using the motion detection and recognition of pigs' resting behavioral patterns. The automated assessment of the boar sperm quality in order to distinguish intact and damaged samples was proposed in [10] using texture features consisting among others of Hu, Legendre, and Zernike moment invariants. Texture classification of mouse liver cell nuclei is proposed in [11]. It is based on invariant moment description of consistent regions—texels. The paper [12] describes the use of Legendre and Zernike moments for texture analysis of MRI T2-weighted images of golden retrievers with muscular dystrophy, the most accurate animal model available for human Duchenne muscular dystrophy.

Six different classes of moment invariants [13] were employed in order to recognize binarized insect images. These moment techniques are geometrical moment invariants, united moment invariants, Zernike moment invariants, Legendre moment invariants, and Chebyshev and Krawtchouk moment invariants. The Krawtchouk MI performed the best. The speckled wood butterfly (*Pararge aegeria*, Linnaeus) and its wing patterns are described by means of Hu moments, and in this way the differences in wing pattern are demonstrated statistically [14].

The rapid mapping and description of plankton is very important for ocean environmental research. In [15] the authors classify large numbers of plankton images detected in real time acquired by a towed underwater video microscope system. The proposed approach combines Hu invariant moment features, Fourier boundary descriptors, and gray-scale morphological granulometries. A learning vector quantization neural network was used to classify acquired samples.

9.2.2   Face and other human parts recognition

Face recognition is a wide area of algorithms, typically used in (but not limited to) security applications. Face recognition systems may be used for person verification/authentication (Is that person actually Mr. X.Y.?) as well as for identification (Who is that person?). Verification is usually carried out in a collaborative manner, and face recognition may be combined with other biometric features such as fingerprints, palmprints, iris patterns, and others (see Forensics applications). Verification has to be performed in real time. Identification is mostly non-cooperative, without a possibility of using other than facial features, and may be requested either real-time or retrospectively. Identification is much more difficult because it must be robust to a head pose, to the presence of a head cover and glasses, and to facial expression. Moreover, a face detection step, in which the person and her face are localized in the scene, must precede the recognition itself. The input of the recognition algorithm is typically a face segmented from the background and bounded by a box, a circle or an ellipse (see Fig. 9.1).

(a) scene with faces   (b) detected face

Figure 9.1   Face recognition: Detection step - the person and its face have to be localized in the scene (a). The output is typically a face segmented from the background (b) and bounded by a box, a circle or an ellipse

There are basically no principal differences between the algorithms used for verification and identification, but usually the success rate of the identification is lower.

The application of moments and moment invariants for face recognition is not a standard solution, and it can be even disputed due to the conflict between the 2D nature of most invariants and 3D human faces and, moreover, due to often large variations between individual images of the same face (hairdress, glasses, to name the most important). Despite this, the moments have been used in face recognition algorithms relatively frequently. Various moment classes have been tested for this purpose: geometric central moments [17], Legendre moments [18], Gaussian-Hermite moments [19], Krawtchouk moments [20], Hahn moments [21], Racah moments [22], Hu, Bamieh [23], Teague-Zernike [23], and complex wavelet moments [24]. Moment invariants to image blurring have been used in scenarios with face recognition on blurred input photographs [25, 26].

The largest group of papers have used Zernike and pseudo-Zernike moments [27–31]. It is a logical choice because the face is often bounded by a circle, and both Zernike and pseudo-Zernike moments can easily be made invariant to its in-plane rotation, as we explained in Chapter 7. There have been also comparative studies showing that the performance of the Zernike and pseudo-Zernike moments in a face recognition experiment is better (or at least equally good) than the performance of other moments, such as Hu moment invariants, Bamieh moment invariants, Zernike, pseudo-Zernike, Teague-Zernike, normalized Zernike, normalized pseudo-Zernike moment invariants and also general moment invariants [23]. An example of an application of the pseudo-Zernike moments in face recognition together with various classifiers can be found in [32]. Performance of k-nearest neighbor algorithm where $k = 1$ and 3, SVM, and HMM classifiers have been studied. The best recognition rate of 91% was achieved with the HMM-based solution. Foon et al. used Zernike moments together with wavelet transformation [33] and provided a comparative study on the moment application [23], too.

While being perfectly invariant to the face rotation around the optical axis and sufficiently robust to noise, the drawback of the moment-based algorithms lies in the intrinsic globality of the moments. This is why the moments are not robust to 3D head pose (the experiments have proved that the tolerable left-right rotation is about 15 degrees) and to the changes of a facial expression. To overcome this limitation, many authors used the moments after they had divided the face into subregions [34] or combined the moments, which capture global characteristics, with local features which are insensitive to the change of a part of the face, such us the presence/absence of glasses or a change of the lips' shape when smiling or speaking. Farokhi et al. proposed using high-pass bands of undecimated wavelet transformation [35] and Hermite kernel filters [16] (see Fig. 9.2) as the local features. The final decision about the person's identity is then done via weighted fusion of the global and local feature decisions. As proven in these papers, the decision fusion yields better performance than individually applied global and local features. The tested scenarios included a variety of facial expressions, and change of the head pose and of the glasses.

Most of the above cited papers work with the images that have been taken in the visible spectrum. However, moments can be applied also to near-infrared face recognition [29, 35, 16] and even to face recognition on thermal images [17]. Using these modalities is advantageous in a dark environment and also provides a higher robustness to skin color, make-up, and similar factors.

While the face recognition systems should be invariant to expression, there exists an opposite task—expression recognition. The aim is not to identify a person, but to estimate his/her disposition by recognizing an expression of the face. Here, common categories are fear, anger, disgust, happiness, surprise, and neutral. Moments can be employed in this task as well, as was demonstrated in [36, 37]. Hu moment invariants were applied by Zhu et al. [38]. They computed them locally, around eyes, nose, and mouth, in order to avoid dependence on changes. Moreover, they used non-centralized versions of moments to be able to capture displacements due to the movements of facial muscles. The recognition was done using HMM. They claim very high recognition rate; however, they put strong limitations on the head movements, glasses

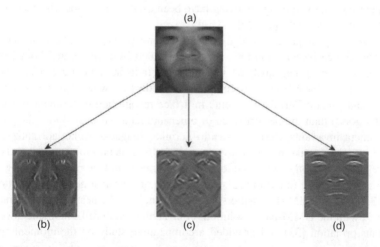

Figure 9.2 Face recognition: Farokhi et al. [16] proposed using Hermite kernel filters as the local features for face recognition in near infrared domain. (a) Original image, (b)–(d) the output of three directional Hermite kernels

and so on. Zhi and Ruan [39] proposed comparison of various moment-based approaches. The applied moments include Hu moments, Zernike moments, wavelet moments, and Krawtchouk moments. Experimental results show that wavelet moments outperform other moment-based methods in facial expression recognition.

As for the recognition of other parts of the human body, we have found two interesting applications of the Zernike moments. In [40] the authors are trying to detect human ears. As stated previously, the Zernike moments are robust in the presence of a noise, and they exhibit rotational invariance. They can also provide non-redundant shape representation because of their orthogonal basis and they reflect the overall shape of the regions. The final verification, when it has to be determined whether a detected object is really an ear, was implemented by means of ZMs. A small set of the lower-order Zernike moments captures the global general shape of an object, which is what is needed here. The magnitudes of the Zernike moments are used for the ear shape representation. If the similarity between the detected candidate and the ear template is high enough, then the claim is validated. In the second application [41], the lip diagnosis for traditional Chinese medicine is proposed. The pipeline starts with the lips' segmentation using the level set approach. Three feature types —color space features, Haralick co-occurrence features, and Zernike moment invariants—are then extracted from the lip image. Finally, SVM is applied for the classification.

9.2.3 Character and logo recognition

The distinct application of the moments, in this case the Zernike moments, is the signature verification [42]. The moments are used for description of individual strokes (curves) of the signature. The signature contour is segmented into a fixed number of small curves, and then the shape features are computed for each curve separately. The magnitude of the Zernike moments captures shape information in a rotationally invariant form, while the relevant complex angles capture rotation with respect to the origin, which is the significant characteristic of signatures. Thus both magnitudes and angles are used as features. The Zernike moments up to the sixth order were evaluated. Proposed signature verification method demonstrates strong invariance among genuine signatures. Lin and Li [43] made use of the Zernike moments too, but in their normalized version [44]. They are verifying Chinese signatures, taking the whole signature as the object to be classified.

In [45] the authors analysed the recognition of the off-line handwriting by means of combination of Hu moment invariants and curvelets and SVM-HMM classifier. The recognition of Arabic handwritten text—written amounts on cheques—is addressed in [46]. They proposed an evaluation of Hu moment invariants on segmented texts. The feature vectors are then fed into the artificial neural network (ANN). Neural network in combination with Hu invariants is used for recognition of Malayalam handwritten text in [47]. Handwritten Devanagari (Marathi) characters are described and successively classified using Zernike moments in [48]. Zernike moments and wavelets are used for recognition of Tamil characters in [49]. MODI, an ancient script of India, is analyzed using Zernike complex moments and Zernike moments with different zoning patterns [50]. Wang et al. [51] employed Krawtchouk and Chebyshev moments as representatives of the discrete orthogonal moments for Chinese handwritten symbols recognition using discrete-time HMM framework. The individual symbols are divided into smaller subparts from which the moment representation is computed. They conclude that the performance of discrete orthogonal moments is much better than that of continuous orthogonal moments (Zernike and Legendre). The Chebyshev moments outperforms Krawtchouk moments in terms of accuracy. Flusser and Suk [52] described an application of affine moment invariants for Latin character recognition. They work with a set of handwritten capital letters. The comparison of recognition ability of four Hu invariants and four AMIs have been described, and the results demonstrated AMI's better robustness with respect to the variations due to handwriting. Matching was realized using the minimum distance classifier.

The very common application of moment descriptors is the recognition of some class of logo images. Logo recognition by moment invariants has already been addressed in [53]. A comparison of methods for identifying the Halal logo is proposed in [54]. The fractionalized principle magnitude is compared with the histogram of oriented gradients (HOG), Hu invariants, Zernike moments, and wavelet co-occurence histogram. Hu moment invariants performance was the lowest in the realized experiments. Zernike moments and the local descriptors defined as the edge-gradient co-occurrence matrices derived from the contor are employed for logo recognition in [55]. Car logos were recognized by means of the radial Chebyshev moments [56]. Hu invariants and simple geometric moments were employed here [57] for recognition of company logos. In [58] radial Chebyshev moment invariants, radial Chebyshev moments, Chebyshev and Krawtchouk moments, Krawtchouk moment invariants, Legendre, pseudo-Zernike, and Zernike moments together with affine moment invariants were evaluated on the Maryland logo database and their comparison published. According to the experiment, Chebyshev moments outperform others in the given task.

9.2.4   Recognition of vegetation and of microscopic natural structures

Moment invariants attracted noticeable attention in the area of tree leaf recognition. The classification of the plant leaves was described in [59]. First, a modified version of the watershed segmentation method combined with pre-segmentation and morphological operation, which can handle partially occluded leaves, is run to segment leaf images. Then, seven Hu geometric moments and sixteen Zernike moments are extracted as shape features from segmented binary images after leafstalk removal. In addition, a moving center hypersphere classifier is designed there. The average correct classification achieved rate was up to 92.6%. Central geometric moments were employed for the leaves' recognition in [60]. Comparison of effectiveness of various moment invariants, namely Zernike moment invariants, Legendre moment invariants, and Chebyshev moment invariants, for the classification of leaf images was performed in [61]. Then, the most effective moment invariants are fed into the general regression neural network (GRNN). Chebyshev moment invariants outperformed others in this test. The paper [62] presents the plants' classification based on their leaf shape using various approaches including Fourier moment technique; however, it was outperformed by SVM in combination with binary decision trees. They tested the methods on a database of 1600 leaf shapes from 32 different classes, where most of the classes have 50 leaf samples of similar kind. Suk et al. applied a set of Zernike, Legendre and Chebyshev moments in order to recognize tree leaves [63].

The second object of interest to be classified using moment invariants is water algae. Thiel et al. [64] proposed applying Hu invariants for recognition of blue-green algae for measuring water quality. Fischer and Bunke [65] classified unicellular algae found in the water by means of affine moment invariants.

Weed visual recognition was presented in [66]. Weeds have to be recognized on images acquired by a camera, moving quickly above the field. The composite motion-defocus blur moment invariants have been proposed for this application. Similarly, in [67] the system for automatic recognition of wood slices is introduced based on moment features invariant to composite motion-defocus blur and to linear motion blur in unknown and arbitrary direction. The wood specimens are placed on a moving platform and scanned by the down-looking camera. In [68], the robust timber cell recognition scheme using the low-quality color timber cell images, often blurred by motion and defocus blur, is described using blur moment invariants. An automated bamboo species recognition system using Fourier and Legendre moment descriptors was proposed in [69].

Moving to a much smaller scale, the moment invariants were incorporated into the automatic analysis of images of the cells and their nuclei [70]. In the paper, a set of features consisting of Zernike moments, Daubechies wavelets, and Gabor wavelets gives the highest reliability for the classification of cells or cell nuclei in fluorescent microscopy images. The methodology for automated detection and classification of microorganisms based on Zernike and Chebyshev moments as well as Haralick texture features is proposed in [71]. The method is processing colony scatter patterns. A set of 1000 scatter patterns representing 10 different bacterial strains was used.

9.2.5   Traffic-related recognition

There are several papers on recognition of aircraft and ships by means of moments. This undertaking commenced with the well-known paper by Dudani et al. [72] using rotational moment invariants. Reeves et al. [73] varied the proposed approach. They presented methods for identification of a 3D object from a 2D image recorded at an arbitrary viewing angle and range. The 3D aircraft models from the database were represented by several 2D projections. They utilized so-called standard moments (normalized moments). Several experiments are described with combinations of silhouette and boundary moments and different normalization techniques. The Mokhtarian and Abbasi method [74] stems from [72]; moreover, they are using the database of aircraft images, too. They are selecting the image subset of optimal views for multi-view free-form object recognition. They address how to choose the optimum number of representative views for each object in a database that can enable the accurate recognition of that object from any single arbitrary view. They propose to represent effectively each object by its boundary and its curvature scale space, Hu moment invariants, and Fourier descriptors. The selection algorithm is described, using the representative subset of the most typical views. The number of retained views varies depending on the complexity of the object and the measure of the expected accuracy. The recognition of landed aircrafts based on a pulse-coupled neural network model and affine moment invariants was recently published in [75]. For the future the authors planned to analyze flying aircraft too.

Alves et al. [76] tried to classify ship types from their infrared silhouette. They compute Hu invariants both from the boundary silhouette and from the region, which were then fed into the neural-network classifier. The proposed method has a problem with the ship's occlusions as a consequence of globally computed invariant features. The ship target recognition is mentioned in [77], where Hu's moments, affine invariant moments, and wavelet invariant moments are combined into one feature vector.

In [78] the method for vehicle detection in high-resolution satellite images was described. First, the image segmentation of the input data was realized, working with multi-spectral images, panchromatic images, and a road network. Then the two-stage classification was evaluated. After primary analysis, the following feature set was chosen: length, width, compactness, elongation, rectangularity, boundary gradient, spatial spread, contrast, smoothness, region mean, gradient mean, variance, and Hu moments. The authors concluded that for the whole system segmentation is the most critical part. Color moment invariants are employed in order to recognize occurrences of buses in the traffic surveillance images [79].

Apatean et al. [80] were involved in the detection of an obstacle in traffic scene situations. Detection of an obstacle (a pedestrian or a vehicle) is often very difficult due to the complex outdoor environment and the variety of appearances of obstacles. The authors intended to achieve classification into the five classes—a standing person, an unknown posture, a motor bike, a tourist car, and a utility car. Visible and infrared images were used as input data and for recognition system features of different types, among them seven geometric moments up to the fourth order, were evaluated and fed to the classification module - k-nearest neighbor algorithm with $k = 1$ and SVM with a radial basis function kernel. Accuracy rates above 92% have been achieved during the experiments.

Zhenghe et al. [81] proposed the modified version of the standard Hu moment invariants and applied them for the recognition of road traffic signs. The feature vectors were classified using a backpropagation neural network. Similar approach, based on Hu moment invariants and backpropagation neural network, appeared recently in [82]. In the novel method for road traffic sign location [83], the proposed solution is described together with following traffic sign recognition by means of affine moment invariants. Zernike moment invariants found their way to traffic sign recognition application, too [84]. Traffic lights were detected by means of Hu's invariants in [85], specifically in the night-time. License plate recognition by means of the Gaussian-Hermite moments was published in [86]. A new method for automotive airbag shape evaluation is proposed in [87]. The boundaries extracted from the airbag images are processed by means of moment invariants, and their similarity is evaluated.

9.2.6 Industrial recognition

SubhiAl-Batah et al. [88] utilized the Hu, Zernike, and affine moment invariants evaluated on the binarized boundary and area to form the feature vector for the description of aggregates in concrete, which are very important for the final quality of concrete. They proposed a new method for the automatic classification of aggregate shapes, based on the described feature vector and artificial neural network. In [89] to automatize visual inspection system the method for surface defection is presented. Copper strips are the target application. The method uses gray level co-occurrence matrices and Hu moment invariants for feature extraction. The adaptive genetic algorithm then processes evaluated feature vectors. Weld radiograms in radiographic testing are the objects of interest in [90]. The authors try to solve the efficient search in the image database in order to find discontinuities similar to some common weld defect types such as crack, lack of penetration, porosity, and solid inclusion. The proposed feature set consists of shape geometric descriptors (SGD), the features based among others on Hu moment invariants and a generic Fourier descriptor. In the paper [91] an automatic detection method for defects in heating, ventilation and air-conditioning ductwork is proposed. The system was set by means of an infrared-CCD device and a method based on cascading seeded k-means and the C4.5 decision tree. The features used in the system are based as well on Hu moment invariants.

Li and Huang [92] proposed a method for recognition of defect patterns in semiconductor manufacturing. The produced circuits have to meet the specifications and must be tested as fast as possible. A method based on self-organizing maps and SVM for wafer bin map classification is proposed. First, they distinguish between the systematic and random wafer bin map distribution. After smoothing the chosen descriptive features – co-occurrence matrix characteristics and Hu moment invariants – are extracted, and the wafer bin maps are then clustered and manufacturing defects classified. The proposed method can transform a large number of wafer bin maps into a small group of specific failure patterns. The experimental results show over 90% classification accuracy, higher than in the case of the back-propagation neural network. In [93] a recognition system for defects of fabricated semiconductor wafers is described. The authors claim up to 95% accuracy (depending on the product type). The chosen intrinsic methods for wafer descriptions are the polar Fourier transformation and rotational moment invariants. To be able to detect geometrical defects of non-silicon Micro-Electro-Mechanical Systems (MEMS) parts, their images are binarized, and their respective skeletons are produced [94]. On them, Hu moment invariants are evaluated and compared with the templates in order to obtain warning that geometrical defects may be present. Here, the skeletonized images outperformed the approach with graylevel images.

In [95] the method based on Hu and affine moment invariants for localization of screws on the tray is described. The detection framework to detect pegs and bits for electric screw drivers is proposed in [96]. The authors utilized Hu moment invariants in order to reliably detect screw tips. Hu moment invariants were applied in [97] to recognize pegs and holes. Hu moment invariants help to detect utility poles in [98]. Here, Hu invariants describe the localized pole segments and then are fed into a neural network for the recognition. The Hu moment invariants are employed in checking the overall dimension of the auto transmission head covers [99]. It helps to check the whole structure earlier before it cools down, which significantly decreases production time.

A new vision-based force measurement method was proposed to evaluate microassembly forces through analyzing pseudo-Zernike moment invariants [100]. The method efficiency is demonstrated on elastic cantilevers. A feature vector is used to describe the deformed cantilever shape in microassembly. A training set describes the relationship between the deformed cantilever under known forces and feature vectors. The recognition is done by means of SVM. In the paper two different methods, Hu image moment invariants and wavelet packet-energy entropy, are compared with respect to their efficiency to extract the pump dynamograph characteristics [101]. It turned out the moment invariants outperformed the wavelet-based approach.

The last to mention is an automatic classification of 3D CAD models[102]. It is based on deep neural networks and Zernike moment descriptors. The latter are employed to represent the rendered 2D images.

9.2.7 Miscellaneous applications

An automated system for recognition of numerals from Auslan—the Australian Sign Language—was proposed in [103] based on the first four Hu invariants and SVM or neural network approach for the classification. Menotti et al. [104] addressed the sign language recognition as well. Their approach is based on SVM classifier and features consisting of Hu and Zernike moment invariants. Their comparison demonstrated the superiority of Zernike moments' features over Hu moments. In [105] a robust vision-based hand gesture recognition system for appliance control in smart homes is demonstrated, based on RTS moment invariants. The gesture recognition is present in [106], addressing the human-robot interaction and cooperation. Using a particular gesture, the person chooses the type of help she wants. The system detects people and then recognizes their gestures. Humans are detected based on face and legs detection. The moments, specifically Hu's moment invariants, are involved in the gesture recognition. General geometric moments are used for classification of human silhouettes in various actions related to shopping [107]. The data are acquired by the Kinect system.

As for the remote sensing applications, Keyes and Winstanley [108] tested the ability of the moment invariants to characterize topographical data. The Hu invariants evaluated on the boundaries of the objects horizontal projections are used for the recognition. Their work is focused on objects on large-scale maps. The final conclusion is that the moments are fairly reliable at distinguishing certain classes of topographic object; however, the authors would recommend fusing them with the results of other recognition techniques. The method for the retrieval of the height of buildings using attribute filters and Hu's moment invariants is introduced in [109]. The main concept of the method relies on the spatial properties of very high-resolution images acquired from different viewing angles. The mutual correspondence of the segmented buildings is then estimated by Hu moment invariants. An underwater mine countermeasure system adapted to operate onboard an aircraft is proposed in [110], using 3D moment invariants.

Pseudo-Zernike moments were used for radar micro-Doppler classification [111] as robust features able to uniquely describe the content of two-dimensional matrices with a small number of coefficients. This ability is very important to exploit the potential of micro-Doppler information for target classification and micro-motion analysis.

An automated shape description of illicit XTC tablets based on the moment invariants with respect to translation, rotation, and scaling is proposed in [112]. The 2D united moment invariants and 3D Zernike moment descriptors are employed in the description of the structure of molecular components of ATS Drugs in [113]. 3D descriptors can be found in [114], where the ability of 3D Zernike moments to capture volumetric shapes is used for the protein surface shape analysis. They are employed in encoding of the molecular structure, which is amenable to fasten further screening.

Regarding a general image/object recognition task, moment invariants were employed, for example, in texture classification. Zernike moments were computed from the correlation functions of individual color channels [115]. The method is independent of rotation, scale, and illumination. The key idea of the method is the relation between illumination, rotation, and scale changes in the scene and a specific transformation of the correlation function of Zernike moment matrices, which can be estimated from a color image. Campisi et al. [116] proposed the use of independent rotation moment invariants based on moments up to the fourth order computed from the texture model. The model is created as an output of a linear system driven by a binary image. The latter retains the morphological characteristics of the texture and is specified by its spatial autocorrelation function (ACF). They show that moment invariants extracted from the ACF of the binary excitation are representative enough to capture the texture for classification purposes. Moreover, the proposed scheme has been shown to be robust with respect to a noise. The geometric moments, Legendre moments, and complex moments were utilized for the texture description and an unsupervised texture segmentation in [117]. All the moments were evaluated from the Gabor decompositions of texture patches.

Legendre moment invariants were used to constrain the evolution of a region-based active contour with respect to a reference shape [118] . A shape prior is here defined as a distance between shape descriptors based on the Legendre moments of the characteristic function. The shape model includes intrinsic invariance with regard to pose and affine deformations. In [119], a denoising method based on nonlocal means (NLM) approach is introduced. Krawtchouk moments are employed to evaluate the level of similarity of the neighborhoods. The method was tested on images degraded by Gaussian, Poisson, and Rician noise.

An application of moment invariants for 3D objects/scenes was described by Xu et al. in [120]. The authors proposed to extract several features (texture and color characteristics, Hu moment invariants, and affine moment invariants) from 2D views of 3D objects. These feature vectors are then presented to the backpropagation neural network for training. The trained network can recognize 3D objects when provided with the feature vectors of unseen views. A 100% correct rate of recognition was achieved when the training views were presented for every 10° change of the viewing angle. In [121], the author proposed the method for selection of the most informative 2D views of 3D objects using web image databases. They directly explore human perception on observing 3D shapes from the relevant web images. Hu moment invariants are used for silhouette comparison here.

As it is apparent from the mentioned approaches, Hu invariants catch the biggest attention for the object description and consecutive classification, even though their choice for given tasks may be questionable. They are invariant to simple geometric deformations (RTS) only; they are not mutually independent; and, moreover, using just a finite set of geometric moments, as is the case in applications with Hu moment invariants, the discriminability can be violated (imagine two objects, whose moment representation differs only in higher-order moments). In spite of all this, the recognition systems often give reasonable success rates. The more complex invariants (AMIs, the invariants based on the orthogonal moments) have already found their position in the pattern recognition awareness and from the latest comparative studies, as in example [23], their bigger efficiency is evident. The choice of the object descriptors is often not directly connected to the choice of the classifier; there are various combinations of many approaches, so the generalization of what is the best pipeline is outside the scope of this book.

9.3   Image registration

Image registration, one of the key image processing objectives, is a process of overlaying two or more images of the same scene taken at different times, from different viewpoints, and/or by different sensors. It aims at the geometrical alignment of the reference and the sensed images. The present distortions between the images are introduced due to different conditions during the data acquisition process. The image registration task appears, for example, in remote sensing, in medicine, and in computer vision, to name a just a few. There are many variants of image registration methods [122] based on various image processing ideas. Their subcategory employing moments or moment invariants, although numerous and significant, is just a fraction of the overall total number. Even so, they form an important subclass of moment-based algorithms.

In general image registration methods can be divided into two categories: land-mark-based and landmark-free. The *landmark-based registration* estimates the correspondence between distinctive objects - landmarks, such as points, closed boundary areas, fixed-sized windows, in the reference and the sensed images. To characterize these features, the methods often employ some kind of invariants for their description. These features override the influence of distortions, introduced by a geometric transformation during the image acquisition process, and also possible discrepancies in the intensity levels. The moment invariants are an example of such invariant representation of the detected objects, which are frequently used for feature matching. The second category of the registration methods, *landmark-free registration* (sometimes called global or area-based methods), utilizes whole images for the estimation of necessary geometrical transformation, which will align the reference and sensed images. Although the moments and moment invariants are more suitable for the landmark-based methods, where they serve as the descriptors of the landmark neighborhoods, their usage in landmark-free registration algorithms can be found in the literature as well. In the following text, both method categories will be discussed from the moment theory point of view.

9.3.1   Landmark-based registration

In this category, two important choices have to be made in order to design a moment-based approach. The first one is about the class of the moment invariants which should be used, depending on the type of the present geometric and even radiometric degradations. For example, in real situations a perspective deformation is present in acquired images, which can often be successfully modeled by an affine transformation, if the distance from the camera to the scene is much larger than the object's size. From the previous statement the usability of the affine moment invariants is substantiated. If radiometric degradations are also expected, it is well-advised to consider the use of combined invariants. For instance, if we expect contrast changes between images, then the invariants to contrast should be used. If blurring may be present in images, we use convolution invariants, which are robust to main prospective kinds of blur, such as the out-of-focus blur, the motion blur, and the atmospheric turbulence blur and which at the same time provide robustness with respect to geometric degradations.

The second important decision which has to made is about the area from which the moments will be calculated. We can use fixed-sized windows localized on the regular mesh over the image or around the detected salient points such as corners, road crossings, line endings, etc. Or even entire images can be used for the moment calculation. The limitations of this approach originate in the relation between the window shape versus the kind of geometric deformation present. First, if the whole image is used for the computation, the method will function only if the reference and the sensed images have a 100% overlap. Any scene occlusion or a difference in the field of view would harm the invariance, so this is not a recommended approach. The simple rectangular window suits the registration of images which locally differ only by translation. If images are deformed by more complex transformations, this type of window is not able to cover the same parts of the scene in the reference and the sensed images (the rectangle can be transformed to some other shape, depending on the deformation). Several authors have proposed using circular windows for mutually rotated images [123].

However, the comparability of such simple-shaped windows is violated as well if more complicated geometric deformations such as similarity or perspective transformations are present. However, recent methods for invariant neighborhood selection can handle even affine deformations; see for example local affinely invariant regions [1]. Another possibility for selecting the area the moments are computed from is not to use the grayvalues but to segment the images first and select several significant closed-boundary regions, such as lakes or fields on remote sensing images (see Fig. 9.4). The moments are then computed only from the binary regions and are insensitive to the particular colors. The robustness of this approach depends largely on the quality of the segmentation algorithm. In the case of a reliable segmentation we usually obtain better results than by calculating moments directly from the graylevels.

The registration methods can be described by the choice of the particular moment descriptors and the regions, from which they are computed. Goshtasby [123] published one of the first algorithms—he used Hu invariants, computed from the circular neighborhood. Moreover, he discussed the matters of the moment value normalization in his paper. When the moment invariants of different scales are used in the similarity estimation by correlation, the feature with the largest scale would dominate the correlation value. He proposed the normalization related to the size of the image and to the order of the used moments. The second improvement he introduced is the speed-up of the registration by splitting the similarity estimation into two stages. First, just the zeroth order moment is used for localization of likely matches, and then only at the selected positions full feature representation is computed. Li et al. [124] proposed to use contours of the objects as salient features; and their correspondence is found by matching their basic geometric features (the perimeter and the longest and shortest distances from boundary to the centroid), then by matching the first two Hu moment invariants, and finally by means of the chain code correspondence. Dai and Khorram [125] proposed an approach to image registration combining the moment-based descriptors with improved chain-code matching. The robustness of the minimum distance classifier is improved by matching of the image, using the rule of root mean-square error. S. X. Hu et al. [126] discussed the Hu invariants based registration algorithm, computed either on detected corners or distinctive edge points. They proposed another way of normalization of the computed invariants values - they use the standard deviation of the absolute value of corresponding moment invariant difference. The improved similarity measure should exclude more outliers than the metric-Euclidean distance. Tuytelaars and Van Gool applied generalized color moments, computed from the local affinely invariant regions [1]. Guo et al. [127] addressed the registration of color images by means of the so called quaternion Fourier-Mellin moments. They can be seen as the generalization of traditional Fourier-Mellin moments for gray-level images. They are meant to be used for matching of the color images differing in the mutual rotation and scaling.

In Shi et al. [128], the moment descriptors with higher-level invariance are exploited. They applied AMIs for the registration of the satellite images. For the computation, they use circular neighborhoods located at the position of the points detected by means of the redundant wavelet transformation. They obtain good results, however it is important to mention here, that the possibility to register affinely deformed images is limited—if the choice of the area for the moment computation is a fixed-sized circle, than in the case of more profound affine deformation it would not cover the same area, which is important for the assurance of moment invariance. This fact was the motivation for the usage of segmented closed boundary regions in [129] for the registration of the satellite images. Here, the AMIs are computed on the segmented areas of reasonable size. Bentoutou et al. [130] amplified the invariance of the descriptors even to blurring, similarly as in the second described experiment. Their registration of remote sensing data uses the feature points from the reference image detected by thresholding the gradient magnitude of the reference image and by applying the improved version of the Harris corner detector [131]. Then the corresponding points in the sensed image are found using the minimum distance of the invariant description, computed from the reference point and tested candidates from the sensed image. They propose the version when even the subpixel accuracy can be achieved.

There is a large group of a methods based on Zernike or pseudo-Zernike moments. Yang and Guo [132] described the registration method based on the Harris corner feature points [131] and pseudo-Zernike moments expressing the neighborhoods centered on feature points. The matching is then performed using RANSAC [133]. Badra et al. [134] utilized the rotational properties of the ZM's. Using the phase change of ZMs and the Fourier shift theorem, they estimate the rotation plus scaling parameters and translation parameters, respectively, in the iterative manner until the final registration is achieved. They present panoramic mosaics that were automatically generated using their approach. They claim that their algorithm can easily handle images with little overlap and scale differences. A family of Mexican-hat wavelets in combination with ZMs were employed in Yasein and Agathoklis [135] and Sarvaiya et al. [136]. Here, the ZMs are computed around the distinctive points found using the scale-interaction of Mexican-hat wavelet decomposition of the registered and the sensed images. In [135] the matching is done by estimating the correlation coefficients of the ZMs' vectors, so opposite to the previous approach when the invariance of the ZMs has been used and not their ability to capture the parameters of the rotation and the scale. In [136], an iterative weighted least squares algorithm is implemented to estimate the transformation parameters. In [137] the authors proposed to use curvature scale space (CSS) to detect distinctive corners, in whose neighborhood the pseudo-Zernike moments are evaluated. The obtained correspondences are then processed by means of RANSAC in order to compute the transformation parameters mapping the sensed image to the reference one. In [138], the normalized regions are represented by a Zernike moment phase-based descriptor. The descriptor is mainly based on the phase components of the feature vector, while the magnitude components are used only as the weighting factors. They use it for an accurate estimation of the rotation angle between two matching regions, too. Finally, the authors included comparison of the proposed ZM phase descriptor and five popular descriptors such as SIFT, steerable filters, or complex moments. Similarly in [139] an approach exploiting the phase information of Zernike moments is proposed in order to estimate the object orientation in underwater sonar images. Zernike moments were used in [140] for optical flow estimation used in the image registration. They defined driving points of the optical flow by means of the Canny edge detector and evaluated Zernike descriptors for them in the multi-scale manner. Zernike moments are employed in an accurate coregistration of the neighborhood frames in noisy video sequences [141].

Other variations of the registration methods are using the ability of low-order moments to capture the orientation and the shape of the objects. Sato and Cipolla [142] computed directly, without correspondence estimation, the parameters of the present geometric deformations (an affine transformation was expected) using the circular moments of distribution of the line features orientation. They combined moments and the scale-space representation of the images. Brivio et al. [143] modeled shadow structures in mountain images by means of their inertia ellipses. The ellipses are here described by their area, inclination of the main axis, and ellipticity. All these attributes are functions of moments. The very matching is realized using multivalue logical tree. Wide baseline image mosaicing [144] based on Maximum stable extremal regions (MSER) and Hessian-Affine detectors is supported by an adjustment of the detected irregular regions to the ellipses which have the same first and second moments as the originally detected regions. In [145] the low-order moments were used to estimate the center of gravity and extremal points of the ellipses. They are further used as control points for the registration of object contours. The applicability of the method is demonstrated on the images of the cement particles from the diffraction topography beam. Time efficiency of the closest iterative point (ICP) algorithm is increased in [146] by the preliminary evaluation of the low-order moments in order to get the initial translation parameters. The rotation angles acquired among others by the second-order central moments are referred to as the initial rotation parameters of the ICP algorithm for image registration. The method was applied on medical data images. In [147] the authors address the problem of the normalization position for general shapes. They introduced so-called Principal moments, defined on the set of N points with reformulated complex coordinates. The Principal moments can be used for estimating the object unique orientation.

There is a large group of registration methods based on pre-normalization of the detected affine invariant distinctive regions, which is realized by the matrix of second-order moments. Then, the matching follows based on the rotationally invariant features. More on this topic can be found in Section 9.2.

There exist applications of the moments even for the registration of 3D data. Jaklič and Solina [148] applied the moments for the 3D registration, in this case the range images were processed. They showed that objects that are compositions of super-ellipsoids can be computed as simple sums of moments of individual parts and that the 3D geometric moment of a globally deformed super-ellipsoid in a general position and orientation can be computed as a linear combination of 3D geometric moments of the corresponding non-deformed super-ellipsoid in the canonical coordinate system. They use a standard technique to find centers of gravity and principal axes in the pairs of range images while the third-order moments are used to resolve the four-way ambiguity. The topic of the registration in R^m by means of the discrete polynomial moments is addressed in [149]. Shen and Davatzikos [150] were solving the 3D registration of brain images, expecting the elastic model of present geometric difference. An attribute vector, formed by a set of geometric moment invariants defined on each voxel is used for description of the tissue maps to reflect the underlying anatomy at different scales. The proposed method, the hierarchical attribute matching mechanism for elastic registration, captures the image similarities using their moment representation, contrary to other volumetric deformation methods, which are typically based on directly maximizing the image similarities. The 3D object alignment system which can be applied to detect objects and to estimate their 3D pose from a depth image containing cluttered background is described in [151]. Here, the relationship of the first-order and second-order Fourier moments of the images is used for the least-squares estimation of the 3D affine parameters. The authors addressed the problem of occluded scenes. They propose a sliding window approach to partition the template and the target images according to their depth maps and perform the matching of the Fourier moments for all partitioned sub-templates. Finally, there is the public Point Cloud Library (PCL) [152] supported by an international community of robotics and perception researchers, which aims at providing support for all the common 3D building blocks that applications need. In addition to many other functionalities, it also contains 3D moment invariants.

Landmark-based blur-invariant registration methods appeared right after the first papers on blur invariants [153, 154] had been published. The basic algorithm is as follows. Significant corners and other dominant points are detected in both frames and considered to be the control point—landmark candidates. To establish the correspondence between them, a vector of blur invariants is computed for each candidate point over its neighborhood, and then the candidates are matched in the space of the invariants. Various authors have employed this general scenario in several modifications, differing from one another by the particular invariants used as the features. The invariants based on geometric moments can only be used for registration of blurred and mutually shifted images [154–156] and for rough registration if a small rotation is present [130, 157, 158, 126, 159]. Nevertheless, they found successful applications in matching and registration of satellite and aerial images [130, 160, 126, 159] and in medical imaging [155, 157, 158]. The discovery of combined blur and rotation invariants from complex moments [161] and Zernike moments [162] led to registration algorithms that are able to handle blurred, shifted, and rotated images, as was demonstrated on satellite images [163, 164], indoor images [165, 166], and outdoor scenes [162, 167]. Zuo [168] even combined moment blur invariants and SIFT features [169] into a single vector with weighted components but without a convincing improvement. In all cases the methods register blurred images directly without necessity of any deblurring. This is their big advantage.

The moments have an unusual role in the image registration in [170]. The registration of the infrared and visual images is proposed here, and possible illumination variations are diminished by means of the so-called moment images. The moment image is defined pixelwise, when for each pixel its value is evaluated as the ratio of two locally computed geometrical moments, with chosen (different) orders. They are then used for the registration process itself.

As can be seen from the above discussion, there are many variations of the landmarks-based image registration scheme using moments. Either they use only low-order moments for the estimation of the main object orientations, or they describe invariantly selected distinctive parts of the scene, to provide input to the matching algorithm and thus to help to estimate the parameters of the present geometric deformation. In all cases, it is extremely important to compute the moments from the regions in the images, which correspond to the same part of the scene.

Before the overview of the landmark-free methods starts, the two illustrative examples of the landmark-based image registration will be described.

The first example sets as its aim to register two satellite images of the same area (Czech Republic, north-west of Prague) taken at different times and using data from different sensors—the first image was acquired by the Landsat 5 satellite while the second image was taken by the French satellite SPOT. They have different spatial as well as spectral resolution—The Landsat image has a spatial resolution of 30 m and 7 spectral bands, while the SPOT image has a resolution of 20 m and 3 spectral bands. In the preprocessing phase the three most important components were computed from the Landsat data set using principal component analysis (PCA) (Fig. 9.3a); the SPOT channel data were used directly (Fig. 9.3b).

The moments were not computed directly from original images but from fifteen segmented salient regions obtained by region growing. The too small and too large regions, which are not favorable for region matching were rejected. Sufficient overlap of these two object sets was achieved, which is vital for the following correspondence estimation. It is important to be aware of the time distance between the reference and the sensed images (the images were acquired in different years and even in different seasons), which causes variations in the region intensities and shapes, and thus some imperfections in the area detection and segmentation may appear.

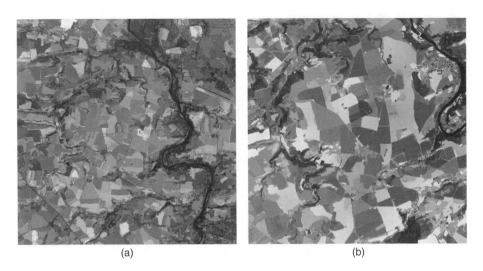

(a)   (b)

Figure 9.3   Image registration: Satellite example. (a) Landsat image – synthesis of its three principal components, (b) SPOT image – synthesis of its three spectral bands

For these reasons the regions were binarized after the segmentation, so contrast normalization was not necessary.

For each detected region from both images, the normalized moments τ_{pq} up to the eighth order were computed, and their counterparts were searched by the minimum distance classifier. The three most forceable pairs of the regions from the reference and sensed images were found. Having the three matched regions, the coordinates of their centroids were used for the computation of the preliminary affine transformation parameters. After this initial alignment, additional matching pairs were searched by minimum distance with thresholding in the spatial domain. Six other region pairs were matched. Finally, the centroids of matching regions were used as control points for the final image registration. The result can be seen on Figure 9.4. The three most forceable regions have numbers in a circle, the other matching regions are just numbered. The final result of the image registration can be seen on Figure 9.5, where the Landsat and the registered SPOT images are superimposed. The normalized moments proved to be robust enough with respect to the geometric deformation present on the satellite data.

In the second described experiment, the application of the moment invariants for image fusion is shown. The aim of the image fusion is to combine several images of the scene of low quality (blurred and noisy images of low resolution) and produce one output image, which has better resolution with the blur and noise removed or diminished (see an example of input images in Fig. 9.7 and the corresponding result in Fig. 9.10b). This can be achieved due to the fact that each input image is degraded in a slightly different way and the image fusion combines them in such way that the qualities are emphasized and the degradations are removed. One condition has to be satisfied before the image fusion can start—the input images have to be registered so that the same objects on the scene are geometrically aligned across the images. Then the fusion analyzes different versions of the same object, captured on the individual input images (see Fig. 9.6 for the image fusion flowchart). As stated previously, a registration

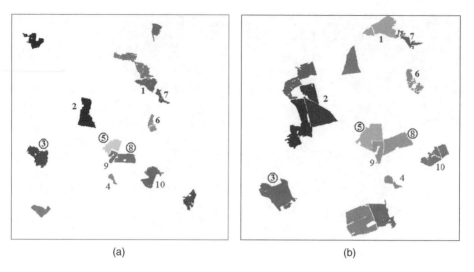

(a)   (b)

Figure 9.4   Image registration: Satellite example. (a) Landsat, (b) SPOT image – segmented regions. The regions with a counterpart in the other image are numbered, the matched ones have numbers in a circle

Figure 9.5   Image registration: Satellite example. The superimposed Landsat and the registered SPOT images

method able to handle blurred and noisy data is required here that is not fulfilled by the most traditional registration techniques. The registration algorithms often expect the noise and blur removal ahead of their main process loop. However, the feature-based method using combined blur invariants can handle such situation thanks to their simultaneous invariance with respect to blur and geometric transformation.

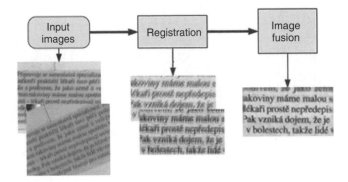

Figure 9.6   Image registration: Image fusion example. Image fusion flowchart

Figure 9.7   Image registration: Image fusion example. Examples of several low quality images of the same scene – they are blurred, noisy, and of low resolution

The proposed registration algorithm first detects feature points in all input images by means of the feature detector [171] (see Fig. 9.8), robust with respect to the geometric and as well to the radiometric degradations—blur. Then, the combined blur invariants are computed for the circular neighborhood of each detected distinctive point. The geometric parameters of necessary alignments among individual input images are found using the same correspondence estimation method as in the previous experiment. After the image registration (see Fig. 9.9) is processed, the image fusion via multichannel deconvolution (for more details see [172]) can be computed, and the resulting image of higher quality is created (Fig. 9.10a). None of the input images shows comparable quality in terms of edge sharpness, noise level, and detail visibility (see Fig. 9.10b).

9.3.2   Landmark-free registration methods

A straightforward idea of a moment-based global registration uses the *moment matching* approach. Having the reference and deformed images and knowing the assumed transformation model, we can deduce how the image moments have been transformed under the

Figure 9.8   Image registration: Image fusion example. Low-quality images of the same scene with detected distinctive points

Figure 9.9   Image registration: Image fusion example. Low-quality images after the registration

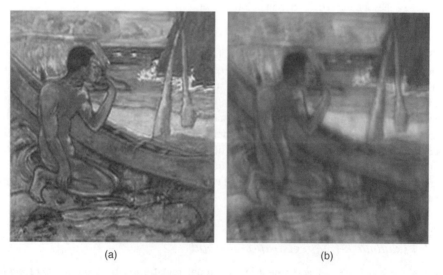

(a)   (b)

Figure 9.10   Image registration: Image fusion example. (a) The output of the image fusion. The resulting image has higher resolution and the blur and noise have been removed or diminished. (b) The scaled version of an input image for comparison. The image does not show comparable quality in terms of edge sharpness, noise level, and details visibility

deformation. This relation depends on the unknown transformation parameters. We search the parametric space for the minimum error (usually in the mean square sense) between the reference and deformed moments. The parameter values, at which this minimum occurs, are then considered as the registration parameters. A serious drawback of this approach lies in the global character of the moments. If the images do not have a 100% overlap, their moments do not match each other, and the method cannot produce correct alignment. This method can be applied only if complete visibility is guaranteed, typically for registering simple shapes or point sets of a small size, extracted from the background. Affine registration by this technique was described, for instance, in [173] and [174]. Similarly, Liu and Ribeiro [175] addressed a more general polynomial model of the mapping function. Elastic registration by moment matching, where the transformation model was supposed to be approximately polynomial, was introduced in [176] and later generalized in [177] where the authors also consider features other than plain moments. In general, any moment-based normalization (such as those introduced in Chapters 3–5) can be viewed as a particular case of moment-matching alignment, where the reference image is the normalized position.

A different group of landmark-free registration methods based on image moments is aimed at registering blurred images. Most of them are variants and modifications of the famous *phase correlation* [178] which has been proven very powerful in registering shifted, rotated, and scaled images of the same modality or of different but similar modalities. Phase correlation is implemented in the Fourier domain, which makes it very fast comparing to traditional image-domain cross-correlation. It calculates a correlation between whitened phase-only images and looks for the correlation peak. The whitening removes intensity-dependent/color-dependent information and works basically with the edges only, which makes it robust to any intensity transformations that preserve edges. However, the phase correlation is not robust to image blurring. If one of the images has been blurred, the correlation matrix is smoothed by the same kernel. When working with whitened images, this leads to a flat peak or even to multiple false peaks, depending on the shape of the blur.

The first attempt to make the phase correlation insensitive to blurring was described by Ojansivu et al. [179, 180]. The authors assumed a centrosymmeric blur, which only changes the image phase by π, and correlated the double of the phase instead of the pure phase itself. This fact actually makes the method robust to blur, as demonstrated in [180] for the case of a linear motion blur and in [179] for some other blurs. In the latter paper, an extension of the method to rotation and scaling was also proposed. Their method was used for example by Tang [181] for registering X-ray images.

Pedone et al. proposed a blur-invariant phase correlation for N-fold symmetric blur [182] and later also for N-fold dihedral blur [183]. The core idea of their methods is to perform the phase correlation of the primordial images, which is implemented in the Fourier domain as a phase correlation of the blur invariants $I(f)$ and $I(g)$ (see Theorem 6.1). Their method does not produce a single correlation peak but a pattern of N peaks ($2N$ in the case of a dihedral blur) distributed along a circle, the center of which determines the between-image shift. This method is particularly suitable for registering consecutive out-of-focus snapshots for a consequent multichannel blind deconvolution, and significantly outperforms the traditional phase correlation as demonstrated by extensive experiments in [182].

There exists also a moment-matching method designed specifically for blurred images. Zhang et al. [184, 185] proposed to estimate the registration parameters by bringing the input images into the same normalized form. Since blur-invariant moments were used to define the normalization constraints, neither the type nor the level of the blur influences the parameter estimation.

9.4   Robot and autonomous vehicle navigation and visual servoing

One of the key issues of robotics is providing a robot/autonomous vehicle (they will be collectively called robots in the following text) with information about the outer world and about position-related tasks, so they can act autonomously without human help. In visual servoing, robot tasks are defined visually and not based on the known coordinates, so the ability to understand the scene is crucial. Here, the vision data directly controls the motion of a robot. This approach is often used for robot arms. The next relevant research area covered in this section is the human-robot communication, how instructions can be passed to the robot in other ways than through the program or some input device. In general, significant parts of the robot navigation tasks are related to the scene understanding and thus object recognition, where moment invariants represent one of possible solutions.

There are several options how to tackle the robot's notion about the surrounding world. If the robot's working space is complicated or is not static, then a *dynamic approach* can be the best solution. The information about the robot trajectory and the previous world representation is used for the estimation of the current robot position. The robot can construct its own world map based on the data acquired by its sensors. After an incorporation of the robot motion trajectory and the sensor data into the evolving world representation, the robot is able to make a decision about its current position and the next action to take.

However, in some tasks, the *landmark-based approach* is more appropriate. Here, the map of the robot's world can be assumed as a priori known (either learned by the robot in the preprocessing phase, or directly given to the robot). The robot makes the decision about its current position and next action by matching the new collected sensor data with the stored reference world representation. So called *landmarks* – objects in the scenes which are found distinctive—are often used for the robot position estimation. Using such an approach, no representation of the surrounding world is necessary, so complex and often memory-demanding descriptions can be avoided. Moreover, landmarks can be used to impart certain types of information to the robot. For example, recovery from failure during robot navigation can be based on landmark localization. The landmarks can be either some natural parts of the world like doors, straight lines, corners, and so on, or specially designed artificial signs, placed into the robot's environment. In both cases the landmarks are detected by the robot's visual sensors during its movement, and compared against the stored database. If the artificial marks are used, we are not limited by the natural layout of objects. Moments and moment invariants are utilized for the landmark classification. The question to be asked is which moment invariants should be used; in other words, what kind of geometric deformation we can expect. The robot equipped with the common type of camera and moving in the usual scene introduces to the landmark acquisition process the projective deformation and an additive noise. The good choices for such situations are the affine moment invariants, which were proven to be robust under this deformation. An applicability of such approach can be demonstrated on the following examples.

The first experiment is demonstrating a view-invariant recognition of circular landmarks for mobile robot navigation based on AMIs. The robustness of the system with respect to an additive random noise and to various viewing angles was tested, and the discriminability and stability of the recognition model was proven in real situations. The proposed mark set consists of patterns formed by two concentric circles with equal outer and different inner radii, allowing easy localization of the marks in the complex environment. The information which has to be imparted to the robot can be encoded into the ratio of the mark's inner and outer radii. The proposed recognition model computes for a given mark image three non-zero AMIs (radial symmetry limitation) and finds the most similar mark from the database using the minimum distance similarity method. Although the AMIs are invariant only under the affine transformation, which is an approximation of the perspective transformation occurring in the robot vision system, their recognition ability was proven to be high enough for the described experiment. The stability of AMIs under a random zero-mean noise present in the acquisition process of mark images leads to good results, too. Even high values of STD had a rather small influence on the recognition results of AMI's.

In the second example, the robustness of the AMIs to even more complex geometric deformation than the perspective transformation is demonstrated. Here, the recognition of artificial landmarks using the AMIs is investigated under the condition that the robot is equipped with a fish-eye lens camera. Such a camera provides more information than the one with a conventional lens. Especially, the robot navigation in indoor environments can be improved using the fish-eye lens because of its wide angle of view (approximately 180° in the diagonal direction) and the ability to obtain information even from very close robot neighborhood [186]. However, such a camera introduces complex geometric deformation of the acquired images (see Fig. 9.11). The landmark set consists of four shapes with different properties of curvature, number of sides, holes, and so on which makes them easy to differentiate. As already discussed, the AMIs are robust under the perspective transformation if the distance between the camera and the object is significantly larger than the size of the object. We assumed

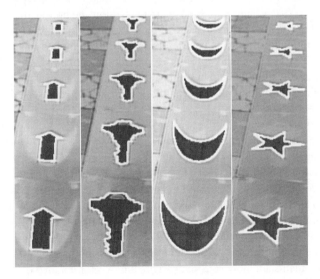

Figure 9.11 Robot navigation: The examples of navigation marks with detected boundaries. Introduced complex geometric deformations introduced by the fish-eye lens camera are apparent

that under similar conditions the AMIs recognition model can be robust enough under the fish-eye lens deformation, too. Experiments confirmed expectations that AMIs can be robust enough even under the perspective and fish-eye lens degradations. All landmarks were classified correctly using the minimum distance classifier even when the introduced degradation is perceptible. Although the AMIs are invariant theoretically only under the affine transformation and we worked with images transformed by much more complex geometric deformations, AMIs recognition ability proved to be high enough even in such cases.

Addressing the general applicability of moment invariants in robot navigation tasks, one of the first papers (Courtney and Jain [187]) proposed using natural landmarks. They fused integrated sonar, vision, and infrared input data for sensor-based spatial representation of the scene. The approximate robot position is determined by comparison of the features extracted from the recently acquired images to the world representation, prepared during the pre-processing phase, when the robot creates grid maps of the distinctive regions that it encounters in its workspace. They compute the area and seven Hu invariants for each modality and each analyzed structure and apply fusion of the multisensory grid maps. The experiment was held in ten different rooms and ten different doorways, and it revealed a 94% recognition rate of the rooms and a 98% recognition rate of the doorways. The navigation of the household service robot by means of the target-based detection is proposed in [188]. Hu moment invariants are used for the target description. The authors claim sufficient robustness with respect to the light conditions and the camera perspective.

Xu et al. [189] proposed to use AMIs for unmanned aerial vehicles (UAVs) navigation—automatic landing on the ship's deck. They use the infrared radiation images of the scene with inserted targets (the T- shaped cooperative objects put on the runway). The infrared radiation images feature high temperature differences between the target and background, which help to precisely segment the target object. After segmentation the four simplest AMIs are computed and used for target identification. The achieved recognition rate was 97.2% even when the present geometric deformation was perspective and not just an affine one. In [190] the authors aimed their navigation method at the small unmanned rotorcrafts, namely on the precision increase of their attitude estimation. The method consists of SVM and Hu moment invariants. The descriptors for the template matching in [191] are exploiting pseudo-Zernike moments in order to provide predefined single-object tracking in an aerial surveillance method. The method proved to be efficient even in the case of noisy data with high illumination changes. The authors proposed to represent the image data in Gabor wavelets coefficient domain.

The natural landmarks are used in the work of Lee and Ko [192], too. They do not insert any new signs into the robot's surrounding; they work directly with distinctive parts of the scene, where the robot should be navigated. They call the method gradient-based local affine invariant feature extraction. The places where the invariants should be computed were identified using the Harris corner detector [131], and then invariants were evaluated in the close neighborhood based on the regular grid of squared windows. Beforehand, the acquired images were transformed from the direct intensity domain into the normalized gradient magnitudes of the image intensities to decrease dependability on the image information change. They tested the lengths of the feature vector from 4 to 10 and proved the assumption that the invariants based on higher-order moments are more sensitive. The feature vector length 6 outperformed others.

The variant solution of the recognition system for robot navigation, based on Zernike moments is described in [193]. Zhe et al. make use of the local distinctive structures for robot navigation. They extract highly robust and repeatable features representing unique structures with the strongest response in both the spatial and scale domains. Instead of the geometric moment invariants, they proposed to compute rotationally invariant weighted Zernike moments on the normalized image patches. They use parts of the scene as the landmarks. The experiment showed good system performance on object recognition as well as on indoor topological navigation, even in the case of severe occlusions and scale changes.

Wells and Torras [194] proposed to use simple functions of geometric moments up to the third order to characterize statistical variations in the object's projection in the acquired image when the camera position is changed. They choose such functions which can express individual geometric deformations, but they are not invariant with respect to them. The very features were computed on binary images corresponding to the thresholded versions of the acquired data (three different thresholds were applied). Feedforward neural networks learned the mapping between the feature variations and the pose displacements, when the robot moved. In this experiment, the geometric moments outperformed the eigenfeatures (Karhunen-Loeve features) and pose-image covariance vectors, even though they use the 2D geometric moments (not the moment invariants) for the description of the projections of 3D movements. They improved the performance of the method by the feature subset selection using mutual information.

Celaya et al. [195] used the geometric moments to normalize detected regions with respect to the patch of the given size to make its description invariant to changes in scale and moderate perspective deformations, so that the recognition algorithm does not need to take these deformations into account. The geometric central moments up to the second order are computed, and the equivalent ellipse axes are found providing the rotation angle to align the region and the scale factors for the x and y dimensions, respectively.

In [196] the authors proposed the solution for a manipulator to correctly grab an object and apply required actions. In order to diminish the effect of the ambient lighting on system performance, Pinto et al. proposed to use a laser range finder coupled in the manipulator. The manipulator by means of exploratory movements creates a representative 2D image of the scene, and using its invariant representation it can classify surrounding objects and thus is able to select appropriate actions. The proposed descriptors are Hu moment invariants.

An intuitive interface for human-robot communication is critical for the development of robots which can serve. Gesture-based interface is one of the possibilities. It can provide reasonable flexibility of interaction with robots. It is in a sense related to the topic of sign language recognition [103]. In [197] the authors proposed a system for manipulating service robots. The recognition is done by means of SVM, working with Hu moment invariants evaluated on the segment image data. They tested the algorithms under various conditions such as complex background, changing lights, and image noise. Several other applications aiming at human-robot communications are described in Section 9.2.7.

Visual servoing (vision-based robot control) [198] makes use of the robot vision system to get feedback for estimating the next steps of the robot to achieve the required aim. The control techniques can be either image-based, when we work directly with image data on which the task function is based, or position-based, when the error function is related to the robot's position in the Cartesian space, which therefore has to be explicitly evaluated. In the image-based approach, which is relevant to the moment descriptors, the task function has been defined using pixel coordinates of some image features such as interest points, and the control law maps the evaluated task onto robot motion directly. In the more recent method, the task function can be based not on the pixel positions themselves but on some derived features such as pixel distances, line orientations, and so on. Chaumette [199] proposed to use the geometrical moments of segmented images for evaluation of the robot progress. They proposed six combinations of moments able to control the six degrees of freedom (DOF) of the system. So called *photometric* moments were proposed to use in [200], which enable us to leave out the segmentation step in the task formulation, because they work directly with image intensities. An application of moments in visual servoing in medical imaging is described in [201], where B-mode ultrasound images are used as the input for the control of the motion of a 2D ultrasound probe held by a medical robot. Considering the industrial application, the alignment of the micropegs and holes is solved by means of the moment-based task function in [202].

To conclude the robot navigation research, applications of moments and moment invariants make use of their recognition abilities even under perspective transformation and stability with respect to an additive random noise. The moments are used here for both the image invariant representation and the image normalization.

9.5   Focus and image quality measure

An interesting application of moments is their use for focus/defocus measuring. In the real world, an image acquisition process faces blurring degradation from various sources such as atmospheric blurring, out of focus blurring, or motion blurring, to name the most common examples. The presence of some level of blur in images is inevitable; however, its amount and its strength can be different. Having a relative focus measure (answering the question which image is the least blurred) is important for ordering an image sequence according to the amount of blur. This is a common task for example in astronomy, where long sequences of images (typically several hundreds or even thousands) of the same object are taken by ground telescopes. The images are blurred by atmospheric turbulence; the amount of the blur may quickly vary as the atmospheric conditions change. Our task is to select several best images that are further used (see Fig. 9.12).[1]

Moreover, the performance of image processing algorithms can be influenced by proper setting of parameters, describing the amount of present blurring, or certain similar characteristics. All these observations are the reason why it is important to know how to evaluate the degree of blur degradation in the acquired images, in both absolute (answering the question how much is the image blurred) and relative manners.

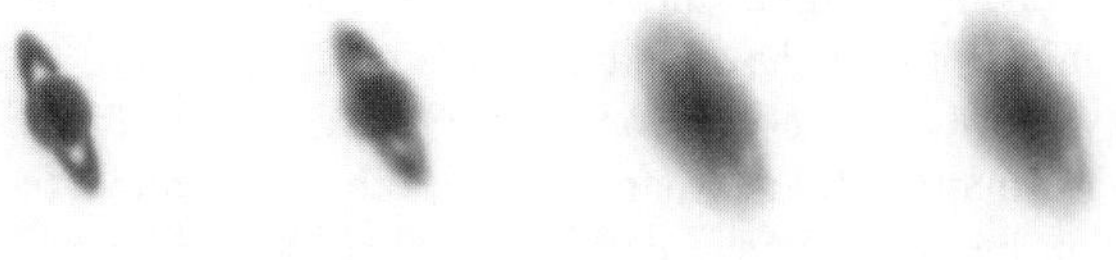

Figure 9.12   Focus measure: Focus measurement by moments. Four sample frames from the Saturn observation sequence, ordered automatically from the sharpest to the most blurred one. The result matches the visual assessment

[1] In astronomy, such selecting of a few good images from a large set of observations is called "lucky imaging".

Various focus measures have been reported in the literature. Most of them try to emphasize high frequencies of the image and measure their quantity. The main idea behind this approach is that the blurring suppresses high frequencies regardless of the particular point spread function. The most popular approaches are based on the variance of image gray-levels or on the image derivatives (ℓ_1 norm and ℓ_2 norm of the image gradient, ℓ_1 norm of the second derivatives, and the energy of image Laplacian) [203].

Moments have been used for the focus measure design in several ways. Zhang et al. [204] proposed a measure based on the second μ_{20} or on the fourth order geometric central moments of an image sequence and on their linear combinations for the measure behavior on the image sequence from Fig. 9.12). They are exploiting the fact that these moments decrease monotonically as the image blur level decreases and reach a minimum when the image is exactly focused. They are even able to express the blur property of the imaging system due to the independence of the method on the imaged objects. The algorithm also returns an explicit expression of the parameter of the point spread function, modeling the present blurring. However, the functionality of the proposed method is violated when the actual blurring is not caused exactly by the convolution. One of the common modification of the model is the cropping on the image boundaries. Wee et al. [205] introduced their full-reference measure (defined with respect to the reference image) based on the symmetric geometric moments (SGM). They divide the tested image into several blocks and for each of them compute all SGMs up to the fourth order. Then they evaluate the score of the image as the average of the quality measures of each block, computed using SGMs of the tested and the reference images. The visual quality assessment method based on the geometric central moments computed on the edge projections of the image is presented in [206]. The edge map of the image is separately projected along horizontal/vertical directions, twenty corresponding central moments are used as features.

Yap and Raveendran [207] based their measure on Chebyshev moments. They define the measure as the ratio of the norm of the high-order moments to that of the low-order ones. In this way, they characterize the low and high spatial-frequency components of an image, respectively. The focus measure is monotonic and unimodal with respect to image blurring, which is important. Additionally, it is invariant to contrast changes due to the differences in the intensities of illumination and robust under Gaussian and salt-and-pepper noise. The paper [208] presents a no-reference (blind) blur evaluation algorithm based on discrete Chebyshev moments. The Chebyshev moments and their energy are calculated locally on the gradient image. The proposed blur score is defined as the variance-normalized moment energy. Thung et al. [209] introduced the content-based image quality metric (a more general task than the focus measure, but strongly related) based on locally applied Chebyshev transformation. The image quality metric is defined as the similarity of the moment vectors evaluated on the test and the reference image blocks combined with the result of the block classification. The content of each reference image block is classified as "plain", "edge", or "texture" patch, according to the energy level and distribution. The overall image quality is then evaluated as an average of each block type. Blockwise evaluated Chebyshev moments were used for an estimation of the blockiness [210], too. It was observed that blocking artifacts have direct impact on the distribution of discrete Chebyshev moments.

To conclude the topic of focus measure and moments, we have to admit that even though the mentioned publications show various ways in which moments can help to estimate the blur level, the idea of using moments for focus measure design is quite minor.

9.6   Image retrieval

Nowadays, falling prices of all kinds of image sensors and widespread of the end user handheld cameras produce a growing stream of images resulting in data organization difficulties especially with respect to the efficient data searching. One approach is to tag the acquired images or introduce some other type of metadata and then ask text-based queries. This methodology is laborious and time consuming, and it can be easily biased by the human factor and an operator's understanding of the image content. Moreover, many large-scale databases do not contain any annotations. The variant approach is based directly on the image data. Since the appearance of the first image databases in the 1980s, image retrieval has been the aim of intensive research. And even now content-based image retrieval (CBIR) methods are among the most important challenges in computer vision. The search engine looks for images which are the most "similar" (in a pre-defined metric) to the given query image; or the query image can be expressed as a collection of basic colored shapes at the desired positions in the scene, representing the rough sketch of the image in question. The image-based solution for a database search has a big advantage over the first tag-based one; there is no loss of information due to the pipeline *image - text tag - image*. Moreover, one image can tell more than a thousand words. The weak point of the image-based search is how to algorithmically express the human notion of the similarity between two images. This is the key issue of the algorithms, belonging to the category of content-based image retrieval (CBIR) applications [211]. Here, each image is described by a set of features (often hierarchical or highly compressive ones), which should reflect the image content characteristics the user prefers—colors, textures, dominant object shapes, and so on. The between-image similarity is then measured by a proper (pseudo)metric in the corresponding feature space. It is important to note that CBIR is a subjective task because there is no "objective" similarity measure between the images. Hence, many CBIR systems aim to retrieve images which are perceived as the most similar to the query image for a majority of users and the users feel this similarity at the first sight without a detailed exploration of the image content.

A significant part of the CBIR methods use moments and moment invariants. They are exploited for object description, as described previously. A small illustration of possible difficulties with the description of image similarities can be seen in Figure 9.13. On the one hand,

(a)   (b)

Figure 9.13   Image retrieval: (a) the mountain (Malá Fatra, Slovakia), (b) the city (Rotterdam, Netherlands). The images are similar in terms of simple characteristics such as color distribution, but they have very different content

they are similar: in the bottom part of both images, there are some trees, in the middle part, there are some solid non-living objects, and in the top part, there is sky. There is agreement (in terms of the correlation) of 94% of the pixel values. On the other hand, the images are completely different: mountain versus city. From the point of view of CBIR, such images would usually be considered as non-similar.

There are several papers comparing efficiency of individual kinds of descriptors utilized for the CBIR systems, depending on the type of data to be searched, on the expected image degradations, or on the author's preferences. In [212] the authors realized and described thorough comparison of the scale, rotation, and translation invariant descriptors for the shape representation and retrieval. Specifically, the authors studied Fourier descriptors, curvature scale space, angular radial transformation, and affine moment invariants. The results showed that the moment descriptors outperformed others in terms of the shape representation quality. Di Ruberto and Morgera [213] compared applicability of other group of moment based features—Hu, Flusser, and Taubin invariants, Legendre and Zernike moments, and generic Fourier descriptors. The experiments were run on the database of binary objects. Similarly, Yadav et al. [214] analyzed the applicability of Fourier descriptors, Legendre moment descriptors, and wavelet Zernike moment descriptors as descriptors for CBIR. The Euclidean distance was evaluated as a similarity measure. The experiments proved that the Zernike-based descriptors outperform other techniques.

The well-known QBIC (IBM) [215] image-retrieval system employs geometric moments up to the fifth order for the description of the object shape variances. Srivastava et al. [216] normalize all images to their graylevel versions sized 256×256 and divide them into 16 blocks 64×64. Then the geometric moments of all blocks are computed. The moments are compared by the Euclidean distance.

Various classes of moment-based descriptors were included in the proposed CBIR algorithms. Banerjee et al. [217] designed a CBIR method based on the normalized geometric moments up to the second order evaluated around visually significant point features. The distinctive point clusters are extracted using a fuzzy set theoretic approach. The moments are computed from the detected regions of interest as well as from the whole image in a normalized RGB model. The authors tried to select an optimal set of features suitable for retrieving a set of images perceptually relevant to the query image. It is shown that a single set of features may not be suitable for all types of queries. The CBIR system based on Legendre moments in combination with SVM is presented in [218]. Krawtchouk moment invariants were chosen as the base CBIR element in [219]. The authors used non-subsampled shearlet transformed image data, represented by the parameters of the distribution of their coefficients and low-order Krawtchouk descriptors capturing the shape. The combination of Gabor filters and Chebyshev moments compared by Euclidian distance was chosen in [220] to form the CBIR system aimed at medical imaging data.

Wei et al. [221] decided to use Zernike moments. They proposed the system for trademark image retrieval in the trademark registration system. They detect edges and realize the shape normalization. Then global and local features are extracted to capture the main shape variations and details, respectively. The curvature and the distance to the object centroid are used as local features, whereas first fourth-order ZMs are employed to capture the global differences. ZMs can be found in [222], too. They investigate texture and shape features for CBIR by combining Gabor filters for the texture extraction and ZMs for the shape description. The combination of the proposed descriptors outperformed their individual application. The pseudo-Zernike moments are employed in [223]. The authors make use of color, texture, and shape information. They applied preliminary search based on the fast color quantization algorithm. Then steerable filter decomposition for the spatial texture features and the pseudo-Zernike moments for the shape description are evaluated. Zernike moments, pseudo-Zernike moments, and orthogonal Fourier-Mellin moments were incorporated into the CBIR system and mutually compared in [224]. The distinctness of this approach is in the moment computation domain. They were evaluated on the edge images. The experiments were performed on various databases to examine the performance of the system in the presence of noise, partial occlusion, distortions, complex structures, and so on.

Tuytelaars and van Gool [225] made use of locally defined AMIs [226], computed on the affinely invariant regions to be able to deal with large changes in viewpoint. The system is robust to occlusions and changes in the background as well as to illumination changes. Karakasis et al. [227] search all salient areas in each image and then compute AMIs $I_1, I_2, I_3, I_6, I_7, I_8, I_9$ and I_{48} from Chapter 5 of each color channel. They combine them pairwise

$$I'_k = \frac{I^R_{p(k)}}{I^B_{p(9-k)}} \cdot \frac{I^G_{p(9-k)}}{I^B_{p(k)}}, \tag{9.1}$$

where $\mathbf{p} = \{1,2,3,6,7,8,9,48\}$ is set of indices, $k = 1,2, \ldots 8$. As a variant, they compute I'_k not directly from RGB, but from chromaticities $C_\ell = \frac{\ell}{R+G+B}$, where $\ell = R,G,B$, respectively. Another variant is an unconventional magnitude normalization $I''_k = \text{sign}(I'_k) \cdot \sqrt[8]{|I'_k|} \cdot e^{\sqrt[8]{|I'_k|}}$, where I'_k are computed from the chromaticities here.

A large group of methods is based on Hu moment invariants. The normalized Hu moment invariants scaled to the same range were applied in [228] together with a histogram of edge directions. In the refinement phase, the edge-deformable models were employed to solve the ambiguities. The proposed system was tested on the trademark image database. Color histograms for color similarities, Gabor filters for the texture-related appearance, and, finally, Hu moment invariants for the shape characteristic of the image are incorporated into the CBIR system for biometric security [229]. Shao and Brady [230] make use of Hu moment invariants, five higher-order invariants from reference [231], and generalized color moment invariants [226]. They applied the salient regions detection followed by their invariant description for CBIR. The detected regions have very high repeatability under various viewpoint and illumination changes. They are detected according to the local entropy and the scale selection. Each region is then described by chosen descriptors. The minimum distance classifier is used for the similarity estimation. The method proved to be effective at retrieving a variety of cluttered images with partial occlusion. The application of the CBIR approach based on Hu moment invariants is presented in [232]. The authors design the system for tree leaves retrieval. The selected feature set consists of the basic geometric descriptors such as eccentricity, aspect ratio, or elongation. The more informative features are Hu invariants, which proved their efficiency during the experiments. The Euclidean distance is used as the similarity measure in k-NN classifier with $k = 10$. Similarly, the tree leaf CBIR was introduced by Suk and Novotný in [233]. They provide the web application for recognition of woody species in Central Europe (see Fig. 9.14). The contours of segmented leaves are described by means of discrete Chebyshev moments normalised to rotation and Fourier descriptors normalised to translation, rotation, scaling, and starting point change, and the classification is done using the k-NN classifier.

Figure 9.14   Image retrieval: Suk and Novotný [233] proposed the CBIR system for recognition of woody species in Central Europe, based on the Chebyshev moments and Fourier descriptors

The efficiency of the method was demonstrated on five available tree leaves datasets, the success rates are between 81% and 96%. The system returns top ten results together with similarity rates.

The CBIR system not for the object but for the texture retrieval is described in [234]. The proposed texture descriptors are modified Zernike moments, invariant to translation, scaling, and rotation. First, the power spectrum of an original texture image is calculated to achieve the translation invariance, and then the power spectrum image is normalized for scale invariance. Finally, modified Zernike moments are calculated for rotation invariance.

In [235] the authors address the issue of 3D retrieval. For feature characteristics, a method combining a distance histogram and geometric moment invariants is proposed to improve retrieval performance. The mutual information and Euclidean metrics estimate the similarity between features, and, finally, the SVM classification is applied. They demonstrate the applicability of the method on the shred database search. 3D object retrieval is also the subject of the paper by Mademlis et al. [236]. They introduce weighted 3D Krawtchouk moments for efficient 3D analyses that are suitable for a content-based search and retrieval applications. Due to the relatively high spatial frequency components of Krawtchouk low-order moments, they can capture sharp shape changes of the object. Thus, proposed weighted 3D Krawtchouk moments can form a very compact and highly discriminative descriptor vector.

The requirement to retrieve the image which is "visually similar" regardless of its particular content, along with the need for rapid system response, has led to a frequent utilization of low-level lossy features based on image colors/graylevels. A typical example is an intensity or color histogram. It is well known that the histogram similarity is a salient property for human vision. Two images with similar histograms are mostly perceived as similar even if their actual content may be very different from each other (see Fig. 9.13). On the other hand, those images that have substantially different histograms are rarely rated by observers as similar. Another attractive property of the histogram is that, if normalized to the image size, it does not depend on image translation, rotation, and scaling, and depends only slightly on elastic deformations. Thanks to this, one does not need to care about image geometry and to look for geometric invariants. Simple preprocessing can also make the histogram insensitive to linear variations of the contrast and brightness of the image. Hence, the histogram established itself as a meaningful image characteristic for CBIR [237–239]. The histogram is rarely used for CBIR directly because it is redundant and inefficient for retrieval purposes. It is sufficient to capture only the prominent features of the histogram and to suppress insignificant details. To do so, some authors compress the histogram from the full color range into just a few bins [240, 241], while others represent the histogram by its coefficients in a proper functional basis. The advantage of the latter approach is that the number of coefficients is a user-defined parameter—we may control the trade-off between a high compression on one hand and an accurate representation on the other hand. It is very natural to be inspired by clear analogy between the histogram of an image and a probability density function (pdf) of a random variable. In probability theory, the pdf is usually characterized by its moments, which has motivated some authors to do the same in the histogram-based CBIR [242, 243].

Stricker and Orengo [244] proposed to compute histograms of individual color channels, Mostafa et al. [245] used the histograms of hue-saturation-brightness (HSB) color channels, and Huang et al. [246] combined the histograms in hue-saturation-value (HSV) color space with Gabor filters for the texture description. They all use moments up to the third order for comparison of the histograms. They are computed either directly from the whole image or, locally, on the regular grid or even on the segmented objects. It is one of the most often used applications of moments in CBIR. Yap and Paramesran in [243] proposed their own color space

$$(rg, yb) = \left(\frac{R - G}{R + G + B + \varepsilon}, \frac{R/2 + G/2 - B}{R + G + B + \varepsilon} \right),$$ (9.2)

where $\varepsilon = 0.03$ prevents division by zero. Then they compute Legendre moments of two-dimensional histogram in this space.

The CBIR methods based on the histogram comparison are sensitive to noise in the images, regardless of the particular histogram representation. Additive noise results in a histogram smoothing, the degree of which is proportional to the amount of noise. Provided that the noise and the image are mutually independent, the histogram of the noisy image is a convolution of the original histogram and of the noise histogram. This immediately leads to a drop of retrieval performance because different histograms tend to be more and more similar to each other due to their smoothing. Although noise in digital photographs is an issue we can neither avoid nor ignore, most authors have either skipped this problem altogether or rely on denoising algorithms applied to all images before they enter the database. Moment blur invariants offer a much more efficient solution, which was proposed by Höschl [247], first for 1D graylevel images and later generalized to color images [248]. If the noise is white Gaussian, the core idea is to represent the histogram by Gaussian blur invariants introduced in Section 6.10 (either by their 1D version in the case of graylevel images or by 3D version in the case of color images and 3D RGB histogram). Since they are invariant w.r.t. convolution, they are – when applied on the histogram – invariant to the amount of noise in the image. As demonstrated in the papers, using just first ten invariants leads to significantly better retrieval rate than using just plain moments, complete histograms, and even than applying image denoising followed by a full histogram matching.

Histograms are not the only variation of the domain the moment descriptors are computed from. For example, Yu et al. [249] convert HS part of the HSV system to the Cartesian coordinates, then they compute the first and the second moments of the 1D Fourier spectrum in 8-neighborhood of each pixel.

There are many variants of CBIR algorithms using the recognition ability of moments. The integral part of the system design is the choice of the classifier, but, as was demonstrated, often the standard minimum distance classifier is utilized.

9.7   Watermarking

In digital watermarking a chosen marker or a random signal is covertly inserted into the image to be protected. Such an application can be regarded as an example of image forensic, aiming at uncovering possible attacks and verifying copyrights, but due to its importance it is discussed separately.

Historically, a watermark was an image printed by water instead of ink, which causes thickness variations in the paper. The watermark then appears as various shades of lightness/darkness when viewed by transmitted light. The symbol used for the watermark referred to the creator/owner of the document. Today's aim of digital watermarking stays the same—we want to encode new information (watermark) into the original signal (host image). The original image is stamped in this way; if the image data are copied, the digital stamp is carried as well. The watermark can be visible, or the brightness change at the place of the embedded watermark can be reduced so it becomes invisible. The watermarks can be used for copyright protection and for active authentication of the images. An example of the watermarked image is in Fig. 9.15.

An important issue for copyright protection is the robustness of watermarks against various attacks including various geometric transformations (at least the similarity transformation), multiple JPEG compressions, cut-outs, and so on, to name a few examples. As in the previous sections, only the solutions based on the moments will be considered. The orthogonal moments—Zernike moments are the most frequent ones—are usually the key components of proposed algorithms. However, other families of moments are mentioned too, such as Krawtchouk (for their better reconstruction properties), Chebyshev, or Hu moment invariants. The moments can be used either for the watermark insertion itself, when the values of

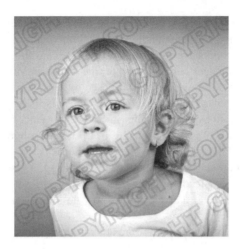

Figure 9.15   Watermarking: The watermarked image. The watermark "COPYRIGHT" is apparent. This is an example of the visible watermark

moment invariants (often Zernike moments) are appropriately changed, or they are applied for the deformation inhibition when the watermarked image is transformed into the normalized position, and thus the influence of present geometric deformations is decreased.

Duan et al. [250] claim Hu moment invariants have the semi-fragile characteristic, meaning they have good robustness to non-malicious manipulations such as JPEG or JPEG2000 compression, additive Gaussian noise, and/or Gaussian filtering, but they provide reasonable fragility to malicious manipulations such as copy-paste action, too. They combined Hu invariants with the wavelet-based contourlet transformation.

Alghoniemy and Tewfik [251] applied the Hu and affine moment invariants directly in the creation of the invariant watermark. They considered the image I as watermarked if a predefined function of the invariants has a certain value N within given tolerance, otherwise the image is declared unwatermarked. They propose to modify the host image in a way that the previous claim may be fulfilled. Various moment invariants can be used, depending on the expected geometric deformations. The predefined function can be any function of the invariants, linear or nonlinear, for example the weighted sum of the invariants can be used. Also, for security purposes, a secret key should be introduced in the function design.

For the image modification, a proper perturbation image $\triangle I$, where $\triangle I = \beta \Pi(I)$ should be constructed and added to the host image I (see Fig. 9.16). Π is the mapping function which serves as a noise added to the host image with a weight β, which is controlled by feedback to ensure N. The authors propose $\log(I)$ as an appropriate choice for Π. For the decoding, the equality of the invariants value to N is simply checked. The decoder does not need the host image for the decoding, which can be advantageous; only the N and the error tolerance has to be known. However, this watermarking approach returns YES – NO answers; thus the method offers 1-bit capacity only, which can be insufficient. Other disadvantages of the methods are its inability to preserve fidelity; that is, the watermarked image will create contrast and brightness variations compared to the host image (see Figs. 9.16(b) and the method vulnerability to the change of the aspect ratio, cropping, and to attacks that would change the geometric moments,

(a)   (b)

Figure 9.16 Watermarking: An approach based on the geometric moments. (a) the original host image to be watermarked by the method [251], (b) the corresponding watermarked image. There is no visible inserted pattern into the image, only intensity variations are apparent. These changes are known disadvantages of the method [251]

such as a histogram equalization. To improve robustness, the image can be normalized a priori. A similar approach was applied in [252], where the Hu moment invariants are computed on a human object detected by means of the skin color distribution.

An example of a more realistic watermark insertion scheme can be found, for instance, in [253], where the dither modulation, a special form of the quantization index modulation, was introduced. They proposed to use invariance properties of circularly orthogonal moments (Zernike moments and pseudo-Zernike moments). In general, Zernike moments are very popular in watermarking applications [254–256]. In the paper, Xin et al. included in the algorithm design the fact that the moment invariance can be corrupted due to the different level of computational accuracy of moments. Due to this the proposed method is robust against geometrical deformations such as image rotation, scaling, and flipping, as well as to other image manipulations – lossy compression, additive noise, and lowpass filtering. Here, the watermark is encoded as a sequence of binary digits, and it is inserted in the host image using the quantization of the ZM's/PZM's magnitudes through the quantization index modulation. Some part of the included binary code contains information about the unit disk used for the watermark encoding. The set of suitable low-order ZMs and PZMs with an appropriate computational accuracy is selected and modified using the watermark binary representation and the dither modulation. The quantization step influences the final quality of a watermarked image – it determines the trade-off between the visibility and the robustness of a watermark. A larger step gives better watermark robustness but makes the watermark more visible, and vice versa. The desired step can be estimated using the required peek signal-to-noise ratio (PSNR) of the image. The watermarked image is then the sum of the original host image and the watermark signal reconstructed from the difference between the modulated and original ZMs and PZMs. For the data extraction, the test image, which can be a distorted version of the original host image with the watermark after some possible manipulations (rotation, scaling, etc.), is analyzed. The unit disk region where the watermark was encoded is located, extracting the first part of the included binary code. Then the binary code is retrieved using the invariant magnitudes of the same subset of ZMs and PZMs, the same way of quantization, and a minimum distance decoder. The comparison of the extracted code with the watermark can indicate if the test image is the distorted version of the host image or not.

In [257], a similar approach is applied, based on the combination of the pseudo-Zernike moments and Krawtchouk moments and support vector regression (SVR). The pseudo-Zernike moments of the host image represented in the polar coordinates are computed. Magnitude values of selected low-order pseudo-Zernike moments are quantized in order to encode the digital watermark. Then, the Krawtchouk moments of the test image form the input to the trained SVR to obtain the estimate of the geometric parameters. They are used for the geometric rectification of the test image, followed by the extraction of the watermark. In this way the robustness against filtering, sharpening, noise adding, and JPEG compression is achieved, and even the geometric attacks such as RST transformation, cropping, and combination of various attacks can be handled successfully.

Zhang et al. [258] proposed to use Chebyshev moments for the watermark embedding and for the decoding. They make use of the independent component analysis (ICA) method for the blind source separation. The watermark is generated randomly independent on the original image and embedded by modifying perceptually invariant Chebyshev moments of the host image. The watermarked image is then created using the inverse Chebyshev transformation on the weighted sum of the Chebyshev moments of the host and of the watermark images. The weight can influence the intensity of the watermark so it should be set according to the requirements of the human visual system. The approach proposed in this paper withstands rotation, scaling, and translation attacks as well as filtering, additive noise, and JPEG compression. Similarly, in [259] Zhang et al. introduced a solution for the local affine degradation with Krawtchouk moment invariants. The watermark is inserted locally into the significant Krawtchouk moment invariants, enabling watermark embedding at the most distinctive part of the image, resulting in great cropping robustness. The watermark detection is realized by means of the ICA.

Papakostas et al. [260] analyzed the possibilities of influencing the performance of the watermarking by means of the parameter setting. They designed the image watermarking methodology, based on Krawtchouk moments, which is optimizing applied parameters using genetic algorithms. They introduced several image quality measures. Similarly, the work described in [261] tries to individually optimize the embedded information quantization for each coefficient, comparing traditional moment families such as Zernike, pseudo-Zernike, or Chebyshev moments and the next generation of moment-based watermarking using polar harmonic transformation (PHT) [262]. The magnitudes of PHT are invariant to image rotation and scaling, and embedding is realized by modification of several PHTs according to the binary watermark sequence. In [263], Papakostas et al. proposed an adaptive moment-based color image watermarking, based on the quaternion radial moments. The method adapts the strength of the embedding with respect to the image morphology.

Zhang et al. [264] addressed the situation where images can be degraded by affine transformation. They have applied orthogonal Legendre moments providing better image description. The watermark is generated from the Legendre moment invariants before being inserted into the invariant domain of the original image. Watermark detection is motivated by [251]. The moments that are used can be exploited for estimating geometric distortion parameters so the watermark can be extracted.

Zhu et al. [265] make use of the Radon transformation and invariance properties of the related complex moment invariants, which they use for the estimation of the geometric distortion factors of analyzed images. Watermark embedding is performed in the discrete cosine transformation (DCT) domain of the original image. Savelonas and Chountasis [266] proposed the watermarking based on the fractional Fourier transformation and related moment representation of images. Embedding watermark sequences into fractional Fourier domain provides more degrees of freedom for the watermark encoding and thus higher security against attackers. Dong et al.'s method is based on Jacobi-Fourier moments [267] applied for the watermark reconstruction. They claim that Jacobi-Fourier moments perform better than Zernike moments for small images. Gao and Jiang utilized Bessel-Fourier moments [268] with proclaimed better performance than other orthogonal moments defined in the polar coordinates, for example Zernike moments. Ahmad et al. applied scale-invariant feature transformation (SIFT) and discrete orthogonal Hahn moment invariants to embed the watermarks [269].

Several authors proposed divergent applications of moments for image watermarking [251, 270–272]. They used central geometric moments for the image normalization with respect to the flipping, translation, scale, and orientation of the image preceding the very watermark encoding and testing. This approach should provide robustness with respect to various geometric attacks; however, it cannot resist, in general, copy and crop attacks. Parameswaran and Anbumani [270] combined such moment normalization with the wavelet-based watermark encoding. Chun-Shien Lu [271] used the DCT decomposition for the watermark insertion after the images were normalized, and in [272] they combined the image normalization with multibit watermarking system based on the spread spectrum watermarking approach [273].

Regarding the watermarking in higher dimensions, namely 3D, Bors [274] introduced an application of moments for the watermarking of 3D graphical objects described by means of polygonal meshes. Here, the first and second-order moments are utilized for the bounding ellipsoids construction, which are then used for the localization of object parts, where the watermark should be embedded.

To conclude watermarking by moments, there is a noteworthy paper by Tsougenis et al. [275]. They provided theoretical analysis of the representative method for watermarking based on moments. Next to the theory overview, they included tests of the performance of individual methods and their sensitivity. They tested the influence of moment order and moment family (Zernike, pseudo-Zernike, wavelet, Krawtchouk, Chebyshev, Legendre, Fourier-Mellin). They concluded that the watermark reconstruction with lower moment orders leads to time saving, but the quality decreases. On the other side, the influence of the moments family selection can be diminished by a smartly designed method in combination with the appropriate tools. They divided methods into two groups, based on the chosen approach. The methods from the first category includes one global watermark inside the host image, resulting in usually low-quality watermarked images with satisfying capacity. Here large number of moments have to be calculated from the whole image. The second category embed information to specific locations, when only a small number of moments need to be calculated. They show that one of the most critical parts of many moment-based watermarking methods is their usage of the Harris detector ([276], utilizing a Harris-Laplace detector for feature points localization and Chebyshev moments for watermark embedding), which can become unstable under attacking conditions. One of the overall best-rated methods is the one by Xin et al. [253], based on pseudo-Zernike moments.

9.8   Medical imaging

A wide application area of the moments and their invariants is in medical imaging. In fact, there can be found similar tasks as in the previous sections (object recognition or image registration, to name a few); however, this domain is often perceived as a specific application category, which is why it is mentioned separately. The main divergences inherent to this kind of methods are caused by the characteristics of the input data. They are often 3D, sometimes with quite low resolution, the noise level can be very high, and the contrast and the data information content are often quite low. Due to the previous statements, the general image processing methods have to be adjusted to fit given limitations.

To illustrate the medical imaging application, the experiment evaluating scoliosis progress (a medical condition in which the spine displays a tendency to take on an S-like shape, and may also be rotated) is described. The rotation moment invariants from Chapter 3 were applied here to analyze medical images shown in Figure 9.17. The aim of the medical project was to measure and evaluate changes to the women's back and spine arising from pregnancy (see [277] for the details of this study). This technique allows 3D measurements from a sequence of 2D images. A template-matching algorithm was used for automatic localization of the landmarks and their centers. Having detected landmarks, the extent of the S-shaped deformation can be evaluated. It should be noted that the application of the Hu invariants failed due to the complexity of present geometric deformation.

As stated before, medical applications do not have any unifying task definition, so the following overview is diverse. Ruggeri and Pajaro [278] applied various moment invariants in order to decide which layer of the eye cornea is seen in the confocal microscope. The proposed

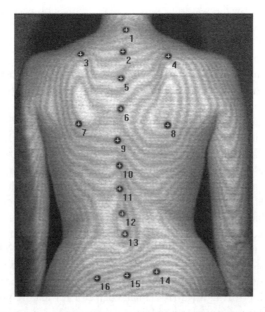

Figure 9.17   Medical imaging: Landmark recognition in the scoliosis study. An example of the human body with attached landmarks and Moire contour graphs. The aim was to detect the landmarks and find their centers

layer recognition is based on cell shapes, which are distinctive for individual layers. They use Hu invariants computed on binarized data and classify them by an artificial neural network. To avoid binarization, which can bias the result, an alternative cell shape description was investigated, based on Zernike moments. The achieved results with ZMs outperformed the ones with Hu moment invariants, and, moreover, no binarization was necessary. A supervised method for blood vessel detection in retinal images [279] was published by Marin et al. Data pixels are classified by neural network, and then the 7D feature vector is generated, consisting of gray-level and moment invariants-based features. In [280] Hu moments are employed for determining the correspondence between vessel bifurcations for retinal image registration.

Hu moment invariants were employed in order to better characterize breast tumors in contrast-enhanced MR images [281]. The automatic segmentation is extended by refining coarse manual segmentation feature and by creating a so-called spatio-temporal model of the segmented tumor. Characterization of the breast tumor in contrast-enhanced MR images by means of the Fourier transformation and by Hu moment invariants was proposed in [282], too. The diagnosis of breast masses by means of Zernike moments used as shape descriptors was proposed in [283]. A novel approach is proposed for the classification of microcalcifications based on Jacobi moments [284], which extract representative orthogonal features at corresponding places.

In [285] the authors proposed to employ Zernike moment invariants, pseudo-Zernike moment invariants, and orthogonal Fourier-Mellin moment invariants to capture the texture characteristics in ultrasound images. A new analysis method for positron emission tomography (PET) data [286] using the spatial characteristics of the radiotracer distribution in consistent regions of interest provides characteristics that may aid in distinguishing between pathological and normal states. The authors proposed to use 3D moment invariants to evaluate spatial changes present in PET images. The features used were found to be correlated with the scores on the united Parkinson's disease rating scale in all studied anatomical regions. Bentoutou et al. [155] applied moment invariants for the image preprocessing for digital subtraction angiography, when blood vessels in a sequence of X-ray images are visualized. The registration of image sequences is based on local similarity detection according to combined moment invariants, followed by thin-plate spline image warping. The proposed technique can handle both slow and sudden motions. Cephalometry is used for estimating absolute and relative measures of the craniofacial skeleton. The paper [287] proposes to apply Zernike moment-based features for an estimation of interest points.

Sommer et al. [288] proposed the method for the comparison of protein binding sites. They use 3D geometric moment invariants as feature vectors for the description of bindings. For many molecules, the coordinates of atoms are available from X-ray crystallography or NMR experiments. They fit a 3D Gaussian onto the center of each atom and sum over all the Gaussians. In this manner, the density of a molecule is approximated, and it is used further as the molecule shape, from which a feature vector is computed. They found the moment invariants to be a robust technique for comparing densities on a global level, useful also for other applications in structural bioinformatics. Similarly, Hung et al. [289] constructed the method for the protein function identification, such as of SARS coronavirus. They apply Hu moment invariants as feature descriptors of protein fragments.

Hu invariants are exploited in [290] for classification of human parasite eggs. The recognition system is based on the multi-class SVM. Banada et al. [291] were looking for the solution of detection and classification of bacterial pathogens, the topic of which is crucial for protecting food supplies. They use a light-scattering sensor together with the recognition system based on the magnitudes of Zernike moments and Haralick texture descriptors. The correspondence itself was estimated using SVM. The detection and identification of the bacterial colonies consisting of the progeny of a single parent cell is achievable in real-time. The success rate for all bacteria categories was over 90%.

Medical image processing often has to deal with 3D data due to the characteristic of the investigated objects. In [292], the authors described the morphometry of cortical sulci (depressions or fissures in the surface of an organ) using 3D geometric moments and their invariants with respect to the translation, rotation, and uniform change of scale. They demonstrate the potential of such an approach for the characterization of the brain activities on the PCA of the 12 first invariants computed for 12 different deep brain structures manually segmented for 7 different brains. This invariant description was used to find correlates of handedness and sex among the shapes of 116 different cortical sulci automatically identified in the database of 142 brains. They plan to apply 3D moment invariants to study the influence of cognitive, genetic, and pathologic features on the shapes of various other cerebral objects. In [293], frequent patterns of the cortical sulci are extracted and captured by means of the 3D moment invariants. Unsupervised agglomerative clustering is performed to define the patterns. The authors in [294] proposed a segmentation of the brain structures in 3D magnetic resonance data by means of the artificial neural network. The 3D moment invariants along with the voxel intensities and the coordinates are used as the input features. In [295] an efficient methodology for the description of the 3D shape of cerebral aneurysms is proposed. Two morphology-oriented methods were considered – geometrical moment invariants and Zernike moment invariants. Zernike moment invariants proved to be more efficient. Ng et al. [296] intended to characterize spatial distribution changes in functional magnetic resonance imaging (fMRI) activation statistics under different experimental conditions. They computed 3D geometric moment invariants with respect to the translation, rotation, and change of scale from the data set recorded from eight healthy subjects performing internally or externally-cued finger tapping sequences. Voxel-based activation statistics were characterized in several regions of interest. The 3D moment invariants analysis revealed the difference between externally and internally cued tasks. Zacharaki et al. [297] proposed to use the aforementioned method [150] to deformably register 3D brain data acquired from magnetic resonance. They limit themselves to the application of low-order geometrical moments as the feature descriptors.

The temporal tracking of coronaries using multi-slice computed tomography dynamic sequence is solved using 3D moment descriptors in [298]. Here, Laguitton et al. proposed to apply the geometric moments and local cylindrical approximation of the vessel. In this way they plan to capture local volume variations and to estimate its displacement along the image sequence. The main idea is to find given points in the successive volumes. The locally defined vessel model, a cylinder, is constructed with parameters, computed from the moments up to the second order. The resulting positions inside the vessel are then adjusted considering neighboring points from the vessel central axis. The cardiac motion from MRI is the topic of the research proposed in [299]. Myocardial motion is estimated within a 4D registration, in which all 3D images are simultaneously registered. An attribute vector (AV) is constructed for each point in the image to represent evaluated data. The AV includes intensity, boundary, and geometric moment invariants. The paper [300] uses Hu moment invariants for automated classification of echocardiography images by means of textural information. An image registration in 4D is proposed for a cardiac motion estimation from MR image sequences [301].

To provide robustness and reliability, a feature vector consisting of intensity, boundary, and geometric moment invariants is instructed for each image voxel.

Hu invariants were employed in recognition of teeth in dental X-ray images [302]. The regions of interest were manually selected. The feature vectors are then compared to the templates. The moment-based object description is used in Morales et al.'s [303] paper, where the authors combined Hu moment invariants together with the Bayes classifier [304] to classify the human embryos according to their quality for its transfer.

An example of content-based image retrieval system (CBIR) for medical purposes is the one proposed by Jones et al. [305]. They concentrate on the database of thermal images, using the idea that a query on images showing symptoms of any given disease will provide visually similar output images, which will usually also represent similarities in medical terms. Their CBIR system is based on the set of Hu moment invariants computed from graylevel thermal images.

An important subgroup of medical imaging algorithms is the class of image reconstruction utilizing data from different modalities, such as X-ray computed tomography or emission tomography. Milanfar et al. [306] proposed a variational framework for the tomographic reconstruction of an image from the maximum likelihood estimates of its orthogonal Legendre moments. They show that in a linear manner they can compute these estimated moments and their error statistics directly from given noisy and possibly sparse projection data. Moreover, they show that this iterative refinement results in superior reconstructions of images from very noisy data as compared with the well-known filtered back-projection algorithm. Basu and Bresler [307] offered solutions for situations where not all viewing angles are known (for example, due to the involuntary motion of the patient). They address the problem of determining the viewing angles directly from the projection data itself in the 2D parallel beam case. They showed that when the projections are shifted by some random amount which must be jointly estimated with the viewing angles, unique recovery of both the shifts and the viewing angles is possible. They exploit the Helgasson-Ludwig consistency conditions, which relate the geometric moments of the object to the moments of its projections. A new method for a reconstruction of medical images with omnipresent noise (image denoising) is presented in [308]. The method is based on pseudo-Zernike, Legendre, Gaussian-Hermite, and Krawtchouk moments. The experiments have shown Legendre moments are superior to all other tested classes.

Moments and moment invariants are used in medical applications in two main roles. They can be employed as the descriptors capable of capturing the shape variations or other structure changes, leading to data classification or data alignment. Alternatively, they are used as tools for data reconstructions from different medical imaging modalities such as magnetic resonance, computed tomography, or emission tomography. Often the 3D moments are applied because of the volumetric nature of the analyzed medical data.

9.9   Forensic applications

Forensic science and analysis of related image-based information for various applications such as person identification or photo forgery detection is another area where the moment theory and especially the moment invariants play an important role. Image analysis helps to provide forensic evidentiary examinations of photographs, documents, handwriting, shoeprints, fingerprints, and so on. The application of moments is based on their invariance (an image verification) and their recognition ability (a person identification by means of various biometrics). Face recognition tasks (see Section 9.2.2) or watermarking (see Section 9.7) can be understood as forensic applications, too, but we paid special attention to them in their dedicated sections due to their popularity.

Nowadays the dominant approach for image capturing is digital imaging; thus images and videos are widespread. Their use can be helpful when they are used in the software solutions for surveillance systems, person identifications, or archiving purposes. However the images can be the subject of a crime, as they are in image forgeries and in intentional data manipulations. Image processing software, which is easily available today, can provide almost anybody with effective tools for image misuse, and so the necessity for detection methods of such image tampering grows. Other specific area where image processing methods can be very useful is biometric identification (by means of fingerprints, iris, etc.) for security applications or for probate proceedings.

A detailed example will introduce the case, when the moment invariants are used for their invariance. Here, the blur invariants help to detect near duplicated image regions in the copy-paste forgeries. Often the naked eye is not able to detect artificially manipulated image data (see Fig. 9.18). Mahdian and Saic [309] proposed the method for automatic detection and localization of near-duplicated regions in digital images. Copy-move forgery detection methods have to take into account that copied parts can be retouched or processed by other image processing tools. During this image tampering, the near-duplicated regions are created in an image. Some regions of the image are used for covering unwanted parts of the original image. They are partially blurred, especially their borders, to conceal traces of pasting copied versions of other parts of the image. Mainly due to this blurring, it is difficult to apply standard similarity measures such as cross-correlation for the tamper localization. The proposed method is based on blur moment invariants and on detection of duplicated

(a)   (b)

Figure 9.18 Forensic application: Detection of near-duplicated image regions. An example of the tampered image with near-duplicated regions – (a) the forged image, (b) the original image

regions in the image, since the duplicated regions may be indicators of this type of forgery. Here, the moment blur invariants allow successful detection of the copy-move forgery even when blur degradation and arbitrary contrast changes are present in the duplicated regions and without the necessity to deblur the image. The lossy format such as JPEG can be treated, too. Moreover, these invariants also work well with the presence of an additive noise, which is important, because the additive noise is often introduced to the image to make detection even more difficult. The proposed method belongs to the category of *passive* (blind) approaches. In contrast to *active* approaches (such as watermarking), the passive methods do not need any a priori information about the tested image, which can be a great advantage in many cases.

Thus, a given image I can contain an arbitrary number of duplicated regions of unknown location and shape, and the task is to determine the presence of such regions in the image and to localize them. The authors suppose that the blurring present can be modeled by convolution with a shift-invariant symmetric energy-preserving point spread function, so that the blur moment invariants can be applied. The proposed copy-move forgery detection method works with the image tiled into the overlapping blocks. For each block the blur moment invariants are evaluated, and then the principal component transformation of all acquired block representations is run to decrease the dimensionality of the feature space. They use twenty-four normalized blur invariants up to the seventh order. Then, the blocks' similarity is analyzed, using the hierarchical k-D tree structure for an efficient identification of all blocks which are in a desired similarity relation with the analyzed block. The k-D tree is a commonly used structure for searching for nearest neighbors. The main idea of the block similarity step is that a duplicated region consists of many neighboring duplicated blocks. If two similar blocks in the analyzed space are found and their neighborhoods are also similar to each other, there is a high probability that they are duplicated.

Figure 9.19 shows an example of a feature space, where black dots represent overlapping blocks. All similar blocks are found for each tested block (similar to the nearest neighbors search), and their neighborhood is verified. In this step, similar blocks with different neighbors are eliminated. Finally a duplicated regions map is created (see Fig. 9.20) showing the image parts, which are likely copied and smoothly pasted in other positions of the image.

In comparison to other methods based mainly on DCT and PCT, the description of duplicated regions by moment invariants has better discriminating properties in cases where the

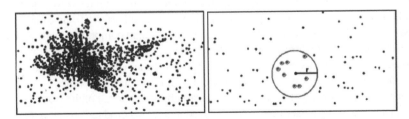

Figure 9.19   Forensic application: Detection of near-duplicated image regions. A two-dimensional feature space. Black dots represent overlapping blocks, which have to be analyzed (left image). The method finds all similar blocks to each block and analyzes their neighborhood (right image). In other words, all blocks inside a circle, which has the analyzed block as centroid are found. The radius r is determined by the similarity threshold

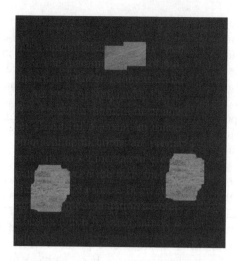

Figure 9.20 Forensic application: Detection of near-duplicated image regions. The estimated map of duplicated regions

regions were modified by blurring or by contrast changes. The results are not affected by an additive zero mean noise. However, the proposed approach, like other existing methods, has a problem with uniform areas since it looks for identical or similar areas in the image from its definition.

There are more papers addressing copy-paste forgeries. Ryu et al. [310] applied Zernike moments; this time they exploited rotation invariance to unveil duplicated regions after arbitrary rotations. Experiments demonstrated high robustness against degradations like JPEG compression, blurring, additive white Gaussian noise, and moderate scaling. J. Yang et al. [311] computes locally Zernike moments in the wavelet domain on individual bands. Huang et al. [312] combined wavelet decomposition with Krawtchouk invariants. Huang et al. [313] preferred Chebyshev moments for tampering detection; they created so-called histogram statistics of reorganized block-based Chebyshev moment features and applied them for analysis of medical data. Review of several moment-based methods for image forgery detection and passive image authentication can be found in [314]. Existing copy-move forgery detection methods were evaluated [315] to analyze their efficiency, both image-wise and pixel-wise. The authors tested the fifteen most prominent feature sets, such as SIFT and SURF, as well as the block-based DCT, wavelet transformation, kernel principal component analysis, PCA, and Zernike features. These feature sets exhibit the best noise robustness, while reliably identifying the copied regions. Results showed that keypoint-based methods are time-wise the best but Zernike moments outperform others in terms of the most precise detection.

The analysis of the origin of the image data is the subject of interest of the work by Chen et al. in [316]. They are trying to estimate if the analyzed image was produced by computer graphics methods or if it is real photograph, and for this they are using the lower-order moments in the image wavelet domain in combination with SVM. Xu et al. [317] utilized the moment ability to capture some characteristic of the data in the statistical sense and use them for the identification of the camera brand and model. The moments were computed from the characteristic function of the image and from the image JPEG representation.

Shabanifard et al. [318] utilized Zernike moments for image classification into four classes: original, contrast modified, histogram-equalized, and noisy images. They estimate absolute values of the first thirty-six Zernike moments of the pixel-pair histogram and its binary form for each image in the polar coordinates. These feature vectors are then fed into the SVM classifier. The digital blind forensics method [319] designed for the medical imaging field tries to answer the question of whether the analyzed medical data has been modified by some image processing. This is realized by the histogram statistics of reorganized block-based discrete cosine transformation coefficients and by the histogram statistics of reorganized block-based Chebyshev moments, substituting DCT coefficients, all classified by SVM. Performance evaluation claims a detection rate greater than 70%.

Invariance of the moment descriptors plays an important role in the very extensive class of applications for biometric identifications, when the final aim is the person's identification. The key components – fingerprints, palmprints, irises, or even shoeprints or ears – are identified after some image enhancement and/or segmentation. Yang et al. in [320] described the Hu invariants based fingerprint recognition. They compute invariants from four sub-images, invariantly selected from the enhanced fingerprints and matched them by means of absolute distance matching and the backpropagation neural network. The modified version of the approach can be found in [321], where the Hu moment invariants are computed for surrounding each detected minutiae point to decrease dependence on sub-image localization (fingerprint minutiae points are local ridge characteristics that occur at either a ridge bifurcation or a ridge ending). Yang et al. [322] combined Hu and Zernike moments in order to achieve higher robustness of the recognition system. Lakshmi Deepika et al. [323] used even more types of moment invariants; in this case they applied Legendre, pseudo-Zernike, and Chebyshev moments. Liu and Yap [324] worked with the moment invariants in the fingerprint analysis in a slightly different manner. They have implemented polar complex moments for extraction of rotation-invariant fingerprint representation. They are describing fingerprint ridge flow structures and creating an orientation field of fingerprints.

Both previous methods apply moment descriptors on already enhanced fingerprint images. The preprocessing of the fingerprints is important due to the low contrast and low quality these images often have. Almansa and Lindeberg [325] capitalized on the properties of the second-order moment to indicate the local image contrast, to measure the degree of anisotropy and to reflect the dominant orientation (the orientation most orthogonal to the gradients in a neighborhood of the point, which will be used for estimating the ridge orientation). They proposed the mechanism for shape-adapted smoothing based on the second moment descriptors and, moreover, an automatic scale selection based on normalized derivatives. The shape adaptation procedure estimates the strength of the fingerprint smoothing according to its local ridge structure, which allows the interrupted ridges to be joined without destroying singularities such as branching points. They produce detailed estimates of the ridge orientation field and the ridge width, as well as a smoothed graylevel version of the input image. The minutiae detection is presented on the post-processed fingerprint which shows good performance.

A multimodal biometric system in which features from the person's face and fingerprint images are extracted using Zernike moments was designed in order to tolerate local variations in the face or fingerprint image of an individual [326]. It makes use of the relevance vector machine (RVM). Testing has demonstrated that using RVM outperforms SVM.

Besides fingerprints, palmprints are typical biometric features for image forensic applications. Once again the moment invariants play the role of the invariant features used for recognition of the palmprint papillary lines structures. Besides the Hu moment invariants [327], applied on the Gabor-filtered images, the application of the Zernike moments is proposed in [328, 329] or [330]. The method described in [331] proposes a palmprint based verification system which uses low-order Zernike moments of palmprint sub-images, and thus it is able to cope with situations when just partial palmprints are available. Pang et al. [332] compared the effectiveness of utilizing three various orthogonal moments, namely the Zernike moments, the pseudo-Zernike moments, and the Legendre moments, in the application of palmprint verification. The orthogonal property of these moments is capable of characterizing independent features of the palmprint image and thus have minimum information redundancy in a moment set, the opposite to the Hu moment invariants, which consist of the geometric moments. Pseudo-Zernike moments have the best performance among all the moments. They compute moments directly on the palmprint images. High-order Zernike moments are used for the description of the hand shape, described in [333]. They compute ZMs for binarized and segmented fingers and palm separately to better capture the geometry of individual parts of the hand. In the latter approach, the papillary lines are not taken into account, just the shape variation of the palm and the fingers are utilized for the person identification.

The combination of the fingerprints and palmprints as biometric tools was studied in [334]. The authors paid attention to the proper selection of features for the tools description in order to achieve the best suitability for the given shape or biometric. They have tested Legendre and pseudo-Zernike moments, and the experiments showed that the nature of Legendre moments makes them more suitable for palmprint images, while Zernike moments are better suited to fingerprint images.

Iris patterns are random, which makes them one of the most reliable biometric feature. On the other hand, due the complex iris image structure and their intra-class variations it is difficult to represent the iris efficiently. Nabti and Bouridane [335] described a multiscale edge detection approach based on a pre-processing localization step followed by filtering with special Gabor filters run on the wavelet-decomposed image. Finally, a feature vector representation using Hu moment invariants is created. The experimental results have shown that the proposed system is comparable to the best iris recognition systems. The distinction of their method is the computation of moments on the results of Gabor wavelet decomposition, which should emphasize the main iris features. In [336] Hu moment invariants were used for description of graylevel segmented iris images, classified then by the k-means algorithm. Zernike moments were utilized in [337], where authors proposed to form feature vectors of eleven features based on Zernike moments computed on the binarized images and k-nearest neighbor classifier. The higher-order Gauss-Hermite moments were extracted from the iris images [338] for their identification. The special codes generated from the zero-crossings of these features capture iris structures. Finally, the performance of Legendre moments, Zernike moments, and pseudo-Zernike moments and their robustness with respect to noise in the iris recognition applications was evaluated in [339]. The authors claim that the recognition rate by all types of features with moments of higher orders are almost the same. The most time-effective representation is the one based on the pseudo-Zernike moments, and the Legendre moment invariants showed more robustness against the white Gaussian noise.

When the biometric features computed on the human head are discussed, we should mention methods based on the comparison of the person's ear shape. Liu et al. [340] combine front and backside view of the ear by extracting features based on the Chebyshev moment descriptors. The ear representation makes use of number of lines, perpendicular to the longest axis in the ear contour. The authors extracted ear shape geometry features and ear high-order invariant moments features based on Chebyshev radial polynomials. The novelty of the paper also consists in the fact they tried multi-view based 3D ear biometrics.

The shape of a hand can be used for a person's identification, too. In [341], high-order Zernike moments were employed for such methodology. In order to facilitate feature computation only the most discriminative features were selected by means of the fast correlation-based filter and sparse Bayesian multinomial logistic regression to reduce the dimensionality of the feature space.

AlGarni and Hamiane [342] proposed to use Hu moment invariants for shoeprint classification against the database in order to link suspects to a crime scene. They work with binarized shoeprint images, variously rotated and with various resolution. Their experiments showed good recognition rates; however, it is important to note that the method would fail in the case of non-complete shoeprints, which is often the case.

The gait is another biometrics feature used for a person's identification. Other biometrics technologies, such as face, hand, and fingerprint recognition, are useless when the tested person is distant. The main idea behind the application of the moments is to divide the silhouette at each frame of the sequence in a predefined way, to compute a set of features based on the moments for each subpart of the image, and then, finally, somehow reduce the acquired feature vector to efficiently represent the gait sequence. Lee and Grimson [343] represented each orthogonal view video silhouettes of human walking motion by a fixed number of ellipses, and they compute the feature vector based on the second-order moments. They aggregate these feature vectors over time and finally process them using Fourier transformation to obtain frequency-based information. Shi et al. [344] proposed the silhouette subdivision in a multiscale manner, where the moments are computed for each resolution. They use a very small number of moments, which limits the characterization ability of their method. The dimensionality of their feature space is reduced using PCA, and the recognition itself is done by SVM. Contour matching in [345] was realized by means of Hu moment invariants. For gender classification by means of the gait, Sudha and Bhavani [346] used extracted moments and two spatial features based on a bounding rectangle of the human silhouette. They applied k-nearest neighbor and SVM classifiers. An innovative gait identification method [347] introduces the 3D radial silhouette distribution transformation and the 3D geodesic silhouette distribution transformation, and it combines different feature extractors by means of the genetic algorithm. Specifically, three feature extraction techniques are involved – the radial and circular integration transformations and one based on the weighted Krawtchouk moments. Moreover, the use of orthogonal moments based on the weighted Krawtchouk polynomials enables reconstruction of the original silhouette.

Video surveillance can also be seen as an image forensic application. In [348] to detect and observe humans appearance, movements and activities the authors subtract the image background, apply appropriate enhancement if necessary, and extract affine moment invariants of the full body and the head-shoulder parts. In this manner, they are able to handle partially occluded images. In [349], a humans and cars identification technique for real-time video surveillance systems is proposed. Here, the feature extraction and classification is done by affine moment invariants. The algorithm can identify humans by extracting the human head-shoulder up to 60–70% occlusion. The accuracy of the human identification of the proposed technique is very good (98.33%). Asaidi et al. in [350] enhance the automatic traffic surveillance system. They proposed an elimination of shadows and the classification of vehicles, based on Hu moments, calculated in such way that the perspective effects are reduced.

As is apparent, the moment ability to invariantly represent the object is quite popular for person identification by means of various biometrics. It has to be mentioned that the efficiency of the moment descriptors is preserved only if the original image and deformed image depict the same scene, so in the case of partial shoeprint or partial fingerprint the algorithms could fail. On the other hand, the near-duplicated region detection uses the invariance with respect to the radiometrical deformations, which ability is rare among descriptors, and due to this the proposed algorithm can achieve very good results even in situations where other methods break down.

9.10   Miscellaneous applications

Besides large categories of moment applications, there are various other image processing domains, where the moment theory has found use and offers new viewpoints. These are not so distinguished as the previous topics covered in this section; however, they surely deserve to be mentioned here. In the following text several examples of such idea upgrading are covered.

9.10.1   Noise resistant optical flow estimation

Optical flow estimation is an important tool to track motion in a sequence of images with many applications; for example, recovering surface structures, change detection, and various medical problems. The objective of optical flow computation is to obtain measurements describing the 2D pixel motion in images, known as the *motion field*. However, such motion is only the projection of the relative 3D motion between the camera and the scene surfaces in 3D space on the image plane. There are strong limitations on the accuracy of such models due to the approximating nature of the optical flow, which is only the approximation of the motion field depending on the observed motion of points in the image. Most optical flow methods are very sensitive to the quality of input images. They prefer images with constant brightness and texture of the image and close sampling (small motion) and noise-free scenes. Unfortunately, in many applications, for example in medical imaging, the images are very noisy, which prevents use of standard methods (for example [351]) relying on local computation of derivatives in each pixel. Moments were utilized to achieve higher robustness with respect to noise and small perturbations. In [352], the authors took the advantage of the stability of rotation-invariant Zernike moments with respect to noise. They modified the method [351] to work with the values of invariants based on Zernike moments, instead of the pixel values. The linear equations system, which forms the solution for optical flow, is then formed over each small local region based on the principle of conservation of those invariant features. They obtain a more robust estimation at the expense of a slight decrease of spatial resolution. Zernike moments-based optical flow is introduced in [140], mentioned already earlier, where driving points of the optical flow are localized by Canny edge detector, and for them Zernike descriptors are evaluated in the multi-scale manner.

Kharbat et al. [353] proposed to apply geometric moments for the optical flow stabilization with respect to intensity changes. They describe the pixel values by the content of their neighborhood using geometric moments. Then, the description of each pixel is normalized and made

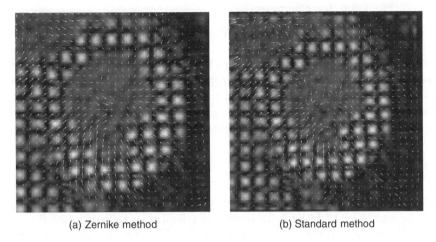

(a) Zernike method   (b) Standard method

Figure 9.21 Optical flow: Noise resistant estimation. The result of (a) Zernike moments; (b) the standard method [351]. Stabilization of optical flow computation in the case of noisy magnetic resonance images of the heart

insensitive to fluctuations of intensity. In this way, resulting optical flow is much less sensitive to visual phenomena like varying illumination, specular reflections, and shadows.

Figure 9.21 shows an application of this method to fast sequences of heart images taken by MR (magnetic resonance) scanner. The images are marked by a harmonic phase imaging technique resulting in a typical grid-like pattern, which helps to improve the precision of the applied optical flow algorithm. It is apparent that the method using Zernike moments exhibits many fewer artifacts and irregularities of the optical flow field, especially in the areas lacking distinctive texture than the standard method [351] (see Fig. 9.21).

One of the possible applications of the optical flow is super-resolution [354], when the image of higher spatial resolution is generated using several low-resolution input images. One such spatio-temporal super-resolution reconstruction model was proposed, employing a robust optical flow estimator and Zernike moments. After the estimation of optical flow, the motion compensation is applied, followed by the multi-frame information fusion scheme to make the spatio-temporal super-resolution reconstruction. The latter is based on Zernike moments and non-local self-similarity. Here, Zernike moments express the level of similarity of the pixel to be restored with the data used for the restoration.

9.10.2   Edge detection

Edge detection is one of base elements of many image processing tasks. Moments found their application even here. In [355], using orthogonal Fourier-Mellin moments, the edge detection with subpixel accuracy was proposed. It is based on the moments with the lower radial orders and on the mutual relation between their different orders and degrees. Experimental results show that high edge location accuracy can be achieved (the authors claim 0.16 pixel for straight lines with noise and 0.23 pixel for curves with noise). The authors of [356] detected step edges with subpixel accuracy, too. Their approach is based on a set of Zernike orthogonal complex moments. Yang and Pei in [357] proposed a combination of Zernike moment-based edge detection with Otsu adaptive threshold algorithm to achieve higher robustness with respect to the threshold selection. Lyvers et al. [358] derived an edge operator based on two-dimensional geometric moments. Precision is achieved by correcting for many deterministic errors caused by non-ideal edge profiles using a pre-generated look-up table to correct the original estimates of edge orientation and location. This table was generated using a synthesized edge which is located at various subpixel locations and various orientations. An experiment with the measurement of metal object is also presented.

9.10.3 Description of solar flares

The moments can be used for evaluation of such complex processes as the behavior of the Sun. In [359] the application of moments for analysis of solar flares is proposed. One of the important flare characteristics is the pre-flare phase (Figure 9.22), when the brightness of the solar surface begins to oscillate. The beginning of the flare and frequencies of these oscillations are the main descriptors of this phenomenon and are the subject of the research. The development of the flare could be described by means of a time curve, and the proposed idea is to use moments of the histogram for the time curve construction.

The moments of the histogram can be expressed by a non-standardized version

$$m_n = \sum_{u=0}^{L} u^n h(u) = \sum_{x=0}^{M-1} \sum_{y=0}^{N-1} f^n(x,y), \quad n = 0, 1, \dots, \tag{9.3}$$

where $h(u)$ is *histogram* of the $M \times N$ image $f(x,y)$ with gray levels from 0 to L, that is, $h(u)$ is the number of pixels with gray level u. The standardized versions of the moments are

$$\mu_0 = m_0, \quad \mu_1 = m_1/m_0, \quad \mu_2 = m_2/m_0 - \mu_1^2,$$

$$\mu_n = \frac{1}{\mu_0} \frac{1}{\mu_2^{n/2}} \sum_{u=0}^{L} (u - \mu_1)^n h(u) = \frac{1}{\mu_0} \frac{1}{\mu_2^{n/2}} \sum_{x=0}^{M-1} \sum_{y=0}^{N-1} (f(x,y) - \mu_1)^n, \quad n = 3, 4, \dots, \tag{9.4}$$

representing the number of data (μ_0), their mean value (μ_1), variation (μ_2), standard deviation ($\sqrt{\mu_2}$), skewness (μ_3), kurtosis μ_4, hyperskewness (μ_5), and hyperkurtosis (μ_6). In our case, μ_0 is the number of pixels in the selected region of interest with the flare, μ_1 is the average

Figure 9.22 Solar flare: An example of the solar flare (the arrow-marked bright object). Dark areas correspond to sunspots

brightness, and $\sqrt{\mu_2}$ corresponds to the contrast. The skewness μ_3 is the lowest moment that could express something about the flare. If the time curve generated by $\mu_3(t)$ is not sufficiently information-rich, we can evaluate moments of higher orders up to, for instance, $\mu_8(t)$, evaluate PCT, and use the time curve generated by the first principal component. Various dynamic ranges of different moments (even the standardized ones) have to be taken into consideration while the PCT is computed.

An example of time curves is in Figure 9.23. In the first half minute, there is a quiet phase, then there are pre-flare oscillations lasting about four minutes. Then, finally, the flare is fully developed. It is convenient to remove the oscillations by a Gaussian filter before the detection of the flare beginning starts (see [359]). Then the beginning can be found as a so-called *turning point*. The turning point is defined here as the next significant point of a function in the row defined by the following rules [$f(x) = 0$ ... zero crossing, $f'(x) = 0$ ... an extreme, $f''(x) = 0$ ... inflection point, $f'''(x) = 0$ ... turning point].

The frequencies of the pre-flare oscillations can be searched either by Fourier transformation [359] or by wavelet transformation [360]. Morlet wavelets have been found the most suitable in this case.

9.10.4 Gas-liquid flow categorization

In [361] the authors presented a novel approach for analyzing two-dimensional flow field data. They employed geometric invariant moments for the purpose of interactive exploration of flow field data. They modified moment invariants for vector fields in order to extract and visualize 2D flow patterns, invariant under translation, scaling, and rotation. They use moments up to the third order and related the invariant values to specific structures in the flow fields. The moment evaluation is done for all positions and for all possible scales in the so-called moment pyramid model, to capture all different structures. Similarly, Zhou et al. [362] applied Hu invariants and co-occurrence matrices for the gas-liquid flow regime classification. They use multi-feature

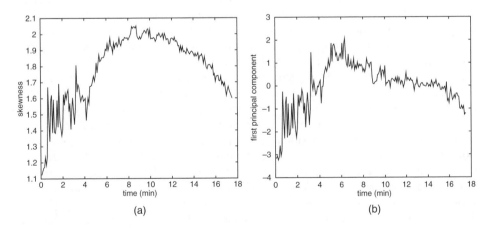

(a)   (b)

Figure 9.23 Solar flare: The time curve of a solar flare – (a) skewness, (b) first principal component. Data from Ondřejov observatory: the scanning began on the 18 December 2003, at 12:25 (time 0 in the graphs)

fusion and SVM recognition kernels. The database of the various kinds of flow was created, feature vectors evaluated, and finally, SVM was trained. The feature sets, reduced using the rough sets theory capture variations between seven typical flow regimes. The generalization of the flow field description for 3D vector fields was introduced by Bujack et al. in [363]. They proposed invariants based on moment normalization, forming a complete and independent set. They search for patterns independent from their position, orientation and scale.

9.10.5   3D object visualization

An interesting application from the computer graphics area is the simplification of meshes [364] for visualization of 3D surfaces. Mesh simplification is important for accurate as well as plausible virtual reality scene generation. The simplification metric is a vital part of the whole topic. The authors proposed using surface geometric moments as well as volume geometric moments for mesh complexity measuring. The objective function is defined as the difference between moments computed on the original mesh and those evaluated on the simplified mesh. These metrics are used in an edge collapse scheme. For a given maximum order and the number of triangles required, the optimal mesh with a minimum moment difference from the original mesh can be determined. An example from material analysis application area is the work of O' Connor et al. [365]. They used second-order moments for 3D modeling of the inner structure of the solid materials based on X-ray scans of individual objects. They first detect regions of interest and then approximate them by best-fit ellipsoids, which provide the main orientation of the inner parts as well.

9.10.6   Object tracking

Person or object tracking is an important research topic; it has already been partially mentioned earlier, where pseudo-Zernike moments were used for predefined single object tracking in aerial surveillance in [191]. In [366] to be able to handle videos with changing scale Hu moment, invariants were employed to modify the bandwidth for object detection. The authors used combined spatial and color information for the object detection in individual frames. Yan et. all [367] addressed the topic of object tracking in the case of a scene with a ceiling-mounted camera, when the person's appearance is very different from the more usual camera position facing the person directly. The challenging part is the significant change of the person's shape while walking around the room. The authors proposed to use particle filters with the input from different visual streams. To enhance the robustness of the tracking system, they proposed to incorporate various cues (motion, shape, shape memory, color memory) to update the probability of the person at the particle position. The shape cue is based on the moments, specifically on the Hu moment invariants computed on the edge images of the scene, to decrease the influence of lighting changes. Multilayer perceptron is used for the classification itself.

To conclude this assortment of various applications the role of moment invariants in image processing can be illustrated by their inclusion into the Point Cloud Library [152] – a large-scale, standalone open project for 2D/3D image and point cloud processing. There are many various algorithms for processing 3D data, their registration and model fitting, extracting features, and handling surfaces. Moment invariants have found a place in their 3D features library.

9.11   Conclusion

In the previous text, the main application areas of moments and moment invariants have been named, together with illustrative examples and many variations. The main properties of moments, which are used during these applications are their invariance under geometric and radiometric degradations, their ability to capture the shape characteristics of individual objects, and their competence in image normalization. The Hu group of geometric moment invariants are often the first choice for many authors, even though other classes of moment invariants can offer better properties in particular tasks. Vital issues of moment applications are proper selection of regions for moment computation, their high sensitivity to segmentation errors, their breakdown in the case of object occlusion, and discrepancy between the facts that the features are defined in 2D but many described objects are defined in the 3D space. In practice, not many applications described in scientific journals are using moment computation acceleration. Despite the mentioned issues, the moments are still the best solution for many aims, as is apparent from the publication dates of many of the listed papers.

References

[1] T. Tuytelaars and L. van Gool, "Wide baseline stereo based on local, affinely invariant regions," in *Proceedings of the British Machine Vision Conference BMVC'00*, pp. 412–422, BMVA, 2000.

[2] J. Matas, O. Chum, M. Urban, and T. Pajdla, "Robust wide baseline stereo from maximally stable extremal regions," in *Proceedings of the British Machine Vision Conference BMVC'02*, pp. 384–396, BMVA, 2002.

[3] K. Mikolajczyk and C. Schmid, "Scale and affine invariant interest point detectors," *International Journal of Computer Vision*, vol. 60, no. 1, pp. 63–86, 2004.

[4] B. Zion, A. Shklyar, and I. Karplus, "In-vivo fish sorting by computer vision," *Aquacultural Engineering*, vol. 22, no. 3, pp. 165–179, 2000.

[5] D. White, C. Svellingen, and N. Strachan, "Automated measurement of species and length of fish by computer vision," *Fisheries Research*, vol. 80, no. 2–3, pp. 203–210, 2006.

[6] C. Gope, N. Kehtarnavaz, G. Hillman, and B. Würsig, "An affine invariant curve matching method for photo-identification of marine mammals," *Pattern Recognition*, vol. 38, no. 1, pp. 125–132, 2005.

[7] N. Kehtarnavaz, V. Peddigari, C. Chandan, W. Syed, G. Hillman, and B. Wursig, "Photo-identification of humpback and gray whales using affine moment invariants," in *Image Analysis* (J. Bigun and T. Gustavsson, eds.), vol. 2749 of *Lecture Notes in Computer Science*, pp. 109–116, Springer, 2003.

[8] C. Beyan and R. B. Fisher, "Detection of abnormal fish trajectories using a clustering based hierarchical classifier," in *Proceedings of the British Machine Vision Conference BMVC'13* (T. Burghardt, D. Damen, W. Mayol-Cuevas, and M. Mirmehdi, eds.), pp. 1–11, BMVA Press, 2013.

[9] B. Shao and H. Xin, "A real-time computer vision assessment and control of thermal comfort for group-housed pigs," *Computers and Electronics in Agriculture*, vol. 62, no. 1, pp. 15–21, 2008. Precision Livestock Farming (PLF).

[10] E. Alegre, V. González-Castro, R. Alaiz-Rodríguez, and M. T. García-Ordás, "Texture and moments-based classification of the acrosome integrity of boar spermatozoa images," *Computer methods and programs in biomedicine*, vol. 108, no. 2, pp. 873–881, 2012.

[11] F. Albregtsen, H. Schulerud, and L. Yang, "Texture classification of mouse liver cell nuclei using invariant moments of consistent regions," in *Computer Analysis of Images and Patterns CAIP'95* (V. Hlaváč and R. Šára, eds.), vol. 970 of *Lecture Notes in Computer Science*, pp. 496–502, Springer, 1995.

[12] G. Yang, V. Lalande, L. Chen, N. Azzabou, T. Larcher, J. de Certaines, H. Shu, and J.-L. Coatrieux, "MRI texture analysis of GRMD dogs using orthogonal moments: A preliminary study," *IRBM*, vol. 36, no. 4, pp. 213–219, 2015.

[13] S. N. Yaakob and L. Jain, "An insect classification analysis based on shape features using quality threshold ARTMAP and moment invariant," *Applied Intelligence*, vol. 37, no. 1, pp. 12–30, 2012.

[14] R. White and L. Winokur, "Quantitative description and discrimination of butterfly wing patterns using moment invariant analysis," *Bulletin of Entomological Research*, vol. 93, no. 04, pp. 361–376, 2003.

[15] X. Tang, W. Kenneth Stewart, L. Vincent, H. Huang, M. Marra, S. M. Gallager, and C. S. Davis, "Automatic plankton image recognition," *Artificial Intelligence Review*, vol. 12, no. 1–3, pp. 177–199, 1998.

[16] S. Farokhi, U. U. Sheikh, J. Flusser, and B. Yang, "Near infrared face recognition using Zernike moments and Hermite kernels," *Information Sciences*, vol. 316, pp. 234–245, 2015.

[17] N. Zaeri, F. Baker, and R. Dib, "Thermal face recognition using moments invariants," *International Journal of Signal Processing Systems*, vol. 3, no. 2, pp. 94–99, 2015.

[18] S. Dasari and I. V. Murali Krishna, "Combined classifier for face recognition using Legendre moments," *Computer Engineering and Applications*, vol. 1, no. 2, pp. 107–118, 2012.

[19] N. Belghini, A. Zarghili, and J. Kharroubi, "3D face recognition using Gaussian Hermite moments," *International Journal of Computer Applications, Special Issue on Software Engineering, Databases and Expert Systems*, vol. SEDEX, no. 1, pp. 1–4, 2012.

[20] J. Sheeba Rani and D. Devaraj, "Face recognition using Krawtchouk moment," *Sādhanā*, vol. 37, no. 4, pp. 441–460, 2012.

[21] F. Akhmedova and S. Liao, "Face recognition using discrete orthogonal Hahn moments," *International Conference on Image Analysis and Processing ICIAP'15*, vol. 2, no. 5, p. 1269–, 2015.

[22] J. P. Ananth and V. Subbiah Bharathi, "Face image retrieval system using discrete orthogonal moments," in *4th International Conference on Bioinformatics and Biomedical Technology ICBBT'12*, vol. 29, pp. 218–223, IACSIT, 2012.

[23] A. Nabatchian, E. Abdel-Raheem, and M. Ahmadi, "Human face recognition using different moment invariants: A comparative study," in *Proceedings of the Congress on Image and Signal Processing CISP'08*, pp. 661–666, IEEE, 2008.

[24] C. Singh and A. M. Sahan, "Face recognition using complex wavelet moments," *Optics & Laser Technology*, vol. 47, pp. 256–267, 2013.

[25] R. Gopalan, P. Turaga, and R. Chellappa, "A blur-robust descriptor with applications to face recognition," *IEEE Transactions on Pattern Analysis and Machine Intelligence*, vol. 34, no. 6, pp. 1220–1226, 2012.

[26] J. Flusser, S. Farokhi, C. Höschl IV,, T. Suk, B. Zitová, and M. Pedone, "Recognition of images degraded by Gaussian blur," *IEEE Transactions on Image Processing*, vol. 25, no. 2, pp. 790–806, 2016.

[27] C. Singh, N. Mittal, and E. Walia, "Face recognition using Zernike and complex Zernike moment features," *Pattern Recognition and Image Analysis*, vol. 21, no. 1, pp. 71–81, 2011.

[28] C. Singh, E. Walia, and N. Mittal, "Discriminative Zernike and pseudo Zernike moments for face recognition," *Pattern Recognition and Image Analysis*, vol. 2, no. 2, pp. 12–35, 2012.

[29] S. Farokhi, S. M. Shamsuddin, J. Flusser, U. U. Sheikh, M. Khansari, and K. Jafari-Khouzani, "Rotation and noise invariant near-infrared face recognition by means of Zernike moments and spectral regression discriminant analysis," *Journal of Electronic Imaging*, vol. 22, no. 1, pp. 1–11, 2013.

[30] J. Haddadnia, M. Ahmadi, and K. Faez, "An efficient method for recognition of human faces using higher orders pseudo Zernike moment invariant," in *Fifth International Conference on Automatic Face and Gesture Recognition FGR'02*, pp. 330–335, IEEE, 2002.

[31] J. Haddadnia, K. Faez, and M. Ahmadi, "An efficient human face recognition system using pseudo Zernike moment invariant and radial basis function neural network," *International Journal of Pattern Recognition and Artificial Intelligence*, vol. 17, no. 1, pp. 41–62, 2003.

[32] A. Nabatchian, I. Makaremi, E. Abdel-Raheem, and M. Ahmadi, "Human face recognition using different moment invariants: A comparative study," in *Proceedings of the 3rd International Conference on Convergence and Hybrid Information Technology ICCIT'08*, pp. 933–936, IEEE, 2008.

[33] N. H. Foon, Y. Pang, A. T. B. Jin, and D. N. C. Ling, "An efficient method for human face recognition using wavelet transform and Zernike moments," in *Proceedings of the International Conference on Computer Graphics, Imaging and Visualization CGIV'04*, pp. 65–69, IEEE, 2004.

[34] E. Sariyanidi, V. Dagli, S. C. Tek, B. Tunc, and M. Gokmen, "Local Zernike moments: A new representation for face recognition," in *19th IEEE International Conference on Image Processing ICIP'12*, pp. 585–588, IEEE, 2012.

[35] S. Farokhi, S. M. Shamsuddin, U. U. Sheikh, J. Flusser, M. Khansari, and K. Jafari-Khouzani, "Near infrared face recognition by combining Zernike moments and undecimated discrete wavelet transform," *Digital Signal Processing*, vol. 31, pp. 13–27, 2014.

[36] S. M. Lajevardi and Z. M. Hussain, "Higher order orthogonal moments for invariant facial expression recognition," *Digital Signal Processing*, vol. 20, no. 6, pp. 1771–1779, 2010.

[37] R. Londhe and V. Pawar, "Facial expression recognition based on affine moment invariants," *International Journal of Computer Science Issues*, vol. 9, no. 6, pp. 388–392, 2012.

[38] Y. Zhu, L. C. de Silva, and C. Ko, "Using moment invariants and HMM in facial expression recognition," *Pattern Recognition Letters*, vol. 23, no. 1–3, pp. 83–91, 2002.

[39] R. Zhi and Q. Ruan, "A comparative study on region-based moments for facial expression recognition," in *Proceedings of the Congress on Image and Signal Processing CISP'08*, pp. 600–604, IEEE, 2008.

[40] S. Prakash, U. Jayaraman, and P. Gupta, "A skin-color and template based technique for automatic ear detection," in *Proceedings of the Seventh International Conference on Advances in Pattern Recognition ICAPR'09*, pp. 213–216, IEEE, 2009.

[41] F. Li, C. Zhao, Z. Xia, Y. Wang, X. Zhou, and G.-Z. Li, "Computer-assisted lip diagnosis on traditional Chinese medicine using multi-class support vector machines," *BMC Complementary and Alternative Medicine*, vol. 12, no. 1, pp. 1–13, 2012.

[42] S. Chen and S. Srihari, "Use of exterior contours and shape features in off-line signature verification," in *Proceedings of the Eighth International Conference on Document Analysis and Recognition ICDAR'05*, p. 5–, IEEE, 2005.

[43] H. Lin and H.-Z. Li, "Chinese signature verification with moment invariants," in *International Conference on Systems, Man, and Cybernetics SMC'96*, vol. 4, pp. 2963–2968, IEEE, 1996.

[44] S. O. Belkasim, M. Shridhar, and M. Ahmadi, "Pattern recognition with moment invariants: A comparative study and new results," *Pattern Recognition*, vol. 24, no. 12, pp. 1117–1138, 1991.

[45] P. Kumawat, A. Khatri, and B. Nagaria, "Comparative analysis of offline handwriting recognition using invariant moments with HMM and combined SVM-HMM classifier," in *International Conference on Communication Systems and Network Technologies CSNT'13*, pp. 140–143, IEEE, 2013.

[46] M. O. A. Albaraq and S. C. Mehrotra, "Recognition of Arabic handwritten amount in cheque through windowing approach," *International Journal of Computer Applications*, vol. 115, no. 10, pp. 33–38, 2015.

[47] P. Raj and A. Wahi, "Zone based method to classify isolated Malayalam handwritten characters using Hu-invariant moments and neural networks," *International Journal of Computer Applications*, vol. Proceedings on International Conference on Innovations In Intelligent Instrumentation, Optimization and Electrical Sciences ICIIIOES'13, pp. 10–14, December 2013.

[48] K. V. Kale, P. D. Deshmukh, S. V. Chavan, M. M. Kazi, and Y. S. Rode, "Zernike moment feature extraction for handwritten Devanagari (Marathi) compound character recognition," *International Journal of Advanced Research in Artificial Intelligence*, vol. 3, no. 1, pp. 68–76, 2014.

[49] P. P. Amitabh Wahi, S. Sundaramurthy, "A comparative study for handwritten Tamil character recognition using wavelet transform and Zernike moments," *International Journal of Open Information Technologies*, vol. 2, no. 4, pp. 30–35, 2014.

[50] A. K. Sadanand, L. B. Prashant, R. M. Ramesh, and L. Y. Pravin, "Offline MODI character recognition using complex moments," *Procedia Computer Science*, vol. 58, pp. 516–523, 2015. Second International Symposium on Computer Vision and the Internet (VisionNet'15).

[51] X. Wang, Y. Yang, and K. Huang, "Combining discrete orthogonal moments and DHMMs for off-line handwritten Chinese character recognition," in *Proceedings of the Fifth International Conference on Cognitive Informatics ICCI'06*, vol. 2, pp. 788–793, IEEE, 2006.

[52] J. Flusser and T. Suk, "Affine moment invariants: A new tool for character recognition," *Pattern Recognition Letters*, vol. 15, no. 4, pp. 433–436, 1994.

[53] A. K. Jain and A. Vailaya, "Shape-based retrieval: A case study with trademark image databases," *Pattern Recognition*, vol. 31, no. 9, pp. 1369–1390, 1998.

[54] K. Saipullah and N. Ismail, "Determining halal product using automated recognition of product logo," *Journal of Theoretical and Applied Information Technology*, vol. 7, no. 2, pp. 190–198, 2015.

[55] F. M. Anuar, R. Setchi, and Y. Lai, "Trademark image retrieval using an integrated shape descriptor," *Expert Systems with Applications*, vol. 40, no. 1, pp. 105–121, 2013.

[56] K.-T. Sam and X.-L. Tian, "Vehicle logo recognition using modest AdaBoost and radial Tchebichef moments," in *International Conference on Machine Learning and Computing ICMLC'12* (Q. Huang, ed.), IACSIT, 2012.

[57] S. Ghosh and R. Parekh, "Automated color logo recognition system based on shape and color features," *International Journal of Computer Applications*, vol. 118, no. 12, pp. 13–20, 2015.

[58] Z. Zhang, X. Wang, W. Anwar, and Z. L. Jiang, "A comparison of moments-based logo recognition methods," *Abstract and Applied Analysis*, pp. 1–8, 2014. Special Issue.

[59] X.-F. Wang, D.-S. Huang, J.-X. Dua, H. Xu, and L. Heutte, "Classification of plant leaf images with complicated background," *Applied Mathematics and Computation*, vol. 205, no. 2, pp. 916–926, 2008.

[60] L. Jiming, "A new plant leaf classification method based on neighborhood rough set," *Advances in Information Sciences and Service Sciences*, vol. 4, no. 1, pp. 116–123, 2012.

[61] Z. Zulkifli, P. Saad, and I. A. Mohtar, "Plant leaf identification using moment invariants & general regression neural network," in *11th International Conference on Hybrid Intelligent Systems HIS'11*, pp. 430–435, IEEE, 2011.

[62] K. Singh, I. Gupta, and S. Gupta, "SVM-BDT PNN and Fourier moment technique for classification of leaf shape," *International Journal of Signal Processing, Image Processing and Pattern Recognition*, vol. 3, no. 4, pp. 67–78, 2010.

[63] T. Suk, J. Flusser, and P. Novotný, "Comparison of leaf recognition by moments and Fourier descriptors," in *Computer Analysis of Images and Patterns* (R. Wilson, E. Hancock, A. Bors, and W. Smith, eds.), vol. 8047 of *Lecture Notes in Computer Science*, pp. 221–228, Springer, 2013.

[64] S. U. Thiel, R. J. Wiltshire, and L. J. Davies, "Automated object recognition of blue-green algae for measuring water quality—A preliminary study," *Water Research*, vol. 29, no. 10, pp. 2398–2404, 1995.

[65] S. Fischer and H. Bunke, "Automatic identification of diatoms using decision forests," in *Proceedings of the Second International Workshop on Machine Learning and Data Mining in Pattern Recognition MLDM'01*, vol. LNAI 4571, pp. 173–183, Springer, 2001.

[66] J. Flusser, T. Suk, and B. Zitová, "Comments on 'Weed recognition using image blur information' by Peng, Z. & Jun, C., Biosystems Engineering 110 (2), p. 198–205," *Biosystems Engineering*, vol. 126, no. 2, pp. 104–108, 2014.

[67] J. Flusser, T. Suk, and B. Zitová, "On the recognition of wood slices by means of blur invariants," *Sensors and Actuators A: Physical*, vol. 198, pp. 113–118, 2013.

[68] P. Zhao, "Dynamic timber cell recognition using two-dimensional image measurement machine," *AIP Review of Scientific Instruments*, vol. 82, no. 8, pp. 083703–1, 2011.

[69] K. Singh, I. Gupta, and S. Gupta, "Classification of bamboo species by Fourier and Legendre moment," *International Journal of Advanced Science and Technology*, vol. 50, pp. 61–70, 2013.

[70] S. Liu, P. A. Mundra, and J. C. Rajapakse, "Features for cells and nuclei classification," in *Engineering in Medicine and Biology Society EMBC'11, Annual International Conference of the IEEE*, pp. 6601–6604, 2011.

[71] W. M. Ahmed, B. Bayraktar, A. K. Bhunia, E. D. Hirleman, J. P. Robinson, and B. Rajwa, "Classification of bacterial contamination using image processing and distributed computing," *IEEE Journal of Biomedical and Health Informatics*, vol. 17, no. 1, pp. 232–239, 2013.

[72] S. A. Dudani, K. J. Breeding, and R. B. McGhee, "Aircraft identification by moment invariants," *IEEE Transactions on Computers*, vol. 26, no. 1, pp. 39–45, 1977.

[73] A. Reeves, R. Prokop, S. Andrews, and F. Kuhl, "Three-dimensional shape analysis using moments and Fourier descriptors," *IEEE Transactions on Pattern Analysis and Machine Intelligence*, vol. 10, no. 6, pp. 937–943, 1988.

[74] F. Mokhtarian and S. Abbasi, "Robust automatic selection of optimal views in multi-view free-form object recognition," *Pattern Recognition*, vol. 38, no. 7, pp. 1021–1031, 2005.

[75] H. Li, X. Jin, N. Yang, and Z. Yang, "The recognition of landed aircrafts based on PCNN model and affine moment invariants," *Pattern Recognition Letters*, vol. 51, no. 1, pp. 23–29, 2015.

[76] J. Alves, J. Herman, and N. C. Rowe, "Robust recognition of ship types from an infrared silhouette," in *Proceedings of the Command and Control Research and Technology Symposium CCRTS'04*, p. 18–, CCRP, 2004.

[77] J. Yu, J. Lv, and X. Bai, "Target recognition research based on combined invariant moments," in *International Workshop on Image Processing and Optical Engineering* (H. Guo and Q. Ding, eds.), vol. 8335, pp. 83350M–1–83350M–6, SPIE, 2012.

[78] L. Eikvil, L. Aurdal, and H. Koren, "Classification-based vehicle detection in high-resolution satellite images," *ISPRS Journal of Photogrammetry and Remote Sensing*, vol. 64, no. 1, pp. 65–72, 2009.

[79] X. Sun, H. Lu, and J. Wu, "Bus detection based on sparse representation for transit signal priority," *Neurocomputing*, vol. 118, pp. 1–9, 2013.

[80] A. Apatean, A. Rogozan, and A. Bensrhair, "Objects recognition in visible and infrared images from the road scene," in *Proceedings of the International Conference on Automation, Quality and Testing, Robotics AQTR'08*, vol. 3, pp. 327–332, IEEE, 2008.

[81] S. Zhenghe, Z. Bo, Z. Zhongxiang, W. Meng, and M. Enrong, "Research on recognition method for traffic signs," in *Proceedings of the Second International Conference on Future Generation Communication and Networking FGCN'08*, pp. 387–390, IEEE, 2008.

[82] H. Gao, X. Liu, Z. Liu, and K. Hong, "Investigation on traffic signs recognition based on BP neural network and invariant moments," in *Advances in Future Computer and Control Systems* (D. Jin and S. Lin, eds.), vol. 159 of *Advances in Intelligent and Soft Computing*, pp. 157–162, Springer, 2012.

[83] J. Zhang, Y. Yu, and X. Liu, "An improved tracking method for automatic location of traffic signs in image-based virtual space," in *Advances in Image and Graphics Technologies, Chinese Conference IGTA'13* (T. Tan, Q. Ruan, X. Chen, H. Ma, and L. Wang, eds.), vol. 363 of *Communications in Computer and Information Science*, pp. 322–330, Springer, 2013.

[84] K. Balasundaram, M. K. Srinivasan, K. Sarukesi, and P. Rodrigues, "Implementation of next-generation traffic sign recognition system with two-tier classifier architecture," in *Proceedings of the International Conference on Advances in Computing, Communications and Informatics ICACCI'12*, pp. 481–487, ACM, 2012.

[85] H.-K. Kim, Y.-N. Shin, S.-g. Kuk, J. H. Park, and H.-Y. Jung, "Night-time traffic light detection based on SVM with geometric moment features," *World Academy of Science, Engineering and Technology (WASET)*, vol. 7, no. 4, pp. 454–457, 2013.

[86] X. Ma, R. Pan, and L. Wang, "License plate character recognition based on Gaussian-Hermite moments," in *Second International Workshop on Education Technology and Computer Science ETCS'10*, vol. 3, pp. 11–14, IEEE, 2010.

[87] X. X. Shao and M. J. Xie, "An automotive airbag detection method based on image processing," *Applied Mechanics and Materials*, vol. 416, pp. 1350–1354, 2013.

[88] M. S. Al-Batah, N. A. M. Isa, K. Z. Zamli, Z. M. Sani, and K. A. Azizli, "A novel aggregate classification technique using moment invariants and cascaded multilayered perceptron network," *International Journal of Mineral Processing*, vol. 92, no. 1–2, pp. 92–102, 2009.

[89] X. Zhang, W. Li, J. Xi, Z. Zhang, and X. Fan, "Surface defect target identification on copper strip based on adaptive genetic algorithm and feature saliency," *Mathematical Problems in Engineering*, pp. 1–10, 2013.

[90] N. Nacereddine, D. Ziou, and L. Hamami, "Fusion-based shape descriptor for weld defect radiographic image retrieval," *International Journal of Advanced Manufacturing Technology*, vol. 68, no. 9–12, pp. 2815–2832, 2013.

[91] Y. Wang and J. Su, "Automated defect and contaminant inspection of HVAC duct," *Automation in Construction*, vol. 41, pp. 15–24, 2014.

[92] T.-S. Li and C.-L. Huang, "Defect spatial pattern recognition using a hybrid SOM–SVM approach in semiconductor manufacturing," *Expert Systems with Applications*, vol. 36, no. 1, pp. 374–385, 2009.

[93] M. P.-L. Ooi, H. K. Sok, Y. C. Kuang, S. Demidenko, and C. Chan, "Defect cluster recognition system for fabricated semiconductor wafers," *Engineering Applications of Artificial Intelligence*, vol. 26, no. 3, pp. 1029–1043, 2013.

[94] X. Cheng, X. Jin, Z. Zhang, and J. Lu, "A geometrical defect detection method for non-silicon MEMS part based on HU moment invariants of skeleton image," in *Fifth International Conference on Graphic and Image Processing ICGIP'13* (Y. Wang, X. Jiang, M. Yang, D. Zhang, and X. Yi, eds.), vol. 9069, pp. 90691R–1–90691R–5, SPIE, 2014.

[95] S. Wanchat, S. Plermkamon, and D. Chetchotsak, "Object's centroid localization using Hu-Flusser's moments invariant," *Applied Mechanics and Materials*, vol. 526, pp. 316–323, 2014.

[96] D. Shukla, Ö. Erkent, and J. Piater, "General object tip detection and pose estimation for robot manipulation," in *Computer Vision Systems, 10th International Conference ICVS'15* (L. Nalpantidis, V. Krüger, J.-O. Eklundh, and A. Gasteratos, eds.), vol. 9163 of *Lecture Notes in Computer Science*, pp. 364–374, Springer, 2015.

[97] Y.-L. Kim, H.-C. Song, and J.-B. Song, "Hole detection algorithm for chamferless square peg-in-hole based on shape recognition using F/T sensor," *International Journal of Precision Engineering and Manufacturing*, vol. 15, no. 3, pp. 425–432, 2014.

[98] A. I. Barranco-Gutiérrez, S. Martínez-Díaz, and J. L. Gómez-Torres, "An approach for utility pole recognition in real conditions," in *the 6th Pacific-Rim Symposium on Image and Video Technology PSIVT'13 Workshops* (F. Huang and A. Sugimoto, eds.), vol. 8334 of *Lecture Notes in Computer Science*, pp. 113–121, Springer, 2014.

[99] X. Li and F. Sha, "Application of moment invariants in checking the overall dimension of the head cover," in *International Conference on Electronic and Mechanical Engineering and Information Technology EMEIT'11*, vol. 2, pp. 932–935, IEEE, 2011.

[100] W. Hu, H. Liu, C. Hu, S. Wang, D. Chen, J. Mo, and Q. Liang, "Vision-based force measurement using pseudo-Zernike moment invariants," *Measurement*, vol. 46, no. 10, pp. 4293–4305, 2013.

[101] W. Wu and Y. Meng, "Comparing different feature extraction methods of pump dynamograph based on support vector machine," in *Advances in Automation and Robotics, Vol. 2, Selected papers from the International Conference on Automation and Robotics ICAR'11* (G. Lee, ed.), vol. 123 of *Lecture Notes in Electrical Engineering*, pp. 501–506, Springer, 2012.

[102] F. Qin, L. Li, S. Gao, X. Yang, and X. Chen, "A deep learning approach to the classification of 3D CAD models," *Journal of Zhejiang University SCIENCE C*, vol. 15, no. 2, pp. 91–106, 2014.

[103] P. Premaratne, S. Yang, Z. Zou, and P. Vial, "Australian sign language recognition using moment invariants," in *9th International Conference on Intelligent Computing, Theories and Technology ICIC'13* (D.-S. Huang, K.-H. Jo, Y.-Q. Zhou, and K. Han, eds.), vol. 7996 of *Lecture Notes in Computer Science*, pp. 509–514, Springer, 2013.

[104] K. C. Otiniano-Rodríguez, G. Cámara-Chávez, and D. Menotti, "Hu and Zernike moments for sign language recognition," in *International Conference on Image Processing, Computer Vision, and Pattern Recognition IPCV'12* (H. R. Arabnia, L. Deligiannidis, and G. Schaefer, eds.), pp. 1–5, CSREA, 2012.

[105] R. Senanayake and S. Kumarawadu, "A robust vision-based hand gesture recognition system for appliance control in smart homes," in *International Conference on Signal Processing, Communication and Computing ICSPCC'12*, pp. 760–763, IEEE, 2012.

[106] F. G. Pereira, R. F. Vassallo, and E. O. T. Salles, "Human-robot interaction and cooperation through people detection and gesture recognition," *Journal of Control, Automation and Electrical Systems*, vol. 24, no. 3, pp. 187–198, 2013.

[107] M. Popa, A. K. Koc, L. J. Rothkrantz, C. Shan, and P. Wiggers, "Kinect sensing of shopping related actions," in *Constructing Ambient Intelligence, Aml 2011 Workshops, Revised Selected Papers* (R. Wichert, K. Van Laerhoven, and J. Gelissen, eds.), vol. 277 of *Communications in Computer and Information Science*, pp. 91–100, Springer, 2012.

[108] L. Keyes and A. Winstanley, "Using moment invariants for classifying shapes on large-scale maps," *Computers, Environment and Urban Systems*, vol. 25, no. 1, pp. 119–130, 2001.

[109] G. A. Licciardi, A. Villa, M. D. Mura, L. Bruzzone, J. Chanussot, and J. A. Benediktsson, "Retrieval of the height of buildings from WorldView-2 multi-angular imagery using attribute filters and geometric invariant moments," *IEEE Journal of Selected Topics in Applied Earth Observations and Remote Sensing*, vol. 5, no. 1, pp. 71–79, 2012.

[110] B. Li, "Airborne mine countermeasures," 2015. US Patent 9092866 B1.

[111] L. Pallotta, C. Clemente, A. D. Maio, J. J. Soraghan, and A. Farina, "Pseudo-Zernike moments based radar micro-Doppler classification," in *Radar Conference*, pp. 0850–0854, IEEE, 2014.

[112] M. Lopatka and W. van Houten, "Automated shape annotation for illicit tablet preparations: A contour angle based classification from digital images," *Science & Justice*, vol. 53, no. 1, pp. 60–66, 2013.

[113] S. F. Pratama, A. K. Muda, Y.-H. Choo, and A. Abraham, "A comparative study of 2D UMI and 3D Zernike shape descriptor for ATS drugs identification," in *Pattern Analysis, Intelligent Security and the Internet of Things* (A. Abraham, A. K. Muda, and Y.-H. Choo, eds.), vol. 355 of *Advances in Intelligent Systems and Computing*, pp. 237–249, Springer, 2015.

[114] V. Venkatraman, L. Sael, and D. Kihara, "Potential for protein surface shape analysis using spherical harmonics and 3D Zernike descriptors," *Cell Biochemistry and Biophysics*, vol. 54, no. 1, pp. 23–32, 2009.

[115] L. Wang and G. Healey, "Using Zernike moments for the illumination and geometry invariant classification of multispectral texture," *IEEE Transactions on Image Processing*, vol. 7, no. 2, pp. 196–203, 1998.

[116] P. Campisi, A. Neri, G. Panci, and G. Scarano, "Robust rotation-invariant texture classification using a model based approach," *IEEE Transactions on Image Processing*, vol. 13, no. 6, pp. 782–791, 2004.

[117] J. Bigün and J. M. Hans du Buf, "N-folded symmetries by complex moments in Gabor space and their application to unsupervised texture segmentation," *IEEE Transactions on Pattern Analysis and Machine Intelligence*, vol. 16, no. 1, pp. 80–87, 1994.

[118] A. Foulonneau, P. Charbonnier, and F. Heitz, "Affine-invariant geometric shape priors for region-based active contours," *IEEE Transactions on Pattern Analysis and Machine Intelligence*, vol. 28, no. 8, pp. 1352–1357, 2006.

[119] A. Kumar, "Nonlocal means image denoising using orthogonal moments," *Applied Optics*, vol. 54, no. 27, pp. 8156–8165, 2015.

[120] X. Sheng and P. Qi-cong, "3D object recognition using multiple features and neural network," in *Conference on Cybernetics and Intelligent Systems CIS'08*, pp. 434–439, IEEE, 2008.

[121] H. Liu, L. Zhang, and H. Huang, "Web-image driven best views of 3D shapes," *The Visual Computer*, vol. 28, no. 3, pp. 279–287, 2012.

[122] B. Zitová and J. Flusser, "Image registration methods: A survey," *Image and Vision Computing*, vol. 21, no. 11, pp. 977–1000, 2003.

[123] A. Goshtasby, "Template matching in rotated images," *IEEE Transactions on Pattern Analysis and Machine Intelligence*, vol. 7, no. 3, pp. 338–344, 1985.

[124] H. Li, B. S. Manjunath, and S. K. Mitra, "Contour-based multisensor image registration," in *26th Asilomar Conference on Signals, Systems and Computers ACSSC'92*, vol. 1, pp. 182–186, IEEE, 1992.

[125] X. Dai and S. Khorram, "A feature-based image registration algorithm using improved chain-code representation combined with invariant moments," *IEEE Transactions on Geoscience and Remote Sensing*, vol. 37, no. 5, pp. 2351–2362, 1999.

[126] S. X. Hu, Y.-M. Xiong, M. Z. W. Liao, and W. F. Chen, "Accurate point matching based on combined moment invariants and their new statistical metric," in *Proceedings of the International Conference on Wavelet Analysis and Pattern Recognition ICWAPR'07*, pp. 376–381, IEEE, 2007.

[127] L.-Q. Guo and M. Zhu, "Quaternion Fourier-Mellin moments for color images," *Pattern Recognition*, vol. 44, no. 2, pp. 187–195, 2011.

[128] A. Shi, M. Tang, F. Huang, L. Xu, and T. Fan, "Remotely sensed images registration based on wavelet transform using affine invariant moment," in *Proceedings of the Second International Symposium on Intelligent Information Technology Application IITA'08*, vol. 1, pp. 384–389, IEEE, 2005.

[129] J. Flusser and T. Suk, "A moment-based approach to registration of images with affine geometric distortion," *IEEE Transactions on Geoscience and Remote Sensing*, vol. 32, no. 2, pp. 382–387, 1994.

[130] Y. Bentoutou, N. Taleb, K. Kpalma, and J. Ronsin, "An automatic image registration for applications in remote sensing," *IEEE Transactions on Geoscience and Remote Sensing*, vol. 43, no. 9, pp. 2127–2137, 2005.

[131] C. Harris and M. Stephens, "A combined corner and edge detector," in *Proceedings of the Fourth Alvey Vision Conference AVC'88*, pp. 147–151, University of Manchester, 1988.

[132] Z.-L. Yang and B.-L. Guo, "Image registration using feature points extraction and pseudo-Zernike moments," in *International Conference on Intelligent Information Hiding and Multimedia Signal Processing IIH-MSP'08*, pp. 752–755, IEEE, 2008.

[133] M. A. Fischler and R. C. Bolles, "Random sample consensus: A paradigm for model fitting with applications to image analysis and automated cartography," *Graphics and Image Processing*, vol. 24, no. 6, pp. 381–395, 1981.

[134] F. Badra, A. Qumsieh, and G. Dudek, "Rotation and zooming in image mosaicing," in *Proceedings of the Fourth Workshop on Applications of Computer Vision WACV'98*, pp. 50–55, IEEE, 1998.

[135] M. S. Yasein and P. Agathoklis, "Automatic and robust image registration using feature points extraction and Zernike moments invariants," in *Proceedings of the Fifth International Symposium on Signal Processing and Information Technology ISSPIT'05*, pp. 566–571, IEEE, 2005.

[136] J. Sarvaiya, S. Patnaik, and H. Goklani, "Image registration using Mexican-hat wavelets and invariant moments," in *Computer Networks and Information Technologies, Second International Conference on Advances in Communication, Network, and Computing CNC'11* (V. V. Das, J. Stephen, and Y. Chaba, eds.), vol. 142 of *Communications in Computer and Information Science*, pp. 574–577, Springer, 2011.

[137] Y. Zhanlong and C. Hang, "Image registration using curvature scale space corner and pseudo-Zernike moments," in *13th International Symposium on Communications and Information Technologies ISCIT'13*, pp. 540–543, IEEE, 2013.

[138] Z. Chen and S.-K. Sun, "A Zernike moment phase-based descriptor for local image representation and matching," *IEEE Transactions on Image Processing*, vol. 19, no. 1, pp. 205–219, 2010.

[139] N. Kumar, A. C. Lammert, B. Englot, F. S. Hover, and S. S. Narayanan, "Directional descriptors using Zernike moment phases for object orientation estimation in underwater sonar images," in *International Conference on Acoustics, Speech and Signal Processing ICASSP'11*, pp. 1025–1028, IEEE, 2011.

[140] Q. Yang and Y. Wen, "Zernike moments descriptor matching based symmetric optical flow for motion estimation and image registration," in *International Joint Conference on Neural Networks IJCNN'14*, pp. 350–357, IEEE, 2014.

[141] M. Favorskaya, D. Pyankov, and A. Popov, "Motion estimations based on invariant moments for frames interpolation in stereovision," *Procedia Computer Science*, vol. 22, pp. 1102–1111, 2013. 17th International Conference in Knowledge Based and Intelligent Information and Engineering Systems KES'13.

[142] J. Sato and R. Cipolla, "Image registration using multi-scale texture moments," *Image and Vision Computing*, vol. 13, no. 5, pp. 341–353, 1995.

[143] P. Brivio, A. Ventura, A. Rampini, and R. Schettini, "Automatic selection of control points from shadow structures," *International Journal of Remote Sensing*, vol. 13, no. 10, pp. 1853–1860, 1992.

[144] Y. Ning, R. Chen, and P. Xu, "Wide baseline image mosaicing by integrating MSER and Hessian-affine," in *4th International Congress on Image and Signal Processing CISP'11*, vol. 4, pp. 2034–2037, IEEE, 2011.

[145] X. Zhao, C. Zhang, Y. Wang, and B. Yang, "A hybrid approach based on MEP and CSP for contour registration," *Applied Soft Computing*, vol. 11, no. 8, pp. 5391–5399, 2011.

[146] M. Pan, J. Tang, Q. Rong, and F. Zhang, "Medical image registration using modified iterative closest points," *International Journal for Numerical Methods in Biomedical Engineering*, vol. 27, no. 8, pp. 1150–1166, 2011.

[147] J. F. P. Crespo and P. M. Q. Aguiar, "Revisiting complex moments for 2D shape representation and image normalization," *IEEE Transactions on Image Processing*, vol. 20, no. 10, pp. 2896–2911, 2011.

[148] A. Jaklič and F. Solina, "Moments of superellipsoids and their application to range image registration," *IEEE Transactions on Systems, Man, and Cybernetics – Part B: Cybernetics*, vol. 33, no. 4, pp. 648–657, 2003.

[149] S. S. Nejhum, Y.-T. Chi, J. Ho, and M.-H. Yang, "Higher-dimensional affine registration and vision applications," *IEEE Transactions on Pattern Analysis and Machine Intelligence*, vol. 33, no. 7, pp. 1324–1338, 2011.

[150] D. Shen and C. Davatzikos, "Hammer: Hierarchical attribute matching mechanism for elastic registration," *IEEE Transactions on Medical Imaging*, vol. 21, no. 11, pp. 1421–1439, 2002.

[151] H.-R. Su, H.-Y. Kuo, S.-H. Lai, and C.-C. Wu, "Fast 3D object alignment from depth image with 3D Fourier moment matching on GPU," in *2nd International Conference on 3D Vision 3DV'14*, vol. 1, pp. 179–186, IEEE, 2014.

[152] R. B. Rusu and S. Cousins, "3D is here: Point cloud library (PCL)," in *International Conference on Robotics and Automation ICRA'11*, pp. 1–4, IEEE, 2011.

[153] J. Flusser, T. Suk, and S. Saic, "Recognition of blurred images by the method of moments," *IEEE Transactions on Image Processing*, vol. 5, no. 3, pp. 533–538, 1996.

[154] J. Flusser and T. Suk, "Degraded image analysis: An invariant approach," *IEEE Transactions on Pattern Analysis and Machine Intelligence*, vol. 20, no. 6, pp. 590–603, 1998.

[155] Y. Bentoutou, N. Taleb, M. Chikr El Mezouar, M. Taleb, and J. Jetto, "An invariant approach for image registration in digital subtraction angiography," *Pattern Recognition*, vol. 35, no. 12, pp. 2853–2865, 2002.

[156] I. Makaremi and M. Ahmadi, "Wavelet domain blur invariants for image analysis," *IEEE Transactions on Image Processing*, vol. 21, no. 3, pp. 996–1006, 2012.

[157] Y. Bentoutou and N. Taleb, "Automatic extraction of control points for digital subtraction angiography image enhancement," *IEEE Transactions on Nuclear Science*, vol. 52, no. 1, pp. 238–246, 2005.

[158] Y. Bentoutou and N. Taleb, "A 3-D space-time motion detection for an invariant image registration approach in digital subtraction angiography," *Computer Vision and Image Understanding*, vol. 97, pp. 30–50, 2005.

[159] Y. Bentoutou, N. Taleb, A. Bounoua, K. Kpalma, and J. Ronsin, "Feature based registration of satellite images," in *Proceedings of the 15th International Conference on Digital Signal Processing DSP 2007*, pp. 419–422, IEEE, 2007.

[160] Z. Liu, J. An, and L. Li, "A two-stage registration algorithm for oil spill aerial image by invariants-based similarity and improved ICP," *International Journal of Remote Sensing*, vol. 32, no. 13, pp. 3649–3664, 2011.

[161] J. Flusser and B. Zitová, "Combined invariants to linear filtering and rotation," *International Journal of Pattern Recognition and Artificial Intelligence*, vol. 13, no. 8, pp. 1123–1136, 1999.

[162] B. Chen, H. Shu, H. Zhang, G. Coatrieux, L. Luo, and J. L. Coatrieux, "Combined invariants to similarity transformation and to blur using orthogonal Zernike moments," *IEEE Transactions on Image Processing*, vol. 20, no. 2, pp. 345–360, 2011.

[163] J. Flusser, B. Zitová, and T. Suk, "Invariant-based registration of rotated and blurred images," in *Proceedings of the International Geoscience and Remote Sensing Symposium IGARSS'99*, pp. 1262–1264, IEEE, 1999.

[164] B. Zitová and J. Flusser, "Invariants to convolution and rotation," in *Invariants for Pattern Recognition and Classification* (M. A. Rodrigues, ed.), pp. 23–46, World Scientific, 2000.

[165] J. Flusser, J. Boldyš, and B. Zitová, "Moment forms invariant to rotation and blur in arbitrary number of dimensions," *IEEE Transactions on Pattern Analysis and Machine Intelligence*, vol. 25, no. 2, pp. 234–246, 2003.

[166] B. Zitová and J. Flusser, "Estimation of camera planar motion from blurred images," in *Proceedings of the International Conference on Image Processing ICIP'02*, pp. 329–332, IEEE, 2002.

[167] X. Dai, H. Zhang, H. Shu, L. Luo, and T. Liu, "Blurred image registration by combined invariant of Legendre moment and Harris-Laplace detector," in *Proceedings of the Fourth Pacific-Rim Symposium on Image and Video Technology PSIVT'10*, pp. 300–305, IEEE, 2010.

[168] X. Zuo, X. Dai, and L. Luo, "M-SIFT: A new descriptor based on Legendre moments and SIFT," in *Proceedings of the 3rd International Conference on Machine Vision ICMV'10*, pp. 183–186, 2010.

[169] D. G. Lowe, "Distinctive image features from scale-invariant keypoints," *International Journal of Computer Vision*, vol. 60, no. 2, pp. 91–110, 2004.

[170] T. Mouats and N. Aouf, "Cross-spectral stereo matching based on local self-similarities and image moments," in *International Conference on Systems, Man, and Cybernetics SMC'13*, pp. 4048–4053, IEEE, 2013.

[171] B. Zitová, J. Kautsky, G. Peters, and J. Flusser, "Robust detection of significant points in multiframe images," *Pattern Recognition Letters*, vol. 20, no. 2, pp. 199–206, 1999.

[172] F. Šroubek, G. Cristóbal, and J. Flusser, "A unified approach to superresolution and multichannel blind deconvolution," *IEEE Transactions on Image Processing*, vol. 16, no. 9, pp. 2322–2332, 2007.

[173] J. Ho, A. M. Peter, A. Rangarajan, and M.-H. Yang, "An algebraic approach to affine registration of point sets," in *12th International Conference on Computer Vision'09*, pp. 1335–1340, IEEE, 2009.

[174] C. Domokos and Z. Kato, "Parametric estimation of affine deformations of planar shapes," *Pattern Recognition*, vol. 43, no. 3, pp. 569–578, 2010.

[175] W. Liu and E. Ribeiro, "Incremental variations of image moments for nonlinear image registration," *Signal, Image and Video Processing*, vol. 8, no. 3, pp. 423–432, 2014.

[176] J. Flusser, J. Kautsky, and F. Šroubek, "Implicit moment invariants," *International Journal of Computer Vision*, vol. 86, no. 1, pp. 72–86, 2010.

[177] C. Domokos, J. Nemeth, and Z. Kato, "Nonlinear shape registration without correspondences," *IEEE Transactions on Pattern Analysis and Machine Intelligence*, vol. 34, no. 5, pp. 943–958, 2012.

[178] E. de Castro and C. Morandi, "Registration of translated and rotated images using finite Fourier transform," *IEEE Transactions on Pattern Analysis and Machine Intelligence*, vol. 9, no. 5, pp. 700–703, 1987.

[179] V. Ojansivu and J. Heikkilä, "A method for blur and affine invariant object recognition using phase-only bispectrum," in *The International Conference on Image Analysis and Recognition ICIAR'08*, vol. LNCS 5112, pp. 527–536, Springer, 2008.

[180] V. Ojansivu and J. Heikkilä, "Image registration using blur-invariant phase correlation," *IEEE Signal Processing Letters*, vol. 14, no. 7, pp. 449–452, 2007.

[181] S. Tang, Y. Wang, and Y.-W. Chen, "Blur invariant phase correlation in X-ray digital subtraction angiography," in *Preprints of the IEEE/CME International Conference on Complex Medical Engineering*, pp. 1715–1719, 2007.

[182] M. Pedone, J. Flusser, and J. Heikkilä, "Blur invariant translational image registration for N-fold symmetric blurs," *IEEE Transactions on Image Processing*, vol. 22, no. 9, pp. 3676–3689, 2013.

[183] M. Pedone, J. Flusser, and J. Heikkilä, "Registration of images with N-fold dihedral blur," *IEEE Transactions on Image Processing*, vol. 24, no. 3, pp. 1036–1045, 2015.

[184] Y. Zhang, C. Wen, and Y. Zhang, "Estimation of motion parameters from blurred images," *Pattern Recognition Letters*, vol. 21, no. 5, pp. 425–433, 2000.

[185] Y. Zhang, C. Wen, Y. Zhang, and Y. C. Soh, "Determination of blur and affine combined invariants by normalization," *Pattern Recognition*, vol. 35, no. 1, pp. 211–221, 2002.

[186] B. Zitová, H. Ríos, J. M. Gutierrez, and A. Marin, "Recognition of landmarks distorted by fish-eye lens," in *Proceedings of the Simposio Iberoamericano de Reconocimiento de Patrones SIARP'99*, pp. 611–622, Instituto Politecnico Nacional, 1999.

[187] J. Courtney and A. Jain, "Mobile robot localization via classification of multisensor maps," in *Proceedings of the International Conference on Robotics and Automation ICRA'94*, vol. 2, pp. 1672–1678, IEEE, 1994.

[188] W. Hao, H. Xiao-rong, F. Bao-fu, G. Wei, and M. Fan-hui, "Target recognition of household service robot based on shape moment invariants," in *International Conference on Computer Application and System Modeling ICCASM'10*, vol. 10, pp. V10–82–V10–85, IEEE, 2010.

[189] G. Xu, Y. Zhang, S. Ji, Y. Cheng, and Y. Tian, "Research on computer vision-based for UAV autonomous landing on a ship," *Pattern Recognition Letters*, vol. 30, no. 6, pp. 600–605, 2009.

[190] C. Miao and J. Li, "Autonomous landing of small unmanned aerial rotorcraft based on monocular vision in GPS-denied area," *IEEE/CAA Journal of Automatica Sinica*, vol. 2, no. 1, pp. 109–114, 2015.

[191] N. İmamoğlu, M. Ö. Efe, A. Eresen, and O. Kaynak, "A novel unmanned aerial surveillance scheme," in *International Conference on Mechatronics and Control ICMC'14*, pp. 26–31, IEEE, 2014.

[192] J. Lee and H. Ko, "Gradient-based local affine invariant feature extraction for mobile robot localization in indoor environments," *Pattern Recognition Letters*, vol. 29, no. 14, pp. 1934–1940, 2008.

[193] Z. Lin, S. Kim, and I. S. Kweon, "Robust invariant features for object recognition and mobile robot navigation," in *Proceedings of the IAPR Conference on Machine Vision Applications MVA'05*, pp. 611–622, Oxford University Press, 2005.

[194] G. Wells and C. Torras, "Selection of image features for robot positioning using mutual information," in *Proceedings of the International Conference on Robotics and Automation ICRA'98*, vol. 4, pp. 2819–2826, IEEE, 1998.

[195] E. Celaya, J.-L. Albarral, P. Jiménez, and C. Torras, "Natural landmark detection for visually-guided robot navigation," in *Proceedings of the Tenth Congress of the Italian Association for Artificial Intelligence AI*IA'07*, vol. LNAI 4733, pp. 555–566, Springer, 2007.

[196] A. M. Pinto, L. F. Rocha, and A. P. Moreira, "Object recognition using laser range finder and machine learning techniques," *Robotics and Computer-Integrated Manufacturing*, vol. 29, no. 1, pp. 12–22, 2013.

[197] Q. Xie, G. Liang, C. Tang, and X. Wu, "Robust gesture-based interaction system for manipulating service robot," in *International Conference on Information Science and Technology ICIST'13*, pp. 658–662, IEEE, 2013.

[198] F. Chaumette and S. Hutchinson, "Visual servoing and visual tracking," in *Springer Handbook of Robotics* (B. Siciliano and O. Khatib, eds.), pp. 563–583, Springer, 2008.

[199] F. Chaumette, "A first step toward visual servoing using image moments," in *IEEE/RSJ International Conference on Intelligent Robots and Systems IROS'02*, vol. 1, pp. 378–383, IEEE, 2002.

[200] M. Bakthavatchalam, F. Chaumette, and É. Marchand, "Photometric moments: New promising candidates for visual servoing," in *International Conference on Robotics and Automation ICRA'13*, pp. 5241–5246, IEEE, 2013.

[201] R. Mebarki, A. Krupa, and F. Chaumette, "Image moments–based ultrasound visual servoing," in *International Conference on Robotics and Automation ICRA'08*, pp. 113–119, IEEE,2008.

[202] J. Wang and H. Cho, "Micropeg and hole alignment using image moments based visual servoing method," *IEEE Transactions on Industrial Electronics*, vol. 55, no. 3, pp. 1286–1294, 2008.

[203] J. Kautsky, J. Flusser, B. Zitová, and S. Šimberová, "A new wavelet-based measure of image focus," *Pattern Recognition Letters*, vol. 23, no. 14, pp. 1785–1794, 2002.

[204] Y. Zhang, Y. Zhang, and C. Wen, "A new focus measure method using moments," *Image and Vision Computing*, vol. 18, no. 12, pp. 959–965, 2000.

[205] C.-Y. Wee, R. Paramesran, and R. Mukundan, "Quality assessment of Gaussian blurred images using symmetric geometric moments," in *Proceedings of the International Conference on Image Analysis and Processing ICIAP'07*, pp. 807–812, IEEE, 2007.

[206] D.-O. Kim, R.-H. Park, and D.-G. Sim, "Reduced-reference visual quality assessment using central moments of edge projections," *Journal of Communication and Computer*, vol. 10, pp. 564–566, 2013.

[207] P.-T. Yap and P. Raveendran, "Image focus measure based on Chebyshev moments," *IEE Proceedings of the Vision, Image and Signal Processing*, vol. 151, no. 2, pp. 128–136, 2004.

[208] L. Li, W. Lin, X. Wang, G. Yang, K. Bahrami, and A. C. Kot, "No-reference image blur assessment based on discrete orthogonal moments," *IEEE Transactions on Cybernetics*, vol. 46, no. 1, pp. 39–50, 2016.

[209] K.-H. Thung, R. Paramesran, and C.-L. Lim, "Content-based image quality metric using similarity measure of moment vectors," *Pattern Recognition*, vol. 45, no. 6, pp. 2193–2204, 2012.

[210] W. Zhang, L. Li, H. Zhu, and D. Cheng, "No-reference quality metric of blocking artifacts," *Journal of Information Hiding and Multimedia Signal Processing*, vol. 5, pp. 564–566, 2014.

[211] Y. Rui, T. S. Huang, and S.-F. Chang, "Image retrieval: Current techniques, promising directions, and open issues," *Journal of Visual Communication and Image Representation*, vol. 10, no. 1, pp. 39–62, 1999.

[212] A. Amanatiadis, V. Kaburlasos, A. Gasteratos, and S. Papadakis, "Evaluation of shape descriptors for shape-based image retrieval," *IET Image Processing*, vol. 5, no. 5, pp. 493–499, 2011.

[213] C. Di Ruberto and A. Morgera, "Moment-based techniques for image retrieval," in *19th International Conference on Database and Expert Systems Application DEXA'08*, pp. 155–159, IEEE, 2008.

[214] R. B. Yadav, N. K. Nishchal, A. K. Gupta, and V. K. Rastogi, "Retrieval and classification of objects using generic Fourier, Legendre moment, and wavelet Zernike moment descriptors and recognition using joint transform correlator," *Optics & Laser Technology*, vol. 40, no. 3, pp. 517–527, 2008.

[215] M. Flickner, H. Sawhney, W. Niblack, J. Ashley, Q. Huang, B. Dom, M. Gorkani, J. Hafner, D. Lee, D. Petkovic, D. Steele, and P. Yanker, "Query by image and video content: The QBIC system," *Computer*, vol. 28, no. 9, pp. 23–32, 1995.

[216] P. Srivastava, N. T. Binh, and A. Khare, "Content-based image retrieval using moments," in *2nd International Conference on Context-Aware Systems and Applications ICCASA'13* (P. C. Vinh, V. Alagar, E. Vassev, and A. Khare, eds.), vol. 128 of *Lecture Notes of the Institute for Computer Sciences*, pp. 228–237, Springer, 2013.

[217] M. Banerjee, M. K. Kundu, and P. Maji, "Content-based image retrieval using visually significant point features," *Fuzzy Sets and Systems*, vol. 160, no. 23, pp. 3323–3341, 2009. Theme: Computer Science.

[218] C. S. Rao, S. S. Kumar, and B. C. Mohan, "Content based image retrieval using exact Legendre moments and support vector machine," *The International Journal of Multimedia & Its Applications*, vol. 2, no. 2, pp. 69–79, 2010.

[219] C. Wan and Y. Wu, "Image retrieval by using non-subsampled shearlet transform and Krawtchouk moment invariants," in *Computer Vision – ACCV'14 Workshops* (C. V. Jawahar and S. Shan, eds.), vol. 9010 of *Lecture Notes in Computer Science*, pp. 218–232, Springer, 2015.

[220] B. Jyothi, Y. MadhaveeLatha, and P. Krishna Mohan, "An effective multiple visual features for content based medical image retrieval," in *9th International Conference on Intelligent Systems and Control ISCO'15*, pp. 1–5, IEEE, 2015.

[221] C.-H. Wei, Y. Li, W.-Y. Chau, and C.-T. Li, "Trademark image retrieval using synthetic features for describing global shape and interior structure," *Pattern Recognition*, vol. 42, no. 3, pp. 386–394, 2009.

[222] X. Fu, Y. Li, R. Harrison, and S. Belkasim, "Content-based image retrieval using Gabor-Zernike features," in *Proceedings of the 18th International Conference on Pattern Recognition ICPR'06*, p. 4, IEEE, 2006.

[223] X.-Y. Wang, Y.-J. Yu, and H.-Y. Yang, "An effective image retrieval scheme using color, texture and shape features," *Computer Standards & Interfaces*, vol. 33, no. 1, pp. 59–68, 2011. Special Issue: Secure Semantic Web.

[224] C. Singh and Pooja, "Local and global features based image retrieval system using orthogonal radial moments," *Optics and Lasers in Engineering*, vol. 50, no. 5, pp. 655–667, 2012.

[225] T. Tuytelaars and L. V. Gool, "Content-based image retrieval based on local affinely invariant regions," in *3rd International Conference on Visual Information Systems VISUAL'99*, vol. LNCS 1614, pp. 493–500, Springer, 1999.

[226] F. Mindru, T. Tuytelaars, L. V. Gool, and T. Moons, "Moment invariants for recognition under changing viewpoint and illumination," *Computer Vision and Image Understanding*, vol. 94, no. 1–3, pp. 3–27, 2004.

[227] E. G. Karakasis, A. Amanatiadis, A. Gasteratos, and S. A. Chatzichristofis, "Image moment invariants as local features for content based image retrieval using the bag-of-visual-words model," *Pattern Recognition Letters*, vol. 55, no. 1, pp. 22–27, 2015.

[228] A. K. Jain and A. Vailaya, "Shape-based retrieval: A case study with trademark image databases," *Pattern Recognition*, vol. 31, no. 9, pp. 1369–1390, 1998.

[229] K. Iqbal, M. O. Odetayo, and A. James, "Content-based image retrieval approach for biometric security using colour, texture and shape features controlled by fuzzy heuristics," *Journal of Computer and System Sciences*, vol. 78, no. 4, pp. 1258–1277, 2012.

[230] L. Shao and M. Brady, "Invariant salient regions based image retrieval under viewpoint and illumination variations," *Journal of Visual Communication and Image Representation*, vol. 17, no. 6, pp. 1256–1272, 2006.

[231] Y. Li, "Reforming the theory of invariant moments for pattern recognition," *Pattern Recognition*, vol. 25, no. 7, pp. 723–730, 1992.

[232] E. J. Pauwels, P. M. de Zeeuw, and E. B. Ranguelova, "Computer-assisted tree taxonomy by automated image recognition," *Engineering Applications of Artificial Intelligence*, vol. 22, no. 1, pp. 26–31, 2009.

[233] P. Novotný and T. Suk, "Leaf recognition of woody species in Central Europe," *Biosystems Engineering*, vol. 115, no. 4, pp. 444–452, 2013.

[234] D.-G. Sim, H.-K. Kim, and R.-H. Park, "Invariant texture retrieval using modified Zernike moments," *Image and Vision Computing*, vol. 22, no. 4, pp. 331–342, 2004.

[235] K. Lu, N. He, and J. Xue, "Content-based similarity for 3D model retrieval and classification," *Progress in Natural Science*, vol. 19, no. 4, pp. 495–499, 2009.

[236] A. Mademlis, A. Axenopoulos, P. Daras, D. Tzovaras, and M. G. Strintzis, "3D content-based search based on 3D Krawtchouk moments," in *Proceedings of the 3rd International Symposium on 3D Data Processing, Visualization, and Transmission DPVT'06*, pp. 743–749, IEEE, 2006.

[237] G. Pass and R. Zabih, "Histogram refinement for content-based image retrieval," in *Proceedings 3rd IEEE Workshop on Applications of Computer Vision WACV'96*, pp. 96–102, IEEE, 1996.

[238] W. Xiaoling, "A novel circular ring histogram for content-based image retrieval," in *First International Workshop on Education Technology and Computer Science ETCS'09*, vol. 2, pp. 785–788, IEEE, 2009.

[239] M. J. Swain and D. H. Ballard, "Color indexing," *International Journal of Computer Vision*, vol. 7, no. 1, pp. 11–32, 1991.

[240] S. Jeong, "Histogram-based color image retrieval," tech. rep., Stanford University, 2001. Psych221/EE362 project report.

[241] G.-H. Liu, L. Zhang, Y.-K. Hou, Z.-Y. Li, and J.-Y. Yang, "Image retrieval based on multi-texton histogram," *Pattern Recognition*, vol. 43, no. 7, pp. 2380–2389, 2010.

[242] M. K. Mandal, T. Aboulnasr, and S. Panchanathan, "Image indexing using moments and wavelets," *IEEE Transactions on Consumer Electronics*, vol. 42, no. 3, pp. 557–565, 1996.

[243] P.-T. Yap and R. Paramesran, "Content-based image retrieval using Legendre chromaticity distribution moments," *IEE Proceedings – Vision, Image and Signal Processing*, vol. 153, no. 1, pp. 17–24, 2006.

[244] M. Stricker and M. Orengo, "Similarity of color images," in *Proceedings of the Storage and Retrieval for Image and Video Databases III*, vol. 2420, pp. 381–392, SPIE, 1995.

[245] T. Mostafa, H. M. Abbas, and A. A. Wahdan, "On the use of hierarchical color moments for image indexing and retrieval," in *International Conference on Systems, Man and Cybernetics*, vol. 7, p. 6–, IEEE, 2002.

[246] Z.-C. Huang, P. P. K. Chan, W. W. Y. Ng, and D. S. Yeung, "Content-based image retrieval using color moment and Gabor texture feature," in *International Conference on Machine Learning and Cybernetics ICMLC'10*, vol. 2, pp. 719–724, IEEE, 2010.

[247] C. Höschl IV, and J. Flusser, "Noise-resistant image retrieval," in *22nd International Conference on Pattern Recognition ICPR'14*, pp. 2972–2977, IEEE, 2014.

[248] C. Höschl IV, and J. Flusser, "Robust histogram-based image retrieval," *Pattern Recognition Letters*, vol. 69, no. 1, pp. 72–81, 2016.

[249] H. Yu, M. Li, H.-J. Zhang, and J. Feng, "Color texture moments for content-based image retrieval," in *International Conference on Image Processing ICIP'02*, vol. 3, pp. 929–932, IEEE, 2002.

[250] G. Duan, X. Zhao, A. Chen, and Y. Liu, "An improved Hu moment invariants based classfication method for watermarking algorithm," in *International Conference on Information and Network Security ICINS'14*, pp. 205–209, IET, 2014.

[251] M. Alghoniemy and A. H. Tewfik, "Geometric invariance in image watermarking," *IEEE Transactions on Image Processing*, vol. 13, no. 4, pp. 145–153, 2004.

[252] P. K. Tzouveli, K. S. Ntalianis, and S. D. Kollias, "Human video object watermarking based on Hu moments," in *Workshop on Signal Processing Systems Design and Implementation SiPS'05*, pp. 104–109, IEEE, 2005.

[253] Y. Xin, S. Liao, and M. Pawlak, "Circularly orthogonal moments for geometrically robust image watermarking," *Pattern Recognition*, vol. 40, no. 12, pp. 3740–3752, 2007.

[254] H. S. Kim and H.-K. Lee, "Invariant image watermark using Zernike moments," *IEEE Transactions on Circuits and Systems for Video Technology*, vol. 13, no. 8, pp. 766–775, 2003.

[255] N. Singhal, Y.-Y. Lee, C.-S. Kim, and S.-U. Lee, "Robust image watermarking using local Zernike moments," *Journal of Visual Communication and Image Representation*, vol. 20, no. 6, pp. 408–419, 2009.

[256] X.-C. Yuan, C.-M. Pun, and C.-L. P. Chen, "Geometric invariant watermarking by local Zernike moments of binary image patches," *Signal Processing*, vol. 93, no. 7, pp. 2087–2095, 2013.

[257] X.-Y. Wang, Z.-H. Xu, and H.-Y. Yang, "A robust image watermarking algorithm using SVR detection," *Expert Systems with Applications*, vol. 36, no. 5, pp. 9056–9064, 2009.

[258] L. Zhang, G. Qian, W. Xiao, and Z. Ji, "Geometric invariant blind image watermarking by invariant Tchebichef moments," *Optics Express*, vol. 15, no. 5, pp. 2251–2261, 2007.

[259] L. Zhang, W.-W. Xiao, and Z. Ji, "Local affine transform invariant image watermarking by Krawtchouk moment invariants," *IET Information Security*, vol. 1, no. 3, pp. 97–105, 2007.

[260] G. A. Papakostas, E. D. Tsougenis, and D. E. Koulouriotis, "Moment-based local image watermarking via genetic optimization," *Applied Mathematics and Computation*, vol. 227, pp. 222–236, 2014.

[261] E. D. Tsougenis, G. A. Papakostas, D. E. Koulouriotis, and V. D. Tourassis, "Towards adaptivity of image watermarking in polar harmonic transforms domain," *Optics & Laser Technology*, vol. 54, pp. 84–97, 2013.

[262] L. Li, S. Li, A. Abraham, and J.-S. Pan, "Geometrically invariant image watermarking using polar harmonic transforms," *Information Sciences*, vol. 199, pp. 1–19, 2012.

[263] E. D. Tsougenis, G. A. Papakostas, D. E. Koulouriotis, and E. G. Karakasis, "Adaptive color image watermarking by the use of quaternion image moments," *Expert Systems with Applications*, vol. 41, no. 14, pp. 6408–6418, 2014.

[264] H. Zhang, H. Shu, G. Coatrieux, J. Zhu, J. Wu, Y. Zhang, H. Zhu, and L. Luo, "Affine Legendre moment invariants for image watermarking robust to geometric distortions," *IEEE Transactions on Image Processing*, vol. 20, no. 8, pp. 2189–2199, 2011.

[265] H. Zhu, M. Liu, and Y. Li, "The RST invariant digital image watermarking using Radon transforms and complex moments," *Digital Signal Processing*, vol. 20, no. 6, pp. 1612–1628, 2010.

[266] M. A. Savelonas and S. Chountasis, "Noise-resistant watermarking in the fractional Fourier domain utilizing moment-based image representation," *Signal Processing*, vol. 90, no. 8, pp. 2521–2528, 2010. Special Section on Processing and Analysis of High-Dimensional Masses of Image and Signal Data.

[267] J.-l. Dong, G.-r. Yin, and Z.-l. Ping, "Geometrically robust image watermarking based on Jacobi-Fourier moments," *Optoelectronics Letters*, vol. 5, no. 5, pp. 387–390, 2009.

[268] G. Gao and G. Jiang, "Bessel-Fourier moment-based robust image zero-watermarking," *Multimedia Tools and Applications*, vol. 74, no. 3, pp. 841–858, 2015.

[269] S. Ahmad, Q. Zhang, Z.-M. Lu, and M. W. Anwar, "Feature-based watermarking using discrete orthogonal Hahn moment invariants," in *Proceedings of the 7th International Conference on Frontiers of Information Technology FIT'09*, pp. 38:1–38:6, ACM, 2009.

[270] L. Parameswaran and K. Anbumani, "A robust image watermarking scheme using image moment normalization," *Transactions on Engineering, Computing and Technology*, vol. 13, no. 5, pp. 239–243, 2006.

[271] C.-S. Lu, "Towards robust image watermarking: Combining content-dependent key, moment normalization, and side-informed embedding," *Signal Processing: Image Communication*, vol. 20, no. 2, pp. 129–150, 2005.

[272] P. Dong, J. G. Brankov, N. P. Galatsanos, Y. Yang, and F. Davoine, "Digital watermarking robust to geometric distortions," *IEEE Transactions on Image Processing*, vol. 14, no. 12, pp. 2140–2150, 2005.

[273] I. J. Cox, J. Kilian, F. T. Leighton, and T. Shamoon, "Secure spread spectrum watermarking for multimedia," *IEEE Transactions on Image Processing*, vol. 6, no. 12, pp. 1673–1687, 1997.

[274] A. G. Bors, "Watermarking mesh-based representations of 3-D objects using local moments," *IEEE Transactions on Image Processing*, vol. 15, no. 3, pp. 687–701, 2006.

[275] E. D. Tsougenis, G. A. Papakostas, D. E. Koulouriotis, and V. D. Tourassis, "Performance evaluation of moment-based watermarking methods: A review," *The Journal of Systems and Software*, vol. 85, no. 8, pp. 1864–1884, 2012.

[276] C. Deng, X. Gao, X. Li, and D. Tao, "A local Tchebichef moments-based robust image watermarking," *Signal Processing*, vol. 89, no. 8, pp. 1531–1539, 2009.

[277] K. Jelen and S. Kůsová, "Pregnant women: Moiré contourgraph and its semiautomatic and automatic evaluation," *Neuroendocrinology Letters*, vol. 25, no. 1–2, pp. 52–56, 2004.

[278] A. Ruggeri and S. Pajaro, "Automatic recognition of cell layers in corneal confocal microscopy images," *Computer methods and programs in Biomedicine*, vol. 68, no. 1, pp. 25–35, 2002.

[279] D. Marín, A. Aquino, M. E. Gegúndez-Arias, and J. M. Bravo, "A new supervised method for blood vessel segmentation in retinal images by using gray-level and moment invariants-based features," *IEEE Transactions on Medical Imaging*, vol. 30, no. 1, pp. 146–158, 2011.

[280] S. S. Patankar and J. V. Kulkarni, "Orthogonal moments for determining correspondence between vessel bifurcations for retinal image registration," *Computer Methods and Programs in Biomedicine*, vol. 119, no. 3, pp. 121–141, 2015.

[281] Y. Zheng, S. Baloch, S. Englander, M. D. Schnall, and D. Shen, "Segmentation and classification of breast tumor using dynamic contrast-enhanced MR images," in *10th International Conference on Medical Image Computing and Computer-Assisted Intervention MICCAI'07* (N. Ayache, S. Ourselin, and A. Maeder, eds.), vol. 4792 of *Lecture Notes in Computer Science*, pp. 393–401, Springer, 2007.

[282] Y. Zheng, S. Englander, S. Baloch, E. I. Zacharaki, Y. Fan, M. D. Schnall, and D. Shen, "STEP: spatiotemporal enhancement pattern for MR-based breast tumor diagnosis," *Medical Physics*, vol. 36, no. 7, pp. 3192–3204, 2009.

[283] A. Tahmasbi, F. Saki, and S. B. Shokouhi, "Classification of benign and malignant masses based on Zernike moments," *Computers in Biology and Medicine*, vol. 41, no. 8, pp. 726–735, 2011.

[284] K. Sankar and K. Nirmala, "Orthogonal features based classification of microcalcification in mammogram using Jacobi moments," *Indian Journal of Science & Technology*, vol. 8, no. 15, pp. 1–7, 2015.

[285] K. Wu, H. Shu, and J.-L. Dillenseger, "Region and boundary feature estimation on ultrasound images using moment invariants," *Computer Methods and Programs in Biomedicine*, vol. 113, no. 2, pp. 446–455, 2014.

[286] M. E. Gonzalez, K. Dinelle, N. Vafai, N. Heffernan, J. McKenzie, S. Appel-Cresswell, M. J. McKeown, A. J. Stoessl, and V. Sossi, "Novel spatial analysis method for PET images using 3D moment invariants: Applications to Parkinson's disease," *NeuroImage*, vol. 68, pp. 11–21, 2013.

[287] A. Kaur and C. Singh, "Automatic cephalometric landmark detection using Zernike moments and template matching," *Signal, Image and Video Processing*, vol. 9, no. 1, pp. 117–132, 2015.

[288] I. Sommer, O. Müller, F. S. Domingues, O. Sander, J. Weickert, and T. Lengauer, "Moment invariants as shape recognition technique for comparing protein binding sites," *Bioinformatics*, vol. 23, no. 23, pp. 3139–3146, 2007.

[289] C.-M. Hung, Y.-M. Huang, and M.-S. Chang, "Alignment using genetic programming with causal trees for identification of protein functions," *Nonlinear Analysis*, vol. 65, no. 5, pp. 1070–1093, 2006.

[290] D. Avci and A. Varol, "An expert diagnosis system for classification of human parasite eggs based on multi-class SVM," *Expert Systems with Applications*, vol. 36, no. 1, pp. 43–48, 2009.

[291] P. P. Banada, K. Huff, E. Bae, B. Rajwa, A. Aroonnual, B. Bayraktar, A. Adil, J. P. Robinson, E. D. Hirleman, and A. K. Bhunia, "Label-free detection of multiple bacterial pathogens using light-scattering sensor," *Biosensors and Bioelectronics*, vol. 24, no. 6, pp. 1685–1692, 2009.

[292] J.-F. Mangin, F. Poupon, E. Duchesnay, D. Rivière, A. Cachia, D. L. Collins, A. C. Evans, and J. Régis, "Brain morphometry using 3D moment invariants," *Medical Image Analysis*, vol. 8, no. 3, pp. 187–196, 2004.

[293] Z. Y. Sun, D. Rivière, F. Poupon, J. Régis, and J.-F. Mangin, "Automatic inference of sulcus patterns using 3D moment invariants," in *Medical Image Computing and Computer-Assisted Intervention MICCAI'07* (N. Ayache, S. Ourselin, and A. Maeder, eds.), vol. 4791 of *Lecture Notes in Computer Science*, pp. 515–522, Springer, 2007.

[294] M. Jabarouti Moghaddam and H. Soltanian-Zadeh, "Automatic segmentation of brain structures using geometric moment invariants and artificial neural networks," in *21st International Conference on Information Processing in Medical Imaging IPMI'09* (J. L. Prince, D. L. Pham, and K. J. Myers, eds.), vol. 5636 of *Lecture Notes in Computer Science*, pp. 326–337, Springer, 2009.

[295] R. D. Millán, L. Dempere-Marco, J. M. Pozo, J. R. Cebral, and A. F. Frangi, "Morphological characterization of intracranial aneurysms using 3-D moment invariants," *IEEE Transactions on Medical Imaging*, vol. 26, no. 9, pp. 1270–1282, 2007.

[296] B. Ng, R. Abugharbieh, X. Huang, and M. J. McKeown, "Characterizing fMRI activations within regions of interest (ROIs) using 3D moment invariants," in *Proceedings of the Computer Vision and Pattern Recognition Workshop CVPRW'06*, pp. 63–70, IEEE, 2006.

[297] E. I. Zacharaki, C. S. Hogea, D. Shen, G. Biros, and C. Davatzikos, "Non-diffeomorphic registration of brain tumor images by simulating tissue loss and tumor growth," *NeuroImage*, vol. 46, no. 3, pp. 762–774, 2009.

[298] S. Laguitton, C. Boldak, and C. Toumoulin, "Temporal tracking of coronaries in multi-slice computed tomography," in *29th Annual International Conference on Engineering in Medicine and Biology Society EMBC'07*, pp. 4512–4515, IEEE, 2007.

[299] H. Sundar, H. Litt, and D. Shen, "Estimating myocardial motion by 4D image warping," *Pattern Recognition*, vol. 42, no. 11, pp. 2514–2526, 2009.

[300] J. Chaki and R. Parekh, "Automated classification of echo-cardiography images using texture analysis methods," in *Handbook of Medical and Healthcare Technologies* (B. Furht and A. Agarwal, eds.), pp. 121–143, Springer, 2013.

[301] D. Shen, H. Sundar, Z. Xue, Y. Fan, and H. Litt, "Consistent estimation of cardiac motions by 4D image registration," in *8th International Conference on Medical Image Computing and Computer-Assisted Intervention MICCAI'05* (J. S. Duncan and G. Gerig, eds.), vol. 3750 of *Lecture Notes in Computer Science*, pp. 902–910, Springer, 2005.

[302] N. Pattanachai, N. Covavisaruch, and C. Sinthanayothin, "Tooth recognition in dental radiographs via Hu's moment invariants," in *9th International Conference on Electrical Engineering / Electronics, Computer, Telecommunications and Information Technology ECTI-CON'12*, pp. 1–4, IEEE, 2012.

[303] D. A. Morales, E. Bengoetxea, and P. Larranaga, "Selection of human embryos for transfer by Bayesian classifiers," *Computers in Biology and Medicine*, vol. 38, no. 11–12, pp. 1177–1186, 2008.

[304] R. O. Duda, P. E. Hart, and D. G. Stork, *Pattern Classification*. Wiley Interscience, 2nd ed., 2001.

[305] B. Jones, G. Schaefer, and S. Zhu, "Content-based image retrieval for medical infrared images," in *26th Annual International Conference on Engineering in Medicine and Biology Society EMBC'04*, pp. 1186–1187, IEEE, 2004.

[306] P. Milanfar, W. C. Karl, and A. S. Willsky, "A moment-based variational approach to tomographic reconstruction," *IEEE Transactions on Image Processing*, vol. 5, no. 3, pp. 459–470, 1996.

[307] S. Basu and Y. Bresler, "Uniqueness of tomography with unknown view angles," *IEEE Transactions on Image Processing*, vol. 9, no. 6, pp. 1094–1106, 2000.

[308] K. M. Hosny, G. A. Papakostas, and D. E. Koulouriotis, "Accurate reconstruction of noisy medical images using orthogonal moments," in *18th International Conference on Digital Signal Processing DSP'13*, pp. 1–6, IEEE, 2013.

[309] B. Mahdian and S. Saic, "Detection of copy-move forgery using a method based on blur moment invariants," *Forensic Science International*, vol. 171, no. 2–3, pp. 180–189, 2007.

[310] S.-J. Ryu, M. Kirchner, M.-J. Lee, and H.-K. Lee, "Rotation invariant localization of duplicated image regions based on Zernike moments," *IEEE Transactions on Information Forensics and Security*, vol. 8, no. 8, pp. 1355–1370, 2013.

[311] J. Yang, P. Ran, D. Xiao, and J. Tan, "Digital image forgery forensics by using undecimated dyadic wavelet transform and Zernike moments," *Journal of Computational Information Systems*, vol. 9, no. 16, pp. 6399–6408, 2013.

[312] H. Zi-Long, Z. Zheng-Bao, W. Jia-Fu, and L. Hui-Ying, "A blind forensic algorithm for detecting copy-paste images based on Krawtchouk invariant moments," *Computer Technology and Development*, vol. 2, 2012.

[313] H. Huang, G. Coatrieux, H. Z. Shu, L. M. Luo, and C. Roux, "Blind forensics in medical imaging based on Tchebichef image moments," in *Engineering in Medicine and Biology Society EMBC'11, Annual International Conference of the IEEE*, pp. 4473–4476, 2011.

[314] S. Mushtaq and A. H. Mir, "Digital image forgeries and passive image authentication techniques: A survey," *International Journal of Advanced Science and Technology*, vol. 73, pp. 15–32, 2014.

[315] V. Christlein, C. Riess, J. Jordan, C. Riess, and E. Angelopoulou, "An evaluation of popular copy-move forgery detection approaches," *IEEE Transactions on Information Forensics and Security*, vol. 7, no. 6, pp. 1841–1854, 2012.

[316] W. Chen, Y. Q. Shi, and G. Xuan, "Identifying computer graphics using HSV color model and statistical moments of characteristic functions," in *Proceedings of the International Conference on Multimedia and Expo ICME'07*, pp. 1123–1126, 2007.

[317] G. Xu, Y. Q. Shi, and W. Su, "Camera brand and model identification using moments of 1-d and 2-d characteristic functions," in *16th International Conference on Image Processing ICIP'09*, pp. 2917–2920, IEEE, 2009.

[318] M. Shabanifard, M. G. Shayesteh, and M. A. Akhaee, "Forensic detection of image manipulation using the Zernike moments and pixel-pair histogram," *IET Image Processing*, vol. 7, no. 9, pp. 817–828, 2013.

[319] H. Huang, G. Coatrieux, H. Shu, L. Luo, and C. Roux, "Blind integrity verification of medical images," *IEEE Transactions on Information Technology in Biomedicine*, vol. 16, no. 6, pp. 1122–1126, 2012.

[320] J. Yang, B. Min, and D. Park, "Fingerprint verification based on absolute distance and intelligent BPNN," in *Proceedings of the Frontiers in the Convergence of Bioscience and Information Technologies FBIT'07*, pp. 676–681, IEEE, 2007.

[321] J. Yang, J. Shin, B. Min, J. Lee, D. Park, and S. Yoon, "Fingerprint matching using global minutiae and invariant moments," in *Proceedings of the Congress on Image and Signal Processing CISP'08*, pp. 599–602, IEEE, 2008.

[322] J. Yang, S. Xie, S. Yoon, D. Park, Z. Fang, and S. Yang, "Fingerprint matching based on extreme learning machine," *Neural Computing and Applications*, vol. 22, no. 3–4, pp. 435–445, 2013.

[323] C. Lakshmi Deepika, A. Kandaswamy, and P. Gupta, "Orthogonal moments for efficient feature extraction from line structure based biometric images," in *Intelligent Computing Theories and Applications, 8th International Conference ICIC'12* (D.-S. Huang, J. Ma, K.-H. Jo, and M. M. Gromiha, eds.), vol. 7390 of *Lecture Notes in Computer Science*, pp. 656–663, Springer, 2012.

[324] M. Liu and P.-T. Yap, "Invariant representation of orientation fields for fingerprint indexing," *Pattern Recognition*, vol. 45, no. 7, pp. 2532–2542, 2012.

[325] A. Almansa and T. Lindeberg, "Fingerprint enhancement by shape adaptation of scale-space operators with automatic scale selection," *IEEE Transactions on Image Processing*, vol. 9, no. 12, pp. 2027–2042, 2000.

[326] T. B. Long, L. H. Thai, and T. Hanh, "Multimodal biometric person authentication using fingerprint, face features," in *PRICAI 2012: Trends in Artificial Intelligence, 12th Pacific Rim International Conference* (P. Anthony, M. Ishizuka, and D. Lukose, eds.), vol. 7458 of *Lecture Notes in Computer Science*, pp. 613–624, Springer, 2012.

[327] S. Wang and Y. Xu, "A new palmprint identification algorithm based on Gabor filter and moment invariant," in *Conference on Cybernetics and Intelligent Systems CIS'08*, pp. 491–496, IEEE, 2008.

[328] J. Kong, H. Li, Y. Lu, M. Qi, and S. Wang, "Hand-based personal identification using k-means clustering and modified Zernike moments," in *Proceedings of the 3rd International Conference on Natural Computation ICNC'07*, pp. 651–656, IEEE, 2007.

[329] R. Gayathri and P. Ramamoorthy, "Automatic palmprint identification based on high order Zernike moment," *American Journal of Applied Sciences*, vol. 9, no. 5, pp. 759–765, 2012.

[330] S. Karar and R. Parekh, "Palm print recognition using Zernike moments," *International Journal of Computer Applications*, vol. 55, no. 16, pp. 15–19, 2012.

[331] G. S. Badrinath, N. K. Kachhi, and P. Gupta, "Verification system robust to occlusion using low-order Zernike moments of palmprint sub-images," *Telecommunication Systems*, vol. 47, no. 3, pp. 275–290, 2011.

[332] Y. Pang, T. Andrew, N. David, and H. F. San, "Palmprint verification with moments," in *Journal of WSCG (Winter School of Computer Graphics)*, vol. 12 (1–3), pp. 325–332, University of West Bohemia, Plzeň, Czech Republic, 2003.

[333] G. Amayeh, G. Bebis, A. Erol, and M. Nicolescu, "Hand-based verification and identification using palm–finger segmentation and fusion," *Computer Vision and Image Understanding*, vol. 113, no. 4, pp. 477–501, 2009.

[334] C. L. Deepika, A. Kandaswamy, C. Vimal, and B. Sathish, "Biometric feature extraction with biometric specific shape descriptors," *International Journal of Biometrics*, vol. 4, no. 3, pp. 246–264, 2012.

[335] M. Nabti and A. Bouridane, "An effective and fast iris recognition system based on a combined multiscale feature extraction technique," *Pattern Recognition*, vol. 41, no. 3, pp. 868–879, 2008.

[336] Y. D. Khan, S. A. Khan, F. Ahmad, and S. Islam, "Iris recognition using image moments and k-means algorithm," *The Scientific World Journal*, pp. 1–9, 2014.

[337] S. Aich and G. M. A. Mamun, "An efficient supervised approach for retinal person identification using Zernike moments," *International Journal of Computer Applications*, vol. 81, no. 7, pp. 34–37, 2013.

[338] S. M. M. Rahman, M. M. Reza, and Q. M. Z. Hasani, "Low-complexity iris recognition method using 2D Gauss-Hermite moments," in *8th International Symposium on Image and Signal Processing and Analysis ISPA'13*, pp. 142–146, IEEE, 2013.

[339] S. J. Hosaini, S. Alirezaee, M. Ahmadi, and S. V.-A. D. Makki, "Comparison of the Legendre, Zernike and pseudo-Zernike moments for feature extraction in iris recognition," in *5th International Conference on Computational Intelligence and Communication Networks CICN'13*, pp. 225–228, IEEE, 2013.

[340] H. Liu and J. Yan, "Multi-view ear shape feature extraction and reconstruction," in *Third International IEEE Conference on Signal-Image Technologies and Internet-Based System SITIS'07*, pp. 652–658, IEEE, 2007.

[341] M. Hussain, A. Jibreen, H. Aboalsmah, H. Madkour, G. Bebis, and G. Amayeh, "Feature selection for hand-shape based identification," in *International Joint Conference CISIS'15 and ICEUTE'15* (Á. Herrero, B. Baruque, J. Sedano, H. Quintián, and E. Corchado, eds.), vol. 369 of *Advances in Intelligent Systems and Computing*, pp. 237–246, Springer, 2015.

[342] G. AlGarni and M. Hamiane, "A novel technique for automatic shoeprint image retrieval," *Forensic Science International*, vol. 181, no. 1, pp. 10–14, 2008.

[343] L. Lee and W. E. L. Grimson, "Gait analysis for recognition and classification," in *Proceedings of the Fifth International Conference on Automatic Face and Gesture Recognition FGR'02*, pp. 155–162, IEEE, 2002.

[344] C.-P. Shi, H.-G. Li, X. Lian, and X.-G. Li, "Multi-resolution local moment feature for gait recognition," in *Proceedings of the International Conference on Machine Learning and Cybernetics ICMLC'06*, pp. 3709–3714, IEEE, 2006.

[345] S. D. Choudhury and T. Tjahjadi, "Silhouette-based gait recognition using procrustes shape analysis and elliptic Fourier descriptors," *Pattern Recognition*, vol. 45, no. 9, pp. 3414–3426, 2012.

[346] L. Sudha and R. Bhavani, "Gait based gender identification using statistical pattern classifiers," *International Journal of Computer Applications*, vol. 40, no. 8, pp. 30–35, 2012.

[347] D. Ioannidis, D. Tzovaras, I. G. Damousis, S. Argyropoulos, and K. Moustakas, "Gait recognition using compact feature extraction transforms and depth information," *IEEE Transactions on Information Forensics and Security*, vol. 2, no. 3, pp. 623–630, 2007.

[348] A. M. Ibrahim, A. A. Shafie, and M. M. Rashid, "Human identification system based on moment invariant features," in *International Conference on Computer and Communication Engineering ICCCE'12*, pp. 216–221, IEEE, 2012.

[349] A. A. Shafie, A. B. M. Ibrahim, and M. M. Rashid, "Smart objects identification system for robotic surveillance," *International Journal of Automation and Computing*, vol. 11, no. 1, pp. 59–71, 2014.

[350] H. Asaidi, A. Aarab, and M. Bellouki, "Shadow elimination and vehicles classification approaches in traffic video surveillance context," *Journal of Visual Languages & Computing*, vol. 25, no. 4, pp. 333–345, 2014.

[351] B. D. Lucas and T. Kanade, "An iterative image registration technique with an application to stereo vision," in *Proceedings of the International Joint Conference on Artificial Intelligence IJCAI'81*, vol. 2, pp. 674–679, Morgan Kaufman, 1981.

[352] S. Ghosal and R. Mehrotra, "Robust optical flow estimation," in *International Conference on Image Processing ICIP'94*, pp. 780–784, IEEE, 1994.

[353] M. Kharbat, N. Aouf, A. Tsourdos, and B. White, "Robust brightness description for computing optical flow," in *Proceedings of the British Machine Vision Conference BMVC'08*, p. 10–, BMVA, 2008.

[354] M. Liang, J. Du, X. Li, L. Xu, H. Liu, and Y. Li, "Spatio-temporal super-resolution reconstruction based on robust optical flow and Zernike moment for dynamic image sequences," in *IEEE International Symposium on Industrial Electronics ISIE'13*, pp. 1–6, IEEE, 2013.

[355] T. J. Bin, A. Lei, C. Jiwen, and K. W. L. Dandan, "Subpixel edge location based on orthogonal Fourier-Mellin moments," *Image and Vision Computing*, vol. 26, no. 4, pp. 563–569, 2008.

[356] S. Ghosal and R. Mehrotra, "Orthogonal moment operators for subpixel edge detection," *Pattern Recognition*, vol. 26, no. 2, pp. 295–306, 1993.

[357] H. Yang and L. Pei, "Fast algorithm of subpixel edge detection based on Zernike moments," in *4th International Congress on Image and Signal Processing CISP'11*, vol. 3, pp. 1236–1240, IEEE, 2011.

[358] E. Lyvers, O. Mitchell, M. Akey, and A. Reeves, "Subpixel measurements using a moment-based edge operator," *IEEE Transactions on Pattern Analysis and Machine Intelligence*, vol. 11, no. 12, pp. 1293–1309, 1989.

[359] S. Šimberová, M. Karlický, and T. Suk, "Statistical moments of active-region images during solar flares," *Solar Physics*, vol. 289, no. 1, pp. 193–209, 2014.

[360] S. Šimberová and T. Suk, "Analysis of dynamic processes by statistical moments of high orders," in *Progress in Pattern Recognition, Image Analysis, Computer Vision, and Applications, Proceedings of the 18th Iberoamerican Congress, CIARP'13, Part I* (J. Ruiz-Shulcloper and G. Sanniti di Baja, eds.), vol. 8258 of *Lecture Notes in Computer Science*, pp. 221–228, Springer, 2013.

[361] M. Schlemmer, M. Heringer, F. Morr, I. Hotz, M.-H. Bertram, C. Garth, W. Kollmann, B. Hamann, and H. Hagen, "Moment invariants for the analysis of 2D flow fields," *IEEE Transactions on Visualization and Computer Graphics*, vol. 13, no. 6, pp. 1743–1750, 2007.

[362] Y. Zhou, F. Chen, and B. Sun, "Identification method of gas-liquid two-phase flow regime based on image multi-feature fusion and support vector machine," *Chinese Journal of Chemical Engineering*, vol. 16, no. 6, pp. 832–840, 2008.

[363] R. Bujack, J. Kasten, I. Hotz, G. Scheuermann, and E. Hitzer, "Moment invariants for 3D flow fields via normalization," in *Pacific Visualization Symposium, PacificVis'15*, pp. 9–16, IEEE, 2015.

[364] H. Tang, H. Z. Shu, J. L. Dillenseger, X. D. Bao, and L. M. Luo, "Moment-based metrics for mesh simplification," *Computers & Graphics*, vol. 31, no. 5, pp. 710–718, 2007.

[365] A. O'Connor, K. F. Mulchrone, and P. A. Meere, "WinDICOM: A program for determining inclusion shape and orientation," *Computers and Geosciences*, vol. 35, no. 6, pp. 1358–1368, 2009.

[366] Y. Kang, B. Hu, Y. Wang, and Z. Shao, "A scale self-adaptive tracking method based on moment invariants," *Journal of Signal Processing Systems*, vol. 81, no. 2, pp. 197–212, 2015.

[367] W. Yan, C. Weber, and S. Wermter, "A hybrid probabilistic neural model for person tracking based on a ceiling-mounted camera," *Journal of Ambient Intelligence and Smart Environments*, vol. 3, no. 3, pp. 237–252, 2011.

10

Conclusion

10.1 Summary of the book

The theory of moments and moment invariants has been covered step by step in this book. Our aim was to address the main categories of 2D and 3D moments and moment invariants, the computational issues, and their applicability in various domains. In this way, we hoped to provide both material for theoretical study and references for algorithm design. We also provided readers with an introduction to a broader context – the shape feature design and the classification algorithms for object recognition.

The individual classes of moments were introduced, comprising geometric moments, complex moments, and a wide variety of orthogonal moments such as Legendre, Zernike, Chebyshev, and many others. They were presented together with the comments on their advantages and relevant weak points. Illustrative examples accompany the theoretical parts. The core section of the book is dedicated to the invariance of moments with respect to various geometric and intensity degradations. We consider these moment properties the most interesting ones from the image analysis point of view. In all cases, we presented a consistent theory allowing a derivation of invariants of any orders; in some cases we even demonstrated several alternative approaches. Certain properties of the recently proposed features such as affine invariants and especially the blur invariants have never been published prior to this book. To make the book comprehensive, widely known invariants to translation, rotation, and scaling are also described in detail.

The moments are valued for their representative object description, as we have seen in Chapter 9 dedicated to applications. For time-optimal applications of moments, methods and algorithms used for their computation were reviewed to ensure the time efficiency of the solutions based on moments. The mentioned application areas make use of the moment ability to express the shape variations and at the same time to stay invariant under common image deformations.

2D and 3D Image Analysis by Moments, First Edition. Jan Flusser, Tomáš Suk and Barbara Zitová.
© 2017 John Wiley & Sons, Ltd. Published 2017 by John Wiley & Sons, Ltd.
Companion website: www.wiley.com/go/flusser/2d3dimageanalysis

10.2   Pros and cons of moment invariants

Here, at the end of the book, we would like to summarize the positive as well as negative issues of moment applicability. Following from the previous chapters, the moments are by definition global features, computed from the whole image or from the selected region of interest (ROI). The global nature of the moments results in their low robustness with respect to unexpected local disturbances in the image or in the individual object shapes, respectively. Object occlusion is a typical example. Even a small local change in the image could affect all image moments. Moreover, the choice of the ROIs and their proper localization is crucial for successful moment utilization. The moments are sensitive to segmentation errors and to data correspondence of the analyzed parts of the scene; thus the choice of the segmentation algorithm has to be carried out carefully. Data correspondence is not handled properly even in many published papers, where the authors are using ROIs of the wrong type with respect to the expected geometric degradations. An example of such an incorrect choice is a fix-sized circular neighborhood in the case of expected scale changes. After that, the ROI's content is not valid any more.

Another issue we would like to mention here is the vulnerability of certain moment invariants with respect to the violence of the degradation model. Convolution invariants are a typical example – if the image blurring is not described by a space-invariant convolution or if the boundary effect is significant, they may become inapplicable.

Fortunately, many of the mentioned shortcomings can be diminished by proper experimental setting.

The negative properties are often counterbalanced by moment *plausible features*. The moments are accompanied by a well-developed mathematical theory providing us with an insight into the behavior of moments under various conditions. As we aimed to show in this book, the moments can be made invariant to many common degradations, not only if they are present separately but even if they act simultaneously, as is often the case when the image is blurred and geometrically transformed. They offer good discriminability and relatively high robustness to additive noise thanks to their definition based on an integration over the whole ROI. Moreover, the moments and their corresponding invariants form complete and independent sets of features for object description, which in the end provides higher discriminability and computational time saving than dependent and/or incomplete systems. Among the positive properties of moments we should also recall many efficient algorithms allowing stable and fast evaluation.

In many papers, the authors have tried to resolve the question of which kind of moments is the best. The term "the best" is related to their discrimination power, reconstruction ability, noise robustness or other criteria. In our opinion, which is supported by numerous experiments we have carried out, this problem cannot be resolved in general and the question itself is inconsequential. Since all polynomial bases of the same degree are theoretically equivalent, there are no significant differences between various moment systems. Of course, for a particular task and for the given test data, we can always find the "optimal" moment set. This solution might be different, however, if we were to change the dataset or other minor parameters such as the generation of noise instances. We believe that looking for an optimal (in some sense) moment set should be coupled not only with the purpose of the processing but also with the data. The moments (more precisely, the basis functions) providing an optimal performance should be adapted to the test set.

We hope this book provides readers with insight into moment theory, gives an idea of how to construct moment invariants, and advises how to utilize moments in object recognition tasks. The survey of applications presented in Chapter 9, along with a number of references we have provided, could be used as an indication of where and how the moments can be used and which image analysis problems can be handled using these techniques.

10.3   Outlook to the future

Regarding open problems and desirable future research directions, we foresee several areas of rapid development in the near future.

We envisage breaking the limitations of fixed polynomial basis that have been mostly used so far. The future "moments" should not be restricted on the polynomials, and the basis functions should be dynamically adapted to the given task and to particular datasets. The functional bases should be designed in such a way that they optimize a criterion, which is important for the given problem, such as minimum information loss, maximum separability of the training sets, sparse object representation, and/or fast computability, to name a few examples.

Since the importance and the availability of 3D imaging technologies has been growing continually, the future development of moments should be more focused on the 3D domain, especially in the aspects where the 2D experience cannot be transferred into 3D in a straightforward manner.

The theory and applications of *vector field invariants* is a recently opened area with a big application potential, particularly in fluid dynamics. Detection of flow singularities such as vortices and sinks is of a great interest in many branches of engineering. These tasks are intrinsically three-dimensional, but 3D vector field invariants have not yet been developed at a sufficient level.

Finally, in moment research, just as in most areas of computer science, the development of theory and applications are intertwined issues. We can imagine the appearance of completely new classes of moment invariants w.r.t. the degradations, the models of which have not been described/utilized so far. Such degradations may appear in new, unexplored application areas and/or in connection with future advanced imaging devices and technologies. This may stimulate theoretical research, which would again, if successful, contribute to opening yet other application areas.

Index

2D and 3D Image Analysis by Moments, First Edition. Jan Flusser, Tomáš Suk and Barbara Zitová.
© 2017 John Wiley & Sons, Ltd. Published 2017 by John Wiley & Sons, Ltd.
Companion website: www.wiley.com/go/flusser/2d3dimageanalysis